Algebra by Example
An Elementary Course

HEATH BASIC MATH SERIES

Algebra by Example

An Elementary Course

Herbert I. Gross

Bunker Hill Community College

D. C. Heath and Company

Lexington, Massachusetts   Toronto

Copyright © 1978 by D. C. Heath and Company.

All rights reserved. No part of this publication may be reproduced or transmitted in any form or by any means, electronic or mechanical, including photocopy, recording, or any information storage or retrieval system, without permission in writing from the publisher.

Published simultaneously in Canada.

Printed in the United States of America.

International Standard Book Number: 0-669-00473-1

Library of Congress Catalog Card Number: 77-77922

To my students, whose perceptive and thought-provoking remarks encouraged me to find better answers to old questions.

Preface

Students entering a course in elementary algebra usually range from those who are taking their first course in algebra to those for whom the course is a review. The textbook must be accessible to all, boring to none, and easily adaptable to any mode of instruction.

Algebra by Example: An Elementary Course is made up of fifteen lessons that progress gradually from arithmetic to quadratic equations and other nonlinear relationships. The progression gives the instructor flexibility in deciding where the course should start and where it should end. The text is structured to help the student become an active participant in all aspects of the course. It begins with the idea that algebra is a natural extension of arithmetic and that being able to use the skills taught in arithmetic provides the connecting link between arithmetic and algebra.

Each of the fifteen lessons is subdivided into sections. A section consists of a single objective, and this objective is stated in the right-hand margin, immediately following the title of the section. To determine whether the student has acquired some understanding of the section, two self-quizzes (accompanied by the answers) appear at the end of every section.

Two comprehensive Exercise Sets, Forms A and B, identical in format and only different in the numbers used, appear at the end of each lesson. They are designed to check the student's mastery of the material covered in each lesson. The answers to every exercise in Exercise Set Form A are given in the textbook; complete solutions to Exercise Set Form A are given in the *Student Solutions Manual*. The answers to every exercise in Exercise Set Form B also appear in the *Student Solutions Manual*.

The Appendix offers important material for the student. It includes an arithmetic review, a glossary of the key terms in pre-algebra mathematics, a table of squares and square roots, and a table of metric conversions. The arithmetic review and glossary are especially useful for the student who needs to review arithmetic occasionally and who will be able to do so without having to go to another textbook.

To aid both the instructor and the student, several separate supplements to the text are available. They are:

The Instructor's Resource Manual with Tests. This supplement consists of asides, hints, topics for enrichment, main points of the text, and ideas that the instructor may wish to dovetail with other teaching strategies. The remarks are made on a section-by-section basis, and the discussion of each lesson ends with yet a third group of exercises, Form C, with the answers immediately following.

The Student Solutions Manual. This manual contains the completely worked-out solutions for every problem in Exercise Set Form A in the text. In many cases, more than one solution is given. The format is the same as that of the text, complete with margin notes to illustrate certain points.

Audio Tapes for the Student. Some students learn better by listening than by reading. For this reason, an audio cassette program is available to talk the student through the course. The program begins with a lecture describing the course and how it was developed. For each of the fifteen lessons in the text, there is a 30-minute review lecture. The *Student Solutions Manual* is organized so that it may be used as the visual complement to the taped lectures.

I am deeply indebted to my colleagues Ralph Mansfield and William Hemmer for their especially helpful comments, suggestions, and criticisms regarding the format and the content of this text. I would also like to express my appreciation to Philip A. Glynn, David C. Mitchell, and John Wisthoff for their detailed criticisms of the complete manuscript, to Lucia McKay and Jerald T. Ball and to my close friend and colleague, Samuel Fox McInroy, for their valuable critiques of portions of the manuscript.

Finally, I want to thank my family. It is a time-consuming endeavor to write a textbook. My family was most gracious in accepting this fact and in lending encouragement to me at the times I needed it most. I am especially grateful to my wife, Louise, for her help and understanding during this project.

HERBERT I. GROSS
Needham, Massachusetts

Contents

unit 1   Introduction to Algebraic Expressions and Equations   1

LESSON 1   Using Arithmetic

OVERVIEW   3

Section 1.1   Using Arithmetic   4
Section 1.2   Mathematical Relationships   7
Section 1.3   Squaring a Number   11
Section 1.4   Finding the Square Root of a Number   14
Section 1.5   The Metric System   17
Exercise Set 1   21

LESSON 2   An Introduction to Algebraic Equations

OVERVIEW   25

Section 2.1   Equations as Open Sentences   26
Section 2.2   Some Additional Vocabulary   31
Section 2.3   More on "Undoing" Equations   34
Section 2.4   Guessing Answers to Equations   38
Exercise Set 2   41

LESSON 3   Signed Numbers

OVERVIEW   43

Section 3.1   The Idea of Less than Zero   44
Section 3.2   Adding Signed Numbers   47
Section 3.3   Subtracting Signed Numbers   52
Section 3.4   Multiplying Signed Numbers   55
Section 3.5   Dividing Signed Numbers   58
Section 3.6   Solving Equations with Signed Numbers   60
Exercise Set 3   62

unit 2   Introductory Topics in Linearity   65

LESSON 4   The Structure of Arithmetic

OVERVIEW   67

Section 4.1   The Associative Property for Addition   68
Section 4.2   The Commutative Property for Addition   71
Section 4.3   The Associative Property for Multiplication   74
Section 4.4   The Commutative Property for Multiplication   76
Section 4.5   The Additive Identity and Additive Inverses   79
Section 4.6   The Multiplicative Identity and Multiplicative Inverses   84
Section 4.7   Removing Grouping Symbols   88
Section 4.8   The Distributive Property   92
Exercise Set 4   97

LESSON 5   An Introduction to Linearity

OVERVIEW   102

Section 5.1   Linear Expressions   103
Section 5.2   Equations of the Form $mx + b = c$ 108
Section 5.3   Solving Equations of the Form $mx + b = cx + d$ 113
Section 5.4   Inconsistent Equations and Identities   116
Section 5.5   Inverting Linear Relationships   119
Exercise Set 5   122

LESSON 6   An Introduction to Word Problems

OVERVIEW   124

Section 6.1   Formulas Leading to Linear Equations   125
Section 6.2   Translating Words into Algebra   127
Section 6.3   Mixture Problems   132
Section 6.4   Application to Constant Speed   136
Section 6.5   Applications to Percents   140
Exercise Set 6   144

unit 3   Additional Topics in Linearity   149

LESSON 7   An Introduction to Graphs of Linear Equations

OVERVIEW   151

Section 7.1   The Number Line   152
Section 7.2   Cartesian Coordinates   156
Section 7.3   Graphing Linear Relationships   163
Section 7.4   The General Straight-Line Equation   173
Section 7.5   Solving Linear Equations by Graphs   180
Exercise Set 7   184

LESSON 8   Simultaneous Linear Equations

OVERVIEW   191

Section 8.1   Introduction to Simultaneous Equations   192
Section 8.2   A Special Case   197
Section 8.3   The More General Case   202

Contents

Section 8.4 Applications to Word Problems   210
Exercise Set 8   214

LESSON 9 An Introduction to Inequalities

OVERVIEW   218

Section 9.1 An Introduction to Inequalities   219
Section 9.2 Solving Linear Inequalities   221
Section 9.3 Linear Inequalities in Two Variables   225
Section 9.4 Simultaneous Inequalities   231
Exercise Set 9   237

unit 4 An Introduction to Nonlinear Equations   243

LESSON 10 Integral Exponents

OVERVIEW   245

Section 10.1 Positive Integral Exponents   246
Section 10.2 Other Integral Exponents   250
Section 10.3 Multiplying Like Bases   254
Section 10.4 Other Arithmetical Operations   257
Section 10.5 Scientific Notation   261
Section 10.6 Significant Figures   266
Exercise Set 10   270

LESSON 11 Introduction to Polynomials

OVERVIEW   275

Section 11.1 Monomial Expressions   276
Section 11.2 Polynomial Expressions   281
Section 11.3 Adding Polynomials   285
Section 11.4 Subtracting Polynomials   288
Section 11.5 Multiplication of Polynomials   292
Exercise Set 11   298

LESSON 12 Division of Polynomials

OVERVIEW   301

Section 12.1 Dividing a Polynomial by a Monomial   302
Section 12.2 An Introduction to Factoring   307
Section 12.3 Dividing Polynomials by Factoring   312
Section 12.4 The Division Algorithm   316
Exercise Set 12   322

unit 5 More on Nonlinear Equations   325

LESSON 13 An Introduction to Quadratic Equations

OVERVIEW   327

Section 13.1 Using Arithmetic to Solve Equations   328
Section 13.2 Disadvantages of Arithmetical Solutions   333
Section 13.3 Divisors of Zero   337
Section 13.4 Factoring Expressions of the Form $x^2 + bx + c$ 340

Section 13.5   Solving Quadratic Equations by Factoring   344
Section 13.6   Some Applications   350
Exercise Set 13   354

LESSON 14   The General Solution of a Quadratic Equation

OVERVIEW   359

Section 14.1   More on Square Roots   360
Section 14.2   Arithmetic Involving Square Roots   364
Section 14.3   Completing the Square   369
Section 14.4   How to Solve Any Quadratic Equation   373
Section 14.5   Graphing a Quadratic Relationship   380
Section 14.6   More about Word Problems   386
Exercise Set 14   389

LESSON 15   Algebraic Fractions

OVERVIEW   395

Section 15.1   Fractions Revisited   396
Section 15.2   Ratio and Proportion   402
Section 15.3   Inverse Proportions and Other Variations   406
Section 15.4   "Sharing the Work" Problems   411
Exercise Set 15   416

Appendix

Arithmetic Review   421
Arithmetic Glossary   427
Squares and Square Roots   430
Metric Conversions   431

Index   433

Introduction to Algebraic Expressions and Equations

unit 1

Lesson 1

Using Arithmetic

Overview This lesson is the "missing link" between arithmetic skills and algebra. Basically, there is a big difference between knowing *how* to do arithmetic and knowing *what* arithmetic to do. For example, if we use a calculator to multiply 3.1416 by 6.78, the calculator will do it — even if the correct answer to the problem required addition of these two numbers.

Perhaps the following story will illustrate the point. Three men go to a hotel and are told that the price of a room is $30. Each man pays $10. Later, the room clerk discovers that the room costs only $25, and gives the bellhop $5 to return to the three men. To make the arithmetic easier, the bellhop keeps $2 and gives each of the men $1. Since each man got a dollar back, each paid $9 for the room. Three 9's are 27 and the 2 the bellhop kept makes 29. What happened to the other dollar?

Obviously, nothing happened to the missing dollar. It is true that $3 \times 9 = 27$ and that $27 + 2 = 29$. *But*, we are *not* supposed to add 2 and 27 in this problem. Rather, since the men got $3 in change, they spent only $27. What we should have said is that three 9's are 27, *of which* 25 went for the room and 2 for the bellhop. The $2 the bellhop kept is already included in the $27 (otherwise the men would have spent $25, not $27).

This brainteaser is based on the fact that just because we know *how* to add and multiply, we don't necessarily know what operation to use.

In this lesson we shall apply arithmetic to real-life problems, in which the solution requires not only knowledge of how to do arithmetic but also knowledge of what arithmetic should be done.

Section 1.1
Using Arithmetic

Arithmetic skills teach us how to perform the various operations of arithmetic. They do not teach us *when* these operations should be used.

For example, the fact that $32 \times 45 = 1{,}440$ doesn't tell *when* we are supposed to multiply 32 by 45.

OBJECTIVE
To understand how arithmetic is used to solve certain real-life problems.

Example 1

A football league has 32 teams with 45 players each. How many players are in the league?

There are 1,440 players in the league. Each team has 45 players, so there are 45 players thirty-two times. This means 45×32, which is 1,440.

The big difference between this example and knowing that $45 \times 32 = 1{,}440$ is to decide that you have to multiply 32 and 45 to get the answer.

NOTE

Remember that you have a nice hint if you keep in mind that "per" means "divided by." That is, "players per team" means $\frac{\text{players}}{\text{team}}$, and since there are 45 players *per* team, you may write

$$\frac{45 \text{ players}}{\text{team}}$$

You then have

$$\frac{45 \text{ players}}{\cancel{\text{team}}} \times 32\,\cancel{\text{teams}} = 45 \text{ players} \times 32$$
$$= (45 \times 32) \text{ players}$$
$$= 1{,}440 \text{ players}$$

From another point of view, it does no good to know that $45 \times 32 = 1{,}440$ if you don't know when you're supposed to multiply 45 and 32 to find the answer.

We use parentheses to tell us that everything inside is one number.

Many different real-life problems may be solved by the same arithmetic problem.

Example 2

An economy car gets 32 miles per gallon of gas. How far will the car go on 45 gallons?

The car will go 1,440 miles.

As the car goes 32 miles on each gallon of gas, it will go 32 miles forty-five times on 45 gallons of gas. That is, the car goes (32×45) miles: $32 \times 45 = 1{,}440$. This is the same arithmetic problem you did in the first example, but the situation is different.

Again, in terms of "per" meaning "divided by," you have

$$\frac{32 \text{ miles}}{\cancel{\text{gal}}} \times 45\,\cancel{\text{gal}} = (32 \times 45) \text{ miles}$$

Example 3

If a merchant buys 32 jackets at \$45 each, how much money does she pay for the jackets?

She pays \$1,440 for the jackets. At \$45 each, she must pay \$45 thirty-two times (or \$45 × 32) for the jackets, but 45×32 is still 1,440.

Section 1.1   Using Arithmetic

In terms of "per," meaning "divided by," you have

$$\frac{\$45}{\text{coat}} \times 32 \text{ coats} = \$45 \times 32 = \$1{,}440$$

Sometimes the real-world problem requires that we use decimal fractions rather than whole numbers. At other times we must use common fractions. Let's look at a few more examples.

Example 4

You want to buy paint at $5.95 per gallon. How much must you pay for 16 gallons of paint?

You must pay $95.20 for the 16 gallons of paint.
   This example is like the earlier ones. Each gallon of paint costs you $5.95, so you must pay $5.95 sixteen times, or ($5.95 × 16), for the paint. Using either the calculator or longhand, you find that $5.95 × 16 = $95.20.
   The important point is that while you already know by arithmetic that 5.95 × 16 = 95.20, in this lesson we are trying to show you *when* you use this fact.

Example 5

How far do you travel if you go at a speed of 60 miles per hour for three-quarters of an hour?

You go 45 miles. In a whole hour, you would travel 60 miles, so in $\frac{3}{4}$ of an hour you traveled $\frac{3}{4}$ of 60 miles. That is,

$$\frac{60 \text{ miles}}{\text{hr}} \times \frac{3}{4} \text{ hr} = 60 \text{ miles} \times \frac{3}{4}$$
$$= \left(60 \times \frac{3}{4}\right) \text{ miles}$$
$$= 45 \text{ miles}$$

Remember, you already know that $60 \times \frac{3}{4} = 45$, but here we are emphasizing *when* to use this result.

Sometimes we must multiply two decimal fractions to solve a real problem.

Example 6

If you travel 45.8 miles per hour, how far will you go in 1.25 hours?

You will go 57.25 miles.
   You still multiply the speed by the time, and 45.8 × 1.25 = 57.25.

Of course, not all problems involve multiplication. In such everyday matters as balancing a checkbook, we often use addition and subtraction.

Example 7

You have a balance in the bank of $53.75. You deposit your paycheck, which is for $127.43. Then you write a check for $37.50 for clothes and another check for $23.46 for food. How much money is now left in your account?

You have $120.22 left in the bank account.
   First you add $127.43 to your balance of $53.75. As $53.75 + $127.43 equals

---

REVIEWING THE MAJOR POINT
We already know that 32 × 45 = 1,440, but in this lesson we are showing that there are many times when we must find 32 × 45 without actually being told to do so.

$$\frac{\$5.95}{\text{gal}} \times 16 \text{ gal} = \$5.95 \times 16$$

Notice that 60 miles per hour does not say that you travel for an hour. All it says is that for each hour you travel you go 60 miles.

As a check, notice that 60 miles per hour is 1 mile per minute (since there are 60 minutes in an hour). And $\frac{3}{4}$ of an hour is 45 minutes. So at a mile a minute, you go 45 miles in 45 minutes.

REVIEW OF TERMS
Balance is the amount of money that you have in the bank.
Deposit is an amount of money that you add to your balance.
Withdrawal is an amount of money you take away from your balance. This happens when you write a check to pay for something.

$181.18, there is now $181.18 in the account. But you are spending $37.50 and $23.46. Altogether, you are spending $37.50 + $23.46, or $60.96.

If you subtract the amount you spent ($60.96) from the amount that's in the bank ($181.18), the difference is your balance. You have

$$\$181.18 - \$60.96 \quad \text{or} \quad \$120.22$$

To solve this problem, you had to do the arithmetic problem

$$(\underbrace{\$53.75 + \$127.43}_{\text{amount in the bank}}) - (\underbrace{\$37.50 + \$23.46}_{\text{amount you spent}})$$

$$\underbrace{}_{\substack{\text{equals} \\ \text{amount left in} \\ \text{the bank}}}$$

> You're spending almost $61 and you had about $54 before you deposited your paycheck, so you've spent about $7 more than you had. Therefore, you should have left about $7 less than the amount of your paycheck. This is about $120, which is close to the exact answer of $120.22.

In summary, this course assumes that you know how to do the arithmetic. Now we are learning *when* to do the arithmetic.

Example 8

You have taken three math tests. You scored 80% on the first, 83% on the second, and 92% on the third test. What is your average score for the three tests?

Your average for the three tests is 85%.

To find the average we add the three scores to obtain (80 + 83 + 92)% or 255%, and we then divide by the number of tests (3). That is,

$$\text{Average (in \%)} = \frac{80 + 83 + 92}{3} = \frac{255}{3}$$
$$= 85$$

> DEFINITION
> To find the **average**, we add the scores and divide this sum by the number of tests.

Again, notice that once you know what "average" means, we can solve Example 8 by the use of arithmetic. In the next section, we shall follow up the idea of how we know when to use arithmetic.

> This average means that if you scored 85% on each of the three tests, you would have scored the same number of points as you did with your scores of 80, 83, and 92.

PRACTICE DRILL

1. What is the total cost of 35 jackets if each jacket costs $40?
2. If an economy car gets 27 miles to a gallon of gas, how far will it go on 15 gallons of gas?
3. At $8.75 per gallon, what is the cost of 10 gallons of paint?
4. At a speed of 80 miles per hour, how far will you travel in

    (a) 3 hours?   (b) 3.2 hours?   (c) 3 minutes?

5. How much money is left in your checking account if you started with $253 and wrote checks for $87 and $66?
6. What is your average if your grades on four quizzes are 80, 90, 83, and 95?

> **Answers**
> 1. $1,400 2. 405 miles
> 3. $87.50 4. (a) 240 miles
> (b) 256 miles (c) 4 miles
> 5. $100 6. 87

CHECK THE MAIN IDEAS

1. This section is concerned with when to do arithmetic rather than with _____ to do arithmetic.
2. One hint comes from the word "per," which suggests the operation _____ .

> **Answers**
> 1. how 2. division
> 3. divided by 4. time
> 5. people (persons) 6. minute

Section 1.2   Mathematical Relationships   7

3. Miles per hour means miles _____ hours.
4. "Miles per hour" suggests that you divide distance by _____ .
5. If you are trying to find "cost per person," you divide the total cost by the total number of _____ .
6. You still must pay attention to labels. For example, if a car travels 3 miles in 3 minutes, its speed is 1 mile per _____ .

Section 1.2
Mathematical Relationships

In the last section, we introduced the idea of trying to know when to use arithmetic, and what arithmetic to use. Often this is not easy to do.

Sometimes it is difficult because we may not know the meaning of a word. For example, we cannot expect to find our average test score if we don't know what the word "average" means. At other times we may know what a word means, but we are missing a piece of information. For example, we may want to find out how many centimeters there are in seven inches. To do this, we have to know the number of centimeters in one inch. (If we don't know, we ask someone or else we look it up.) In fact, there are *2.54 centimeters per inch.* Once we know this, we can proceed as we did in the last section.

OBJECTIVE
To be able to read a mathematical relationship and to use the information in problem-solving.

Example 1
How many centimeters are there in 7 inches?

There are 17.78 centimeters in 7 inches.
  Once you know that there are 2.54 centimeters per inch, you know that there are 7 times 2.54 cm in 7 inches. That is,

$$\frac{2.54 \text{ cm}}{\text{in.}} \times 7 \text{ in.} = 2.54 \text{ cm} \times 7$$
$$= (2.54 \times 7) \text{ cm} = 17.78 \text{ cm}$$

The parentheses are used to emphasize that 2.54×7 is really a single number 17.78.

Example 2
How many centimeters are there in 1 foot?

There are 30.48 centimeters in 1 foot.
  What you need to know is that there are 12 inches in a foot and 2.54 centimeters in an inch. Then

$$\frac{2.54 \text{ cm}}{\text{in.}} \times \frac{12 \text{ in.}}{\text{ft}} = \frac{(2.54 \times 12) \text{ cm}}{\text{ft}}$$
$$= 30.48 \frac{\text{cm}}{\text{ft}}$$
$$= 30.48 \text{ cm } per \text{ foot}$$

That is, we have 2.54 centimeters twelve times.

The basic idea is that there is a *relationship* between inches and centimeters. This relationship is given by the rule:

> To convert inches to centimeters, multiply the number of inches by 2.54.

We often write this relationship in the form

$$\text{number of cm} = 2.54 \times \text{number of in.} \qquad [1]$$

When the relationship is given in the form of [1], we call it a **formula**.

We can now use [1] whenever we want to find the number of centimeters in a given number of inches. All we have to do is put the number of inches into formula [1] and do the arithmetic.

DEFINITION
A **formula** is a rule that tells us how two or more quantities (amounts) are related.

Example 3

Use [1] to find the number of centimeters in 36 inches.

There are 91.44 centimeters in 36 inches. Replace "number of inches" in [1] by 36 to obtain

number of centimeters = 2.54 × 36

Because there are 36 inches in a yard, there are 91.44 centimeters in 1 yard.

Then all you have to do is multiply 36 by 2.54. Note that Formula [1] doesn't even require that we have a whole number of inches.

Example 4

How many centimeters are there in 34.77 inches?

There are 88.3158 centimeters in 34.77 inches. All you do is replace "number of inches" by 34.77 in [1] to get

number of centimeters = 2.54 × 34.77

To summarize, a formula tells what to do and when to do it, but it's arithmetic that tells us *how* to do it. For example, in this lesson formula [1] tells us that 2.54 × 36 is the number of centimeters in 36 inches, *but* it's arithmetic that tells us that 2.54 × 36 = 91.44.

Once we know the formula, we don't have to worry about knowing what to do (if we believe the formula). This is what makes using arithmetic easier for most of us. In most situations where we need to use mathematics, we shall either have the formula or we shall be able to look up the formula somewhere.

This explanation of formulas may seem to suggest that we use formulas only when we don't know how two or more amounts are related and we have to look up the information. *This is not true.*

Even when we know what to do and when to do it, we still use formulas. For example, in the last section we saw that if a merchant bought 32 coats at $45 per coat, the total cost was $1,440. We said, "Just multiply $45 by 32." But we were actually using the formula:

Total price for the coats = Price per coat × Number of coats   [2]

This formula tells us more than how to find the cost of 32 jackets at $45 each. It tells us how to find the price of any number of jackets at any given price.

In other words, a formula describes the mathematical relationship between two or more quantities. But it doesn't matter whether we know the relationship already or whether we have to look it up. What [2] does is relate the three quantities "total price," "price per coat," and "number of coats." Once we choose values for two of these quantities, we are in a position to find the value of the third. We shall learn more about this in the next lesson.

Example 5

Use [2] to determine the cost of 72 coats at $34.79 per coat.

The cost of these coats is $2,504.88.

Using [2], replace "price per coat" by $34.79 and "number of coats" by 72. That is,

Total price = Price per coat ($34.79) × Number of coats (72)
            = $34.79 × 72

and $34.79 × 72 = $2,504.88.

Section 1.2   Mathematical Relationships

Example 6

You find the area of a rectangle by multiplying length by width. What is the area of a rectangle whose length is 15 feet and whose width is 9 feet?

The area of the rectangle is 135 *square* feet.
  Here you were given the formula verbally rather than by an equality. That is, you are told in different words that the formula is:

$$\text{Area of rectangle} = \text{Length of rectangle} \times \text{Width of rectangle} \quad [3]$$

All you did was to replace "length of rectangle" in [3] by 15 (feet) and "width of rectangle" by 9 (feet) to get

$$\begin{aligned}\text{Area of rectangle} &= 15 \text{ (feet)} \times 9 \text{ (feet)} \\ &= (15 \times 9) \text{ (feet} \times \text{feet)} \\ &= 135 \text{ square feet}\end{aligned}$$

Even when the length and width of a rectangle are not whole numbers, [3] is still the formula we use to find the area of a rectangle.

Example 7

Find the area of a rectangle whose length is 3.45 meters and whose width is 2.37 meters.

The area of the rectangle is 8.1765 *square* meters. You still use [3], but now you replace "length of rectangle" by 3.45 (meters) and "width of rectangle" by 2.37 (meters). This gives

$$\begin{aligned}\text{Area of rectangle} &= 3.45 \text{ (meters)} \times 2.37 \text{ (meters)} \\ &= (3.45 \times 2.37) \text{ } square \text{ meters} \\ &= 8.1765 \text{ square meters}\end{aligned}$$

Sometimes, using [3] may require that we know *how* to handle mixed numbers or common fractions. The "what to do" part of [3] stays the same, but the "how to do it" part requires the various skills of arithmetic.

Example 8

Find the area of a rectangle whose length is $8\frac{2}{3}$ feet and whose width is $4\frac{1}{5}$ feet.

The area is 36.4 $\left(\text{or } 36\frac{2}{5} \text{ or } \frac{182}{5}\right)$ square feet.

All you do is replace the length by $8\frac{2}{3}$ (feet) and the width by $4\frac{1}{5}$ (feet) in formula [3] to obtain

$$\begin{aligned}\text{Area of rectangle} &= 8\frac{2}{3} \text{ (feet)} \times 4\frac{1}{5} \text{ (feet)} \\ &= \left(8\frac{2}{3} \times 4\frac{1}{5}\right) \text{ square feet} \\ &= \left(\frac{26}{3} \times \frac{21}{5}\right) \text{ square feet} \\ &= \left(\frac{26 \times 21}{3 \times 5}\right) \text{ square feet} \\ &= \frac{182}{5} \text{ square feet}\end{aligned}$$

A NOTE ON AREA

Suppose we pick a unit of length, say 1 foot. Then a **unit square** is a square each of whose sides is 1 foot. The area of such a square is called a **square foot**.

Suppose we have a rectangle that is 3 feet long and 2 feet wide.

This rectangle can be divided into 6 (3 × 2) unit squares.

Because each unit square has an area of one square foot, the area of the rectangle is 6 *square feet, not* 6 feet. In other words, 3 feet × 2 feet is 6 (feet × feet).

Notice that with a calculator, the work necessary to use [3] is about the same whether you have whole numbers or decimal fractions.
  As a check, notice that since 3.45 is between 3 and 4 and 2.37 is between 2 and 3, 3.45 × 2.37 is more than 3 × 2 but less than 4 × 3. That is, the product must be between 6 and 12. This verifies that the decimal point is correctly placed.

Mathematically, all three forms of the answer are correct, but common practice is to use either the decimal or the mixed number rather than an improper fraction.

  Aside from the fact that $\frac{182}{5}$ can be confused with $18\frac{2}{5}$ when we write rapidly, it is usually harder to get a feeling for the size of $\frac{182}{5}$ than to get a feeling for, say, 36.4, or $36\frac{2}{5}$, which we know at once to be between 36 and 37.

Don't be misled into believing that formulas take the place of our having to think.

Example 9

Find the area of a rectangle whose length is 16 feet and whose width is 15 inches.

The area is 20 square feet or 2,880 square inches.

Did you avoid the trap of multiplying 16 by 15 in [3]? If you multiplied 16 by 15, you were multiplying feet by inches. The answer is neither square feet nor square inches. That is, you'd have

$$\text{Area of rectangle} = 16 \text{ (feet)} \times 15 \text{ (inches)}$$
$$= (16 \times 15) \text{ feet} \times \text{inches}$$
$$= 240 \text{ (feet} \times \text{inches)}$$

That is, to use *square* units, we must multiply a given unit by that same unit. If we multiply feet by feet, we get square feet; inches by inches, square inches; miles by miles, square miles, and so on.

You can change 16 feet to 192 (16 × 12) inches, in which case [3] gives the answer as 192 (inches) × 15 (inches) or (192 × 15) square inches; that is, 2,880 square inches. Or you can change 15 inches into $\frac{15}{12}$ or 1.25 feet and then use [3] to see that the area is 1.25 (feet) × 16 (feet) or (1.25 × 16) square feet; that is, 20 square feet.

Notice that while there are 12 inches in a foot, there are 144 (12 × 12) square inches in a square foot.

SOME SHORTHAND

Formula [3] may seem "bulky". It takes a lot of words and a lot of space to write

$$\text{Area of Rectangle} = \text{Length of rectangle} \times \text{Width of rectangle} \quad [3]$$

For this reason we use abbreviations, usually the first letter of the word. In [3], we may let the letter "A" stand for "Area of the rectangle." We may let the letter "L" stand for "Length" of the rectangle." And we may let "W" stand for "Width of the rectangle."

If we do this, then [3] becomes

$$A = L \times W \quad [4]$$

We "may," but we don't have to. The idea is that "A" suggests Area, and this makes it easier for us to recall what each symbol means.

For example, we could have written

$$Q = Z \times V$$

but these letters do not suggest area, length, and width.

Formula [4] is easier to read and to write than [3], *provided* that we know what A, L, and W stand for.

When formulas are written as in [4] rather than as in [3], we begin to think in terms of **algebra** rather than in terms of arithmetic. In many ways, algebra is just a symbolic form of arithmetic.

We shall have more to say about this in the next lesson, but note how closely algebra and arithmetic are related — and why it is important to understand arithmetic before you start studying algebra.

PRACTICE DRILL

1. If there are 2.2 pounds in 1 kilogram, how many pounds are in
   (a) 10 kilograms   (b) 10.5 kilograms
2. What is the area of a rectangle if the rectangle is
   (a) 13 yards long and 6 yards wide   (b) 13 yards long and 6 feet wide
3. If the formula is Number of pounds = 2.2 × Number of kilograms, how would you write this if you let p stand for the number of pounds and let k stand for the number of kilograms?
4. In the formula $G = 454 \times P$, what is the value of G when P is replaced by 3?

Answers
1. (a) 22 (b) 23.1
2. (a) 78 square yards (b) 26 square yards (or 234 square feet)
3. $p = 2.2 \times k$ 4. 1,362

Section 1.3 Squaring a Number

CHECK THE MAIN IDEAS

1. A mathematical relationship expressing an equality is often called a _____ .

2. There are 12 inches in a foot, so an example of a formula is

    Number of inches = _____ × Number of feet

3. If you let I stand for the number of inches and F for the number of feet, the formula becomes

    $I = 12 \times$ _____ .

4. If $F = 6$, then $I =$ _____ .

5. There are 72 (a) _____ in 6 (b) _____ .

Answers
1. formula   2. 12   3. F
4. 72   5. (a) inches   (b) feet

Section 1.3
Squaring a Number

Simply stated, **to square** a number means to multiply the number by itself.

Example 1

What is the square of 3?

The square of 3 is 9.
  The square of a number means the product of the number (in this case, 3) and itself. Because $3 \times 3 = 9$, the square of 3 is 9. Sometimes people say that 9 is 3 **squared** (instead of the square of 3).

  We use the word "square" because when we multiply a number by itself, we are finding the area of the square whose side is the given number.

Example 2

What is the area of a square if the length of a side of the square is 6.7 centimeters?

The area of the square is 44.89 square centimeters.
  You find the area by multiplying the length of the square (6.7 centimeters) by the width of the square (6.7 centimeters). In other words to find the number of square centimeters, you must square 6.7; that is, you must multiply 6.7 by itself.

  Remember, no matter what the number is, its square is always the *product* of the number and *itself*.

Example 3

What is the square of 134.7?

The square of 134.7 is 18,144.09. You find the answer by computing 134.7×134.7.
  There is a special way of indicating the square of a number to save space and time. When we want to multiply a number by itself (that is, to square the number), we place a small 2 *above* and *to the right* of the number.
  For example, 3^2 means 3×3, or 9.

OBJECTIVE
To know the meaning of squaring a number, to recognize the symbol that tells us to square a number, and to be able to use a formula in which a number is being squared.

A **square** is a rectangle whose length and width happen to be equal. So we can still say that the area of a square is given by the formula

$$A = L \times W$$

But because the length and width are equal in a square, we may use L (or W) for both the length and the width. That is:

$$A = L \times L$$

If we want to be impartial, we can let S stand for the "side of the square" and write

$$A = S \times S$$

To find the value of A we *square* the value of S.

Computations like this are easier to do using a calculator.

Example 4

What number is named by

(a) 6^2 (b) 6.54^2

The number named by 6^2 is 36, and the number named by 6.54^2 is 42.7716. All you have to remember is that 6^2 means 6×6 and 6.54^2 means 6.54×6.54.

Don't confuse 6^2 with 6×2. 6^2 is 6 times 6. 6×2 is 6 plus 6, or 12.

DEFINITION
If n stands for any number, then n^2 means $n \times n$.

We read n^2 as "n squared," "the square of n," or as "n to the second power." In this lesson, we shall use the terms "n squared" and "the square of n."

NOTE

The idea of writing $n \times n$ as n^2 leads to yet another shorthand symbol. In the last section, we said that when we multiply feet by feet, we get square feet. If we used the new notation on the "label" feet, we have

$$\text{feet} \times \text{feet} = \text{feet}^2$$

People in science and technology do indeed use feet^2 as an abbreviation for "square feet."

We usually abbreviate "feet" by "ft," so we often write "ft^2" as an abbreviation for "square feet."

Although the idea of squaring a number is easy to visualize as finding the area of a square, this is not the only time we square a number. In the seventeenth century, the famous Italian physicist and astronomer Galileo Galilei found a mathematical relationship between the distance a freely falling body travels (in the absence of air resistance) and the time it takes to fall that distance.

Galileo showed that a freely falling object fell a distance of d feet in a time of t seconds, according to the formula (rule)

$$d = 16 \times (t^2) \qquad [1]$$

Galileo was born in Pisa in 1564 and died in Arcetri in 1642. Many of the relationships he discovered are still taught in physics classes.

That is, distance = $16 \times$ (time $\times$ time)

The parentheses in [1] tell us first to square t — that is, (t^2) is one number — and then to multiply this result by 16.

Formula [1] should not be frightening. You may have to study physics to learn how the formula came to be. But once formula [1] is given, you can use it to get answers even if you don't know for sure where it comes from or why it's true.

Let's try a few examples in which we use [1].

This says that if we take the time (in seconds) that the object falls and multiply this time by itself, and then multiply this product by 16, we find the distance (in feet) that the object falls during this time.

Example 5

Find the value of d in [1] if t is 3.

When t is 3, d is 144. You find this answer by writing

$$d = 16 \times (t^2) \qquad [1]$$

and replacing t by 3. This yields

$$\begin{aligned} d &= 16 \times (3^2) \\ &= 16 \times 9 \\ &= 144 \end{aligned}$$

Remember, 3^2 still means 3×3.

Notice that to use [1], we don't have to know what d and t mean. Of course, if we do know what d and t mean in [1], we can use [1] to find the distance a freely falling object travels in 3 seconds.

Section 1.3 Squaring a Number

Example 6

According to [1], how far will a freely falling body fall in 3 seconds?

It will fall 144 feet. All you do is replace t by 3 in [1], and find the value of d just as you did in Example 5.

The only difference is that in this example, you are using [1] to get an answer to a real-life problem, while in Example 5 you just had to know how to read [1] and do the arithmetic.

Of course, this assumes that the body is at least 144 feet above the ground, because once it hits the ground, it's no longer freely falling.

Again, notice that the formula is used in the same way when the numbers are not whole numbers.

Example 7

According to [1], how far does a freely falling body fall in 6.8 seconds?

It falls 739.84 feet in 6.8 seconds. Just as before, you use [1], but now you replace t by 6.8. This gives

$$d = 16 \times (6.8^2) = 16 \times (6.8 \times 6.8)$$
$$= 16 \times 46.24 = 739.84$$

Be careful when grouping the numbers. Note the following.

Example 8

Suppose d and t are related by the formula

$$d = (16 \times t)^2 \qquad [2]$$

What is the value of d in [2] when $t = 3$?

Using [2], you see that $d = 2{,}304$ when $t = 3$. That is, you first replace t by 3 in $(16 \times t)$ to find that $(16 \times t) = (16 \times 3) = 48$. Then

$$d = (16 \times 3)^2 = 48^2 = 48 \times 48 = 2{,}304$$

In other words, in Example 6 you computed the value of $16 \times (3^2)$, and in this example you computed the value of $(16 \times 3)^2$. So it makes a difference how you group the numbers. The parentheses tell you how the terms should be grouped.

Always make sure that you have the right units (or "labels") when you use a formula.

Notice that [1] and [2] do not give the same relationship. In [1], we first multiplied t by itself (t^2) and then we multiplied this by 16. In [2], the parentheses name a single number. *The* number being squared in [2] is $(16 \times t)$. That is, in [2] we first multiply t by 16, *then* we square the entire product.

This also happens in nonmathematical situations, but instead of parentheses we use hyphens or voice inflection. For example, does

"the high school building"

mean a tall elementary school (in which case we might write: the high school-building) or does it mean a one-story building that has grades 9 through 12 (in which case we would write the high-school building)?

If we used parentheses, the difference is whether we mean

the high (school building)

or

the (high school) building

Everything inside the parentheses is treated as if it were one word.

Example 9

Use [1] to determine how far a freely falling body falls in 1 minute.

In one minute, the freely falling body falls 57,600 feet.

The trick is that in [1] the time must be in *seconds* (if you use minutes, the relationship between distance and time is different). Since 1 minute is 60 seconds, you replace t in [1] by 60, *not by 1*. Then the answer is $d = 16 \times (60^2) = 16 \times 3{,}600$ or 57,600 feet.

Because there are 5,280 feet in 1 mile, if you wanted to, you could divide 57,600 by 5,280 to convert the distance to miles. If you did, the answer would be almost 11 miles $\left(10\frac{10}{11}\right)$.

PRACTICE DRILL

1. Find the value of n^2 if
   (a) $n = 5$ (b) $n = 5.5$ (c) $n = 55$
2. What is the area of a square if the length of a side of the square is
   (a) 5 inches (b) 5.5 feet (c) 55 yards
3. In the formula $d = 16 = (t^2)$, what is the value of d when
   (a) $t = 6$ (b) $t = 6.6$
4. In the formula $d = (16 \times t)^2$, what is the value of d when $t = 6$?
5. If a freely falling body falls a distance of d feet in t seconds according to the formula $d = 16 \times (t^2)$, how far does the body fall in 6 seconds?

Answers
1. (a) 25 (b) 30.25 (c) 3,025
2. (a) 25 square inches (b) 30.25 square feet (c) 3,025 square yards
3. (a) 576 (b) 696.96
4. 9,216 5. 576 feet

CHECK THE MAIN IDEAS

1. Squaring a number involves the operation of _____ .
2. To square a number means that you _____ the number by itself.
3. You square a number when you find the area of a _____ .
4. When you square 4, the answer is _____ .
5. Do not confuse squaring with doubling. If you double 4, you get _____ .

Answers
1. multiplication 2. multiply
3. square 4. 16 5. 8

Section 1.4
Finding the Square Root of a Number

Have you noticed that you are able to do subtraction as soon as you know how to do addition? This is how we usually "make change." There is a similar situation for squaring a number.

In the last section, we worked with finding the area of a square if we knew the length of a side of the square. Let's review quickly.

OBJECTIVE
To know the meaning of square root.

In fact, to check subtraction, we add. For example, the check that $33 - 19 = 14$ is that $14 + 19 = 33$. In other words, $33 - 19$ is the number that must be *added* to 19 to give 33. This is what we mean by "making change." We add on to 19 the amount necessary to give 33.

Example 1

Find the area of a square if the length of one of its sides is 10.7 yards.

The area of the square is 114.49 yd² (that is, 114.49 square yards). You find the answer by multiplying 10.7 (yards) by itself.

In this section, we ask roughly the "opposite" question. Instead of starting with the length of a side and finding the area of the square, we start with the area of the square and see if we can find the length of the side of the square.

Example 2

If the area of a square is 100 square feet, what is the length of a side of the square?

The length of a side is 10 feet.
You want a number that when multiplied by itself is 100. Because $10 \times 10 = 100$, the number you want is 10. Moreover, because the units are square feet, you are multiplying feet by feet. So, the answer is 10 feet.

Do you see the basic difference between what we did in Example 1 and what we did in Example 2? In Example 2, we don't want to multiply 100 by itself. We

Pictorially:

10 feet

10 feet | 100 square feet

Section 1.4 Finding the Square Root of a Number

want to find the number that has to be multiplied by itself to give 100 as the product.

There is an abbreviation for this. When we write $\sqrt{100}$, we are asking for the number that when multiplied by itself (that is, the number whose square) equals 100.

$\sqrt{100}$ is read as "the **square root** of 100."

In other words,

$$\sqrt{100} \times \sqrt{100} = 100$$

Because $10 \times 10 = 100$, $\sqrt{100} = 10$.

Example 3

What number is named by $\sqrt{64}$?

The square root of 64 is 8.
Remember, $\sqrt{64}$ means the number whose square is 64. That is, $\sqrt{64} \times \sqrt{64} = 64$. Because the square of 8 (8×8) is 64, you see that $\sqrt{64} = 8$.

When we start with the length of a side and find the area of the square, we are squaring a number. When we have the area and want to find the length of the side, we take the square root. For example, $8^2 = 64$ says that if the length of a side is 8, the area is 64. $\sqrt{64} = 8$ says that if the area is 64, then the length of a side is 8.

One of the hardest things about finding square roots is that we often have to estimate the answer in fraction form.

For example, suppose we want the square root of 95 (that is, $\sqrt{95}$). This is *exactly* the number whose square is 95. But what does this number look like?

$$9 \times 9 = 81 \quad \text{and} \quad 10 \times 10 = 100$$

This gives us a hint that the square root of 95 is greater than 9 but less than 10.

Is the square root of 95 closer to 9 or closer to 10? To answer this, we see what happens if we square the average of 9 and 10. That is, we square 9.5. Since $9.5^2 = 90.25$, what we have so far is

$$\sqrt{95} \rightarrow \begin{array}{rcl} 9^2 & = & 81 \\ 9.5^2 & = & 90.25 \\ ? & = & ? \\ 10^2 & = & 100 \end{array} \leftarrow 95$$

In terms of area, we want to know the length of the side of a square if the area is 95 square feet. If one side were 9 feet, then the area would be 81 ft². So since the area is 95 ft², the length of a side must be more than 9 ft.

Similarly, if the length of a side were 10 ft, the area would be 100 ft², which is more than 95 ft². So the length must be less than 10 ft.

Therefore, the length of the side of the square is between 9 feet and 10 feet.

In other words, if we square 9.5, the answer is less than 95, but if we square 10 the answer is more than 95. So the number whose square is 95 is between 9.5 and 10; therefore, it is closer to 10 than to 9.

Visually:

```
─────────────────●──────────●×××××××●───────
                 9         9.5         10
```

Anything to the right of 9.5 is closer to 10 than to 9.

We can now say that rounded off to the nearest whole number, the square root of 95 is 10.

If we want a more accurate answer, we can find that

$$\sqrt{95} \rightarrow \begin{array}{rcl} 9.5^2 & = & 90.25 \\ 9.6^2 & = & 92.16 \\ 9.7^2 & = & 94.09 \\ 9.8^2 & = & 96.04 \end{array} \leftarrow 95$$

That is, $\sqrt{95}$ is between 9 and 10 but closer to 10.

This tells us that the square root of 95 is between 9.7 and 9.8. We can then look at 9.71^2, 9.72^2, 9.73^3, and so on, to get an even better estimate for the square root of 95.

Some hand calculators have a square-root key on them. All you have to do is put in the number and then press the key labeled "$\sqrt{}$." Then the square root

appears on the calculator display. The answer is
$$\sqrt{95} = 9.746794$$
As a check, a calculator may show that
$$9.746794 \times 9.746794 = 94.999993$$
and
$$9.746795 \times 9.746795 = 95.000012$$
⤡ 95 ⤢

Table 1 at the end of the book lists the square roots of several numbers. This should be helpful to you if you don't have a calculator with a square-root key.

Example 4

Given the formula
$$m = \sqrt{n+3} \qquad [1]$$
what is the value of m if n is 22?

When n is 22, m is 5. If you replace n in [1] by 22, you get
$$m = \sqrt{22 + 3}$$
$$= \sqrt{25}$$
$$= 5, \text{ as } 5 \times 5 = 25$$

Remember, $5^2 = 25$ and $\sqrt{25} = 5$ are two different ways of saying the same thing (just as $3 + 2 = 5$ and $5 - 3 = 2$ are different ways of saying the same thing).

Again, make sure that you read the problem correctly and use the right grouping.

Example 5

Given the formula
$$m = \sqrt{n} + 3 \qquad [2]$$
what is the value of m (to the nearest tenth) when n is 22?

When n is 22, m is 7.7 (to the nearest tenth).
In [2], first you take the square root of 22 (n), and *then* you add 3. To the nearest tenth, $\sqrt{22} = 4.7$. Therefore
$$m = \sqrt{22} + 3 = 4.7 + 3 = 7.7$$

That is

$\sqrt{22}$ is in here ⟶
$4.6^2 = 21.16$
$4.65^2 = 21.6225$ ⟵ 22
$4.7^2 = 22.09$

In the next lesson we shall go into formulas in more detail; this will begin our study of algebra. For now, we want to make sure that we know the difference between a "how to" and a "what and/or when to" problem, and that we are able to work with formulas. The exercise set in this lesson provides plenty of opportunity to practice what we've been learning.

PRACTICE DRILL

1. What is the square root of
   (a) 81   (b) 121   (c) 12,100
2. Between what two consecutive whole numbers is the square root of
   (a) 5   (b) 50   (c) 500
3. If $B = \sqrt{A} + 7$, find the value of B when A is 9.
4. If $B = \sqrt{A + 7}$, find the value of B when A is 9.

Answers
1. (a) 9 (b) 11 (c) 110
2. (a) 2 and 3 (b) 7 and 8 (c) 22 and 23   3. 10   4. 4

Section 1.5 The Metric System   17

CHECK THE MAIN IDEAS

1. In visual terms, when you square a number, you are finding the _____ of a square.
2. When you take the square root of a number, you are finding the length of the (a) _____ of the square, once you know the (b) _____ of the square.
3. The (a) _____ of 9 is 81, and the (b) _____ of 81 is 9.
4. The symbolic way to write the square root of 81 is _____ .
5. The square root of 100 is _____ .
6. Since 85 is between 81 and 100, the square root of 85 is between the consecutive whole numbers (a) _____ and (b) _____ .

Answers
1. area 2. (a) side (b) area
3. (a) square (b) square root
4. $\sqrt{81}$ 5. (a) nine (b) ten

Section 1.5
The Metric System

Example 1

How many inches are there in one mile?

There are 63,360 inches in one miles.
 You want inches per mile, so you have

$$12\,\frac{\text{in.}}{\text{ft}} \times 3\,\frac{\text{ft}}{\text{yd}} \times 1{,}760\,\frac{\text{yd}}{\text{mi}} = (12 \times 3 \times 1{,}760)\,\frac{\text{in.}}{\text{mi}}$$
$$= 63{,}360 \text{ inches per mile}$$

You also could have used the fact that there are 5,280 feet in one mile to obtain

$$12\,\frac{\text{in.}}{\text{ft}} \times 5{,}280\,\frac{\text{ft}}{\text{mi}} \times (12 \times 5{,}280)\,\frac{\text{in.}}{\text{mi}}$$

and this also tells you that there are 63,360 inches in one mile.

   Look what happened in Example 1. First we had to know such numbers as 12, 3, and 1,760. Then we had to multiply these numbers to get the answer.
   On the other hand, see how much easier it is in terms of money, where we use a decimal system.

Example 2

How many mills are there in $1?

There are 1,000 mills in $1.
 Because we want mills per dollar, we write

$$10\,\frac{\text{mils}}{\text{cent}} \times 10\,\frac{\text{cents}}{\text{dime}} \times 10\,\frac{\text{dimes}}{\text{dollar}} = (10 \times 10 \times 10)$$
$$= 1{,}000 \text{ mills per dollar}$$

   See how Example 2 compares with Example 1? Instead of having to memorize such numbers as 3, 12, 5,280, or 1,760, all we have to do is remember the number 10. We don't have to multiply numbers like 3, 12, and 5,280; we need only to find the product of 10's.
   The English system might be easier to learn if it, too, used a decimal system. There is nothing very bad in definitions like the following:

OBJECTIVE
To learn the metric system and to see how it is related to our English system.

REVIEW
$$12 \text{ inches} = 1 \text{ foot}$$
$$3 \text{ feet} = 1 \text{ yard}$$
$$1{,}760 \text{ yards} = 1 \text{ mile}$$

Remember how we cancel units?

$$\frac{\text{in.}}{\cancel{\text{ft}}} \times \frac{\cancel{\text{ft}}}{\cancel{\text{yd}}} \times \frac{\cancel{\text{yd}}}{\text{mi}} = \frac{\text{in.}}{\text{mi}}$$

This tells us the same thing as

$$1 \times 1{,}760 = 1{,}760 \text{ yd/mi}$$
$$1{,}760 \times 3 = 5{,}280 \text{ ft/mi}$$
$$5{,}280 \times 12 = 63{,}360 \text{ in./mi}$$

REVIEW
A mill is one-tenth of a cent.
$$10 \text{ mills} = 1 \text{ cent}$$

$$\frac{\text{mills}}{\cancel{\text{cent}}} \times \frac{\cancel{\text{cent}}}{\cancel{\text{dime}}} \times \frac{\cancel{\text{dime}}}{\text{dol}} = \frac{\text{mills}}{\text{dol}}$$

Remember that multiplying and dividing by 10's only involves moving a decimal point. For example,

$$2.16 \times 1{,}000 = 2160.$$
$$\phantom{2.16 \times 1{,}000 = }{}^{\wedge}{\scriptstyle 1\,2\,3}$$

$$2.16 \div 1{,}000 = 0.00216$$
$$\phantom{2.16 \div 1{,}000 = 0.00}{\scriptstyle 3\,2\,1}{}^{\wedge}$$

$$10 \text{ inches} = 1 \text{ foot}$$
$$10 \text{ feet} = 1 \text{ yard}$$
$$1{,}000 \text{ yards} = 1 \text{ mile}$$

The idea of using a decimal system was carried out in the metric system. The metric system uses the *meter* as a basic unit of length. Other units of length are then defined as:

$$10 \text{ millimeters} = 1 \text{ centimeter}$$
$$10 \text{ centimeters} = 1 \text{ decimeter}$$
$$10 \text{ decimeters} = 1 \text{ meter}$$
$$10 \text{ meters} = 1 \text{ dekameter}$$
$$10 \text{ dekameters} = 1 \text{ hektameter}$$
$$10 \text{ hektameters} = 1 \text{ kilometer}$$

Some of these units are more commonly used than others. Usually we have:

$$10 \text{ millimeters} = 1 \text{ centimeter}$$
$$100 \text{ centimeters} = 1 \text{ meter}$$
$$1{,}000 \text{ meters} = 1 \text{ kilometer}$$

SOME ABBREVIATIONS
mm = millimeter(s)
cm = centimeter(s)
m = meter(s)
km = kilometer(s)

As you may know, since most of the rest of the world (about 95%) uses the metric system, there is a serious attempt being made to have America convert to the metric system. This text is not the place to debate the merits of such an attempt, but it is a good reason to study the metric system in this section.

Example 3

How many millimeters are there in one meter?

There are 1,000 millimeters in 1 meter.
 This is exactly the same problem (with slightly different words) as Example 2. Both mill and millimeter indicate one thousandth. More specifically:

$$10 \tfrac{\text{mm}}{\text{cm}} \times 10 \tfrac{\text{cm}}{\text{dm}} \times 10 \tfrac{\text{dm}}{\text{m}} = 1{,}000 \tfrac{\text{mm}}{\text{m}}$$

or

$$10 \tfrac{\text{mm}}{\text{cm}} \times 100 \tfrac{\text{cm}}{\text{m}} = 1{,}000 \tfrac{\text{mm}}{\text{m}} = 1{,}000 \text{ millimeters per meter}$$

Even if the English system were more convenient and no changes were useful, you can see that having two systems causes problems. As a result, it is helpful to be able to change measurements from one system to another.
 We've already seen in this lesson that one relationship between English lengths and metric lengths is given by

$$2.54 \text{ centimeters} = 1 \text{ inch}$$

We showed that although it may be convenient to know more interrelationships, one is all we really need.
 It may be difficult to keep track of units of length in the English system, but weights and volumes are much more difficult. How would you like to memorize the following facts?

$$3 \text{ teaspoons (tsp)} = 1 \text{ tablespoon (tbsp)}$$
$$2 \text{ tablespoons} = 1 \text{ fluid ounce (fl oz)}$$
$$8 \text{ fluid ounces} = 1 \text{ cup (c)}$$
$$2 \text{ cups} = 1 \text{ pint (pt)}$$
$$2 \text{ pints} = 1 \text{ quart (qt)}$$
$$4 \text{ quarts} = 1 \text{ gallon (gal)}$$
$$2 \text{ gallons} = 1 \text{ peck (pk)}$$
$$4 \text{ pecks} = 1 \text{ bushel (bu)}$$
$$3\tfrac{15}{16} \text{ bushels } (31\tfrac{1}{2} \text{ gallons}) = 1 \text{ barrel (bl)}$$

In the metric system, the basic unit of volume is a **cubic centimeter (cc)**. 1,000 cubic centimeters make a **liter**.

You may have noticed the problem if you've been in another country. It is often very complicated and confusing to exchange our money for money used in the other country. We have to know how the dollar is related to pounds or lire or pesos or rubles.

For example, in some books we are told that 1 foot is the same length as 30.48 centimeters. But we can determine this just by computing 12 × 2.54 (see Example 2 in Section 1.2).

Don't memorize these facts (unless you'd like to). You can look them up if you need them.

All we're trying to show here is how much there is to memorize in the English system.

A liter is a little more than a quart. It is 1.06 quarts. Table 2 at the end of the book lists several other English–metric conversions.

Section 1.5 The Metric System

Example 4

How many cubic centimeters are there in one cubic inch?

One cubic inch is (about) 16.4 cc.
  A cubic inch means a cube that is 1 inch on each side. That means that the volume is 1 inch × 1 inch × 1 inch = 1 cubic inch. An inch is about the same length as 2.54 centimeters, so the dimensions of the cube are also 2.54 centimeters on each side. Hence, the volume is also given by

$$2.54 \text{ cm} \times 2.54 \text{ cm} \times 2.54 \text{ cm} = 16.387064 \text{ cubic centimeters}$$

  The fact that there are about 16.4 cubic centimeters in one cubic inch allows us to convert any metric volume into an English equivalent. We just convert the metric volume into cubic centimeters and use the fact that there are about 16.4 cubic centimeters in 1 cubic inch.

We rounded off our answer. $2.54 \times 2.54 \times 2.54$ is exactly 16.387064, but there are only approximately 2.54 centimeters in 1 inch.

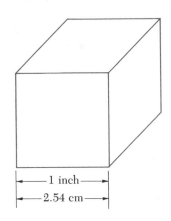

Example 5

How many cubic inches are there in a liter?

There are about 61 cubic inches in 1 liter.
  You want cubic inches per liter, which is $\frac{\text{cu in.}}{\text{lit}}$. This is the same as

$$\frac{\text{cu in.}}{\text{cc}} \times \frac{\text{cc}}{\text{lit}}$$

If you put in the correct numbers, you have

$$\frac{1 \text{ cu in.}}{16.4 \text{ cc}} \times \frac{1000 \text{ cc}}{1 \text{ lit}} = \frac{(1 \times 1000)}{(16.4 \times 1)} \text{ cubic inches per liter}$$

And as $\frac{1000}{16.4}$ is about 61, there are about 61 cubic inches per liter.

  Do you see that the metric system is more "orderly" than the English system, and that you can convert from a unit in one system to a unit in the other just by knowing one interrelationship?
  Let's look at one more example.

The cube contains the same amount of space (it's the same cube) regardless of whether the measurements are in inches or centimeters.

We don't have to do the problem this way. We can simply divide 1,000 by 16.4, because every 16.4 cc is 1 cubic inch. The unit analysis just makes it easier for us to remember whether we should multiply or divide; and if we divide, which number is divided by which.

Example 6

In American cars, the engine capacity is usually measured in cubic inches. In other countries, this capacity is measured in liters. Approximately how many cubic inches are there in an engine whose capacity is 5 liters?

A 5 liter engine has about 325 cubic inches.
  We have already found out in the last example that there are about 61 cubic inches in 1 liter. So in 5 liters there are five times 61 cubic inches, or 325 cubic inches.

  The English system of weights is no simpler than the English system of lengths and volumes. For example, we have the following facts:

$$16 \text{ ounces} = 1 \text{ pound (lb)}$$
$$2{,}000 \text{ pounds} = 1 \text{ ton}$$

  In the metric system, the basic unit of weight is a **gram,** and 1,000 grams is called 1 **kilogram.**

This is a more practical example of where we convert between English and metric units.

$$61 \frac{\text{cu in.}}{\text{lit}} \times 5 \text{ lit} = (61 \times 5) \text{ cu in.}$$

Notice how the language is also easy to learn in the metric system? For example, "kilo" always means 1,000. In the same way, "milli" always means one thousandth. So, for example, a milligram (mg) is one-thousandth of a gram.

Example 7

There are about 454 grams in 1 pound. About how many grams are in 1 ounce?

To the nearest gram, there are 28 grams in 1 ounce.
   If there are 16 ounces in 1 pound, there are 454 grams *per* 16 ounces. Therefore, there are (454 ÷ 16) grams per ounce; and 454 ÷ 16 = 28.375, which is closer to 28 than it is to 29.

> Remember that "per" means "divided by."

   You have probably noticed the effects of the metric system on temperature readings. It is now very common to give the reading in both degrees Fahrenheit (F) and degrees Celsius (C). Fahrenheit is associated with the English system, and Celsius is associated with the metric system. There is a formula that relates degrees Celsius to degrees Fahrenheit. The formula is

$$C = \frac{5}{9} \times (F - 32) \quad [1]$$

> Before the word "Celsius" was used, the temperature unit was called "centigrade." Notice that "centi" suggests 100 ("grad" means degree). Celsius comes from the name of the person who invented the centigrade scale. The main point is that centigrade indicates a decimal system of measurement.

   Equation [1] is a shortcut for telling us how to change a Fahrenheit reading into Celsius. First we subtract 32 from the Fahrenheit reading. Then we multiply this result by $\frac{5}{9}$.

> That's what F − 32 means. The fact that F − 32 is in parentheses tells us that F − 32 is being treated as a single number.

Example 8

What is the Celsius reading when the Fahrenheit reading is 68?

When the Fahrenheit reading is 68, the Celsius reading is 20.
   All you do is replace F by 68 in the formula

$$C = \frac{5}{9} \times (F - 32) \quad [1]$$

We get

$$C = \frac{5}{9} \times (68 - 32) = \frac{5}{9} \times 36 = 20$$

C stands for the Celsius reading, so 68°F is equivalent to 20°C.

> That is, 68°F is the same temperature as 20°C.

PRACTICE DRILL

1. How many centimeters are there in 2 meters?
2. If there are 2.54 centimeters in 1 inch, how many centimeters are there in 100 inches?
3. If there are 454 grams in 1 pound, how many pounds are 4,540 grams?

> **Answers**
> 1. 200   2. 254   3. 10

CHECK THE MAIN IDEAS

1. The _____ system is the measuring system that most of the world uses.
2. The system that is currently used in America is the _____ system.
3. One nice feature of the metric system is that _____ of one denomination always equals one of the next higher denomination.
4. This is much easier than having to remember that there are (a) _____ inches in a foot and (b) _____ feet in a yard.

> **Answers**
> 1. metric   2. English
> 3. ten (10)   4. (a) 12
> (b) 5,280   5. one   6. 12

Exercise Set 1 (Form A)

5. If you want to convert from one system to the other, all you have to know is _____ interrelationship.
6. If you know that there are 2.54 centimeters in 1 inch, you do not have to memorize how many centimeters there are in 1 foot. All you have to do is multiply 2.54 by _____ to see that there are 30.48 centimeters per foot.

EXERCISE SET 1 (Form A)

Section 1.1

1. If each of 58 people donates $65 to charity, what is the total amount of their donation?
2. Of 80 people in a plant, 58 each donate $65 to charity and the other 22 each donate $85. What is the total amount donated?
3. In Exercise 2, what was the average donation per person to charity?
4. A merchant buys 42 jackets at $37 each and 27 jackets at $57 each. If his budget for jackets is $3,500, how much money does he still have left for the purchase of jackets?
5. If a car travels 31.6 miles on each gallon of gas, how many miles will it travel on 8.9 gallons of gas?
6. You have $234 in the bank. You make deposits of $34.50, $118.75, and $54.65. You also make withdrawals of $56,87, $23.75, $24.15, and $32.45. How much money do you then have left in the bank?
7. The price of a radio is $38. In addition, there is a 7% sales tax on the radio. What is the cost of the radio, including the tax?
8. Suppose you went 240 miles on a trip and it took you $3\frac{3}{4}$ hours. What was your average speed for the trip?
9. Suppose a car went $\frac{2}{5}$ of a mile in $\frac{3}{7}$ of a minute.

    (a) What was the average speed of the car in miles per minute?
    (b) What was the speed of the car in miles per hour?
10. What is the price per pound of ground beef if you can buy $3\frac{3}{4}$ pounds for $4.50?
11. If you can buy $3\frac{3}{4}$ pounds of ground beef for $4.50, how much will it cost you to buy 12 pounds?
12. If you can buy $3\frac{3}{4}$ pounds of ground beef for $4.50, how many pounds can you buy for $12?

Section 1.2

13. What is the area of a rectangle (in square feet) if its length is 4.5 feet and its width is 2.3 feet?
14. What is the area of a rectangle (in square meters) if its length is 4.5 meters and its width is 2.3 meters?
15. What is the area of a rectangle (in square feet) if its length is 3 yards and its width is 4 feet?
16. Find the value of h when $k = 12$ if

    (a) $h = 5 \times (k + 6)$ (b) $h = (5 \times k) + 6$ (c) $h = 5 \times (k - 6)$

Answers: Exercise Set 1

Form A
1. $3,770   2. $5,640
3. $70.50   4. $407
5. 281.24 miles   6. $304.68
7. $40.66   8. 64 mph
9. (a) $\frac{14}{15}$ (b) 56 mph
10. $1.20   11. $14.40
12. 10 lbs.   13. 10.35 square feet   14. 10.35 square meters
15. 36 square feet   16. (a) 90 (b) 66 (c) 30   17. (a) 64 (b) 66.0969 (c) $\frac{4}{9}$ (d) $6\frac{1}{4}$ (6.25)
18. 66.0969 cm² (square cm)
19. $\frac{4}{9}$ 20. $6\frac{1}{4}$ or 6.25 square feet
21. (a) 144 feet   (b) 256 feet (c) 196 feet   22. 112 feet
23. 25   24. 26   25. 6
26. 30 meters (m)   27. 30.8 meters   28. 79   29. 198.6
30. 15 pounds, 6 ounces

Section 1.3

17. What is the value of n^2 if

    (a) $n = 8$ (b) $n = 8.13$ (c) $n = \frac{2}{3}$ (d) $n = 2\frac{1}{2}$

18. What is the area of a square if the length of each side is 8.13 cm?
19. What is the area of a square, in square yards, if the length of each side of the square is 2 feet?
20. What is the area of a square, in square feet, if the length of each side is 2 feet, 6 inches?
21. An object falls a distance of d feet at the end of t seconds according to the rule $d = 16 \times (t^2)$. How far has the object fallen at the end of

    (a) 3 seconds   (b) 4 seconds   (c) $3\frac{1}{2}$ seconds

22. Use the facts in Problem 21 to find how far the object falls during the fourth second.

Section 1.4

23. What is the square root of 625?
24. Find the value of $\sqrt{670}$ rounded off to the nearest whole number.
25. Suppose m and n are related by the formula $m = \sqrt{3 \times n}$. What is the value of m when $n = 12$?
26. What is the length of each side of a square if the area of the square is 900 square meters?

Section 1.5

27. What is the length of each side of a square if the area of the square is 950 square meters? Write your answer correct to the nearest tenth of a meter (decimeter).
28. If there are 2.54 centimeters in one inch, how many inches are there in 2 meters? Round off your answer to the nearest inch.
29. If there are 454 grams in one pound, how many grams are there in 7 ounces? Write your answer correct to the nearest tenth of a gram (there are 16 ounces in one pound).
30. If there are 454 grams in one pound, how many pounds are there in 7 kilograms? Write your answer correct to the nearest ounce.

EXERCISE SET 1 (Form B)

Section 1.1

1. If each of 62 men contributes $70 to a fund, what is the total amount of their contribution?
2. Of 95 women in a company, 62 contribute $70 each to a charity and the other 33 contribute $50 each. What is the total amount they contributed to the charity?
3. In Exercise 2, what was the average donation per person to charity?
4. A merchant buys 53 coats for $27 each and 34 coats for $75 each. If her budget for coats is $5,000, how much money does she still have left to buy coats?

Exercise Set 1 (Form B)

5. If a car travels 17.4 miles on each gallon of gas, how far will it travel on 11.7 gallons of gas?
6. You have $453 in the bank. You make deposits of $45.63, $78.53, and $117.94. You also make withdrawals of $193.47, $187.54, and $82.51. How much money do you then have left in the bank?
7. The price of a radio is $53. In addition there is a 9% sales tax. What is the price of the radio, including the 9% tax?
8. Suppose you went 340 miles in $4\frac{1}{4}$ hours on a trip. What was your average speed for this trip?
9. Suppose a car travels $\frac{2}{5}$ of a mile in $\frac{4}{9}$ of a minute.
    (a) What is the average speed of the car in miles per minute?
    (b) What is the average speed of the car in miles per hour?
10. What is the price per pound of candy if you can buy $2\frac{1}{7}$ pounds for $3.75?
11. If you can buy $2\frac{1}{7}$ pounds of candy for $3.75, how much will it cost to buy 14 pounds of candy?
12. If you can buy $2\frac{1}{7}$ pounds of candy for $3.75, how many pounds can you buy for $14?

Section 1.2
13. What is the area of a rectangle (in square feet) if its length is 8.5 feet and its width is 6.1 feet?
14. What is the area of a rectangle (in square centimeters) if its length is 8.5 centimeters and its width is 6.1 centimeters?
15. What is the area of a rectangle (in square inches) if its length is 2 feet and its width is 5 inches?
16. Find the value of m when $c = 5$ if
    (a) $m = 14 \div (2 + c)$ (b) $m = (14 \div 2) + c$ (c) $m = 14 \div (7 - c)$

Section 1.3
17. What is the value of n^2 if
    (a) $n = 11$ (b) $n = 11.7$ (c) $n = \frac{7}{12}$ (d) $n = 2\frac{1}{3}$
18. What is the area of a square if the length of each side is 11.7 centimeters?
19. What is the area of a square (in square feet) if the length of each side of the square is 7 inches?
20. What is the area of a square (in square feet), if the length of each side of the square is 2 feet, 4 inches?
21. An object falls d feet at the end of t seconds according to the rule $d = 16 \times (t^2)$. How far has the object fallen at the end of
    (a) 5 seconds (b) 6 seconds (c) $5\frac{1}{2}$ seconds
22. According to Exercise 21, how far does the object fall during the sixth second?

Section 1.4
23. What is the square root of 1,225?
24. Find the square root of $\sqrt{1,290}$ rounded off to the nearest whole number.
25. Suppose b and c are related by the formula $b = \sqrt{c - 50}$. What is the value of b when $c = 66$?
26. What is the length of each side of a square if the area of the square is 2,500 square feet?

Section 1.5

27. What is the length of each side of a square, to the nearest tenth of a meter (decimeter), if the area of the square is 2,575 square meters?
28. If there are 2.54 centimeters in one inch, how many inches are there in 12 meters? Round off your answer to the nearest inch.
29. If there are 454 grams in one pound, how many grams are there in 19 ounces? Write your answer correct to the nearest tenth of a gram.
30. If there are 454 grams in one pound, how many pounds are there in 10 kilograms? Write your answer correct to the nearest ounce.

Lesson 2

An Introduction to Algebraic Equations

Overview   In this lesson we introduce algebra. The approach we shall use was set up by the work we did with formulas in Lesson 1.

Recall that in Lesson 1 we applied our knowledge of arithmetic to mathematical relationships called formulas. To review, suppose we have the formula

$$m = (3 \times n) + 5 \qquad [1]$$

and we want to find the value of m when n is 4. We replace n by 4 in [1] to obtain:

$$m = (3 \times 4) + 5 \qquad [2]$$

from which we conclude that when n is 4, m is 17.

In this lesson, our goal is to use formula [1] to show the relatively simple connection between algebra and arithmetic. When we are given n and asked to find m, the procedures of Lesson 1 allow us to solve this problem with arithmetic. Now suppose we are given the value of m and are asked to find the value of n.

If we are given [1] and asked to find the value of n when m is 17, we would replace m by 17 in [1] to obtain:

$$17 = (3 \times n) + 5 \qquad [3]$$

Both [2] and [3] are examples of "fill in the blank," where we use m and n rather than blanks. The difference is that we can find the value of m in [2] just by doing arithmetic: we multiply 3 by 4 and add 5 to the result.

But in [3], n does not stand alone, expressed in specific numbers. Instead, we have to apply arithmetic to n itself. In other words, while both [2] and [3] require the number fact that $17 = (3 \times 4) + 5$, it is more difficult to find the value of n in [3] than the value of m in [2].

In this lesson, we introduce a technique for learning how to find the value of n in [3] using logical reasoning rather than relying on trial-and-error guessing. The technique we use is one way to introduce algebra. Essentially, finding n when m is given in [1] begins our study of algebra. We were still doing arithmetic when we were given n and asked to find the value of m.

Section 2.1
Equations as Open Sentences

OBJECTIVE
To be able to understand that algebraic equations are a form of "fill in the blank."

We have all, at one time or another, answered questions of the form "Fill in the blank."

Example 1

Fill in the blank:

    There are _____ months in a year.   [1]

Assuming you know that there are twelve months in a year, you write the word "twelve" where the blank appears (or you may write the numeral "12"). That is,

    There are <u>twelve</u> months in a year.

Sometimes the instructions are worded so that you do not have to rewrite the expression [1]. For example, [1] may be reworded as follows.

Example 2

What word (or words) should replace the "blank" to make [1] a true statement?

    There are _____ months in a year.   [1]

Now you would simply write the word "twelve" as the answer.

Notice that [1] is not a sentence until the blank is replaced by a word (in this case, a number). If you replaced the blank by "fourteen," [1] would become

    There are fourteen months in a year.   [2]

[2] is a complete sentence that happens to be a *false* statement.

> **DEFINITION**
> An expression such as [1] is called an **open sentence**. An open sentence is any sentence in which a blank is used in place of a word.

So what "fill in the blank" really means is "Replace the blank(s) by the word or words that convert the open sentence into a *true* statement."

See how much simpler it is to say "Fill in the blank"?
But we must know the more exact meaning to be able to do it.
For example, if we took "fill in the blank" literally, we might fill in the blank in [1] by writing
    There are <u>////</u> months in a year.
We certainly "filled in the blank," but we suspect that wasn't what we were supposed to do.

Example 3

Fill in the blank:

    _____ is the capital of California.   [3]

You write:

    <u>Sacramento</u> is the capital of California.

You could have converted the open sentence [3] into a statement by replacing the blank with the name of any city. For example, you could have written

    <u>Boston</u> is the capital of California.   [4]

[4] is indeed a statement, but it happens to be false. That's why [4] wouldn't be the right answer in this example. The right answer must not only be a statement — it must be a *true* statement.

Section 2.1   Equations as Open Sentences

So to get the right answer in Example 3, you have to know that Sacramento is the capital of California.

Sometimes the same fact can be written as an open sentence in different ways.

Example 4

Fill in the blank:

Sacramento is the capital of _____.   [5]

The answer is:

Sacramento is the capital of California.

In other words, _____ is the capital of California [3] and: Sacramento is the capital of _____ [5] are different open sentences, but the answers to both require the single fact that *Sacramento is the capital of California.*

The blank really acts as a "place holder." For example, in [3] the blank holds the place of the name of a city. In [5] the blank holds the place of the name of a state.

We can use a different symbol as a place holder if we want to.

Example 5

What city is named by n if

"n is the capital of California"   [6]

is to be a true statement?

n must name Sacramento if [6] is to be a true statement.

It is important for you to notice that [3] and [6] are the same open sentence, with different symbols for the "placeholder."

What does all this have to do with algebraic equations? From the point of view of this lesson, the answer is:

> Algebraic **equations** are a special case of open sentences. They are open sentences that describe equality between two (or more) quantities (numbers).

What does this really mean?

Example 6

Fill in the blank: _____ $= 3 \times 2$ [7]

The answer is $\underline{6} = 3 \times 2$. All you have to know to solve this problem is that $3 \times 2 = 6$.

_____ $= 3 \times 2$ and $n = 3 \times 2$ are called equations because they are open sentences that involve the equality of numbers.

The example could have been worded differently.

Example 7

What number is named by n if $n = 3 \times 2$?

In this case you would simply say n is 6, or you may write $n = 6$.

In Examples 6 and 7, all we had to do was multiply two given numbers to answer the question. Sometimes things get a little more complicated.

Example 8

What number is named by n if

$$6 = 2 \times n \qquad [8]$$

Of course, since $6 = 2 \times 3$, n is 3. In terms of a "blank"

$$6 = 2 \times \underline{3}$$

[8] Notice that Examples 6, 7, and 8 are different open sentences based on the single fact that $6 = 3 \times 2$.

The point is that [8] is also an equation (open sentence), but now the placeholder n is not standing by itself. It is being multiplied by 2. That is, you are asked not to multiply 6 and 2 but rather to find out what you must multiply 2 by in order to get 6.

Sometimes the arithmetic may be harder, but the idea is always the same.

Example 9

What number is named by C if $C = 2.54 \times 6$?

Because $2.54 \times 6 = 15.24$, C is 15.24.

Why should we want to find the value of C for which $C = 2.54 \times 6$? What does this have to do with formulas? Perhaps the next example will tell us.

Example 10

The number of centimeters (C) in a given length is related to the number of inches (I) in that same length by the formula

$$C = 2.54 \times I \qquad [9]$$

Use [9] to find the number of centimeters in 6 inches.

There are 15.24 centimeters in 6 inches.
You find the answer by replacing I (the number of inches) in [9] by 6 to get

$$C = 2.54 \times 6 \qquad [10]$$

[10] is an open sentence with C holding the place of the correct number of centimeters that makes [10] a true statement.
In other words, once you replace I by 6 in [9], the problem is the same as saying

Fill in the blank: _____ $= 2.54 \times 6$ [11]

To solve [9] or [11] correctly, you do not need an almanac or an encyclopedia. All you have to know is how to find the product of 2.54 and 6. You have to know (or be able to work out) only that $2.54 \times 6 = 15.24$.

When we say "solve" an equation, we mean to replace the place holder by the number(s) that converts the open sentence into a true statement.

As we mentioned before, there are different open sentences that use the same information. For example, we used formula [9] to find the number of centimeters in a given number of inches. We can also use formula [9] to find the number of inches in a given number of centimeters.

Section 2.1 Equations as Open Sentences

Example 11

Use formula [9] to find the number of inches in 10 centimeters. Round your answer off to the nearest tenth of an inch.

There are 3.9 inches in 10 centimeters.
 Formula [9] stays the same. The difference is that because you know you have 10 centimeters and that C stands for centimeters, you now replace C by 10 in [9] to get

$$10 = 2.54 \times I \qquad [12]$$

[12] is now an open sentence in which I is the place holder. You could have written instead

$$10 = 2.54 \times \underline{\qquad} \qquad [13]$$

> Because [12] (and [10]) is an open sentence involving the equality of two quantities [10 and $(2.54 \times I)$] we call [12] an *equation*.

 The problem with [13] (or [12]) is that even though you're supposed to multiply, one of the factors is *missing*. It has been replaced by the place holder.
 This didn't happen in [9], when you had

$$\underline{\qquad} = 2.54 \times 6$$

because the two factors being multiplied were both known numbers (2.54 and 6).
 Now you can begin to see why it is important to understand arithmetic when you want to solve equations. Namely, in [13] the place holder names the number that you must multiply by 2.54 to get 10. The number that when multiplied by 2.54 gives 10 is written as $10 \div 2.54$, and by dividing these two decimal fractions you obtain 3.9 (to the nearest tenth).

> A calculator might give the answer 3.9370078.

 If you didn't remember how division and multiplication are related, there is a way of finding the relationship from [12]. The idea is that you want I by itself. But in [12], I is being multiplied by 2.54. To "undo" multiplying by 2.54, you divide by 2.54.
 Note that [12] is an equation. It tells you that both sides must name the same number. So if you divide one side by 2.54, you must also divide the other side by 2.54 if you want the amounts to remain equal. In the language of common fractions, what you do is

> This means the following: Pick any number. Multiply this number by 2.54. Then divide your answer by 2.54. You end up with the number you started with.

STEP 1: Write the equation.

$$10 = 2.54 \times I$$

STEP 2: Divide *both* sides of the equation by 2.54.

$$\frac{10}{2.54} = \frac{(2.54 \times I)}{2.54} \qquad [14]$$

STEP 3: Cancel 2.54 as a common factor of the numerator and denominator on the right side of [14].

$$\frac{10}{2.54} = \frac{\cancel{2.54} \times I}{\cancel{2.54}} = I \qquad [15]$$

> Think in terms of a seesaw. If the seesaw is balanced and we double the weight on one side, then the seesaw won't stay balanced unless we double the weight on the other side.

 You then find the value of I in [15] by dividing 10 by 2.54, as indicated by the meaning of $\frac{10}{2.54}$.

Example 12

Solve the equation:

$$n - 3 = 4 \qquad [16]$$

by "undoing" the 3.

The solution of the equation $n - 3 = 4$ is $n = 7$. That is, $7 - 3 = 4$ is a true statement.

You want to find the value of n by itself. But n appears on the left side of [16], where 3 is being subtracted from n. To "undo" subtracting 3, you add 3. To keep the equality in balance, if you add 3 to one side, you must also add 3 to the other side. You then get

$$\begin{array}{r} n - 3 = 4 \\ +3 +3 \\ \hline n = 7 \end{array}$$

The idea of equations and "fill in the blanks" gives us a rather easy way to see at least one important connection between arithmetic and algebra. To review:

1. We start with a formula in which certain quantities, usually named by letters of the alphabet, are mathematically related.
2. We read the problem in which the formula is going to be used, and we replace each letter by the numerical value given to it in the problem.
3. When we get through, we will have an open sentence (equation) in which only one letter remains as a place holder. (All the other letters have been replaced by numerical values.)
4. Suppose we let n stand for the place holder. Then if n is by itself on one side of the equation, we solve the equation simply by the use of arithmetic. [For example, we might have $n = 2.7 \times 3.9$ or $2.79 \times (2.34 + 5.67) = n$, where all it takes is arithmetic to solve for n.] But if n is not by itself on one side of the equation, we have to "undo" things to get n by itself. The process of "undoing" is what algebra is all about.

Of course, if you remember how addition and subtraction are related, you know right away that $n - 3 = 4$ has the same meaning as $4 + 3 = n$, so that $7 = n$. But the "undoing" method gets you the right answer even if you forget how addition and subtraction are related.

This section is not meant to teach how to solve equations. It is designed to help you make sure that you know what an equation is, how it often comes from a formula, and how arithmetic and algebra are related in terms of formulas and equations.

Section 2.3 deals a little more with the idea of "undoing," but first Section 2.2 will establish a little more vocabulary that's used to deal with equations.

PRACTICE DRILL

1. Find the value of n that makes each of the following open sentences true statements.

    (a) There are n days in a week.   (b) $n = 5 \times 3$
    (c) $n = 5 \times (3 + 4)$ (d) $n = (5 \times 3) + 4$
    (e) $3 + n = 9$ (f) $3 \times n = 18$

2. In the formula $D = 5 + C$, find the value of

    (a) D when $C = 8$ (b) C when $D = 8$

3. In the formula $P = 2.2 \times K$, find the value of K when $P = 13.2$.

Answers
1. (a) 7 (b) 15 (c) 35 (d) 19 (e) 6 (f) 6   2. (a) 13 (b) 3
3. 6

CHECK THE MAIN IDEAS

1. "There are 7 days in a _____" and "There are _____ days in a week" are forms of _____ sentences.
2. When correctly completed, these two open sentences state the fact that there are _____ .
3. An open sentence that involves the equality of numbers is called an _____ .

Answers
1. open   2. 7 days in a week
3. equation   4. formula
5. multiply   6. divide
7. false   8. true

Section 2.2 Some Additional Vocabulary   31

4. You often get equations by replacing letters with numbers in a mathematical _____ .
5. To solve the equation $n = 3 \times 4$, _____ 3 by 4.
6. To solve the equation $n \times 5 = 30$, _____ 30 by 5.
7. If you replace n by 150 in $n \times 5 = 30$, you get a _____ statement.
8. You can always check an equation by seeing if you get a _____ statement when you replace n by a number.

Section 2.2
Some Additional Vocabulary

Have you noticed that numbers occur in two ways in a formula? When you see a formula like

$$C = 2.54 \times I \qquad [1]$$

I or C may stand for different values. That is, you may be asked to find the value of C when I is 5; or you may be asked to find the value of C when I is 17; or you may be asked to find the value of I when C is 10.

Because we may vary the values of C and I in formula [1], we call C and I **variables**.

OBJECTIVE
To become more familiar with various terms and symbols used in equations, with special emphasis on the meaning of constants and variables.

Example 1

What are the variables in the formula

$$p = 2.2 \times k \qquad [2]$$

The variables are p and k. You may use [2] with different values for p or k. For example, if you let $p = 7$, [2] becomes the equation $7 = 2.2 \times k$; but there is no general reason why p must be 7.

On the other hand, no matter what C and I are in [1], the number 2.54 is always the same. To find C for a given value of I, we *always* multiply the value of I by 2.54. Because the number 2.54 always remains the same in the formula, we call it a **constant**.

Example 2

In the formula

$$p = 2.2 \times k \qquad [2]$$

what is the constant?

The constant is 2.2. That is, for any value of k, you always find the corresponding value of p by multiplying k by 2.2.
If you were given a value for p, you would *divide* this by 2.2 to get k.

REVIEW
If you divide both sides of [2] by 2.2, you get

$$\frac{p}{2.2} = \frac{(2.2 \times k)}{2.2} = k$$

Example 3

In the formula

$$A = L \times W \qquad [3]$$

what are the variables and what are the constants?

There are no constants. *A*, *L*, and *W* are all variables.

Remember that we get equations by replacing variables in a formula by specific numbers.

Example 4

What equation do you get if the variable p is replaced by 44 in the formula

$$p = 2.2 \times k \qquad [2]$$

You get the equation

$$44 = 2.2 \times k \qquad [4]$$

If you want [4] to be a true statement, k can no longer be a variable. In fact, since $44 \div 2.2 = 20$, [4] can be true only when k is 20. In this sense, k is no longer a variable in [4]. In an equation such as [4], it is customary to refer to k as an **unknown** rather than as a variable.

What we mean when we call k an unknown in

$$44 = 2.2 \times k \qquad [4]$$

is that we want to *know* the value of k that makes [4] a true statement.

Example 5

In the equation

$$56 = 37 + m \qquad [5]$$

what is the unknown, and for what value of the unknown does [5] become a true statement?

m is the unknown in [5], and [5] is true if m is 19. That is, $56 = 37 + 19$ is a true statement.
Another way of saying this is that the solution of equation [5] is $m = 19$.

Letting letters of the alphabet denote variables can cause small problems. One such problem occurs with the letter "x" — x is a letter of the alphabet, but x is also used to mean multiplication. If we are not careful, we can be confused if we see x. It may be a variable or it may mean "times."
To avoid this problem, we invent a shortcut. If we want to indicate that we are taking the product of two variables, we simply write the variables side by side.

Example 6

How do you read $A = LW$?

You read it as if it were $A = L \times W$ (*L* times *W*). By writing LW we no longer need to use $\times$.

Even if we have a constant times a variable, we leave out the × as a multiplication symbol. Instead, we write the constant and the variable side by side.

Example 7

In the formula $I = 12F$, how are I and F related?

OPTIONAL NOTE

In $A = L \times W$, the A stands alone (as opposed to L and W, which share the same side of the equation). For this reason, we often say that L and W are **independent** variables and that A is the **dependent** variable.
In other words, if we want to use arithmetic, we may pick values for L and W, and then the value of A *depends* on the values of L and W that we picked independently of one another.

That is, $44 = 2.2 \times \underline{20}$ is a true statement.

We see that $m = 19$ either (1) by trial and error or (2) by subtracting 37 from both sides of [5] to "isolate" m.

Section 2.2   Some Additional Vocabulary

The formula really says that $I = 12 \times F$.

To find the value of I for a given value of F, you multiply the given value of F by 12. In other words, $12F$ means $12 \times F$ (12 times F).

On the other hand, if we are multiplying two constants, it would be confusing to put them side by side. For example, if we want to multiply 3 by 2, it would be very confusing to write 32, because this looks like the numeral that stands for thirty-two. Therefore, we write 3×2.

In the formula
$$P = 2(L + W) \qquad [6]$$
what is the value of P when L is 3 and W is 4?

P is 14.

Because $L + W$ is enclosed in parentheses, you treat $L + W$ as a single number; and 2 is multiplying that single number. That is,
$$P = 2 \times (L + W) \qquad [7]$$
If L is 3 and W is 4, then $L + W$ is $3 + 4$, or 7. So [7] becomes
$$P = 2 \times (L + W)$$
$$= 2 \times 7$$
$$= 14$$

> This formula is actually the relationship between feet (F) and inches (I). To change feet to inches, we multiply the number of feet by 12.
>
> There is an exception. Sometimes we put the 3 and 2 in separate parentheses and we then let $(3)(2)$ stand for 3×2.
>
> Or, as shown below, we may put only one of the factors in parentheses, and write $3(2)$ to mean 3×2.

PRACTICE DRILL

1. Name the variables in each of the following.
   (a) $F = \left(\frac{9}{5}C\right) + 32$ (b) $V = LWH$
   (c) $Y = 361$ (d) $C = (5p) + 23$

2. Name the constants in each of the following.
   (a) $P = 2.2K$ (b) $A = 7 + B$

3. In the equation $3m = 7$
   (a) what is the unknown?   (b) what value of the unknown is a solution of the equation?

> **Answers**
> 1. (a) F and C (b) V, L, W, and H (c) Y and I (d) C and p
> 2. (a) 2.2   (b) 7   3. (a) m
> (b) $\frac{7}{3}$

CHECK THE MAIN IDEAS

1. Formulas consist of constants and _____ .
2. Variables are usually represented by _____ .
3. Because x is both a letter and a symbol for multiplication, we usually do not use it to mean "times" in an equation. Instead of writing $I = 12 \times F$, we usually write instead $I =$ _____ .
4. In a formula, if you replace all the variables but one with numbers, the remaining variable is usually called an _____ .
5. In the formula $A = 2(L + W)$, the variables are L, W, and _____ .
6. If you replace L by 7 and W by 5, A is then called an (a) _____ and the formula becomes an (b) _____ .

> **Answers**
> 1. variables   2. letters (of the alphabet)   3. $12F$
> 4. unknown   5. A
> 6. (a) unknown   (b) equation

Section 2.3
More on "Undoing" Equations

In Section 2.1, we learned that equations are special kinds of open sentences, or "fill in the blank" types of sentences. We noticed that in some cases the "blank," or place holder, was already by itself on one side of the equation. When that is the case, we can solve the equation directly by arithmetic.

Example 1

Solve the equation

$$n = 3.14 \times 7.2 \qquad [1]$$

The solution of [1] is 22.608. That is, $22.608 = 3.14 \times 7.2$ is a true statement.

In other words, you find the value of n that converts the open sentence (equation) [1] into a true statement. All you have to do to solve [1] is the direct arithmetic problem 3.14×7.2.

A more difficult case exists when the "blank" is combined with various constants. Let's begin with examples in which the arithmetic is easy. This helps us keep track of the important ideas without worrying about computation.

Example 2

Solve the equation

$$12 = n + 3 \qquad [2]$$

The solution of [2] is $n = 9$. If you replace n by 9 in [2], you get the true statement $12 = \underline{9} + 3$.

In this example, n is not by itself. It appears on the right side of the equation, and 3 is being added to it.

To "undo" adding 3, you want to subtract 3. But to keep the equation balanced, if you subtract 3 from one side of the equality, you must also subtract 3 from the other side. This gives

$$\begin{array}{r} 12 = n + 3 \\ -3 \quad -3 \\ \hline 9 = n \end{array}$$

It is important to see not just what keeps us from having n by itself on one side of the equation, but also to see *how* this is happening.

Example 3

Solve the equation

$$12 = 3n \qquad [3]$$

The solution is $n = 4$.

Here, just as in Example 2, the constant 3 keeps you from having n by itself on one side of the equation. But this time, n is being *multiplied by* 3. (Remember, in Example 2, 3 was being added to n.)

To "undo" multiplying by 3, you divide by 3. But if you divide one side of the equation by 3, you must also divide the other side of the equation by 3 to keep the equation balanced. This leads to

$$\frac{12}{3} = \frac{3n}{3} = \frac{\cancel{3}n}{\cancel{3}} \qquad \text{or} \qquad 4 = n$$

Of course, we don't have to use n as the "blank" every time.

OBJECTIVE
To be able to use some simple principles of arithmetic to solve certain equations.

Of course if you have a calculator, the computational details won't be difficult, because the calculator works quickly.

Notice that even if you have trouble "undoing," you can still see whether an answer is correct by filling in the blank and seeing if you get a true statement.
  For example, if you replaced n by 4 in $5 + n = 20$, you get

$$5 + 4 = 20$$

This is a false statement, so you know that $n = 4$ is not a solution of the equation

$$5 + n = 20$$

even if you don't know that the right answer is ($n = 15$).

How we "undo" 3 depends on how n and 3 are being combined.

Notice how we are using the point of the last section: $3n$ means $3 \times n$.

Section 2.3 More on "Undoing" Equations

Example 4

Solve the equation

$$12 = r \div 3 \qquad [4]$$

The solution of [4] is $r = 36$. If you replace r by 36 in [4], you get $12 = 36 \div 3$, which is a true statement.

In this example, r is being divided by 3, so you "undo" this by multiplying by 3. But if you multiply the right side of [4] by 3, you must also multiply the left side of [4] by 3. The left side becomes 36, and the right side becomes r (since you have divided r by 3 and then multiplied the result by 3).

So far we have been able to "undo" an equation to find the solution with only one operation per problem. Sometimes we may have to do more.

Example 5

Solve the equation

$$2(n + 4) = 20 \qquad [5]$$

The solution of [5] is $n = 6$. As a check replace n by 6 in [5] to obtain $2(6 + 4) = 20$. That is

$$2 \times (6 + 4) = 20$$

which is a true statement.

The left side of [5] tells you that 4 is being added to n [$(n + 4)$] and that the sum is multiplied by 2 [$2(n + 4)$]. The point is that because 2 is a *factor* of the left side of [5], you can "undo" the 2 by dividing by 2.

If you divide both the left and right sides of [5] by 2, you get

$$\frac{2(2n + 4)}{2} = \frac{20}{2} \quad \text{or} \quad n + 4 = 10 \qquad [6]$$

We solve [6] by subtracting 4 from both sides to get $n = 6$.

NOTE

Don't be tempted to solve [5] by first subtracting 4 from both sides of [5]. The expression $2(n + 4)$ means $(n + 4)$ twice, or $(n + 4) + (n + 4)$. That is, the 4 occurs twice, not once.

Putting it another way don't confuse Example 5 with the next example.

Example 6

Solve the equation

$$(2n) + 4 = 20 \qquad [7]$$

The solution of [7] is $n = 8$. That is, $(2 \times 8) + 4 = 20$ is a true statement.

The parentheses in [7] tell us that 4 is being added to $2n$ because everything inside the parentheses is treated as one number. So 4 is being added to $2n$, and you can now "undo" adding 4 by subtracting 4. Subtracting 4 from both sides of [7], you have

$$\begin{array}{r} (2n) + 4 = 20 \\ -4 \quad -4 \\ \hline (2n) \quad = 16 \end{array} \quad \text{or} \quad 2n = 16 \qquad [8]$$

Now you can find the value of n from [8] by dividing both sides of [8] by 2.

You do not have to do this problem the way it's done here. There are usually several different strategies that work.

For example, you may notice that $2(n + 4)$ means twice $(n + 4)$; that is,

$$\begin{array}{r} n + 4 \\ + \; n + 4 \\ \hline (2n) + 8 \end{array}$$

So you can rewrite [5] as

$$(2n) + 8 = 20$$

Then you can subtract 8 from both sides to get

$$\begin{array}{r} (2n) + 8 = \; 20 \\ -8 \quad -\; 8 \\ \hline 2n \quad = \; 12 \end{array}$$

From this you also see that $n = 6$.

Do you see the difference? $2(n + 4)$ means that first you add 4 to n, and then you double the sum. But $(2n) + 4$ means that first you double n and then you add 4. These are quite different.

For example, you have seen that when $n = 6$, $2(n + 4)$ is 20. But when $n = 6$, $(2n) + 4$ is $(2 \times 6) + 4 = 12 + 4 = 16$.

NOTE

It is a nuisance to have to keep writing parentheses. So we make an agreement. We agree that when we write $2n + 4$, it means $(2n) + 4$. If we mean that first we want to add 4 to n and then double the result, we will use parentheses and write the problem $2(n + 4)$.
Let's see if you get the point.

Example 7

What is the value of $3n + 6$ when $n = 5$?

When $n = 5$, $3n + 6$ is 21. Remember, since no parentheses appear, $3n + 6$ means $(3n) + 6$ or $(3 \times n) + 6$. With n equal to 5, you get $(3 \times 5) + 6$, or $15 + 6$.

On the other hand, note the following.

$3n + 6$ is not an equation, because no equality is given. $3n + 6$ is called an **expression**. It has different values for different values of n.
 We don't *solve* expressions (we solve equations). We *evaluate* expressions.

Example 8

Evaluate the expression $3(n + 6)$ when $n = 5$.

When $n = 5$, the value of $3(n + 6)$ is 33. You get this by replacing n by 5 in $3(n + 6)$. This gives you $3(5 + 6) = 3(11) = 3 \times 11$.

In this section, we are practicing with equations. *But*, in most real-life situations, the equation will come from a formula.

The point is that we will not write $3n + 6$ if we mean $3(n + 6)$. Rather, $3n + 6$ will always mean $(3n) + 6$. There is no logic or magic to this choice. It's just an agreement we accept so that we can cut down on the use of parentheses.

Example 9

The cost C (in dollars) of n books is given by the formula

$$C = 2n + 4 \quad\quad [9]$$

Use [9] to decide how many books you can buy for $20.

[9] tells you that you can buy 8 books for $20.
 To get the answer, replace the cost (C) by 20 in [9], to get

$$20 = 2n + 4 \quad\quad [10]$$

This is just the same as equation [7] in Example 6.
 In other words, the equation you began with in Example 6 could have come from trying to do the "real" problem stated in Example 9.

Of course, formula [9] may not seem too realistic, but this isn't important. The important thing is that equations often come from formulas.
 For a more practial example that's a little more advanced, we can return to the study of freely falling bodies that we talked about in the last lesson.
 To review, Galileo showed that a freely falling body fell a distance of d feet in t seconds according to the formula

$$d = 16(t^2) \quad\quad [11]$$

In the last lesson, we converted [11] into an equation by choosing a value for t.

Remember that this means

$$C = (2n) + 4$$

That is, to find C, you first double n and then add 4 to the product.

Note that here you have $20 = 2n + 4$, while in [7] the sides of the equation were reversed: that is, $2n + 4 = 20$.
 But this causes no problem: if the number on the left equals the number on the right, it's also true that the number on the right equals the one on the left. In other words, if b and c are numbers, $b = c$ means the same as $c = b$.

In general, we shall omit the parentheses with the understanding that when we write $16t^2$, it means $16(t^2)$. If we want to square $16t$, we shall write $(16t)^2$.

Example 10

Use formula [11] to find the distance a freely falling body falls in 5 seconds.

It falls 400 feet. All you have to do is replace t by 5 in [11] to obtain

Section 2.3 More on "Undoing" Equations

$$d = 16(5^2) = 16(5 \times 5)$$
$$= 16(25) = 16 \times 25 = 400$$

That is, to solve the equation

$$d = 16(5^2)$$

we need only to use arithmetic.

Formula [11] becomes an algebra problem if we pick the value of d in [11] and then try to solve the resulting equation for t.
Let's try an example this way.

Example 11

Use the formula

$$d = 16t^2 \qquad [11]$$

to find out how long it takes for a freely falling body to fall 100 feet.

It takes 2.5 $\left(2\frac{1}{2} \text{ or } \frac{5}{2}\right)$ seconds to fall 100 feet.

You find the answer by first replacing d by 100 in [11]. This gives

$$100 = 16t^2 \qquad [12]$$

One way to solve equation [12] is to notice that t^2 is being multiplied by 16. To "undo" multiplying by 16, divide by 16. But to keep the equation balanced, you must divide *both* sides of [12] by 16. This gives

$$\frac{100}{16} = t^2 \qquad [13]$$

So far, this is no different from what we've done in other examples.

or in lowest terms

$$\frac{25}{4} = t^2 \qquad [13']$$

Unfortunately, you want the value of t, not the value of t^2. So you must "undo" squaring. To undo squaring, you take the square root. You must take the square root of *both* sides of [13']. The right side becomes t and the left side becomes $\frac{5}{2}\left(\text{because } \frac{5}{2} \times \frac{5}{2} = \frac{25}{4}\right)$. So you see that $\frac{5}{2} = t$.

Do you understand? For example, pick a number, say 5. If you square 5, you get 25, and if you then take the square root of 25, you get 5 back again.

As the course continues you will get into more practical and more advanced examples. For now you should just get the idea of what we mean by "undoing" and see how arithmetic and algebra are related.

That is, if you have to "undo" the equation to solve for n, it is algebra; but if the equation already has n by itself on one side of the equation, it is arithmetic.

PRACTICE DRILL

Solve each of the following equations for n.
1. $21 = 7 + n$ 2. $21 = 7n$ 3. $21 = n \div 7$ 4. $21 = n - 7$
5. $21 = 3(n - 4)$ 6. $21 = 2n - 3$ 7. $n^2 = 25$ 8. $n^2 - 3 = 33$

Answers
1. $n = 14$ 2. $n = 3$
3. $n = 147$ 4. $n = 28$
5. $n = 11$ 6. $n = 12$
7. $n = 5$ 8. $n = 6$

CHECK THE MAIN IDEAS

1. The method of "undoing" is used when the _____ does not appear by itself on one side of the equation.
2. To undo adding a number, _____ that number.
3. To undo multiplying by a number, _____ by that number.

Answers
1. unknown   2. subtract
3. divide   4. add   5. add
6. divide

4. To undo subtracting a number, _____ that number.
5. If you "undo" one side of an equation by adding a number, then you must _____ that same number to the other side of the equation.
6. If you divide one side of an equation by a number, you must _____ the other side of the equation by the same number.

Section 2.4
Guessing Answers to Equations

OBJECTIVE
To learn how to estimate solutions of equations just by doing "regular" arithmetic.

The connection between arithmetic and algebra cannot be stressed enough. In an emergency, you can solve any equation with one unknown just by having patience and using arithmetic. That is, if you don't know a formula or if you don't understand how to "undo," you can still solve equations by guessing — by using trial and error. Let's start with an easy problem.

Example 1

Solve the equation

$$15 + k = 28 \qquad [1]$$

The solution of [1] is $k = 13$. Because 15 is being added to k, you can "undo" this by subtracting 15. If you subtract 15 from both sides of [1], you have

$$\begin{array}{rr} 15 + k = & 28 \\ -15 & -15 \\ \hline k = & 13 \end{array}$$

Suppose we want to *guess* the solution of equation [1]. We just pick any number to replace k. For example, suppose we let $k = 10$. With $k = 10$, the expression $15 + k$ is $15 + 10$, or 25. We don't want $15 + k$ to be 25. According to [1], we want the sum to be 28. We need something greater. Let's try, for example, $k = 16$.

If we evaluate the expression $15 + k$ with $k = 16$, we get $15 + k = 15 + 16$, or 31. According to [1], we want the sum to be 28, so 31 is too great. In other words, when we guessed that k may equal 16, we guessed too high. The value of k must be less than 16.

What do we have so far? We know that 10 is too little and 16 is too great. No matter what the right answer is, the solution of [1] must be between 10 and 16. We can now make another guess, letting k be any number between 10 and 16. Eventually, we shall discover that the solution of [1] is $k = 13$.

Now let's look at a more complicated example. Suppose we want to find a value of t for which

$$t^2 + t = 17 \qquad [2]$$

How should we begin?

One way is to try to find a value for t that makes $t^2 + t$ less than 17 and another value for t that makes $t^2 + t$ more than 17. Then one solution of [2] would have to exist between these two values of t. Let's proceed step by step.

Remember you did things like this in the last lesson? For example, when you wanted the square root of 95, you kept guessing. You found estimates that were too great and estimates that were too little.

Example 2

What is the value of the expression

$$t^2 + t \qquad [3]$$

when $t = 3?$

Do you see the difference? $t^2 + t = 17$ is an equation. $t^2 + t$ is an expression. It is the left side of the equation $t^2 + t = 17$.

Section 2.4   Guessing Answers to Equations

When $t = 3$, expression [3] is 12.
  You replace t by 3 in [3] to get

$$3^2 + 3$$
or $$9 + 3$$

  This tells you that $t = 3$ is not a solution of

$$t^2 + t = 17$$

In fact, because $t = 3$ makes $t^2 + t$ equal to 12, we see that $t = 3$ is too small to be a correct answer (solution) of $t^2 + t = 17$.

Keep track of the fact that $t = 3$ is too small.

  Can we find a value of t that makes $t^2 + t$ greater than 17?

Example 3

What is the value of the expression

$$t^2 + t \qquad [3]$$

when $t = 4$?

Examples 2 and 3 combine to tell us that $t = 3$ is too small while $t = 4$ is too great to be a solution of [2]. But if 3 is too small and 4 is too great, what should our next guess be?

When $t = 4$, $t^2 + t$ is 20.
  All you do is replace t by 4 in [3] to get

$$4^2 + 4 \quad \text{or} \quad 16 + 4 \quad \text{or } 20$$

But 20 is more than 17. Therefore, $t = 4$ is too great to be a solution of $t^2 + t = 17$.

  Visually, what we have so far is

t	t^2	$t^2 + t$
3	9	12
		17?
4	16	20

As 17 is between 12 and 20, it's reasonable to assume that one value of t we want is between 3 and 4.

  Our next guess could be to "split the difference." If $t = 3$ is too small and $t = 4$ is too big, let's try $t = 3.5$.

Example 4

Evaluate $t^2 + t$ when $t = 3.5$

When t is 3.5, $t^2 + t$ is 15.75. That is, you evaluate $t^2 + t$ when t is 3.5 by replacing t with 3.5 in $t^2 + t$. When you do this, you get $3.5^2 + 3.5$, or $12.25 + 3.5$, which is 15.75.

  Since 15.75 is less than 17, we now know that $t = 3.5$ is also too small to be a solution of the equation $t^2 + t = 17$.
  In fact, if we want to take the time to make a more extensive chart, we can show

Now, instead of being able to say only that t is between 3 and 4, we can say that t is between 3.5 and 4; and this means that t is closer to 4 than to 3.

t	t^2	$t^2 + t$	
3.6	12.96	16.56	too small
3.7	13.69	17.39	too big

39

We now know that the answer is between 3.6 and 3.7. We can split the difference again to get

t	t^2	$t^2 + t$	
3.6	12.96	16.56	
3.65	13.3225	16.9725	too small
3.7	13.69	17.39	too big

Do you see how much of what you are doing requires nothing more advanced than some plain arithmetic? By using arithmetic, you already know that a value of t for which $t^2 + t = 17$ is between 3.65 and 3.7

We can keep on going like this, getting a better and better approximation as we go along. In fact, if $t = 3.66$, we have

$$\begin{aligned} t^2 + t &= 3.66^2 + 3.66 \\ &= (3.66 \times 3.66) + 3.66 \\ &= 13.3956 + 3.66 \\ &= 17.0556 \end{aligned}$$

This tells us that $t = 3.66$ is just a little too large to be a solution of $t^2 + t = 17$. We already know that $t = 3.65$ is a little too small to be a solution of $t^2 + t = 17$.

What you should remember from this lesson is that no matter how advanced the material gets in this course, you can always estimate a solution of an equation by using arithmetic and resorting to trial-and-error methods. Often in very advanced problems, the only way you get answers is by trial and error.

If $t = 3.65$ is too small and $t = 3.66$ is too big, then for some value of t between 3.65 and 3.66, there is a solution of the equation

$$t^2 + t = 17$$

Using a calculator and a few other techniques that we'll study later in the course, we find that, to six decimal places, the value of t we want is $t = 3.653312$.

PRACTICE DRILL

1. What is the value of each of the following expressions when $t = 4$?

    (a) $3t$ (b) $3(t-3)$ (c) $3t - 3$ (d) t^2 (e) $t^2 + 3$

2. (a) What is the value of $t^2 + t$ when $t = 7$?
   (b) What is the value of $t^2 + t = 8$?
   (c) Use (a) and (b) to decide between what two consecutive whole numbers there is a solution to the equation $t^2 + t = 60$.
3. Find the square root of 59 to the nearest tenth.

Answers
1. (a) 12 (b) 3 (c) 9 (d) 16 (e) 19   2. (a) 56 (b) 72 (c) 7 and 8   3. 7.7

CHECK THE MAIN IDEAS

1. If you don't know how to "undo" an equation, you can always try to solve the equation by _____ .
2. You may try to guess a solution of the equation $t^2 + t = 33$ by evaluating $t^2 + t$ for different values of _____ .
3. If t is 5, then $t^2 + t$ is _____ than 33.
4. If $t = 6$, then $t^2 + t$ is _____ than 33.
5. From (3) and (4), you may conclude that there is a value of t between 5 and 6 for which $t^2 + t$ is _____ 33.

Answers
1. guessing   2. t 3. less   4. greater (more)   5. equal to

Exercise Set 2 (Form A)

EXERCISE SET 2 (Form A)

Section 2.1

1. Fill in the blank in the following.
   (a) $34.87 + 52.66 =$ _____ (b) $34.87 +$ _____ $= 87.53$
2. For what value of n is each of the following true?
   (a) $n = 56.013 - 27.23$ (b) $28.783 = n - 27.23$
3. In the formula $V = L \times W \times H$, find the value of V when $L = 6$, $W = 4$, and $H = 3$.
4. In the formula $V = L \times W \times H$, find the value of H when $V = 144$, $L = 8$, and $W = 2$.

Section 2.2

5. In the formula $A = bh$, what is the value of A when $b = 5$ and $h = 4$?
6. In the formula $A = bh$, what is the value of A when $b = 5.36$ and $h = 4.03$?
7. In the formula $P = 2(b + h)$, what is the value of P when $b = 5$ and $h = 4$?
8. In the formula $P = 2(b + h)$, what is the value of P when $b = 5.36$ and $h = 4.03$?
9. Find the value of the unknown in each of the following
   (a) $m = 27.875 - 16.578$ (b) $34.56 = q + 22.78$
   (c) $r + 3.14 = 2(5.23 + 4.17)$

Section 2.3

10. For what value of k does $k + 3.7 = 8.9 + 14.4$?
11. For what value of d does $3.14d = 6.28$?
12. For what value of y does $3.14(y - 7) = 6.28$?
13. For what value of y does $3.14y - 7 = 14.98$?
14. For what whole number t does
    (a) $21 = t^2 - 4$ (b) $25 = (t - 4)^2$
15. Given the formula $P = 2(L + W)$, what is the value of W if $P = 52$ when $L = 15$?
16. In the formula $A = (3s)^2$, find the value of
    (a) A when $s = 36$ (b) s when $A = 36$

Section 2.4

17. (a) Evaluate $t^2 + 2t$ when $t = 4$.
    (b) Evaluate $t^2 + 2t$ when $t = 5$.
    (c) Between what two consecutive whole numbers is there a solution of the equation $t^2 + 2t = 30$?
18. (a) Evaluate $(t^2 + 2)t$ when $t = 2$.
    (b) Evaluate $(t^2 + 2)t$ when $t = 3$.
    (c) Between what two consecutive whole numbers is there a solution of the equation $(t^2 + 2)t = 30$?
19. Use trial-and-error methods to find a solution of the equation $t^2 + 2t = 50$ correct to the nearest whole number.
20. A freely falling body falls d feet in t seconds according to the rule $d = 16t^2$.
    (a) To the nearest second, how long does it take the body to fall 480 feet?
    (b) To the nearest tenth of a second, how long does it take the body to fall 480 feet?
    (c) To the nearest second, how long does it take for the body to fall one mile? (There are 5,280 feet in 1 mile.)

Answers: Exercise Set 2
Form A
1. (a) 87.53 (b) 52.66
2. (a) 28.783 (b) 56.013
3. 72   4. 9   5. 20
6. 21.6008   7. 18
8. 18.78   9. (a) 11.297
(b) 11.78 (c) 15.66   10. 19.6
11. 2   12. 9   13. 7
14. (a) 5 (b) 9   15. 11
16. (a) 11,664 (b) 2
17. (a) 24 (b) 35 (c) 4 and 5
18. (a) 12 (b) 33 (c) 2 and 3
19. 6   20. (a) 5 (b) 5.5
(c) 18

EXERCISE SET 2 (Form B)

Section 2.1
1. Fill in the blank
   (a) $86.23 + 91.89 = $ _____
   (b) $86.23 + $ _____ $ = 178.12$

2. For what value of n is each of the following true?
   (a) $n = 87.03 - 29.86$ (b) $57.17 = n - 29.86$

3. In the formula $V = L \times W \times H$, find the value of V when $L = 7$, $W = 6$, and $H = 2$.
4. In the formula $V = L \times W \times H$, find the value of H if $V = 500$, $L = 5$, and $W = 10$.

Section 2.2
5. In the formula $A = bh$, what is the value of A when $b = 7$ and $h = 11$?
6. In the formula $A = bh$, what is the value of A when $b = 7.03$ and $h = 11.02$?
7. In the formula $P = 2(b + h)$, what is the value of P when $b = 7$ and $h = 11$?
8. In the formula $P = 2(b + h)$, what is the value of P when $b = 7.03$ and $h = 11.02$?
9. Find the value of the unknown in each of the following
   (a) $m = 47.013 - 17.238$ (b) $86.231 = q + 32.446$
   (c) $r + 5.17 = 3(4.73 + 3.31)$

Section 2.3
10. For what value of k does $k + 7.32 = 8.93 + 6.41$?
11. For what value of d does $6.07d = 18.21$?
12. For what value of y does $6.07(y - 4) = 18.21$?
13. For what value of y does $6.07y - 2.35 = 28$?
14. For what whole number t does
    (a) $48 = t^2 - 1$ (b) $49 = (t - 1)^2$
15. Given the formula $P = 2(L + W)$, find the value of W if $P = 1{,}024$ and $L = 400$.
16. In the formula $A = (4s)^2$, find the value of
    (a) A when $s = 10$ (b) s when $A = 144$

Section 2.4
17. (a) Evaluate $t^2 + 3t$ when $t = 6$.
    (b) Evaluate $t^2 + 3t$ when $t = 7$.
    (c) Between what two consecutive whole numbers is there a solution of the equation $t^2 + 3t = 60$?
18. (a) Evaluate $(t^2 + 1)t$ when $t = 3$.
    (b) Evaluate $(t^2 + 1)t$ when $t = 4$.
    (c) Between what two consecutive whole numbers is there a solution of the equation $(t^2 + 1)t = 50$?
19. Use trial and error methods to find a solution of the equation $t^2 + 3t = 100$ correct to the nearest whole number.
20. A freely falling body falls d feet in t seconds according to the rule $d = 16t^2$.
    (a) To the nearest second, how long does it take the body to fall 800 feet?
    (b) To the nearest tenth of a second, how long does it take the body to fall 800 feet?
    (c) To the nearest second, how long does it take the body to fall half a mile? (There are 5,280 feet in 1 mile.)

Lesson 3

Signed Numbers

Overview In arithmetic we learn that while the quotient of two whole numbers need not be a whole number, there are times when such a quotient still makes sense. In this lesson, we shall study the case in which the difference of two whole numbers need not be a whole number. That is, when we deal with whole numbers, we can take 3 people from a room that contains 5 people. But we cannot take 5 people from a room in which there are only 3 people.

There is an old joke about a tailor who sold coats for $25 although the coats cost him $30 to make. When he was asked how he made a living, the tailor said, "I have to sell lots and lots of coats." This joke tells us something important: if we think of "breaking even" as a 0 point (that is, it is neither a profit nor a loss), there are worse things than 0. In fact, taking a loss means that we made less than 0.

In this lesson, we want to learn about the idea of signed numbers — numbers that can be less than 0 as well as numbers that can be greater than 0.

We shall interpret signed numbers in terms of directed distances, profit and loss, temperature readings, and so on. Our goal is to become comfortable with signed numbers. Signed numbers become very important when we deal with algebraic equations — a topic we cover in the last section of this lesson.

When we are not involved with algebra, it seems artificial to talk about signed numbers. For example, why talk about $7 less than $0 when all we have to do is talk about a $7 *loss*? In reality, we use things like profit and loss only to make it easier to see the meaning of less than 0 and greater than 0.

Section 3.1
The Idea of Less than Zero

We know that the sum and product of whole numbers are also whole numbers; but while the quotient of two whole numbers can be a whole number (6 ÷ 2), there are other times when the quotient of two whole numbers is not a whole number (5 ÷ 2 or 2 ÷ 5).

There are times when the quotient of two whole numbers makes sense even when it is not a whole number. For example, we can divide the cost of a $5 gift equally between two people (so that each person pays $2.50). We can also divide the cost of a $2 gift equally among five people (so that each person pays $0.40)

In this lesson, we shall make the same kind of study about what happens when we subtract one whole number from another.

When we subtract one whole number from another, sometimes the difference will also be a whole number (for example, 5 − 3 = 2). But what about an expression such as 3 − 5? In terms of whole numbers, the idea is that we can take 3 pieces of candy from a candy dish that has 5 pieces of candy in it.

But how can we take 5 pieces of candy from a candy dish which has only 3 pieces of candy?

The answer is that we can't.

Yet there are times when 0 and none don't mean the same thing. If the temperature reads 0°C, that doesn't mean that there is no temperature. Remember that 0°C is the Celsius temperature at which water freezes. Any temperature that is less than the freezing point of water is less than 0°C.

OBJECTIVE
To understand the meaning of and the need for negative numbers.

REVIEW
By (first no.) − (second no.) we mean the number that must be added to the second to give the first as the sum. For example, 5 − 2 = 3 because 3 is the number that must be added to 2 to give 5 as the sum. 5 − 2 is also read as "2 subtracted from 5." When we subtract one number from another, the correct answer is called the **difference.**

Don't think that the dish still owes you 2 pieces. The reason the dish "owes" you is that you took only 3 pieces (because that's all there were) instead of 5.

Example 1

At midnight in Boston, the temperature is 3°C. During the next hour, the temperature decreases 5°C. What is the temperature then?

The temperature is 2°C *below* 0°C. In this case, unlike taking candy from a dish, when you subtract 3°C from 3°C and get 0°C, there is still temperature left. You then subtract two more degrees to get the answer. That is,

$$3°C - 5°C = 2°C \text{ below } 0°C \qquad [1]$$

If you leave off the label (°C) in [1] and remember that "below" means "less than," you get

$$3 - 5 = 2 \text{ less than } 0 \qquad [2]$$

In other words, if we are willing to use the descriptions "less than 0" and "greater (more) than 0," we can always subtract whole numbers. For example,

$$5 - 3 = 2 \text{ more than } 0$$
and
$$3 - 5 = 2 \text{ less than } 0$$

Because it is lengthy to write 2 more than 0 or 2 less than 0, we use the symbols + and − as abbreviations.

We use + to mean "more than 0," so that (+2) means "2 more than 0"; and (−2) means "2 less than 0."

When we write 5 − 3 = 2, it is understood that 2 means 2 *more than 0*.

We use the parentheses when we write (+2) to emphasize that + is part of the description of the number. We don't want to confuse "2 more than 0" with adding 2 to a number.

Example 2

Use the above abbreviations to find each of the following differences:

(a) 6 − 2   (b) 2 − 6

You write 6 − 2 = (+4) to indicate that when you subtract 2 from 6, the result is 4

Section 3.1 The Idea of Less than Zero

more than 0. You write $2 - 6 = (-4)$ to indicate that when you subtract 6 from 2, the result is 4 *less than* 0.

Did you notice that $(+4)$ and 4 mean the same thing? That is, 4 is four more than zero — exactly what $(+4)$ means.

That is, $(+4)$ means 4 *more than* 0. (-4) means 4 *less than* 0. If the meaning is clear, we may omit the parentheses and write -4 or $+4$.

NOTE

We read $(+4)$ as "*positive* four" and (-4) as "*negative* four."
Just as we give a special name to the quotient of two whole numbers (a rational number), we also give a special name to the *difference* of two whole numbers.

We do *not* say "*plus 4*" because "plus" is already used to mean addition. We do not say "*minus 4*" because "minus" already is used to mean subtraction.

> **DEFINITION**
> When we subtract one whole number from another, we call the answer an **integer**. That is, an integer is the *difference* of any two whole numbers.

The word "integer" has the same base as the word "integral," which means "whole." In other words, an integer is a whole number that may be 0, less than 0, or greater than 0.

Example 3

What integer is named by $6 - 5$?

$6 - 5$ is the integer $(+1)$.
Of course, you already knew that $6 - 5 = 1$. That is $(+1)$ and 1 name the same number. For this reason, the *whole number* 1 is also called a *positive* integer.

1 and $(+1)$ both mean 1 more than 0.

Example 4

What integer is named by $3 - 7$?

$3 - 7$ is (-4). That is, if you subtract 7 from 3, you have 4 less than 0. (-4) is called a *negative integer*.

Example 5

What integer is named by $6 - 6$?

$6 - 6$ is the integer 0. Remember, an integer is the difference of whole numbers; in the expression $6 - 6$, you are subtracting a whole number (6) from a whole number (6).

That is, 0 is an integer, but it is neither positive nor negative. Zero is neither more than nor less than itself.

The idea of less than 0 and more than 0 applies to things other than temperature.
If we think in terms of profit and loss, we may use 0 to indicate "breaking even": that is, as neither a profit nor a loss. In this case, "more than 0" (positive) indicates a *profit* while "negative" indicates a loss.

A profit is more than "breaking even," while a loss is less than "breaking even."

Example 6

In terms of profit and loss what do we mean by $(+\$5)$?

This means a \$5 profit. The positive sign stands for profit.

Example 7

What is meant by $(-\$7)$?

It means a loss of \$7. The negative sign means loss.

We can also use negative and positive in terms of describing direction. For example, draw a line.

On the line, pick a point as a starting point, and name it 0.

Call the left-to-right direction positive, and the right-to-left direction negative.

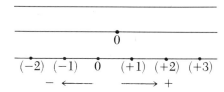

Example 8

What point is named by the integer (+8)?

It is the point on the line that is 8 units to the right of 0.
  Remember: in this situation, positive means to start at 0 and to move to the right. (−8) means that you start at 0 and move 8 units to the left.
  In this case, + and − tell you *direction*. For this reason, the integers are sometimes called *directed* whole numbers.

Of course, there is no need to limit our study to positive and negative whole numbers. For example, we know that a merchant can make a profit of $2.83 or that the temperature can be 2.5 degrees below 0. We are using integers only to keep the arithmetic as simple as possible. If our number is not a whole number, we use the term *signed numbers* rather than integers. In other words, integers are a special class of signed numbers. They are the signed numbers whose sizes are whole numbers.

In summary, a signed number is any magnitude (size) combined with either the positive or the negative signature or sign (unless the size is 0), which tells us whether the signed number is less than 0 or greater than 0.

Now that we know what signed numbers are, in the next sections we shall turn our attention to learning how to add, subtract, multiply, and divide these numbers.

OPTIONAL VOCABULARY
The size is often called the **magnitude**. For example, the magnitude of both (+6) and (−6) is 6. In other words, the magnitude names the size of the number while the sign or signature names the "direction" of the number.
  Sometimes we refer to the **absolute value** of a signed number rather than to its magnitude. The absolute value means that we are going to ignore the sign of the number. There is a special symbol that is used to denote the absolute value of a signed number. We enclose the number between a pair of vertical lines.
  For example, we write |−6| = 6 to indicate that the magnitude or absolute value of −6 is 6. We also write |+6| = 6 because the size of (+6) is also 6.

PRACTICE DRILL

1. What is the signature of each of the following signed numbers:
   (a) −6   (b) +8   (c) +2.34   (d) −3.45   (e) 0
2. What is the magnitude of each signed number in (1)?

Answers
1. (a) negative   (b) positive
(c) positive   (d) negative   (e) no signature (we consider 0 to be neither positive nor negative)
2. (a) 6   (b) 8   (c) 2.34
(d) 3.45   (e) 0

CHECK THE MAIN IDEAS

1. When you write 5 − 3 = 2, you mean that 5 − 3 is 2 _____ than 0.
2. You didn't say "2 more than 0" because before this lesson, no number could be _____ than 0.
3. If you think in terms of directed distance, temperature, profit or loss, you see that it is possible for a number to be _____ 0.
4. The abbreviation for "2 more than 0" is the signed number _____ .
5. In the signed number (+2), you read the + as (a) _____ , and it is called the (b) _____ of (+2).
6. The 2 in (+2) is called the _____ of the signed number.
7. Similarly, the sign of the signed number (−2) is called (a) _____ and its (b) _____ is 2.
8. If you want to subtract 7 from 5, you write 5 − 7 = (a) _____ which means that 5 − 7 is 2 (b) _____ .

Answers
1. more (greater)   2. less
3. less than   4. (+2)
5. (a) positive   (b) sign or signature   6. size (magnitude, absolute value)
7. (a) negative   (b) size (magnitude, absolute value)
8. (a) (−2)   (b) less than 0

Section 3.2
Adding Signed Numbers

Until this lesson, there were no signed numbers. That is, all the numbers we used were at least as great as zero. In other words, prior to this lesson, all *non-zero* numbers were *assumed to be positive*.

Example 1

Find the sum of $(+8)$ and $(+5)$.

The sum is $(+13)$.
$(+8)$ is the same as 8 and, $(+5)$ is the same as 5. Hence

$$(+8) + (+5) = 8 + 5$$

But you already know that $8 + 5 = 13$. Therefore

$$(+8) + (+5) = 13 \qquad [1]$$

13 means the same as $(+13)$, so you may replace 13 by $(+13)$ in [1] to obtain:

$$(+8) + (+5) = (+13)$$

We can easily picture the result of Example 1 in terms of profit and loss. We have already agreed that the positive sign stands for a profit and the negative sign stands for a loss. Suppose that we read "+" ("plus") to mean "followed by." In this way, we would think of $(+8) + (+5)$ as meaning an \$8 profit *followed by* a \$5 profit.
Visually:

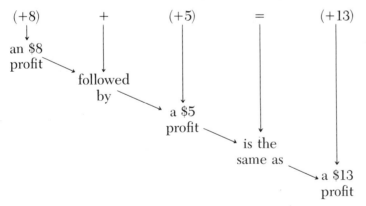

Pictured as a bookkeeping system, we have

Debit (Loss)	Credit (Profit)
	\$8
	\$5
	\$13 Balance

OBJECTIVE
To be able to find the sum of two or more signed numbers.

The ancient Greeks viewed numbers as lengths. Since a length couldn't be less than "nothing," they assumed that all numbers were greater than zero.

Do you see how + has two meanings? In the expression $8 + 5$ it is "plus" and means to add. But in $(+8)$ it is a sign and means "positive." So $(+8) + (+5)$ is read as positive 8 plus positive 5. That is, it means the sum of $(+8)$ and $(+5)$.

This keeps the spirit of more than and less than 0. Whether or not it's profit or loss, the idea remains that 5 more than 0 plus 8 more than 0 is 13 more than 0. As profit and loss, an \$8 profit followed by a \$5 profit is a \$13 profit.

To find the balance (sum) of two credits, we add the two credits (\$8 + \$5) and place the sum in the "Credit" column.

Example 2

Find the sum of $(+15)$ and $(+11)$, and explain the answer in terms of profit and loss.

The sum is (+26). A $15 profit followed by a profit of $11 is a total profit of $26.

Without referring to profit and loss, you see that (+15) + (+11) means 15 + 11, or 26; and 26 and (+26) mean the same thing. In terms of bookkeeping, you have

Debit	Credit
	15
	11
	26 Balance

Do you see a pattern?

> To add two positive numbers, add the two magnitudes and keep the positive signature

This is the same as saying that to find the balance of two credits, add the two credits and put the sum in the credit column.

Example 3

Find the sum of (+6) and (+3).

The sum is (+9). The magnitudes are 6 and 3, so the sum of the magnitudes is 9. If you keep the positve signature, you get (+9) as the sum.

This is just another way of saying that a $6 profit followed by a $3 profit is the same as a $9 profit.

Suppose at least one of the numbers we're adding is not positive. Why not continue to use the bookkeeping or profit-and-loss idea?

For example, we could read

$$(+8) + (-5)$$

as meaning an $8 profit followed by a $5 loss. But if we make $8 and then lose $5, we are still $3 ahead. That is,

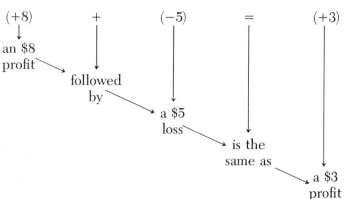

In terms of bookkeeping

Debit	Credit
	8
5	
	3 Balance

Here's what we did. We subtracted the smaller magnitude (5) from the greater (8) and put the balance in the column where the greater magnitude appeared (the credit column).

Section 3.2 Adding Signed Numbers

Let's try another example.

Example 4

Find the sum of (-7) and $(+4)$.

The sum is (-3).
   The problem may be written as

$$(-7)+(+4)$$

In terms of profit and loss, you have $7 *loss* followed by a $4 *profit*. The net result is a $3 loss. In terms of bookkeeping:

Debit	Credit
7	
	4
3	Balance

We subtracted the smaller amount (4) from the greater amount (7), and put the balance in the column that had the greater amount (debit).

NOTE

We could also have written the example as

$$(+4)+(-7)$$

in which case the bookkeeping would be the same but the amounts would have been entered in a different order. That is,

Debit	Credit
	4
7	
3	Balance

In terms of profit and loss, it isn't difficult to see why $(-7)+(+4)$ and $(+4)+(-7)$ mean the same thing.

If we don't want to use the idea of profit and loss, we can follow these steps.

1. Subtract the lesser magnitude from the greater (in this case $7-4$, or 3).
2. Keep the sign of the greater magnitude (in this case the greater magnitude is 7, and its sign is negative).
3. The answer is (-3).

As long as we subtract the lesser from the greater, we are working only with positive numbers. That is, the difference is negative only if we try to subtract more than we have.

   The only case still left to study is the one in which both signed numbers are negative.

Example 5

Use profit and loss to find the sum $(-6)+(-3)$.

The sum is (-9).
   In terms of profit and loss, the problem asks you to find the result of a $6 loss followed by a $3 loss. This is the same as a $9 loss.

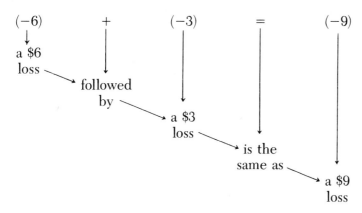

In terms of credit-debit

Debit	Credit
6	
3	
9	Balance

The two amounts are in the same column (debit), so we add them and put the sum in the same column.

Notice that while our discussion was in terms of debit and credit, it could just as well have been in terms of temperature change or directed distance. The important thing is

> To add two negative numbers, add the two magnitudes and keep the negative signature

We can summarize these results for addition as follows:

RULE FOR ADDING TWO SIGNED NUMBERS

To add two signed numbers:

1. If the numbers have the same sign (that is, either both positive or both negative), add the two magnitudes and keep the common sign.
2. If the numbers have different signs (that is, one is positive and the other is negative), subtract the smaller magnitude from the greater and keep the sign of the greater magnitude (if the magnitudes are equal, the sum is 0 and there is no sign).

If we want to add more than two signed numbers, we add them two at a time, just as we did with other numbers.

Example 6

Find the sum

$$[(+6) + (-9)] + (+7)$$

The sum is $(+4)$.

Begin by noticing that $(+6) + (-9)$ is (-3). So

$$[(+6) + (-9)] + (+7) =$$
$$(-3) + (+7) = (+4)$$

The brackets are also used as grouping symbols. They tell us that $(+6) + (-9)$ is being treated as one number, to which $(+7)$ is being added.

Section 3.2 Adding Signed Numbers

Example 7

Find the sum

$$(+6) + [(-9) + (+7)]$$

The sum is $(+4)$. This time you add (-9) and $(+7)$ first to find that the number in brackets is (-2). So:

$$(+6) + [(-9) + (+7)] =$$
$$(+6) + \underbrace{(-2)} =$$
$$(+4)$$

Remember, everything inside the brackets is treated as one number. In this case the number $[(-9) + (+7)]$ is being added to $(+6)$.

We use parentheses as grouping symbols. In other words, everything within a set of parentheses is treated as a single number. Remember, we agreed to write $(+6)$ or (-6) to emphasize the fact that the signature is part of the signed number. In Example 6, we want to indicate that $(+6) + (-9)$ is being treated as *one* signed number. We could enclose $(+6) + (-9)$ in parentheses to get

$$((+6) + (-9)) \qquad [2]$$

In fact, we showed that $(+6) + (-9) = (-3)$, and we then added $(+7)$ to this to get $(+4)$.

The double set of parentheses in [2] is somewhat hard on the eyes. To make it easier to see what we mean, we use a different set of symbols to indicate the fact that we want to group $(+6) + (-9)$ as a single number. We enclose this sum in *brackets* and write

$$[(+6) + (-9)] \qquad [3]$$

In other words, [2] and [3] are different ways of writing the same number, but [3] is easier to read.

To help eliminate the need for so many different sets of grouping symbols, we often indicate the signature of a signed number by raising it slightly. We may write $^+6$ instead of $(+6)$ or $^-9$ instead of (-9). In this way, we can write $(+6) + (-9)$ as $^+6 + ^-9$. If we do this, we can write $(^+6 + ^-9)$ without using another set of grouping symbols. In this lesson, we shall continue to write $(+6)$ rather than $^+6$. When we have worked more with signed numbers, we shall begin to use whichever notation seems the better in a particular problem.

Sometimes we use more than two sets of different grouping symbols. For example, if we want to treat

$$[(+6) + (-9)] + (+7)$$

as one number, we can use *braces* and write

$$\{[(+6) + (-9)] + (+7)\}$$

In this case, everything inside { } is treated as one number.
Our point is that it is easier to read

$$\{[(+6) + (-9)] + (+7)\}$$

than to read

$$(((+6) + (-9)) + (+7)).$$

PRACTICE DRILL

Find each of the following sums.
1. $(+9) + (+5)$
2. $(+9) + (-5)$
3. $(-9) + (+5)$
4. $(-9) + (-5)$
5. $(+2.45) + (+3.47)$
6. $(+2.45) + (-3.47)$

CHECK THE MAIN IDEAS

1. In terms of profit and loss, you may think of $(+9) + (-5)$ as meaning a \$9
    (a) _____ followed by a (b) _____ loss.
2. $(-3) + (+7)$ means a \$3 loss (a) _____ a \$7 (b) _____ .
3. Because a \$3 loss followed by a \$7 profit is a \$4 profit, you may say that $(-3) + (+7) =$ _____ .
4. Without referring to profit and loss, you can compute $(-3) + (+7)$ by subtracting (a) _____ from (b) _____ and keeping the sign of (c) _____ .

Answers
1. $(+14)$ 2. $(+4)$
3. (-4) 4. (-14)
5. $(+5.92)$ 6. (-1.02)

Answers
1. (a) profit (b) \$5
2. (a) followed by (b) profit
3. $(+4)$ 4. (a) 3 (b) 7 (c) $(+7)$

Section 3.3
Subtracting Signed Numbers

OBJECTIVE
To understand how to subtract signed numbers.

Suppose we keep the definition that says

$$(\text{first number}) - (\text{second number})$$

means the number we must add to the second number to get the first number as the sum. Then, for example,

$$(-5) - (-3)$$

means the number we have to add to (-3) to get (-5) as the sum.

From the last section, we know that (-2) is the number we must add to (-3) to get (-5). Therefore

$$(-5) - (-3) = (-2)$$

because

$$(-5) = (-3) + (-2)$$

Example 1

What number is named by $(+6) - (-4)$?

$(+6) - (-4) = (+10)$. Remember, $(+6) - (-4)$ means the number you must add to (-4) to get $(+6)$. From the last section, you know that

$$(-4) + (+10) = (+6)$$

In terms of "change-making," you want to figure out what you have to add to (-4) to get $(+6)$. First you must add $(+4)$ to (-4) to get 0 as the sum. Then you must add $(+6)$ to 0 to get $(+6)$. In other words

$$(-4) \xrightarrow{(+4)} 0 \xrightarrow{(+6)} (+6)$$
$$ \quad (+4) \quad + \quad (+6) \quad = (+10)$$

As a check, notice that

$$\begin{array}{r} (+6) \\ -\underline{(-4)} \\ (+10) \end{array}$$

Example 2

What number is named by $(-6) - (-5)$?

$(-6) - (-5) = (-1)$. Remember that $(-6) - (-5)$ means the number you must add to (-5) to get (-6) as the sum. In other words, if you let $n = (-6) - (-5)$, then

$$n + (-5) = (-6) \qquad [1]$$

From the previous section you already know that

$$(-1) + (-5) = (-6) \qquad [2]$$

If you now compare [1] and [2], you see that $n = (-1)$.

Do we have to go through all of this every time we want to subtract two signed numbers? There *is* a quicker way. It involves knowing what we mean by the "opposite" of a signed number.

> **DEFINITION**
> Given any signed number, the **opposite** of that signed number is the signed number that has the same magnitude as the given number, but the *opposite* sign.

Section 3.3   Subtracting Signed Numbers

Example 3

What is the opposite of $(+3)$?

The opposite of $(+3)$ is (-3). You find this by keeping the magnitude of $(+3)$ [which is 3] and changing the sign. The sign of $(+3)$ is positive, so the sign of its opposite is negative.

Example 4

What is the opposite of (-5)?

The opposite of (-5) is $(+5)$. Keep the magnitude of (-5) but change the negative signature to a positive signature.

If you look at Examples 3 and 4, you may notice that when you add a number and its opposite, you get 0 as the sum. So there is another way of defining the opposite.

> The opposite of a signed number is the number you must add to the given number to get 0 as the sum.

In terms of Examples 3 and 4,
$$(+3) + (-3) = 0$$
$$(-5) + (+5) = 0$$

Example 5

What is the opposite of 0?

The opposite of 0 is 0 itself. That is, the number you must add to 0 to get 0 is 0. In other words, $0 + 0 = 0$.

Zero is its own opposite.

We are now ready to give an easy rule for finding the difference of two signed numbers.

A RULE FOR SUBTRACTING TWO SIGNED NUMBERS

To compute

(first number) − (second number)

1. Leave the first number alone.
2. Change subtraction to addition. That is, replace the minus sign by a plus sign.
3. Replace the second number by its opposite.
4. Perform the resulting addition problem.

Let's follow the rule step by step in an example.

Suppose we want to evaluate $(+8) - (-3)$.

STEP 1: Leave the first number alone [$(+8)$].

STEP 2: Replace the minus sign by a plus sign [$(+8) +\ \ $].

STEP 3: Replace the second number by its opposite [$(+8) + (+3)$].

STEP 4: Perform the addition $(+8) + (+3)$ to obtain $(+11)$.

Therefore:

$$(+8) - (-3) = (+11)$$

This is not a new answer, just a new method. If we used the earlier method, we would still get $(+11)$ as the answer. That is, $(+8) - (-3)$ still means the number we must add to (-3) to get $(+8)$; and $(-3) + (+11) = (+8)$.

Example 6

Use "adding the opposite" to evaluate $(-9)-(-11)$.

$(-9)-(-11)=(+2)$.   In other words,

$$(-9) - (-11) =$$
$$\downarrow \quad \downarrow \quad \downarrow \quad \downarrow$$
$$(-9) + (+11) = (+2)$$

Remember that $(-9)-(-11)$ means the number that must be added to (-11) to give (-9) as the sum. So, no matter what method you use

$$(-9)-(-11)=(+2)$$

because

$$(-9)=(-11)+(+2)$$

Perhaps you are bothered by the number of sets of parentheses used to write signed numbers. This can be changed. Remember that before you had signed numbers, it was understood that all non-zero numbers mean "more than 0." For example, 2 means 2 *more than 0*. Therefore:

> **AGREEMENT**
> When a signed number has a positive signature, we agree to leave out *both* the signature and the parentheses.

This agrees with the notion that 5, for example, is thought of as meaning 5 more than 0, even though we don't write "more than 0."

Example 7

What signed number does 7 stand for?

7 stands for $(+7)$. When a number is positive, you omit the positive sign and the parentheses.

Example 8

What problem with signed numbers is represented by $5+3=8$?

$5+3=8$ means $(+5)+(+3)=(+8)$. The agreement means that 5 is the same as $(+5)$, 3 is the same as $(+3)$, and 8 is the same as $(+8)$. That is,

$5+3=8$ means
$(5)+(3)=(8)$

and this means $(+5)+(+3)=(+8)$

Example 9

Use *addition* of signed numbers to find the value of $3-5$.

First think of $3-5$ as $(3)-(5)$.

This is the same as $(+3)-(+5)$.

By the "add the opposite" rule, you get $(+3)+(-5)$.

By the rule for adding signed numbers, this is (-2).

This gives us another way of looking at subtraction as being a form of addition. Before we learned about signed numbers, we knew that subtraction is a form of addition by using the "change-making" idea. Using signed numbers, we can say that subtracting any number is the same as *adding* its *opposite*.

We already know that the answer is -2. What we want to show is that once signed numbers are invented, the difference of any two whole numbers can be represented as the *sum* of two signed numbers. The "add the opposite" rule is very convenient to use, but it is important to remember the fact that the difference of two numbers means the number you must add to the second to get the first. That is, $3-5 = -2$ because $^-2+5=3$.

Section 3.4 Multiplying Signed Numbers

Example 10

Use the "add the opposite" rule to find the value of $(-3)-(-7)$.

Instead of subtracting (-7), add the opposite of (-7), which is $(+7)$. That is, $(-3)-(-7)$ is the same number as $(-3)+(+7)$. Because $(-3)+(+7)=(+4)$, you know that $(-3)-(-7)=(+4)$.

By the "add the opposite" rule we are emphasizing the fact that every subtraction problem can be done by the *addition* of signed numbers.

PRACTICE DRILL

Find the signed number named by each of the following.
1. $(-9)-(-3)$ 2. $(-9)-(+3)$ 3. $(+9)-(+3)$ 4. $(+9)-(-3)$

Answers
1. (-6) 2. (-12)
3. $(+6)$ 4. $(+12)$

CHECK THE MAIN IDEAS

1. $(-8)-(-3)$ means the number you must add to (a) _____ to get (b) _____ as the sum.
2. If $(-3)+(-5)=(-8)$, then $(-8)-(-3)=$ _____ .
3. Another way of computing $(-8)-(-3)$ is to change the subtraction to (a) _____ and replace (b) _____ by its opposite.
4. In either case, $(-8)-(-3)=$ _____ , which is (-5).

Answers
1. (a) (-3) (b) (-8)
2. (-5) 3. (a) addition
(b) (-3) [the second number]
4. $(-8)+(+3)$

Section 3.4
Multiplying Signed Numbers

The fact that we may omit the parentheses and the sign when we deal with positive numbers is a good start for finding the product of two integers.

Example 1

What is the product of $(+3)$ and $(+2)$?

The product is $(+6)$.
Because you can omit the sign and the parentheses when we deal with positive numbers, you have

$$(+3)\times(+2) = 3\times 2 = 6$$

Because 6 is the same as $(+6)$,

$$(+3)\times(+2)=(+6)$$

Check the procedure in Example 1. To multiply the two positive numbers, we multiplied the two magnitudes and kept the positive sign. More generally:

> To find the product of two positive numbers, multiply the two magnitudes and keep the positive sign.

Now suppose that one of the factors is negative and one is positive. Suppose we want to find the product of (-3) and $(+2)$.
The idea is that we may write $(+2)$ as 2, so

$$(-3)\times(+2)$$

means

$$(-3)\times 2$$

OBJECTIVE
To be able to find the product of any two signed numbers.

Once we have the rule for multiplying integers, we shall extend the result to *all* signed numbers.

In terms of profit and loss $(-3)\times(+2)=(-3)\times 2$ may be thought of as indicating a $3 loss, two times.

but $(-3) \times 2$ means (-3), two times. That is, $(-3) \times 2 = (-3) + (-3)$; and by our rules for adding signed numbers, this sum is (-6). In other words

$$(-3) \times (+2) = (-6)$$

Let's try another problem.

Do you understand the method? You multiply the two magnitudes to get 6, but now the answer has a negative signature.

Example 2

Express as a signed number

$$(+4) \times (-7)$$

$(+4) \times (-7) = (-28)$. Again, begin by replacing $(+4)$ with 4. In this way, $(+4) \times (-7)$ becomes $4 \times (-7)$. In turn, this means (-7), four times, or

$$(-7) + (-7) + (-7) + (-7)$$

But this sum is (-28):

Again, do you notice that to get the answer, you multiply the two magnitudes and keep the negative signature?

$$
\begin{array}{rl}
(-7) & \longleftarrow \text{one } (-7) \\
+ \ (-7) & \\
\hline
(-14) & \longleftarrow \text{two } (-7)\text{'s} \\
+ \ (-7) & \\
\hline
(-21) & \longleftarrow \text{three } (-7)\text{'s} \\
+ \ (-7) & \\
\hline
(-28) & \longleftarrow \text{four } (-7)\text{'s}
\end{array}
$$

This leads us to accept the following rule:

> To find the product of a negative number and a positive number, multiply the two magnitudes and keep the negative sign.

The problem that still faces us is what to do if both factors are negative. If at least one of the factors is positive, we can think in terms of addition. For example, given $(+3) \times (+2)$, we can think of the problem as meaning either $(+3)$, two times, or $(+2)$, three times. Both ways give the answer by addition. If the problem is $(-3) \times (+2)$, we can think of it as $(-3) \times 2$ or $(-3) + (-3)$. And when the problem is $(+3) \times (-2)$, we can think of it as (-2), three times; that is, $3 \times (-2)$, or $(-2) + (-2) + (-2)$.

What can we do with $(-3) \times (-2)$? It doesn't seem to make much sense to say, "Add (-3) *negative two* times." And it makes no more sense to say, "Add (-2) *negative three* times."

Perhaps the following pattern will help.

$$
\begin{array}{l}
(-3) \times (+4) = (-12) \\
(-3) \times (+3) = (-9) \\
(-3) \times (+2) = (-6) \\
(-3) \times (+1) = (-3) \\
(-3) \times \ \ 0 \ \ = \ \ 0
\end{array}
$$

As we read down the list, we see that to get from one product to the next, we add $(+3)$. That is, $(-12) + (+3) = (-9)$; $(-9) + (+3) = (-6)$; and so on.

We've already agreed to the rule involving factors with different signs.

We know that any number times 0 is still 0.

Reading down this column, we get from one number to the next by adding (-1).
That is, $(+4) + (-1) = (+3)$; $(+3) + (-1) = (+2)$; $(+2) + (-1) = (+1)$; $(+1) + (-1) = 0$

Section 3.4 Multiplying Signed Numbers

Why don't we just continue this pattern? Let's keep adding (+3) to each product, and keep adding (−1) to the second factor. If we do this, we get

$$(-3) \times (+4) = (-12)$$
$$(-3) \times (+3) = (-9)$$
$$(-3) \times (+2) = (-6)$$
$$(-3) \times (+1) = (-3)$$
$$(-3) \times 0 = 0$$
$$(-3) \times (-1) = (+3)$$
$$(-3) \times (-2) = (+6)$$
$$(-3) \times (-3) = (+9)$$

From this chart, we see a pattern indicating that $(-3) \times (-2) = (+6)$. Because of this pattern, we agree to the following:

> To find the product of two negative numbers, multiply the two magnitudes and make the sign *positive*.

In other words, if we accept the pattern, we see that we multiply the two magnitudes but make the sign positive (even though both factors are negative).

Putting the pieces together, we have the following general rule:

> **RULE FOR MULTIPLYING SIGNED NUMBERS**
>
> 1. If the two factors have the same sign, multiply the magnitudes and make the sign of the product positive.
> 2. If one factor is positive and the other is negative, multiply the two magnitudes and make the sign of the product negative.
> 3. If at least one of the factors is 0, the product is 0.

Let's try a few problems.

Example 3

What is the product of (−3) and (−5)?

The product is (+15) [or 15]. In this case, the numbers have the same sign, so you multiply the two magnitudes and use the positive signature.

CAUTION

Try to avoid memorizing such rules as "double negative." For example, while it's true that the product of two negative numbers is positive, the sum of two negative numbers is negative. Note that

$$(-3) \times (-5) = (+15)$$

but

$$(-3) + (-5) = (-8)$$

Example 4

What signed number is named by $(-5) \times (+7)$?

$(-5) \times (+7) = (-35)$. In this case, you still multiply the two magnitudes, but because the two factors have different signs, the product is negative.

Example 5

What is the value of $(-6)^2$?

$(-6)^2 = 36$ [or (+36)]. Remember, $(-6)^2$ means $(-6) \times (-6)$. Because you are multiplying two negative numbers, the product is positive. That is, $(-6) \times (-6) = (+36)$.

This is a special case of a more general result. If we multiply any negative number by itself, the product is positive. The product of any two negative numbers is positive. Remember, $(-6)^2$ means $(-6) \times (-6)$.

NOTE

We know that $0 \times 0 = 0$. We also know that any positive number times itself is positive. Example 5 shows us that the product of any negative number and itself is positive. In other words, the square of any signed number is nonnegative. If the square of a number is nonnegative, we call that number a **real number**. These are the only numbers we shall use in this course, so we shall often say "number," although what we mean is "real number."

Example 6

What is the value of $[(-3) \times (+4)] \times (-2)$?

The product is 24.
Remember, the brackets are grouping symbols. In this case, they tell you that $[(-3) \times (+4)]$ is treated as a single number.
Because $(-3) \times (+4) = (-12)$, you have

$$[(-3) \times (+4)] \times (-2)$$
$$= [-12] \times (-2)$$
$$= (+24)$$

You may omit a positive signature, so you may write the product as 24.

The definition of a real number is:

A number b is said to be real if its square is nonnegative.

In more advance courses, there is a need to invent nonreal numbers — that is, numbers whose square is negative. But there is no need for such numbers in our course.

REVIEW
In multiplying two signed numbers, we always take the product of the magnitudes. If the factors have like signs, the product is positive, if the factors have opposite signs, the product is negative. If at least one of the factors is 0, then the product is 0.

PRACTICE DRILL

Find each of the following products.
1. $(-8) \times (-6)$ 2. $(-8) \times (+6)$ 3. $(-6) \times (+8)$ 4. $(+6) \times (+8)$
5. $(-1) \times (-1)$ 6. $(-1) \times [(-1) \times (-1)]$

Answers
1. $(+48)$ or 48   2. (-48)
3. (-48) 4. $(+48)$ or 48
5. 1 or $(+1)$ 6. (-1)

CHECK THE MAIN IDEAS

1. To find the product of (-8) and $(+4)$, you may think of an $8,
   (a) _____, (b) _____ times.
2. This is the same as a $32 loss, so $(-8) \times (+4) =$ _____.
3. In general, when you multiply two signed numbers, you always _____ the two magnitudes.
4. You use the positive sign only when the two factors have (a) _____ and the negative sign when the two factors have (b) _____ .
5. You avoid saying that two negatives make a positive because when you add two negative numbers, the sum is always _____ .

Answers
1. (a) loss (b) four
2. (-32) 3. multiply
4. (a) the same sign (b) different (unequal) signs   5. negative

Section 3.5
Dividing Signed Numbers

If we continue to accept the idea that

first number ÷ second number

means the number we must multiply by the second number to get the first, then division of signed numbers is practically a footnote to multiplying signed numbers.

OBJECTIVE
To find the quotient of two signed numbers.

Section 3.5 Dividing Signed Numbers

Example 1

What number is named by $(-6) \div (-2)$?

$(-6) \div (-2)$ is $(+3)$.

The question asks you to find the number that must be multiplied by (-2) to give (-6). In other words, you want to "fill in the blank" in

$$(-2) \times \underline{\qquad} = (-6) \qquad [1]$$

Working with the magnitudes first, you know that $2 \times 3 = 6$. So the magnitude of the number you are looking for is 3. You also know that if one factor is negative, the other factor must be positive for the product to be negative. Hence the sign of your number must be positive. This means the number is $(+3)$.
As a check,

$$(-2) \times (+3) = (+6)$$

Let's try another example.

Example 2

What is the value of $(+12) \div (-3)$?

The value is (-4).
You are looking for the number that must be multiplied by -3 to give $(+12)$. When you multiply signed numbers, you multiply the magnitudes. Hence you want a number whose magnitude is 4, because $3 \times 4 = 12$.
You also want the number to be negative. The only way you can make a product positive if one of the factors is negative is to make the other factor negative.

After doing these two examples, perhaps you have noticed a shortcut — a shortcut similar to the rule for multiplying signed numbers.

> **RULE FOR DIVIDING SIGNED NUMBERS**
>
> To find the quotient of two signed numbers:
>
> 1. The magnitude of the quotient is the quotient of the two magnitudes.
> 2. The sign (signature) of the quotient is positive if the two numbers have the same sign, and negative if they have opposite signs.

Example 3

What is the quotient when (-15) is divided by $(+3)$?

The quotient is (-5). Using the rule, you know that $15 \div 3 = 5$. Because the two numbers have different signs, the sign of the answer must be negative. The answer is (-5).

What the rule actually means is that we must multiply $(+3)$ by 5 to get a number whose magnitude is 15. To make sure that the product is negative (-15), we must multiply $(+3)$ by a negative number. So the answer is -5.

REVIEW OF MULTIPLICATION
To find the product of two signed numbers:
1. Multiply the two magnitudes.
2. Use the positive sign if the two factors have the same sign and the negative sign if they have different signs.

If either factor is 0, the product is 0.

That is, negative × negative is positive.

That is

$$(-3) \times ? = (+12)$$

REVIEW OF EXAMPLES 1 AND 2

$$(-6) \div (-2) = (-3)$$

Divide the magnitudes to get $6 \div 2 = 3$; the two numbers have the same sign, so the quotient is positive. The answer is $(+3)$.

$$(+12) \div (-3)$$

Divide magnitudes to get $12 \div 3 = 4$; because the two numbers have opposite signs, the quotient is negative. The answer is (-4).
The rule is just a different way of getting the same answer.

NOTE

Division by 0 is still excluded. For example, $(-6) \div 0$ still means the number we must multiply by 0 to get (-6); but any number times 0 is 0.

Because division is the inverse of multiplication, and because we've already studied how to multiply signed numbers, we may still view division as a special case of multiplication.

But $0 \div (-6) = 0$. That is, $0 \div (-6)$ means the number we must multiply by -6 to get 0; and $-6 \times 0 = 0$. Even with signed numbers, if we divide 0 by any non-zero number, the quotient is 0.

PRACTICE DRILL

Find each of the following quotients.
1. $(-12) \div (+3)$
2. $(-12) \div (-3)$
3. $(+12) \div (-3)$
4. $(+12) \div (+3)$
5. $0 \div (-6)$
6. $(-6) \div 0$

Answers
1. (-4)
2. $(+4)$ or 4
3. (-4)
4. $(+4)$ or 4
5. 0
6. no quotient

CHECK THE MAIN IDEAS

1. In dealing with signed numbers, (first number) ÷ (second number) still means the number you must multiply by the (a) _____ to get the (b) _____ .
2. $(-15) \div (-3)$ means the number you must _____ by (-3) to get (-15).
3. $(-3) \times (+5) = (-15)$, so $(-15) \div (-3) = $ _____ .
4. A way to remember this result is to remember that you divide the two magnitudes and, because the two numbers are negative, use the _____ signature.

Answers
1. (a) second number (b) first number
2. multiply
3. $(+5)$ or 5
4. positive

Section 3.6
Solving Equations with Signed Numbers

In this section, we want to apply our new knowledge about signed numbers to expressions and equations. For example, suppose we have the expression $3t$. To evaluate $3t$ means to find the value of $3t$ for a given value of t.

OBJECTIVE
To be able to solve equations and to evaluate expressions which use signed numbers.

Example 1

Evaluate the expression $3t$ when $t = 4$.

When $t = 4$, $3t = 12$.
All you do is replace t by 4 in the expression $3t$ to get $3(4)$ or 12.

There is no reason that t must be a positive number.

If t stands for temperature, the fact that the temperature can be below 0 means that t can be negative.

Example 2

Evaluate the expression $3t$ when $t = -4$.

When $t = -4$, $3t = (-12)$. This time you replace t in $3t$ by (-4) to get $3(-4)$ which is the same as $(+3)(-4)$. By the rule for multiplying signed numbers, this product is (-12).

Of course, we may think of $3 \times (-4)$ as meaning (-4), three times. That is, $(-4) \times 3 = (-4) + (-4) + (-4) = (-12)$.

Example 3

Evaluate $(t-5)(t-6)$ when $t = 3$.

When $t = 3$, $(t-5)(t-6) = 6$.

Section 3.6   Solving Equations with Signed Numbers

Replace t by 3 in the expression $(t-5)(t-6)$ to obtain

$$(3-5)(3-6) \qquad [1]$$

You know that $3-5=-2$ and that $3-6=-3$. Hence [1] becomes

$$(-2)(-3) \qquad [2]$$

By the rule for multiplying signed numbers, the value of [2] is 6, or $(+6)$.

Example 4

Find the value of $(L-M)N$ when $L=3$, $M=-2$, and $N=4$.

The value is 20. Replace L, M, and N by the given values to obtain

$$(3-[-2])4 = (3+[+2])4$$
$$= (5)4 = 20$$

Just as before, we replace each letter by its given value and then do the arithmetic. The only "new" thing is that here the arithmetic involves signed numbers.

Example 5

Use the "undoing" method to find the value of n for which

$$n + 7 = 3 \qquad [3]$$

The value of n is (-4). That is, $(-4)+7=3$.

On the left side of [3], 7 is being added to n. You undo this by subtracting 7, or by adding (-7), to both sides of [3]. That is,

$$\begin{array}{rl} n + (+7) = & (+3) \\ + (-7) & + (-7) \\ \hline n \quad = & (+3) + (-7) \quad \text{or} \quad (-4) \end{array}$$

The fact that $3-7$ is negative does not change the fact that we "undo" adding 7 by subtracting 7. In other words, the "undoing" method does not depend on whether we have signed numbers.

Example 6

For what value of n does

$$2(n-3) = -16 \qquad [4]$$

The value of n is -5.

As a check, replace n by -5 in the equation to get

$$\begin{array}{rl} & 2[(-5)-3] \quad = -16 \\ \text{or} & 2[(-5)-(+3)] = -16 \\ \text{or} & 2[(-5)+(-3)] = -16 \\ \text{or} & 2[(-8)] \quad\quad = -16 \\ & -16 \quad\quad\quad = -16 \end{array}$$

To use the "undoing" method, notice that as 2 is a factor of the left side of [4], you may undo it by dividing both sides of [4] by 2. This gives

$$\frac{2(n-3)}{2} = \frac{-16}{2} \quad [=-16 \div 2]$$

Because $\frac{-16}{2} = -8$, you have

$$n - 3 = -8 \qquad [5]$$

Now add 3 $(+3)$ to both sides of [5] to get

$$\begin{array}{rl} n - 3 = & -8 \\ + 3 & + 3 \\ \hline n \quad = & -8 + (+3) = -8 + 3 = -5 \end{array}$$

So far, we have not tried a real-life equation. Now let's look at a problem involving Celsius and Fahrenheit temperature readings.

Example 7

Use the formula $C = \frac{5}{9}(F - 32)$ to find the Celsius reading that corresponds to 23°F.

When the temperature is 23°F, the Celsius reading is −5°C.

All you do is replace F by 23 in the formula $C = \frac{5}{9}(F - 32)$ to get

$$C = \frac{5}{9}(23 - 32)$$
$$= \frac{5}{9}(-9) = \frac{-45}{9} = -5$$

In this context, negative numbers are quite real. Example 7 also shows us that we must learn the arithmetic of signed numbers if we want to use the formula $C = \frac{5}{9}(F - 32)$ in the most general way, such as when the Fahrenheit temperature reading is less than 32°.

PRACTICE DRILL

1. Find the value of *n* in each of the following equations.
   (a) $n = (-2)(-3) + 4$ (b) $n = (-2)[(-3) + 4]$
   (c) $n = (-2)(-3) + (-5)(+7)$ (d) $n + 3 = 2$ (e) $7 = n + 11$
   (f) $n - 3 = -7$ (g) $(-3)n = -21$ (h) $2n + 13 = 3$
   (i) $2(n + 13) = 20$

Answers
1. (a) 10 (b) −2 (c) (−29) or −29 (d) (−1) or −1 (e) (−4) or −4 (f) −4 (g) 7 (h) −5 (i) −3

CHECK THE MAIN IDEAS

1. The relationship between Celsius (C) and Fahrenheit (F) is $C = \frac{5}{9}(F - 32)$.

   If you start with the Fahrenheit temperature, first you _____ and then you multiply by $\frac{5}{9}$ to find Celsius.

2. If the Fahrenheit reading is 14°, you subtract 32 to get _____ .

3. Then you multiply −18 by $\frac{5}{9}$ to obtain C = _____ .

4. When the temperature is 14°F, the Celsius reading is _____ .
5. It is necessary to understand the meaning of negative numbers in order to use a mathematical _____ correctly.

Answers
1. subtract 32   2. −18
3. −10   4. −10°C, or 10° below 0 on the Celsius scale
5. formula

EXERCISE SET 3 (Form A)

Section 3.1

What is the value of each of the following?

1. $13.7 - 5.8$ 2. $5.8 - 13.7$ 3. $5.8 - 5.8$

Exercise Set 3 (Form B)

Section 3.2
Find each of the following sums.

4. $(+3.45) + (-2.37)$ 5. $(-3.45) + (-2.37)$ 6. $(-3.45) + (+1.873)$
7. $\left(-3\frac{2}{5}\right) + \left(+8\frac{1}{3}\right)$

Section 3.3
8. What is the opposite of each of the following signed numbers?

   (a) $+3.45$ (b) -3.45 (c) 0

9. What number must be added to -3.45 to give $+5.23$ as the sum?
10. Find each of the following differences.

    (a) $(+4.56) - (+2.87)$ (b) $(+4.56) - (-2.87)$
    (c) $(-4.56) - (+2.87)$ (d) $(-4.56) - (-2.87)$

11. What is the value of $[(+3.67) - (-2.54)] - (+4.12)$?
12. What is the value of $(+3.67) - [(-2.54) - (+4.12)]$?

Section 3.4
Find the value of each of the following.

13. $(-2.34) \times (-3.1)$ 14. $(-3)^2$ 15. $-(3^2)$ 16. $(-3)^2(-4)$
17. $[(-2) \times (-5)] \times (-6)$ 18. $[(-2) \times 0] \times (-6.3)$
19. $(-2)[(-3) - (-7)]$ 20. $[(-3)(-7)] + [(-2)(+5)]$

Section 3.5
Perform the indicated operations.

21. $(-0.0004) \div (-0.00002)$ 22. $(-8) \div [(-6) - (-2)]$
23. $[(-18) \div (-6)] \times (-3)$ 24. $(-18) \div [(-6) \times (-3)]$

Section 3.6
25. Find the value of n in each of the following equations.

    (a) $n + 16.7 = 3.9$ (b) $n - 12.5 = -19.3$
    (c) $2.1(n + 7) = 4.2$ (d) $2.1n + 7 = 2.8$

26. For what value of F does $-10 = \frac{5}{9}(F - 32)$?
27. (a) Evaluate $t^2 + 2t$ when $t = -8$.
    (b) Evaluate $t^2 + 2t$ when $t = -9$.
    (c) Find, correct to the nearest integer, a negative value of t for which $t^2 + 2t = 50$.

Answers: Exercise Set 3
Form A
1. $(+)7.9$ 2. -7.9
3. 0 4. $+1.08$ 5. -5.82
6. -1.577 7. $+4\frac{14}{15}$
8. (a) -3.45 (b) $+3.45$ (c) 0
9. $+8.68$ 10. (a) $+1.69$
(b) $+7.43$ (c) -7.43 (d) -1.69
11. $+2.09$ 12. $+10.33$
13. $+7.254$ 14. 9
15. -9 16. -36
17. -60 18. 0 19. -8
20. $+11$ 21. $+20$
22. $+2$ 23. -9 24. -1
25. (a) -12.8 (b) -6.8 (c) -5
(d) -2 26. 14 27. (a) 48
(b) 63 (c) -8

EXERCISE SET 3 (Form B)

Section 3.1
What is the value of each of the following?

1. $15.12 - 7.56$ 2. $7.56 - 15.12$ 3. $7.56 - 7.56$

Section 3.2
Find each of the following sums.

4. $(+8.76) + (-4.34)$ 5. $(-8.76) + (-4.34)$ 6. $(-8.76) + (+3.95)$
7. $\left(+8\frac{2}{3}\right) + \left(-2\frac{3}{4}\right)$

Section 3.3
8. What is the opposite of each of the following signed numbers?
   (a) -7.89 (b) $+7.89$ (c) 0
9. What number must be added to 8.23 to give -3.27 as the sum?
10. Find each of the following differences.
    (a) $(+7.63)-(+4.97)$ (b) $(+7.63)-(-4.97)$
    (c) $(-7.63)-(+4.97)$ (d) $(-7.63)-(-4.97)$
11. What is the value of $[(+8.53)-(-7.26)]-(+9.37)$?
12. What is the value of $(+8.53)-[(-7.26)-(+9.37)]$?

Section 3.4
Find the value of each of the following.
13. $(-4.56)\times(+5.1)$ 14. $(-8)^2$ 15. $-(8^2)$ 16. $(-8)^2(-2)$
17. $[(-7)\times(-1)]\times(-3)$ 18. $[(-3.2)\times 0]\times 4.37$
19. $(-6)[(-7)-(-8)]$ 20. $[(+8)(-3)]+[(-9)(-4)]$

Section 3.5
Perform the indicated operations.
21. $(-0.0012)\div(-0.0003)$ 22. $(-24)\div[(-4)-(-10)]$
23. $[(-60)\div(-10)]\times(+3)$ 24. $(-60)\div[(-10)\times(+3)]$

Section 3.6
25. Find the value of n in each of the following equations.
    (a) $n+22.9=11.7$ (b) $n-13.4=-8.7$ (c) $4.7(n+3)=9.4$
    (d) $4.7n+7=2.3$
26. For what value of F does $-15=\frac{5}{9}(F-32)$?
27. (a) Evaluate t^2+3t when $t=-4$.
    (b) Evaluate t^2+3t when $t=-5$.
    (c) Find, correct to the nearest integer, a negative value of t for which $t^2+3t=5$.

Introductory Topics
in Linearity

unit 2

Lesson 4   The Structure of Arithmetic

Overview

    First man: Did you ever see a man eating shark?
    Second man: No, but once in a restaurant I saw a man eating tuna.

    This joke depends on how words are grouped. It makes a difference whether we say "a man eating tuna" or "a man-eating tuna."
    There are other times when the order in which things are done makes a difference. It matters whether first we undress and then we shower or first we shower and then we undress.
    These concepts are very important in the study of arithmetic — especially when we deal with algebraic expressions such as $x+3$ and $4-(x+3)$. For example, when we add two numbers, the sum does not depend on the order: $3+4$ is the same as $4+3$. But when we subtract two numbers, the difference does depend on the order, so $5-3$ is 2 but $3-5$ is -2.
    When we add three numbers, the sum does not depend on how the three numbers are grouped. For example, $3+4+5$ is the same sum whether we think of it as $(3+4)+5$ or as $3+(4+5)$. On the other hand, $9-5-3$ has different answers for different groupings. Thus $(9-5)-3$ means $4-3$ or 1, but $9-(5-3)$ means $9-2$ or 7.
    In this lesson we shall explore the various structural properties of numbers, including rules about how numbers should be grouped. In particular, we shall try to see how these properties apply to the study of algebra. In many respects, the study of algebra is based on the material in this lesson.

Section 4.1
The Associative Property for Addition

Have you noticed that when we want to find the sum of three numbers such as 2, 3, and 4, we write

$$2 + 3 + 4 \qquad [1]$$

There are two ways the numbers in [1] may be grouped. We may write

$$2 + (3 + 4) \qquad [2]$$

or we may write

$$(2 + 3) + 4 \qquad [3]$$

Although these are two different groupings, the sum is the same in both cases. That is

$$2 + (3 + 4) = 2 + 7 = 9$$

and

$$(2 + 3) + 4 = 5 + 4 = 9$$

The sum of 2, 3, and 4 in [1] comes out the same with either grouping, so it is usual not to bother writing the grouping symbols (the parentheses) in this case.

Example 1

Compute the sum $3 + 6 + 5$ first by writing it as $3 + (6 + 5)$ and then by writing it as $(3 + 6) + 5$

In both cases, the sum is 14. That is,

$$3 + (6 + 5) = 3 + 11 = 14$$

and

$$(3 + 6) + 5 = 9 + 5 = 14$$

So the sum $3 + 6 + 5$ does not depend on how the summands are grouped. You can write $3 + 6 + 5$ without the parentheses because the sum is 14 with either grouping.

The fact that the sum of three numbers does not depend on the grouping of the summands is very helpful in solving algebraic equations. For this reason, we give this fact a special name. We say that addition is **associative**. In more exact language:

> **DEFINITION**
> The **associative principle for addition** means that for any three numbers b, c, and d:
> $$b + (c + d) = (b + c) + d$$

In everyday language, the associative principle for addition says that the sum of three numbers doesn't depend on "voice inflection" (how we group the symbols). For this reason, there is no need to use grouping symbols in a problem that involves only addition.

Be very careful. All we have shown so far is that addition has the associative property. In subtraction, it *does* make a difference how terms are grouped.

OBJECTIVE

To learn that the sum of three or more numbers does not depend on the way the numbers are grouped; and to see how this fact is used in algebra.

The reason is easy to see if we use tally marks. The grouping in [2] looks like

| | | | | | | | |

The grouping in [3] looks like

| | | | | | | | |

In either case, there are nine tally marks. (Remember that when we use tally marks, the number does not depend on the spacing).

REVIEW

If an expression is grouped inside parentheses, it means the expression is being treated as a single number. For example, $b + (c + d)$ means *the* number $c + d$ is being added to b. On the other hand, $(b + c) + d$ means that d is added to *the* number $b + c$.

By "voice inflection" we refer, for example, to the joke in the lesson overview. The meaning of "man eating shark" depends on how we pronounce it. It could mean a person who eats sharks, or a shark who eats people.

Section 4.1 The Associative Property for Addition

Example 2

Group the numbers in the expression $9 - 5 - 4$ in two different ways, and show that you get two different answers.

One grouping is $(9 - 5) - 4$, which is the same as $4 - 4$, or 0. The other grouping is $9 - (5 - 4)$, which is the same as $9 - 1$, or 8.

Without the parentheses, the two *different* numbers $(9 - 5) - 4$ and $9 - (5 - 4)$ look like the single expression
$$9 - 5 - 4$$

How does the fact that addition obeys the associative principle help us to simplify algebraic expressions? Suppose that we have the expression

$$(x + 4) + 3 \qquad [4]$$

Because x stands for a number, and because the sum of three numbers does not depend on grouping, we may rewrite [4] as

$$x + (4 + 3) \qquad [5]$$

$4 + 3 = 7$, so [5] may be rewritten as

$$x + 7 \qquad [6]$$

In other words, the associative principle for addition allows us to write

$$(x + 4) + 3 = x + (4 + 3) = x + 7$$

That is,

$$(x + 4) + 3 = x + 7$$

The fact that the expression $x + 7$ is simpler than the expression $(x + 3) + 4$ is a good reason for replacing $(x + 3) + 4$ by $x + 7$ in any equation. (This idea will be followed in more detail in the next lesson.)

But the expression $(x - 4) - 3$ is *not* equal to the expression $x - (4 - 3)$ or $x - 1$. For example, if we replace x by 5 in the expression $(x - 4) - 3$, we get $(5 - 4) - 3$ or $1 - 3$ or -2. But if we replace x by 5 in the expression $x - 1$, we get $5 - 1$ or 4. As -2 is not the same as 4, the two expressions $(x - 4) - 3$ and $x - (4 - 3)$, are not equal.

REVIEW
When we write
$$(x + 4) + 3 = x + 7$$
we mean that for any given value of x, these two expressions have the same value. For example, when $x = 5$,
$$(x + 4) + 3 = (5 + 4) + 3$$
$$= 9 + 3 = 12$$
and
$$x + 7 = 5 + 7 = 12$$

Example 3

When $x = 6$, what is the value of each of the two expressions $(x + 7) + 4$ and $x + (7 + 4)$?

When $x = 6$, both expressions equal 17. That is, when $x = 6$

$$(x + 7) + 4 = (6 + 7) + 4 = 13 + 4 = 17$$

and

$$x + (7 + 4) = 6 + (7 + 4) = 6 + 11 = 17$$

In other words, if we are given the expression $x + 7 + 4$, we may write it as either $(x + 7) + 4$ or $x + (7 + 4)$. Because $7 + 4 = 11$, we probably would group $x + 7 + 4$ as $x + (7 + 4)$ and then rewrite this in simpler form as $x + 11$. When we do this, we say we have **simplified** the expression $x + 7 + 4$.

Example 4

Evaluate the expressions $(x - 7) - 4$ and $x - (7 - 4)$ when $x = 6$.

When $x = 6$

$$(x - 7) - 4 = (6 - 7) - 4 = -1 - 4$$
$$= -1 - (+4) = -1 + (-4) = -5$$

and

$$x - (7 - 4) = 6 - (7 - 4) = 6 - 3 = 3$$

REVIEW OF SIGNED NUMBERS

To subtract signed numbers, we leave the first alone, change subtraction to addition, and change the sign of the second number. When the sign is omitted from a signed number, we assume the number is positive.

Every subtraction problem may be rewritten as an addition problem, so we may use the associative principle in subtraction problems provided that we first change the problem to addition.

For example, $x - 7$ is the same as $x + (-7)$, and $(x - 7) - 4$ is the same as $(x - 7) + (-4)$. So we may rewrite $(x - 7) - 4$ as $(x + [-7]) + (-4)$, in which case we are adding x, (-7) and (-4). By the associative principle for addition we can say that

$$\begin{aligned}(x - 7) - 4 &= (x + [-7]) + (-4) \\ &= x + [(-7) + (-4)] \\ &= x + [(-11)] \\ &= x - 11\end{aligned}$$

This is just another case of using the rule for subtracting signed numbers.

The brackets tell us that first we add x and -7. Then we add this sum to -4.

Example 5

Evaluate the expressions $(x - 7) - 4$ and $x - 11$ when $x = 15$.

When $x = 15$, each expression is equal to 4. Namely,

$$\begin{aligned}(x - 7) - 4 &= (15 - 7) - 4 \\ &= 8 - 4 = 4\end{aligned}$$

and

$$x - 11 = 15 - 11 = 4$$

In other words, we are trying to show that $(x - 7) - 4$ is equal to the simpler expression $x - 11$; and that we get this result by rewriting $(x - 7) - 4$ as an addition problem and using the associative principle.

Example 6

Rewrite $(x - 3) - 5$ as the sum of three terms.

$(x - 3) - 5$ is the same as $(x + [-3]) + (-5)$. That is, $x - 3 = x + (-3)$ and subtracting 5 is the same as adding (-5). Hence

$$(x - 3) - 5 = (x + [-3]) + (-5)$$

In other words

$$(x - 3) - 5 = x + (-3) + (-5)$$

Using raised signs, we may omit the parentheses and write $x + {}^-3 + {}^-5$.

Example 7

Show that the expressions $(x - 3) - 5$ and $x - 8$ are equal.

From Example 6, you know that $(x - 3) - 5$ is equal to $(x + [-3]) + (-5)$. By the associative principle for addition, $(x + [-3]) + (-5)$ is equal to $x + [(-3) + (-5)]$ or $x + (-8)$ or $x - 8$.

Although we have been talking about the associative principle of addition with respect to the sum of three numbers, the principle holds for the sum of any number of terms. For example the sum

$$1 + 2 + 3 + 4$$

can be grouped as

$$(1 + 2) + (3 + 4)$$

or as

$$[(1 + 2) + 3] + 4$$

For the two expressions to be equal, the value of one for any given value of x must equal the value of the other for that same value of x. As a partial check, if we let $x = 12$, $x - 8$ becomes $12 - 8$ or 4. At the same time $(x - 3) - 5$ becomes $(12 - 3) - 5$ or $9 - 5$, which is also 4.

This grouping says that we add the sum of 1 and 2 to the sum of 3 and 4.

This grouping says that we first add 1 and 2, then add 3 to it, and then add 4 to that sum. We remove grouping symbols starting with the innermost and working our way out.

or $[1 + (2 + 3)] + 4$

This grouping says that we add 2 and 3, then add 1 to this, and then add 4.

or $1 + [2 + (3 + 4)]$

This says we first add 3 and 4, then add 2, and then add 1.

or $1 + [(2 + 3) + 4]$

This says we first add 2 and 3, then add 4, and then add 1.

In each case the sum is 10.
That is,

$$(1 + 2) + (3 + 4) = 3 + 7 = 10$$
$$[(1 + 2) + 3] + 4 = [3 + 3] + 4 = 10$$
$$[1 + (2 + 3)] + 4 = [1 + 5] + 4$$
$$= 6 + 4 = 10$$
$$1 + [2 + (3 + 4)] = 1 + [2 + 7]$$
$$= 1 + 9 = 10$$
$$1 + [(2 + 3) + 4] = 1 + [5 + 4]$$
$$= 1 + 9 = 10$$

Later in this lesson we shall talk more about the way we remove grouping symbols, but for now the important point is that when any number of terms are added, the grouping does not affect the final sum.

PRACTICE DRILL

1. Compute each of the sums (a) $9 + (3 + 5)$ (b) $(9 + 3) + 5$
2. Compute each of the differences

    (a) $(9 - 3) - 1$ (b) $9 - (3 - 1)$

3. Write $(9 - 3) - 1$ as a sum.
4. Simplify the expression $(x + 7) + 6$.
5. Simplify the expression $(x - 7) - 6$.
6. By evaluating both expressions when $x = 15$, show that $(x - 7) - 6$ and $x - (7 - 6)$ are unequal.

Answers
1. Each of the sums is 17
2. (a) 5 (b) 7
3. $(9 + [-3]) + (-1)$, or $9 + (-3) + (-1)$ 4. $x + 13$
5. $x - 13$ 6. When $x = 15$, $(x - 7) - 6 = (15 - 7) - 6 = 8 - 6 = 2$; but $x - (7 - 6) = 15 - (7 - 6) = 15 - 1 = 14$

CHECK THE MAIN IDEAS

1. Generally the _____ to an arithmetic problem depends on how you group the terms.
2. For example, $(9 - 3) - 1$ and _____ are different numbers even though both look like $9 - 3 - 1$ when the parentheses are left out.
3. However, when you add numbers, your _____ does not depend on how the numbers are grouped.
4. For this reason, in addition problems you may omit the _____ .
5. For example, $(9 + 3) + 1$ names the same number as does _____ . You don't need parentheses to tell you that $9 + 3 + 1$ must be 13.

Answers
1. answer   2. $9 - (3 - 1)$
3. sum (or answer)
4. parentheses (or grouping symbols)   5. $9 + (3 + 1)$

Section 4.2
The Commutative Property for Addition

At the beginning of this lesson, we reviewed the fact that sometimes the order in which we do things is important. We used the example of undressing and showering. The same idea is true in the study of arithmetic. Sometimes the answer to an arithmetic problem depends on the order in which we write the terms. At other times, the order does not matter. For example, $5 + 3$ and $3 + 5$ name the same number. But $5 - 3$ and $3 - 5$ do not name the same number.

OBJECTIVE
To know that the sum of two numbers does not depend on the order in which the summands appear; and to be able to use this fact in evaluating algebraic expressions.

Example 1

Find the following sums: (a) $188 + 2$ (b) $2 + 188$

In either case, the sum is 190.

Example 2

Find the following differences: (a) $15 - 3$ (b) $3 - 15$

$15 - 3$ is 12. That is, $12 + 3 = 15$. But $3 - 15$ is (-12), because $15 + (-12) = 3$.

> Did you sense that there is a psychological difference to parts (a) and (b)? Usually it seems easier to add 2 to 188 than to add 188 to 2. Solving $188 + 2$ seems to suggest the series 188, 189, 190; but solving $2 + 188$ seems like saying "3, 4, 5, . . . ," which is a much longer process.

The fact that the sum of two numbers does not depend on the order in which the two numbers are written is very helpful in solving certain algebra problems. We give this property of addition a special name; we say that addition is **commutative**. More exactly:

> **DEFINITION**
> We say that addition is **commutative**. This means that for any two numbers b and c:
> $$b + c = c + b$$

Do you understand the difference between order and grouping? For example, to get from $(3 + 4) + 5$ to $3 + (4 + 5)$, all you do is change the grouping. In either case, the order is $3 + 4 + 5$. But the order in the expression $(4 + 3) + 5$ is different from the order $(3 + 4) + 5$. The order in which 3 and 4 appear has been changed.

Sometimes we use the associative principle and the commutative principle in the same problem.

> We often use *rule* and *principle* to mean the same thing.

Example 3

Simplify the expression

$$4 + (m + 5) \qquad [1]$$

In simplified form, $4 + (m + 5)$ is the same as $m + 9$.

METHOD 1: Because $m + 5 = 5 + m$, you may rewrite [1] as

$$4 + (5 + m) \qquad [2]$$

Addition is associative, so you may rewrite [2] as

$$(4 + 5) + m \qquad [3]$$

This is the same as $9 + m$. In summary:

$$4 + (m + 5) = 4 + (5 + m) = (4 + 5) + m = 9 + m$$

Addition is commutative, so this is the same as $m + 9$.

METHOD 2: Because $(m + 5)$ is one number, by the commutative principle you may rewrite [1] as

$$(m + 5) + 4 \qquad [4]$$

Then by the associative principle, you may rewrite [4] as

$$m + (5 + 4) \qquad [5]$$

This is the same as $m + 9$. By the commutative rule, $m + 9$ and $9 + m$ are equal.

> We are showing that $4 + (m + 5)$ and $9 + m$ are two different expressions that give the same answer for a given value of m. For example, if we let $m = e$, $4 + (m + 5)$ is $4 + (3 + 5)$ or $4 + 8$, which is 12. And $9 + m$ is also 12 in this case, because when $m = 3$, $9 + m = 9 + 3$.

> Until now we have emphasized that $3 + 5$ names the same number as $5 + 3$. Now we're saying that for any number m, $5 + m$ and $m + 5$ name the same number. As an equation, $5 + m = m + 5$ becomes a true statement for any value of m.

Section 4.2 The Commutative Property for Addition

Addition is commutative, but subtraction isn't. Hence, before we can apply the commutative principle to a subtraction problem, we must first rewrite the problem as an addition problem.

Example 4

Show that $3 - 5$ is the same number as $(-5) + 3$.

$3 - 5$ means $3 - (5)$ or $3 - (+5)$. By the rule for subtracting signed numbers, $3 - (+5)$ equals $3 + (-5)$. But $3 + (-5)$ is the *sum* of two numbers. Addition is commutative, so $3 + (-5)$ is equal to $(-5) + 3$.

Example 5

Simplify the expression $3 + (q - 4)$.

$3 + (q - 4)$ is the same as $q - 1$. You can rewrite $q - 4$ as $q + (-4)$. By the commutative property of addition, $q + (-4)$ is equal to $(-4) + q$. So

$$3 + (q - 4) = 3 + (q + [-4])$$
$$= 3 + ([-4] + q)$$
$$= (3 + [-4]) + q$$
$$= (-1) + q$$
$$= q + (-1)$$
$$= q - 1$$

We could have solved Example 5 by writing

$$3 + (q - 4) = (q - 4) + 3$$
$$= (q + {}^-4) + 3$$
$$= q + ({}^-4 + 3)$$
$$= q + ({}^-1)$$
$$= q + {}^-1$$
$$= q - 1$$

It is important to remember that addition is both associative and commutative. Any expression containing only addition may be simplified by using these two properties. We often use these two properties in arithmetic without realizing it.

For example, we often like to add by grouping in tens. So given $7 + 8 + 2$, we may prefer to group the numbers as $7 + (8 + 2)$, which is $7 + 10$, or 17. If we do this, we are using the associative property of addition, which tells us that $(7 + 8) + 2$ and $7 + (8 + 2)$ have the same value. Another way to group by tens is by rewriting $2 + 7 + 8$ as $2 + 8 + 7$. This is changing the order of the 7 and 8, which means that we are using the commutative property.

We have used some of the properties of addition. We treat $(q - 4)$ as a single number, so we use the commutative property to change the order of the terms in $3 + (q - 4)$ to $(q - 4) + 3$. Then we use the "add the opposite" rule to rewrite $q - 4$ as $q + {}^-4$.

Next we use the associative property to rewrite $(q + {}^-4) + 3$ as $q + ({}^-4 + 3)$. We then replace ${}^-4 + 3$ by ${}^-1$, and finally, we rewrite $q + ({}^-1)$ as $q + {}^-1$ or $q - 1$.

The net result is that by the properties of addition we are able to show that for any given value of q, $3 + (q - 4)$ and $q - 1$ have the same value. In other words, the expression $3 + (q - 4)$ can be replaced by the simpler expression $q - 1$.

PRACTICE DRILL

1. For what value(s) of h will $h + 8 = 8 + h$?
2. Evaluate each of the following when $h = 5$.

    (a) $h - 8$ (b) $8 - h$

3. Simplify the expression $5 + (t + 11)$.
4. Simplify the expression $5 + (t - 11)$.
5. Simplify the expression $5 + (11 - t)$.

Answers
1. Because addition is commutative, $h + 8 = 8 + h$ for all values of h.   2. (a) (-3) (b) 3
3. $16 + t$ or $t + 16$ 4. $t - 6$
5. $16 - t$

CHECK THE MAIN IDEAS

1. In many arithmetic problems, both grouping and _____ make a difference.
2. The commutative rule for addition states that the sum of two numbers does not depend on the _____ of the two numbers.
3. The cummutative principle means that $6 + 7$ is the same number as _____.
4. Subtraction is not commutative. In other words, $4 - 7$ is not equal to _____.
5. You use the commutative rule when you write that $(3 + 4) + 7 =$ _____ $+ 7$.
6. When you write $(3 + 4) + 7 = 3 + (4 + 7)$, you are using _____.

Answers
1. order   2. order
3. $7 + 6$ (Do not write 13. $6 + 7$ is 13, but not by the commutative rule.)   4. $7 - 4$ 5. $(4 + 3)$
6. the associative principle (rule) for addition.

Section 4.3
The Associative Property for Multiplication

Just as we don't need grouping symbols in addition problems, we don't need them in multiplication problems.

Example 1

What is the value of (a) $2 \times (3 \times 4)$? (b) $(2 \times 3) \times 4$?

In both cases, the product is 24. That is, $2 \times (3 \times 4) = 2 \times 12 = 24$ and $(2 \times 3) \times 4 = 6 \times 4 = 24$.

But just as grouping is important in subtraction, it is also important in division.

Example 2

What is the value of (a) $12 \div (6 \div 2)$? (b) $(12 \div 6) \div 2$?

The answer to (a) is 4 and the answer to (b) is 1. That is,

$12 \div (6 \div 2) = 12 \div 3 = 4$
and $(12 \div 6) \div 2 = 2 \div 2 = 1$

Both (a) and (b) look like $12 \div 6 \div 2$ if the parentheses are left out. This means that in division, the answer *does* depend on how the terms are grouped.

The fact that the product of three or more numbers doesn't depend on grouping is given a special name. As in addition, we say that multiplication is *associative*.

DEFINITION
For any three numbers b, c, and d:
$$b(cd) = (bc)d$$
This is known as the *associative principle for multiplication*.

The product bcd can be grouped only in two ways, $b(cd)$ or $(bc)d$, so the associative property of multiplication tells us that the product bcd doesn't depend on how the factors are grouped.

OBJECTIVE

To learn that the product of three or more numbers does not depend on how the numbers are grouped; and to see how this fact is used in algebra.

The property of being associative extends to any operation in which grouping doesn't change the answer. Arithmetic is associative with respect to both addition and multiplication, but not with respect to either subtraction or division.

REVIEW

When we work with letters instead of numbers, we write bc rather than $b \times c$.

In other words, in a problem that involves only multiplication, we may omit the grouping symbols.

Section 4.3 The Associative Property for Multiplication

Example 3

Simplify the expression $4(3m)$.

A simpler way of expressing $4(3m)$ is to write $12m$. This is seen from the associative principle:

$$4(3m) = (4 \times 3) \times m = 12m$$

That is,
$$4(3m) = 4 \times (3 \times m)$$
$$= (4 \times 3) \times m$$

If we use the fact that, for example, $12 \div 6$ means $12 \times \frac{1}{6}$, we can rewrite all division problems in terms of multiplication and use the associative principle for multiplication.

To divide two fractions, we leave the first one alone, change the division to multiplication, and invert the second fraction. (Invert means we interchange numerator and denominator.)

Example 4

Simplify the expression $(m \div 6) \div 3$.

$(m \div 6) \div 3$ is the same as $m \div 18$. That is,

$$(m \div 6) \div 3 = \left(m \times \frac{1}{6}\right) \times \frac{1}{3} = m \times \left(\frac{1}{6} \times \frac{1}{3}\right)$$
$$= m \times \frac{1}{18} = m \div 18$$

Notice that Example 4 tells us that division is not associative. If we regroup $(m \div 6) \div 3$, we get $m \div (6 \div 3)$, or $m \div 2$. But $m \div 2$ is quite different from $(m \div 6) \div 3$. In fact, we saw in Example 4 that $(m \div 6) \div 3$ is the same as $m \div 18$.

In summary, both addition and multiplication obey the associative principle. But if we want to apply this principle to either subtraction or division, we must rewrite each subtraction in terms of addition and each division in terms of multiplication.

For example, if we replace m by 36 we find that,
$$(m \div 6) \div 3 = (36 \div 6) \div 3 = 2$$
$$m \div 18 = 36 \div 18 = 2$$
But
$$m \div (6 \div 3) = 36 \div (6 \div 3)$$
$$= 36 \div 2$$
$$= 18$$

PRACTICE DRILL

1. Compute the products
   (a) $9 \times (2 \times 4)$ (b) $(9 \times 2) \times 4$

2. Compute the quotients
   (a) $(40 \div 10) \div 2$ (b) $40 \div (10 \div 2)$

3. Simplify the expression $6(8m)$. (That is, rewrite $6(8m)$ without grouping symbols.)

4. Simplify the expression $(p \div 8) \div 4$.

5. (a) Evaluate $(p \div 8) \div 4$ when $p = 64$.
   (b) Evaluate $p \div (8 \div 4)$ when $p = 64$.
   (c) Are the expressions $(p \div 8) \div 4$ and $p \div (8 \div 4)$ equal?

Answers
1. (a) 72 (b) 72   2. (a) 2 (b) 8   3. $48m$ 4. $p \div 32$
5. (a) 2 (b) 32 (c) no

CHECK THE MAIN IDEAS

1. The associative principle applies to both addition and _____ .

2. For example, $7 \times (6 \times 8)$ names the same number as _____ . Therefore, you do not need grouping symbols in a multiplication problem.

3. On the other hand, even though they both look like $72 \div 6 \div 3$ when you omit the parentheses, $(72 \div 6) \div 3$ is not the same as _____ .

Answers
1. multiplication
2. $(7 \times 6) \times 8$ 3. $72 \div (6 \div 3)$

Section 4.4
The Commutative Property for Multiplication

Just as the sum of two numbers does not depend on the order in which the two numbers appear, the product of two numbers doesn't depend on the order of the two factors. The reason for this is easily seen in terms of the area of a rectangle.

Example 1

One side of a rectangle is 6 feet and another side is 4 feet. What is the area of the rectangle?

The area of the rectangle is 24 square feet.
  To find the area, you multiply the length and width. In this example, you are not told which is the length and which is the width. Thus, you may write either 4×6 or 6×4. In either case, the product is 24. In the diagram on this page, you can see that there are 24 unit squares in each rectangle.

> **DEFINITION**
> Multiplication is said to possess the **commutative property**. That is, for any two numbers b and c
>
> $$b \times c = c \times b \qquad \text{or } bc = cb$$

Example 2

Compute the following two products in the order in which the factors appear.

  (a) 3.46×7.89 (b) 7.89×3.46

In either case the product is 27.2994. This is simply one illustration of the commutative property of multiplication.

  Sometimes we prefer to emphasize the general meaning rather than to show a particular example. Perhaps the following problem will be more helpful.

Example 3

For what value of h is it true that

$$8 \times h = h \times 8 \qquad [1]$$

$h \times 8 = 8 \times h$ for every numerical value of h. This example is a restatement of the commutative property of multiplication. If you think of [1] as a "fill in the blank" problem with h as the blank, you mean that any number you put into the blank will make [1] a true statement.

  It is more conventional to write $8h$ rather than $8 \times h$. But it is not generally the practice to write $h8$ in place of $h \times 8$. Normally, people write the numerical factor to the left of the literal (letter) factor.
  We can write $8h$ to abbreviate $h \times 8$ because of the commutative property of multiplication. For example, $h \times 8 = 8 \times h$, which is $8h$.
  There is a special vocabulary used to discuss the factors in an expression such as $8h$. When we have the product of two factors, one of the factors is called the **coefficient** of the other. For example, in the expression $8h$, 8 is the coefficient of h, and h is the coefficient of 8. It is more customary to talk about the coefficient of the literal factor than the coefficient of the numerical factor. That is, we talk more often about 8 being the coefficient of h than we do about h being the coefficient of 8.

OBJECTIVE
To learn that the product of two numbers does not depend on the order of the two factors; and to use this fact in the study of algebra.

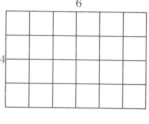

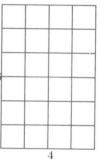

In this example, you can still use the area of a rectangle. A rectangle whose dimensions are 3.46 cm by 7.89 cm has the same area regardless of which dimension you call the length.

An equation that becomes a true statement for *each* value of the unknown is called an **identity**. For example, $8 \times h = 8 \times h$ is an identity.
  In an identity, the expression on one side of the equation is a synonym (different name) for the expression on the other side. That is, for each value of h, $8 \times h$ is another name for $h \times 8$.

Section 4.4 The Commutative Property for Multiplication

Example 4

What is the coefficient of m in the expression $2.54m$?

The coefficient of m is 2.54, because the expression means that you are multiplying 2.54 and m.

Example 5

What is the coefficient of m in the expression $3m + 4$?

The coefficient of m in the expression $3m + 4$ is 3.
 The coefficient of m is the number that multiplies m. Looking at the expression $3m$, you see that this number is 3. The 4 is being *added* to $3m$; therefore, by the definition of a coefficient, 4 is not part of the coefficient of m.

Example 6

What is the coefficient of m in the expression $(3 + 4)m$?

In $(3 + 4)m$, the coefficient of m is 7. You see that $(3 + 4)$ is multiplying m; since $3 + 4 = 7$, you may write that the coefficient of m is 7.

> Notice the use of the grouping symbols. 3 is not being added to $4m$. Rather, m is being multiplied by $(3 + 4)$. In the previous example, the grouping was understood to be $(3m) + 4$, which meant that 4 was being added to the product $3m$.

Example 7

What is the coefficient of m in the expression $(b + c)m$?

The coefficient of m is $b + c$.
 Since $b + c$ is enclosed within parentheses, it is a single number. Hence m is being multiplied by $b + c$. This means that $b + c$ is the coefficient of m.

> This shows the difference between numbers and letters. When we had $3 + 4$, we could replace it by 7. But since b and c are symbols for numbers, we cannot simplify the expression $b + c$. This is one reason that the structure of arithmetic is so important. We need the rules, especially when we deal with letters rather than numbers.

Example 8

What is the coefficient of m in the expression $3(4m)$.

In the expression $3(4m)$, the coefficient of m is 12. The definition of coefficient requires that you have the product of two terms. This can always be done by the associative property. For example,

$$3(4m) = [(3)(4)]m = 12m$$

Keep track of the difference between the associative and the commutative property of multiplication. When we keep the order the same but change the grouping, we are using the associative property. When we change the order and keep the grouping the same, we are using the commutative property.
 For example, when we rewrite $1 \times (2 \times 3)$ as $(1 \times 2) \times 3$, we are using the associative property of multiplication. But when we rewrite $1 \times (2 \times 3)$ as $1 \times (3 \times 2)$, we are using the commutative property of multiplication.
 Sometimes we may have to use both properties in a single problem.

Example 9

Simplify the expression $(7 \times k) \times 5$.

$(7 \times k) \times 5$ can be written more simply as $35k$.
 By the commutative property, $(7 \times k) \times 5$ equals $5 \times (7 \times k)$. By the associative property, $5 \times (7 \times k)$ equals $(5 \times 7) \times k$ or $35 \times k$, which is the same as $35k$.

> That is $(7 \times k) \times 5$ and $35k$ have the same value for a given value of k.

In terms of coefficients, you can also see in this example that the coefficient of k in $(7 \times k) \times 5$ is 35.

Although division is not commutative, we can still simplify division problems by rewriting each division as an equivalent multiplication problem. That is, we may always rewrite $b \div c$ as $b \times \frac{1}{c}$.

Example 10

Rewrite $10 \div 5$ in terms of multiplication.

$10 \div 5$ is the same as $10 \times \frac{1}{5}$. All you do is "invert and multiply."

Example 11

Simplify $(12 \times p) \div 6$.

$(12 \times p) \div 6$ is the same as $2p$.

$$
\begin{aligned}
(12 \times p) \div 6 &= (12 \times p) \times \tfrac{1}{6} && \text{(def. of division)} \\
&= (p \times 12) \times \tfrac{1}{6} && \text{(mult. is commutative:} \\
& && 12 \times p = p \times 12) \\
&= p \times \left(12 \times \tfrac{1}{6}\right) && \text{(mult. is associative)} \\
&= p \times 2 && \text{(replacing } 12 \times \tfrac{1}{6} \text{ by 2)} \\
&= 2 \times p && \text{(mult. is commutative)} \\
&= 2p && \text{(another way of writing } 2 \times p\text{)}
\end{aligned}
$$

Example 11 is useful for the algebra that we do later in the course. It tells us we can replace $(12 \times p) \div 6$ by the simpler $2p$. That is, for a given value of p, it is easier to evaluate $2p$ than $(12 \times p) \div 6$.

Example 11 shows the basic difference between the associative and commutative properties.

1. Whenever we leave the numbers in the same order but change the grouping, we are using the associative property.
2. Whenever we leave the grouping alone but change the order in which the numbers are written, we are using the commutative property.
3. If we change both the grouping and the order, we are using a combination of both properties. For example, we used both properties in Example 11 when we simplified $(12 \times p) \div 6$ to obtain $2p$.

For example, $(4 \times 7) \times 5 = 4 \times (7 \times 5)$.

For example, $(4 \times 7) \times 5 = (7 \times 4) \times 5$.

$(4 \times 7) \times 5 = 4 \times (5 \times 7)$. The order is not the same because in one case we have $4 \times 7 \times 5$ and in the other we have $4 \times 5 \times 7$. The grouping is also different, in one case the first two factors are grouped, and in the other, the last two are grouped.

PRACTICE DRILL

1. For what value(s) of t will $7 \times t = t \times 7$?
2. Evaluate each of the following expressions when $b = 10$.
   (a) $b \div 4$ (b) $4 \div b$
3. Simplify each of the following expressions.
   (a) $(5 \times p) \times 6$ (b) $(5 \times p) \div 6$ (c) $(p \div 5) \div 6$
4. Find the coefficient of m in each of the following.
   (a) $9m$ (b) $9m + 4$ (c) $(9 + 4)m$ (d) $9(4m)$

Answers
1. All numerical values of t
2. (a) 2.5 (or $\frac{5}{2}$ or $2\frac{1}{2}$)   (b) 0.4 (or $\frac{2}{5}$, and so on)
3. (a) $30 \times p$ (or $30p$)   (b) $\frac{5}{6}p$ (c) $p \div 30$ (or $\frac{p}{30}$)
4. (a) 9   (b) 9   (c) 13   (d) 36

Section 4.5 The Additive Identity and Additive Inverses

CHECK THE MAIN IDEAS

1. The commutative property for multiplication states that the product of two or more numbers does not depend on the _____ of the factors.
2. The commutative property of multiplication tells you that 7×5 is the same as the product _____ .
3. On the other hand, since $5 \div 10$ and $10 \div 5$ are not equal, division does not possess the _____ property.
4. When you rewrite $(2 \times 3) \times 4$ as $2 \times (3 \times 4)$, you are using the (a) _____ property. But when you rewrite $(2 \times 3) \times 4$ as $(3 \times 2) \times 4$, you are using the (b) _____ property.

Answers
1. order   2. 5×7
3. commutative
4. (a) associative
   (b) commutative

Section 4.5
The Additive Identity and Additive Inverses

In Lesson 2, we introduced equations as a form of "fill in the blank," and we saw how we could solve equations by "undoing" certain operations. In this section, we shall see how the "undoing" idea is tied to the structure of arithmetic. Let's begin with a short review.

OBJECTIVE
To reexamine the facts that for any number b

(1) $b + 0 = b$
(2) $b + {}^-b = 0$

and to see how these properties are used in solving equations.

Example 1

Use the "undoing" idea to solve the equation

$$m + 3 = 7 \qquad [1]$$

The solution of $m + 3 = 7$ is $m = 4$.

The approach is that 3 is being added to m, so you can "undo" it by subtracting 3. To keep the equation balanced, if you subtract 3 from one side, you must also subtract 3 from the other side. You get

$$\begin{array}{r} m + 3 = 7 \\ -3 \quad -3 \\ \hline m \quad = 4 \end{array}$$

We add (-3) to both sides of [1] to obtain

$$[m + 3] + (-3) = 7 + (-3)$$
$$m + [3 + (-3)] = 4$$
$$m + 0 = 4$$
$$m = 4$$

In doing Example 1, we used two important facts. They are:

1. For any number m, $m + 0 = 0$.
2. If we subtract any number from itself, the answer is 0.

The first fact is given a special name.

DEFINITION
0 is called the **additive identity**. That is, for any number n:
$$n + 0 = n$$

This means that we can always replace $n + 0$ by n, for any number n. And, if we wish, we can always replace n by $n + 0$.
In other words, n and $n + 0$ are synonyms.

Example 2
Simplify $-3 + 0$.

$-3 + 0$ is -3. You do not change the identity of any number by adding 0 to it.

This problem asks us to see that when added to any number, the additive identity produces that number as the sum.
This is generalized in Example 3.

Example 3

Simplify the expression $p + 0$.

$p + 0$ is p. This is a generalization of Example 2. Adding 0 to any number, in this case p, doesn't change the number.

Example 4

Simplify the expression $-3 + 0 + 4$.

$-3 + 0 + 4$ is the same as 1. That is, by the associative property of addition

$$-3 + 0 + 4 = (-3 + 0) + 4$$
$$= -3 + 4$$
$$= 1$$

$-3 + 0 + 4$ is also $-3 + (0 + 4)$ or $-3 + 4$, which is still 1.

The second important fact leads to the idea of the **additive inverse.** When we subtract a number from itself, we are really doing a form of addition. Let's review.

Example 5

Rewrite $3 - 3$ as an addition problem.

$3 - 3$ means the same as $3 + (-3)$. In other words,

$$3 - 3 = 3 - (3)$$
$$= 3 - (+3)$$
$$= 3 + (-3)$$

Look at what Example 5 tells us. We already know that $3 - 3 = 0$. The answer tells us that $3 + (-3) = 0$. That is, if we add -3 to 3, we get 0, or if we add 3 to -3, we get 0. This leads to the definition:

> **DEFINITION**
> If the sum of two numbers is 0, each of the numbers is called the **additive inverse** of the other.

Let's try using the definition.

Example 6

What is the additive inverse of 7?

The additive inverse of 7 is $^-7$.
You know from your study of signed numbers that $7 + {^-7} = 0$. Hence $^-7$ is the number you must add to 7 to get 0. By the definition of the additive inverse, the additive inverse of 7 is $^-7$.

Remember that the additive inverse of 7 means the number you must add to 7 to get 0.

Example 7

What is the additive inverse of $^-7$?

The additive inverse of $^-7$ is 7. You want the number that must be added to $^-7$ to give 0 as the sum. This number also comes from the fact that $7 + {^-7} = 0$.

There is a pattern. The additive inverse of $^+7$ is $^-7$, and the additive inverse of $^-7$ is $^+7$. You find the additive inverse of a number by changing its sign.

Section 4.5   The Additive Identity and Additive Inverses

What we call the additive inverse in this section is a new name for something we already learned about in Lesson 3. Look at Examples 6 and 7. In each case, we found the additive inverse of the given number by changing the sign of the number. In Lesson 3, we mentioned that the opposite of a signed number is the number we get when we just change the sign of the number.

In other words, the additive inverse of a number is the opposite of the number.

NOTATION

If b is any given number, we denote the additive inverse of b by ^-b. In other words, ^-b is defined as the number such that $b + {}^-b = 0$.

Example 8

Simplify $^-(^-7)$.

$^-(^-7) = 7$. Because $^-7$ is enclosed in parentheses, it is considered to be a single number. Therefore, $^-(^-7)$ means the additive inverse of $^-7$. That is, you want the number that must be added to $^-7$ to get 0. $^-7 + 7 = 0$, so the additive inverse of $^-7$ is 7: that is, $^-(^-7) = 7$.

In this way, ^-b no longer has to be a negative number, but rather the opposite or additive inverse of b. In Example 8, notice that $^-(^-7)$ is the positive number 7.

Example 9

What is the additive inverse of $(5 - 3)$?

The additive inverse of $5 - 3$ is -2.
   Another name for $5 - 3$ is 2. Therefore, you want the additive inverse of 2. This means that you want the number that must be added to 2 to give 0 as the sum. $2 + (-2)$ is 0, so the additive inverse of $5 - 3$ is -2.

   There is a simple way of getting the additive inverse of a difference. All we have to do is change the order of the numbers we are using in the subtraction problem. For example, if we change the order of the numbers in $5 - 3$, we get $3 - 5$ or -2. As we saw in Example 9, -2 is the additive inverse of $5 - 3$.
   To see why this works, we notice that in terms of addition, $5 - 3$ is $5 + (-3)$ while $3 - 5$ is $3 + (-5)$. If we add $5 - 3$ and $3 - 5$, we have

$$[5 - 3] + [3 - 5] = [5 + (-3)] + [3 + (-5)]$$
$$= 5 + [(-3) + 3] + (-5) \longrightarrow \text{By the associative property.}$$
$$= 5 + \underbrace{0} + (-5) \longrightarrow \text{Because 3 and } -3 \text{ are additive inverses.}$$
$$= [5 + 0] + (-5) \longrightarrow \text{Again by the associative property.}$$
$$= 5 + (-5) \longrightarrow \text{Because 0 is the additive identity.}$$
$$= 0 \longrightarrow \text{Because 5 and } -5 \text{ are additive inverses.}$$

   In other words, the sum of $(5 - 3)$ and $(3 - 5)$ is 0. By definition, these two numbers are the additive inverses of one another.

Example 10

What subtraction problem gives the additive inverse of $2 - 9$?

The additive inverse of $2 - 9$ is $9 - 2$. All you do is reverse the numbers in the expression $2 - 9$.
   As a check, notice that $2 - 9$ is -7 and $9 - 2$ is 7. The sum of 7 and -7 is 0.

   When we are working with numbers like $2 - 9$, we can always subtract to obtain -7. Then it is easy to see that the additive inverse is 7. We do not have to rewrite $2 - 9$ as $9 - 2$, even though we can if we want to.

But when we deal with letters that stand for numbers, we can't replace $b - c$ by a single letter. However, we can reverse the b and c to write $c - b$, which is the additive inverse of $b - c$.

Example 11

What is the additive inverse of $m - 4$?

The additive inverse of $m - 4$ is $4 - m$. To get this result, just reverse the m and the 4.

As a check, notice that

$$\begin{aligned}(m - 4) + (4 - m) &= (m + [-4]) + (4 + [-m]) \\ &= m + ([-4] + 4) + [-m] \\ &= m + 0 + [-m] \\ &= (m + 0) + (-m) \\ &= m + (-m) \\ &= 0\end{aligned}$$

In other words, the opposite or the additive inverse of $m - 4$ is $4 - m$. Another way of writing this is as $-(m - 4) = 4 - m$.

The fact that we can find the additive inverse (the "opposite") of an algebraic expression helps us to simplify some expressions.

Example 12

Simplify the expression $7 - (m - 4)$.

$7 - (m - 4)$ is equal to $11 - m$.

Remember that in Lesson 3, you learned how to subtract signed numbers. You leave the first number alone, change the subtraction to addition, and change the sign of the second number. Changing the sign means to take the additive inverse. Therefore

$$\begin{aligned}7 - (m - 4) &= 7 + {}^-(m - 4) \\ &= 7 + (4 - m) \\ &= 7 + (4 + {}^-m) \\ &= (7 + 4) + {}^-m \\ &= 11 + {}^-m \\ &= 11 - m\end{aligned}$$

As we shall see in more detail in the next lesson, the main use of additive inverses in algebra is to simplify equations.

Example 13

Make use of the additive inverse to solve the equation $m + (-8) = 7$.

Add 8 to both sides of the equation. That is, the additive inverse of -8 is 8. So adding 8 to both sides of the equation gives

$$\begin{aligned}[m + (-8)] + 8 &= 7 + 8 = 15 \\ m + [(-8) + 8] &= 15 \\ m + 0 &= 15 \\ m &= 15\end{aligned}$$

The same idea can be used even if the equation is a bit more complicated.

Namely,

$$\begin{aligned}(b - c) + (c - b) &= \\ (b + {}^-c) + (c + {}^-b) &= \\ b + (c + {}^-c) + {}^-b &= \\ \underbrace{b + 0}_{b} + {}^-b &= \\ + {}^-b &= 0\end{aligned}$$

That is, for any number m, ${}^-(m - 4) = 4 - m$. More generally, for any two numbers m and n

$${}^-(m - n) = n - m$$

Again, notice that ${}^-(m - 4)$ need not be a negative number. For example, if we let $m = 1$, ${}^-(m - 4) = {}^-(1 - 4) = {}^-({}^-3)$, which is 3, not $^-3$.

As we said before, ${}^-b$ is always the number we add to b to get 0. So ${}^-b$ will be negative only if b is positive.

Previously we evaluated expressions such as $7 - (m - 4)$ for various values of m. Here we want to express $7 - (m - 4)$; more simply, but still in terms of m.

Remember that in Example 11, we showed that $4 - m$ is the additive inverse of $m - 4$.

This means that $11 - m$ and $7 - (m - 4)$ will give the same value for any chosen value for m. For example, if $m = 10$,

$$11 - m = 11 - 10 = 1$$

and $\quad 7 - (m - 4) = 7 - (10 - 4)$
$$= 7 - 6 = 1$$

The equation may be rewritten as $m - 8 = 7$; then we can solve it as in Lesson 2 by adding 8 to both sides. But we are trying to get better acquainted with the structure of arithmetic, because later in the course it will be helpful to be able to use this method.

Section 4.5   The Additive Identity and Additive Inverses   83

Example 14

For what value of m does $7 - (2 - m) = 6$?

$7 - (2 - m) = 6$ when $m = 1$.
   The problem may be viewed as
$$7 + {}^-(2 - m) = 6$$
so that
$$7 + (m - 2) = 6$$ Remember that ${}^-(2 - m) = m - 2$.

Because $m - 2 = m + {}^-2$, you have
$$7 + (m + {}^-2) = 6$$

Therefore
$$7 + ({}^-2 + m) = 6 \text{ (by the commutative property)}$$

This is the same as
$$(7 + {}^-2) + m = 6 \text{ (by the associative property)}$$

As $7 + {}^-2 = 5$, you have
$$5 + m = 6$$

You can now subtract 5 from both sides to get $m = 1$.
   As a check, if you let $m = 1$, you get
$$7 - (2 - 1) = 7 - 1 = 6$$

Once we learn the rules of algebra we usually combine several steps. For example, we might write

$$\begin{aligned} 7 - (2 - m) &= 6 \\ 7 + (m - 2) &= 6 \\ (7 - 2) + m &= 6 \\ 5 + m &= 6 \\ -5 & -5 \\ \hline m &= 1 \end{aligned}$$

If we are right, then $7 - (2 - m) = 6$ should be a true statement when m is replaced by 1.

   We shall drill more on this in the exercise sets, but now let's conclude by making sure that we see the relationship between the additive identity and the additive inverse. As shown below,

$$b \quad + \quad {}^-b \quad = \quad 0$$

with arrows labeling ${}^-b$ as "additive inverse (of b)" and 0 as "additive identity".

PRACTICE DRILL

1. Find the additive inverse of
   (a) 10   (b) −10   (c) $9 - 11$
2. Simplify
   (a) $-(-8)$ (b) $-(3 - 11)$ (c) $9 - (3 - 11)$
3. What is the additive inverse of
   (a) $(m - 6)$ (b) $(6 - m)$
4. Simplify
   (a) $9 - (m - 6)$ (b) $9 - (6 - m)$

Answers
1. (a) −10   (b) 10   (c) 2
2. (a) 8   (b) 8   (c) 17
3. (a) $6 - m$ (b) $m - 6$
4. (a) $15 - m$ (b) $3 + m$ (or $m + 3$)

CHECK THE MAIN IDEAS

1. Adding _____ to a sum doesn't change the sum.
2. For this reason 0 is called the _____ identity.

Answers
1. 0   2. additive
3. inverse   4. opposite
5. sign (or signature)
6. simplify

3. The number you add to a given number to get 0 as a sum is called the additive _____ of the given number.
4. In your earlier study of signed numbers, the additive inverse of a given number was called the _____ of the given number.
5. To get the additive inverse (or opposite) of a given number, you change the _____ of the given number.
6. Using the additive inverse helps to _____ algebraic expressions.

Section 4.6
The Multiplicative Identity and Multiplicative Inverses

In the last section, we saw that 0 played a very special role in addition. When we add 0 to a number, we don't change the sum. We described this property of 0 by calling 0 the additive identity.

In this section, we want to notice that 1 does for multiplication what 0 does for addition. That is, if we multiply a given number by 1, the product is the given number.

For this reason, 1 is called the **multiplicative identity**. More precisely:

OBJECTIVE

To reexamine the facts that for any number b

(1) $b \times 1 = b$

(2) $b \times \dfrac{1}{b} = 1$ (unless $b = 0$)

and to see how these properties are used in solving equations.

> For any number n:
> $$1 \times n \text{ (or } 1n) = n$$

In other words, whenever we see a letter like n that stands for a number, we may think of it as being $1n$ (that is, $1 \times n$). We can always replace $1n$ by n, or n by $1n$.

Example 1

Perform the given operations

    (a) $8 + 0$ (b) 8×1 (c) 8×0

(a) Because 0 is the additive identity, $8 + 0 = 8$.
(b) Because 1 is the multiplicative identity, $8 \times 1 = 8$.
(c) 0 eight times is still 0. That is, $8 \times 0 = 0$.

While we're on the subject of multiplying by 1, we should also look at the effect of multiplying a number by -1.

The point of (c) is to help you remember that 0 is the identity only for addition. 1 is the identity for multiplication. In our study of algebra, it is very important to remember that *any number times 0 is 0.*

Example 2

Compute the products

    (a) -1×4 (b) -1×-4

(a) $-1 \times 4 = -4$
(b) $-1 \times -4 = 4$
You get these answers by the rule for multiplying signed numbers.

Did you notice what happened in Example 2? In each case, multiplying a number by -1 changed the sign of the number. But when you change the sign of a number, you get the additive inverse of the number. In other words, the general result is

REVIEW

To find the product of two signed numbers, multiply their magnitudes and call the sign positive if both factors have the same sign; call the sign negative if the factors have different signs. Thus

$$-1 \times 4 = -1 \times +4 = -(1 \times 4)$$
$$= -4$$
$$-1 \times -4 = (1 \times 4) = 4$$

> For any number n:
> $$-1 \times n = -n$$

In other words we can always replace ^-n by ^-1n and ^-1n by ^-n.

Section 4.6 The Multiplicative Identity and Multiplicative Inverses

Example 3

Write -8 as a product of two numbers, one of which is 8.

$-8 = -1 \times 8$, or $-1(8)$.

Example 4

Write 8 as a product of two numbers, one of which is -8.

$8 = -1 \times -8$.

These examples illustrate the fact that we can get the additive inverse of a number just by multiplying the number by -1. This will be an important aid to us later in our study of simplifying various algebraic expressions.

In the last section, we explained that the additive inverse of a number is what we had to add to a number to get the additive identity (0). Is there a similar thing in multiplication? If we are given a number, is there a number we can multiply it by to get the multiplicative identity (1) as a product? Let's try to answer this by looking at a few examples.

Example 5

What number can you multiply $\frac{2}{3}$ by to get 1 as the product?

If you multiply $\frac{2}{3}$ by $\frac{3}{2}$, you get 1 as the product.

Remember, you multiply fractions by multiplying numerators and denominators. Therefore

$$\frac{2}{3} \times \frac{3}{2} = \frac{2 \times 3}{3 \times 2} = \frac{6}{6} = 1$$

Notice that the number you wanted is the *reciprocal* of the given number. The only number that doesn't have a reciprocal is 0.

If neither b nor c is 0, the reciprocal of $\frac{b}{c}$ is $\frac{c}{b}$. That is,

$$\frac{b}{c} \times \frac{c}{b} = \frac{b \times c}{c \times b} = \frac{b \times c}{b \times c} = 1$$

In other words, the multiplicative inverse of a number is the reciprocal of that number if the number is written as a common fraction.

Example 6

What number can you multiply 0 by to get 1 as the product?

There is no number that when multiplied by 0 is 1. Remember, any number multiplied by 0 gives 0 as the product.

The point is that except for 0, every number has a reciprocal. The reciprocal of a number is the number we multiply the number by to get 1 as the product. The more formal name given to the reciprocal of a number is the multiplicative inverse of the given number.

In summary:

> If n is any non-zero number, there exists a number denoted by $\frac{1}{n}$ such that:
>
> $$n \times \frac{1}{n} = 1$$

Remember, multiplicative inverse means the same as the reciprocal — except that the term "reciprocal" is usually used only when we use common fractions. No other numerals have numerators and denominators.

$\frac{1}{n}$ is called the **multiplicative inverse** of n.

Just as we used the additive inverse to solve certain types of equations, we use the multiplicative inverse to solve other types of equations.

To review, remember that in Lesson 2 we solved an equation like

$$2p = 36 \qquad [1]$$

by dividing both sides of [1] by 2. That is, we wrote

$$\frac{2p}{2} = \frac{36}{2} \qquad [2]$$

which gives us

$$p = 18 \qquad [3]$$

If we recall that dividing by 2 is the same as multiplying by $\frac{1}{2}$, we can re-examine equation [1] in terms of the multiplicative inverse and the multiplicative indentity.

That is, first we multiply both sides of equation [1] by $\frac{1}{2}$ to get

$$\tfrac{1}{2}(2p) = \tfrac{1}{2}(36) = 18 \qquad [4]$$

Then by the associative property for multiplication, we may rewrite [4] as

$$\left[\tfrac{1}{2}(2)\right]p = 18 \qquad [5]$$

Because $\frac{1}{2}$ is the multiplicative inverse of 2, we know that $\frac{1}{2} \times 2 = 1$. Hence [5] becomes

$$1p = 18 \qquad [6]$$

Finally, $1p = p$, as any number times 1 is still the same number. Therefore [6] becomes

$$p = 18$$

Using the multiplicative inverse allows us to rewrite any division problem in terms of multiplication (just as using the additive inverse allows us to rewrite subtraction problems in terms of addition).

Example 7

Write $m \div 8$ as a multiplication problem.

$m \div 8$ is the same as $m \times \frac{1}{8}$ or $\frac{m}{8}$. Use the usual "invert and multiply" rule, which in this case means that dividing by 8 is the same as multiplying by the reciprocal of 8.

Example 8

Simplify $3 \div (m \div 8)$.

$3 \div (m \div 8)$ is the same as $24 \div m$ or $\frac{24}{m}$. From Example 7, $m \div 8$ is $\frac{m}{8}$. Hence

$$3 \div (m \div 8) = 3 \div \frac{m}{8} = 3 \times \frac{8}{m} = \frac{24}{m}$$

In other words, just as adding the additive inverse to a number "undoes" that number, multiplying a number by its multiplicative inverse "undoes" that number. In more precise terms, we see that for any number b

$$b + {}^-b = 0$$

and if b is not 0, then

$$b \times \tfrac{1}{b} = 1$$

↓ multiplicative identity

↓ multiplicative inverse

Section 4.6 The Multiplicative Identity and Multiplicative Inverses

We must take one precaution here that didn't worry us in addition: 0 has no multiplicative inverse. *We cannot divide by 0.* In our answer to Example 8, we must state that the result holds *except when m is 0.*

Example 9

For what value of k doesn't $(k-1) \div 8$ have a multiplicative inverse?

$(k-1) \div 8$ doesn't have a multiplicative inverse when $k = 1$. When $k = 1$, $k - 1$ is 0 and $0 \div 8$ is 0. But 0 is the only number that doesn't have a multiplicative inverse.

In Example 9, the important thing to notice is that the "wrong" value for k is *not* 0 but 1. We can't take the reciprocal of 0, and $(k-1) \div 8$ is 0 only when $k = 1$.

Example 10

For what value of g doesn't $(g+3) \div 6$ have a multiplicative inverse?

$(g+3) \div 6$ does not have a multiplicative inverse when $g = -3$. That is, the only time you need worry is when $g + 3$ is 0; and this happens only if $g = -3$.
If you replace g by 0 in the expression $(g+3) \div 6$, you get $(0+3) \div 6$, or $3 \div 6$, which is $\frac{1}{2}$. The reciprocal of $\frac{1}{2}$ is 2. Therefore, $(g+3) \div 6$ does have a multiplicative inverse when g is 0. It doesn't have a multiplicative inverse when $g = -3$, because when g is -3, the value of the *entire expression* is 0.

PRACTICE DRILL

1. Find the multiplicative inverse of
   (a) 12 (b) $\frac{1}{12}$ (c) $18 \div 3$
2. For what value of m does the expression $\frac{m}{6}$ not have a multiplicative inverse (reciprocal)?
3. Simplify $4 \div (m \div 6)$, indicating for what values of m, if any, the simplification does not apply.

Answers

1. (a) $\frac{1}{12}$ (b) 12 (c) $\frac{1}{6}$
2. 0
3. $24 \div m$ or $\frac{24}{m}$, provided that m is not equal to 0.

CHECK THE MAIN IDEAS

1. Multiplying a product by _____ doesn't change the product.
2. Multiplying a number by -1 gives you the _____ of the number.
3. The number you must multiply a given number by to get 1 is called the multiplicative _____ of the given number.
4. The only number that doesn't have a multiplicative inverse is _____ .
5. Any number times 0 is _____ .

Answers

1. 1 2. additive inverse (or opposite) 3. inverse 4. 0
5. 0

Section 4.7
Removing Grouping Symbols

OBJECTIVE
To learn the agreements that have been accepted in algebra to allow us to omit grouping symbols without giving the expression more than one meaning.

We have already seens that we may omit grouping symbols in problems involving only addition or only multiplication.

For example, $3 + 4 + 5 + 2$ is 14 regardless of how the numbers are grouped. $3 \times 4 \times 5 \times 2$ is 120 regardless of how the numbers are grouped.

But if we have both addition and multiplication in the same problem, grouping does make a difference.

Example 1

Show that $2 \times 3 + 4$ can be either 10 or 14, depending on how you group the numbers.

If you write $2 \times 3 + 4$ as $(2 \times 3) + 4$, you have $6 + 4$, or 10. But if you write $2 \times 3 + 4$ as $2 \times (3 + 4)$, you have 2×7, or 14.

Now addition and multiplication are in the same expression, and as a result, grouping makes a difference. For example, we may group $2 \times 3 + 4$ as $(2 \times 3) + 4$ and get 10; or we may group it as $2 \times (3 + 4)$ and get 14. It is not a case of which is right or wrong. Rather, different groupings give different answers.

In other words, given an expression like $2 \times 3 + 4$, we must write it as $2 \times (3 + 4)$ or as $(2 \times 3) + 4$ so that other people will know what the expression means. The grouping symbols are important because different groupings give different answers.

To help cut down on the amount of grouping symbols, mathematicians agree to accept certain rules. The rules are not self-evident (any more than it's self-evident that 1 foot equals 12 inches; we learn that there are 12 inches in a foot by memorizing and accepting the definition).

We agree to use the plus signs to break expressions up into smaller pieces called **terms**. For example, in an expression such as

$$3 \times 4 \times 5 + 5 \times 3 + 9 \qquad [1]$$

there are *three* terms. The three terms are $3 \times 4 \times 5$, 5×3, and 9. These are smaller pieces that lie between the plus signs.

In other words, given [1] without further grouping symbols, we agree to read it as if it were written as

$$(3 \times 4 \times 5) + (5 \times 3) + (9) \qquad [2]$$

This is the same as $(60) + (15) + (9)$, or 84

We can get from [1] to [2] by a series of small steps:

STEP 1: Start a set of parentheses at the beginning of the expression. $(3 \times 4 \times 5 + 5 \times 3 + 9$

STEP 2: End this set of parentheses just to the left of the first plus sign in the expression. $(3 \times 4 \times 5) + 5 \times 3 + 9$

STEP 3: Start the next set of parentheses just to the right of this plus sign. $(3 \times 4 \times 5) + (5 \times 3 + 9$

STEP 4: Repeat the same procedure. In the present example, close the set of parentheses just before the next plus sign. $(3 \times 4 \times 5) + (5 \times 3) + 9$

STEP 5: Start a new set of parentheses to the right of the next plus sign. $(3 \times 4 \times 5) + (5 \times 3) + (9$

STEP 6: Continue this process until you reach the end of the expression. $(3 \times 4 \times 5) + (5 \times 3) + (9)$

This explains how we get from

$$3 \times 4 \times 5 + 5 \times 3 + 9 \qquad [1]$$

to

$$(3 \times 4 \times 5) + (5 \times 3) + (9) \qquad [2]$$

We can now simplify each set of parentheses because only multiplication is used; and we can then simplify this result because our expression becomes a sum of numbers.

Section 4.7 Removing Grouping Symbols

Example 2

Use the above method to find the number named by

$$4 \times 3 + 2 \times 2 \times 3 \times 4 + 1 \times 2$$

The number is 62. With the plus signs to separate the expression into terms, the implied grouping is

$$(4 \times 3) + (2 \times 2 \times 3 \times 4) + (1 \times 2)$$
$$= (12) + (48) + (2)$$
$$= 12 + 48 + 2$$
$$= 62$$

> **NOTE**
>
> If we want to indicate any grouping other than the given one, we must use parentheses to tell us what grouping we want.

Example 3

What number is named by the expression

$$4 \times (3 + 2) \times (2 \times 3) \times (4 + 1) \times 2$$

The number is 1,200. That is, you have

$$4 \times (3 + 2) \times (2 \times 3) \times (4 + 1) \times 2$$
$$= 4 \times (5) \times (6) \times (5) \times 2$$
$$= 1,200$$

Everything within a set of parentheses is a single number. For example, $(3 + 2)$ means 5.

There is no trouble here because outside the grouping symbols, the only operation is multiplication, and multiplication is associative.

   Notice that the expressions in Examples 2 and 3 are the same except for the grouping symbols. Changing the grouping symbols does change the value of the expression.
   The key point is that we agree to omit grouping symbols according to the understanding that plus signs break expressions up into terms. If we want any other type of grouping, we must use the grouping symbols.
   Sometimes expressions use minus signs and division signs in addition to the plus sign and the times sign. When this happens, we use the minus sign in the same way as we did the plus sign. That is, either plus signs or minus signs separate expressions into terms. Division signs, like times signs, do not separate expressions into terms. Let's look at a few more examples.

Example 4

What is the value of

$$3 \times 4 - 2 \times 5 + 7 \times 6 \div 3$$

The value is 16.
   Using plus and minus signs to get the terms, you have

$$(3 \times 4) - (2 \times 5) + (7 \times 6 \div 3)$$
$$= (12) - (10) + (14)$$
$$= 12 - 10 + 14 = 12 + (-10) + 14$$
$$= 2 + 14$$
$$= 16$$

Remember, $7 \times 6 \div 3$ means

$$7 \times 6 \times \frac{1}{3} = 42 \times \frac{1}{3} = 14.$$

Example 5

What is the value of

$$3 \times (4-2) \times (5+7) \times (6 \div 3)$$

The value is 144. That is,

$$3 \times (4-2) \times (5+7) \times (6 \div 3)$$
$$= 3 \times \quad (2) \quad \times \quad (12) \quad \times \quad (2)$$
$$= 144$$

Again, notice that in Examples 4 and 5 we have expressions that differ only in how the numbers are grouped. The fact that we get different answers shows us that grouping does make a difference. However, unless we are given directions telling us to do otherwise, the expression

$$3 \times 4 - 2 \times 5 + 7 \times 6 \div 3$$

is interpreted as we did in Example 4.

We have to be careful when a term contains the operation of division. For example, we have already seen that the expression

$$12 \div 6 \div 2$$

has two different values depending upon whether we write

$$(12 \div 6) \div 2 = 2 \div 2 = 1$$
$$\text{or } 12 \div (6 \div 2) = 12 \div 3 = 4$$

For this reason, we agree that when multiplication and division appear in the same term, we perform the operations in the *given* order, from left to right. If we want any other type of grouping, we must use the grouping symbols.

For example,

$$\underbrace{12 \div 6}_{\downarrow} \div 2 \text{ means}$$
$$2 \div 2 \text{ or } 1$$

If we wanted it to mean $12 \div (6 \div 2)$, then we would have to use the grouping symbols. As a second example, if we don't use grouping symbols, $24 \div 4 \times 2$ means 6×2, or 12. If we wanted the answer to be 3, we would have to write $24 \div (4 \times 2)$, which would mean $24 \div 8$, or 3.

Example 6

What is the value of

$$12 \div 6 \div 2 \times 5 + 24 \div 6 \times 2$$

The value is 13.
You have

$$(12 \div 6 \div 2 \times 5) + (24 \div 6 \times 2) \qquad [3]$$
$$\uparrow$$

If you evaluate each term in the left-to-right order that each number occurs in, you have

$$12 \div 6 \div 2 \times 5 =$$
$$\underbrace{2 \div \ 2}_{} \times 5 =$$
$$\underbrace{1 \quad \times 5}_{} = 5$$

Similarly

$$24 \div 6 \times 2 =$$
$$\underbrace{4 \quad \times 2}_{} = 8$$

So [3] becomes

$$(5) + (8), \text{ or } 13$$

The greatest use of our agreement is for algebraic expressions.

We do not read this as

$$12 \div (6 \div 2) \times 5$$

If we meant this, we would have to use the grouping symbols.

Again, we do not mean

$$24 \div (6 \times 2) = 24 \div 12 = 2$$

If we meant this, we would use the parentheses.

Section 4.7   Removing Grouping Symbols

Example 7

Evaluate $bc + de$ when $b = 2$, $c = 3$, $d = 4$, and $e = 5$.

The value is 26. No grouping symbols appear, so the terms are determined by the plus (or minus) signs. In this example, you have:

$$(bc) + (de)$$
↑

or $(b \times c) + (d \times e)$

If you replace b, c, d, and e by their given values, you get

$$(2 \times 3) + (4 \times 5) = 6 + 20 = 26$$

Example 8

Evaluate the expression $b(c + d)e$ when $b = 2$, $c = 3$, $d = 4$, and $e = 5$.

The value is 70. Using the grouping symbols that are given, you have

$$2(3 + 4)5 = 2(7)5 = [2(7)]5 = [14]5 = 70$$

Example 9

Simplify $2 + 3 \times b + 4 \times 3$.

This is another way of saying $14 + 3b$. That is, using the plus signs to find the terms, you have

$$(2) + (3 \times b) + (4 \times 3) = 2 + (3b) + 12$$
  ↑   ↑

$$= 2 + [(3b) + 12]$$
$$= 2 + [12 + (3b)]$$
$$= [2 + 12] + (3b)$$
$$= [14] + (3b)$$
$$= 14 + 3b$$
↑

Remember, this means that we add 14 to $3b$. That is, if b is 5, then $3b$ is 15 and $14 + 3b$ is 29. In other words,

$$14 + 3b = (14) + (3b)$$
$$= 14 + (3 \times b)$$

So when $b = 5$,
$$14 + 3b = 14 + (3 \times 5)$$
$$= 14 + 15$$
$$= 29$$

   More informally, look at $2 + 3b + 12$ and see that $2 + 12 = 14$, and so on. That is, you often use the associative and commutative properties without actually saying them each time.

PRACTICE DRILL

What number is named by each of the following?
1. $4 \times 5 + 7$ 2. $4 \times 5 + 7 \times 3$ 3. $4 \times 5 - 7 \times 4$
4. $8 \div 4 \div 2 - 5 \times 3$ 5. $12 \div 6 \times 3 - 7 \times 6 + 15$

Answers
1. 27   2. 41   3. −8
4. −14   5. −21

CHECK THE MAIN IDEAS

1. The portion of an expression that lies between consecutive plus or minus signs is called a _____ .
2. Plus and minus signs help to break expressions into smaller parts called _____ .
3. If no grouping symbols appear, you assume that a set of parentheses exists around each _____ .

Answers
1. term   2. terms
3. term   4. order
5. associative

4. If a term contains both multiplication and division, carry out the arithmetic in the _____ the numbers appear.
5. If only multiplication appears in a term, you can do the arithmetic in any order you want, because multiplication obeys both the commutative and the _____ rules.

Section 4.8
The Distributive Property

In the last section, we pointed out that unless grouping symbols are used, it is understood that an expression such as

$$2 \times 3 + 4$$

is evaluated as if it were written

$$(2 \times 3) + (4) \quad \text{or} \quad 10$$
↑

In this section, we are interested in the special expression

$$2 \times (3 + 4)$$

That is, we are interested in multiplying a number by the sum of two numbers.

Example 1
John has $3 and Bill has $4. How much will they have if their money is doubled?

John will have $6 and Bill will have $8, or a total of $14 for both boys.
There are two ways of doing this problem. First you can double 3 and get 2×3, or 6; then you can double 4 and get 2×4, or 8. Then you add 6 and 8 to get 14. But you also can add 3 and 4 first to get 7, and then double this amount.
Symbolically, one way is to compute $2(3 + 4)$ and the other is to compute $2(3) + 2(4)$. You get the same answer either way, so this example shows you that

$$2(3 + 4) = 2(3) + 2(4) \qquad [1]$$

Expression [1] is a special case of the more general result known as the **distributive property**; or more precisely, the **distributative property of multiplication over addition**.

> The **distributive property of multiplication over addition** states that for any numbers b, c, and d:
> $$b(c + d) = bc + bd$$

Remember, if no grouping symbols appear in the expression $bc + bd$, we have agreed to interpret it as

$$(bc) + (bd)$$

Visually:

	c	d
b	$b \times c$ or bc	$b \times d$ or bd

In this diagram the lengths b, c, and d do not have to be whole numbers. This means that the distributive property should apply even if we aren't working with whole numbers. This is illustrated in Example 3.

$b(c + d) =$ the area of the whole rectangle
$bc + bd =$ the sum of the area of the two pieces

OBJECTIVE
To learn how to remove the grouping symbols from an expression such as $b(c + d)$.

$2 \times 3 \times 4$ and $2 + 3 + 4$ do not require grouping symbols because both multiplication and addition are associative. But this doesn't apply to $2 \times 3 + 4$ because the operations are neither purely additive nor purely multiplicative.

This can be seen in terms of areas, as the following diagram shows.

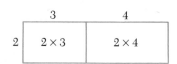

On the one hand, the area of the rectangle is $2 \times (3 + 4)$. On the other hand, it is equal to the sum of the areas of the smaller parts; that is, $2 \times 3 + 2 \times 4$.

Section 4.8 The Distributive Property

The term "distributive" implies that the "outside" (first) number is distributed with each of the two numbers in the sum that makes up the second factor. That is,

$$\text{first} \times (\text{second} + \text{third}) = (\text{first} \times \text{second}) + (\text{first} \times \text{third})$$

Example 2

Evaluate each of the following expressions when $b = 3$, $c = 5$, and $d = 2$.

(a) $b(c + d)$ (b) $bc + bd$

The value of each expression is 21. That is,

$$b(c + d) = 3(5 + 2) = 3(7) = 21$$
$$bc + bd = (3)(5) + (3)(2)$$
$$= 15 + 6 = 21$$

This illustrates another situation in which $b(c + d) = bc + bd$.

Example 3

Evaluate each of the following expressions when $b = 2.5$, $c = 3.6$, and $d = 4.4$.

(a) $b(c + d)$ (b) $bc + bd$

The value of each expression is 20. That is, $2.5(3.6 + 4.4) = 2.5(8) = 20$; and $bc + bd$ becomes $(2.5)(3.6) + (2.5)(4.4)$, or $9 + 11$, or 20.

Example 4

Evaluate each of the following expressions when $b = 3$, $c = 2$, and $d = -4$.

(a) $b(c + d)$ (b) $bc + bd$

The value of each expression is -6. That is,

$$b(c + d) = 3(2 + {}^-4)$$
$$= 3(-2) = -6$$

and
$$bc + bd = 3(2) + 3(-4)$$
$$= 6 + {}^-12 = -6$$

This illustrates that the distributive property applies even when you deal with negative numbers.

NOTE
This example is a special case of a more general result. That is, if we write $2 + {}^-4$ in the more compact form $2 - 4$, we have

$$3(2 - 4) = 3(2) - 3(4)$$

More generally,

$$b(c - d) = bc - bd$$

As the following examples show, the distributive property allows us to simplify many more algebraic expressions.

Example 5

Simplify $5(b + 3) + 8$.

$5(b + 3) + 8$ is the same as $5b + 23$.
By the distributive property, $5(b + 3)$ is $5b + 5(3)$ or $5b + 15$. Therefore

$$[5(b + 3)] + 8 = [5b + 15] + 8$$
$$= 5b + [15 + 8]$$
$$= 5b + 23$$

Remember that we may use either brackets or parentheses as grouping symbols. However, if only one set is needed, we usually choose parentheses; but we don't have to.

In an expression like $5(b + 3) + 8$, we *do not* add 3 and 8. The 3 is being multiplied by 5, so we must use the distributive property first.

Remember, this shows that $5b + 23$ means the same thing as does $5(b + 3) + 8$. For example, when $b = 7$, $5b + 23 = 5(7) + 23 = 35 + 23 = 58$; and $5(b + 3) + 8 = 5(7 + 3) + 8 = 50 + 8 = 58$.

Everything within a set of grouping symbols is treated as a single number. Where we break the expression up into terms, the plus sign in $b + 3$ is not used. That is, in $5(b + 3) + 8$, we first place a grouping symbol before the 5 to form

$$[5(b + 3) + 8$$

We used [rather than (to make it obvious that we have more than one set of grouping symbols surrounding b.

Then $b + 3$ is a single number because it is enclosed in the parentheses. In a way, we don't even look inside the parentheses. When we form terms, we act as if we have $5(\ \ ) + 8$.

Since the entire parentheses is multiplying 5, we don't come to a "true" plus sign until just before the 8. In other words,

$$[5(b + 3)] + 8$$
$$\uparrow$$

REMINDER

To remove grouping symbols, we work from the inside to the outside. In this example, we first remove the parentheses to obtain

$$[5(b + 3)] + 8$$
$$= [5b + 15] + 8$$
$$= 5b + 15 + 8$$

(We can remove the brackets because addition is associative).

Example 6

Simplify

$$5(b + 3) + 2(b + 4) \qquad [2]$$

This is the same as $7b + 23$. Remember, [2] is read as

$$[5(b + 3)] + [2(b + 4)]$$
$$\uparrow$$

This is the same as

$$[5b + 15] + [2b + 8]$$
$$= 5b + 15 + 2b + 8$$
$$= (5b + 2b) + (15 + 8)$$
$$= \quad 7b \quad + \quad 23$$

We now have another way of seeing why $5b + 2b = 7b$, other than by adding like denominations. By the distributive property

$$5b + 2b = (5 + 2)b = 7b$$

NOTE

If you made a mistake and wrote [2] as, for example, $5b + 3 + 2b + 4$ or $7b + 7$, you can see that you made a mistake if you evaluate the two expressions for given values of b. For example, when $b = 10$, $7b + 7 = 7(10) + 7 = 70 + 7 = 77$; but when $b = 10$:

$$5(b + 3) + 2(b + 4) = 5(10 + 3) + 2(10 + 4)$$
$$= 5(13) \quad + 2(14)$$
$$= 65 \quad\quad + 28$$
$$= 93$$

Example 7

Simplify

$$2[5 + 3(m + 4)] \qquad [3]$$

$2[5 + 3(m + 4)] = 6m + 34$.
    Starting with the inner grouping symbols (parentheses), you have

$$3(m + 4) = 3m + 12.$$

"Simplifying" means that it's easier to work with $6m + 34$ than with $2[5 + 3(m + 4)]$ although both expressions will have the same value for a given value of m.

Section 4.8   The Distributive Property   95

So you may think of [3] as

$$2[5 + (3m + 12)] \qquad [4]$$

Addition is associative, so you may rewrite [4] as

$$2[5 + 3m + 12] \qquad [5]$$

Because addition is both associative and commutative, you may rewrite [5] as

$$2[3m + 17]$$

By the distributive property, this is the same as $2(3m) + 2(17)$, or $6m + 34$.

Example 8

Simplify

$$7 - 2(3 - m) \qquad [6]$$

$7 - 2(3 - m) = 1 + 2m$.   By the rules for grouping, you may read [6] as

$$\begin{aligned} &7 - [2(3-m)] \\ &= 7 - [2(3) - 2m] \\ &= 7 - [6 - 2m] \end{aligned} \qquad [7]$$

At this stage, you can simplify [7] by using the result in Section 4.5 that $-(a - b) = b - a$. That is,

$$\begin{aligned} 7 - [6 - 2m] &= 7 + [2m - 6] \\ &= 7 + 2m - 6 \\ &= 7 + 2m + {}^-6 = 1 + 2m \end{aligned}$$

You can also use the fact that $-[6 - 2m]$ is the same as $({}^-1)[6 - 2m]$. In this case, the distributive property tells you that $({}^-1)[6 - 2m] = ({}^-1)[6 + {}^-2m] = {}^-1(6) + {}^-1({}^-2m)$, or ${}^-6 + 2m$. That is,

$$\begin{aligned} 7 - [6 - 2m] &= 7 + {}^-1[6 + {}^-2m] \\ &= 7 + {}^-1[6] + {}^-1({}^-2m) \\ &= 7 + {}^-6 + 2m = 1 + 2m \end{aligned}$$

The technique of removing grouping symbols, either in problems that require the distributive property or problems that do not, requires much practice. So additional practice is left for the Exercise Set.

However, there is one more use of the distributive property that I'd like to discuss with you in this Section. It involves solving equations which contain fractions.

Example 9

For what value of x does

$$\frac{x}{3} + \frac{x}{2} = 10 \qquad [8]$$

$\frac{x}{3} + \frac{x}{2} = 10$ when $x = 12$.

METHOD 1: $\frac{x}{3}$ means $\frac{1}{3}x$ and $\frac{x}{2}$ means $\frac{1}{2}x$. Therefore

$$\frac{x}{3} + \frac{x}{2} = \frac{1}{3}x + \frac{1}{2}x = \left(\frac{1}{3} + \frac{1}{2}\right)x = \frac{5}{6}x$$

---

[4] If we want to use the rules of mathematics step-by-step, we have

$$\begin{aligned} &2[5 + 3(m + 4)] \\ &= 2[5 + (3m + 12)] \\ &= 2[5 + (12 + 3m)] \\ &= 2[(5 + 12) + 3m] \\ &= 2[17 + 3m] \\ &= 2[17] + 2[3m] \\ &= 34 + [2(3)]m \\ &= 34 + 6m \\ &= 6m + 34 \end{aligned}$$

In general, in a course like ours we don't insist on using the rules word for word. But it's important to know that whenever we're in doubt, we can use the rules to justify each step.

In other words, as we gain experience we find that there is often more than one way to develop the same result. But whatever method we use, we've shown that for any given value of m, $7 - 2(3 - m)$ and $1 + 2m$ have the same value.

CHECK: $\frac{12}{3} + \frac{12}{2} = 4 + 6 = 10$

REVIEW

$\frac{n}{3}$ means $n \div 3$ or $n \times \frac{1}{3}$ or $\frac{1}{3} \times n$. If we leave out the "times sign" we have $\frac{1}{3}n$.

Hence we may rewrite [8] as $\frac{5}{6}x = 10$, and then multiply both sides of this equation by 6 to get

$$5x = 60, \text{ or } x = 12$$

But if we had used the distributive property sooner, we could have avoided having to work with fractions.

That is,

$$6\left(\frac{5}{6}x\right) = \left[6\left(\frac{5}{6}\right)\right]x = 5x$$

METHOD 2: Multiply both sides of [8] by 6 to obtain

$$6\left(\frac{x}{3} + \frac{x}{2}\right) = 6(10) = 60 \qquad [9]$$

Then by the distributive property, [9] becomes

$$6\left(\frac{x}{3}\right) + 6\left(\frac{x}{2}\right) = 60, \quad \text{or} \quad 2x + 3x = 60$$

That is,

$$6\left(\frac{x}{3}\right) = 6\left(\frac{1}{3}[x]\right)$$
$$= \left[6\left(\frac{1}{3}\right)\right]x = 2x$$
$$6\left(\frac{x}{2}\right) = 6\left(\frac{1}{2}[x]\right)$$
$$= \left[6\left(\frac{1}{2}\right)\right]x = 3x$$

A NOTE ON THE GENERALIZED DISTRIBUTIVE PROPERTY

Using the properties of arithmetic, we can show that the distributive rule exists in more general forms, such as

$$a \times (b + c + d) = (a \times b) + (a \times c) + (a \times d)$$

or as

$$(a + b)(c + d) = ac + ad + bc + bd$$

Rather than use the rules of mathematics, it is often easier to picture these results as areas of rectangles.
For example:

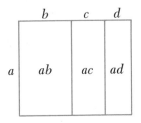

More generally, we form all possible terms where the first factor comes from the first set of grouping symbols and the second factor comes from the second set of grouping symbols. If you want a demonstration in terms of the properties of arithmetic, we have

$$(a + b)(c + d)$$
$$= (a + b)c + (a + b)d$$
$$= c(a + b) + d(a + b)$$
$$= [ca + cb] + [da + db]$$
$$= ca + (cb + da) + db$$
$$= ca + (da + cb) + db$$
$$= ac + (ad + bc) + bd$$
$$= ac + ad + bc + bd$$

On the one hand, the area of the rectangle is $a(b + c + d)$. On the other hand, it is $ab + ac + ad$. But since the rectangle has only one area, then

$$a(b + c + d) = ab + ac + ad$$

As we mentioned before, there are other ways in which the proper steps could have been carried out.

	c	d
a	ac	ad
b	ab	bd

In this case, the area of the rectangle is given both by $(a + b)(c + d)$ and by $ac + ad + bc + bd$.

One place we use the generalized distributive property is in multiplication.

Exercise Set 4 (Form A)

Example 10

Use the distributive property to compute 765×111 by writing 111 as $100 + 10 + 1$.

The product is 84,915. You see that

$$765 \times 111 = 765 \times (100 + 10 + 1)$$
$$= 765 \times 100 + 765 \times 10 + 765 \times 1$$
$$= 76{,}500 + 7{,}650 + 765$$

This is how we usually multiply. That is,

```
    765
  × 111
    765    (765 × 1)
   7650    (765 × 10)
  76500    (765 × 100)
  84915    (765 × 111) or
           765 × (1 + 10 + 100)
```

PRACTICE DRILL

Simplify each of the following expressions.

1. $2(b + 4) + 7$
2. $2(b + 4) + 7b$
3. $2(b + 4) + 7(b + 2)$
4. $7(b - 3)$
5. $2(b + 4) - 7(b - 3)$
6. $8 - 3(b - 4)$
7. $9 - (4 - b)$
8. $3[7 + 2(4 - b)]$

Answers
1. $2b + 15$
2. $9b + 8$
3. $9b + 22$
4. $7b - 21$
5. $29 - 5b$ (or $-5b + 29$)
6. $20 - 3b$
7. $5 + b$ (or $b + 5$)
8. $45 - 6b$ (or $-6b + 45$)

CHECK THE MAIN IDEAS

1. The fact that $b(c + d) = bc + bd$ is called the _____ property.
2. In the expression $4(m + 2) + 5$, you can't add the 2 and 5 because you must first multiply the 2 by _____.
3. Because of the commutative property of multiplication, the distributive property can also be written in the form $(c + d)b =$ _____.

Answers
1. distributive
2. 4
3. $cb + db$

EXERCISE SET 4 (Form A)

Section 4.1

1. What number is named by $4 + \{[2 + (3 + 1)] + (2 + 3)\}$?
2. When $m = 7$, what is the value of the expression
   (a) $(m - 2) - 4$
   (b) $m - (2 - 4)$
3. Do the expressions $(m - 2) - 4$ and $m - (2 - 4)$ have the same meaning?
4. For what value of m does
   (a) $(m + 7) + 4 = 35$
   (b) $(m - 7) - 4 = 35$
   (c) $m - (7 - 4) = 35$

Section 4.2

5. When $q = 15$, what is the value of
   (a) $5 + (q + 11)$
   (b) $5 + (11 + q)$
6. When $q = 15$ what is the value of
   (a) $5 + (q - 11)$
   (b) $5 + (11 - q)$
   (c) $5 - (11 - q)$
7. For what value of r does
   (a) $8 + (r + 3) = 22$
   (b) $8 + (r - 3) = 22$
   (c) $8 + (3 - r) = 22$
8. What reason justifies each of the following equalities?
   (a) $3 + m + 4 = 3 + (m + 4)$
   (b) $3 + (m + 4) = 3 + (4 + m)$
   (c) $3 + (4 + m) = (3 + 4) + m$
9. Simplify $[3 + (4 + m) + 1] + 2$.

Answers: Exercise Set 4 Form A

The following abbreviations are used in the answers to this exercise set: APA = associative property of addition; CPA = commutative property of addition; APM = associative property of multiplication; and CPM = commutative property of multiplication.

1. 15
2. (a) 1 (b) 9
3. no
4. (a) 24 (b) 46 (c) 38
5. (a) 31 (b) 31
6. (a) 9 (b) 1 (c) 9
7. (a) 11 (b) 17 (c) −11
8. (a) APA (b) CPA (c) APA
9. $10 + m$ or $m + 10$

Continued

Lesson 4 The Structure of Arithmetic

Section 4.3

10. Evaluate each of the following expressions when $c = 90$.
    (a) $(c \times 15) \times 3$ (b) $c \times (15 \times 3)$

11. For what values of c does $(c \times 15) \times 3 = c \times (15 \times 3)$?

12. Evaluate each of the following expressions when $c = 90$.
    (a) $(c \div 15) \div 3$ (b) $c \div (15 \div 3)$

13. For what value of c does
    (a) $(c \times 15) \times 3 = 135$ (b) $(c \div 15) \times 3 = 135$

Section 4.4

14. In the expression $(3 + 4)m$, what is the coefficient of m?
15. In the expression $3(4m)$, what is the coefficient of m?
16. In the expression $(b + c)m$, what is the coefficient of m?
17. What reason justifies each of the following steps?
    (a) $(150 \times p) \div 10 = (150 \times p) \times \frac{1}{10}$
    (b) $(150 \times p) \times \frac{1}{10} = 150 \times \left(p \times \frac{1}{10}\right)$
    (c) $150 \times \left(p \times \frac{1}{10}\right) = 150 \times \left(\frac{1}{10} \times p\right)$
    (d) $150 \times \left(\frac{1}{10} \times p\right) = \left(150 \times \frac{1}{10}\right) \times p$

18. Evaluate $(150 \times p) \div 10$ when $p = 6\frac{1}{3}$.
19. For what value of p does $(150 \times p) \div 10 = 90$?

Section 4.5

20. What is the value of $-(-32)$?
21. If $c + {}^-c = 0$, what is the value of ${}^-c + c$?
22. What is the value of ${}^-c$ if $c = {}^-5$?
23. Supply the reason for each of the following steps.
    (a) $(m - n) + (n - m) = (m + {}^-n) + (n + {}^-m)$
    (b) $(m + {}^-n) + (n + {}^-m) = m + ({}^-n + n) + {}^-m$
    (c) $[m + ({}^-n + n)] + {}^-m = [m + 0] + {}^-m$
    (d) $[m + 0] + {}^-m = m + {}^-m$

24. What is the additive inverse of $m - n$? (Write without grouping symbols.)

Section 4.6

25. For what value of g does $g + 3$ not have a multiplicative inverse?
26. What is the multiplicative inverse of $g + 3$ when $g = \frac{2}{3}$?
27. To solve the equation $3p = 18$, you begin by multiplying both sides of the equation by $\frac{1}{3}$ to get $\frac{1}{3}(3p) = \frac{1}{3}(18) = 6$. Justify the following steps:
    (a) $\left[\frac{1}{3}(3)\right]p = 6$ (b) $1p = 6$ (c) $p = 6$

Section 4.7

28. What number is named by
    (a) $(9 - 5) - (3 - 1)$ (b) $[9 - (5 - 3)] - 1$ (c) $9 - [(5 - 3) - 1]$
    (d) $9 - 5 - 3 - 1$

---

10. (a) 4,050
    (b) 4,050   11. all values
    12. (a) 2 (b) 18   13. (a) 3
    (b) 675   14. 7   15. 12
    16. $b + c$ 17. (a) def. of division (b) APM (c) CPM
    (d) APM   18. 95   19. 6
    20. 32   21. 0   22. 5
    23. (a) def. of subtraction
    (b) APA (c) additive inverse
    (d) additive identity
    24. $n - m$ 25. -3
    26. $\frac{3}{11}$ 27. (a) APM
    (b) mult. inverse (c) mult. identity
    28. (a) 2 (b) 6 (c) 8
    (d) 0   29. (a) 70
    (b) 46 (c) 26   30. 20
    31. 5   32. (a) -9 (b) 9
    (c) -9 33. 26
    34. (a) $5b - 7$ (b) $10b - 22$
    35. 13   36. 4   37. 10
    38. 7   39. 240   40. 9

Exercise Set 4 (Form B)

29. Evaluate each of the following when $a = 2$, $b = 3$, $c = 4$, and $d = 5$.
    (a) $a(b + c)d$ (b) $a(b + cd)$ (c) $ab + cd$
30. Simplify $5(-4) \div 2(6 - 8)$
31. Simplify $5(-4) \div [2(6 - 8)]$.
32. Simplify
    (a) $-(3^2)$ (b) $(-3)^2$ (c) -3^2

Section 4.8

33. Simplify $5 + \{2[(6 - 4) + 7] + 3\}$.
34. Simplify
    (a) $5(b - 3) + 8$ (b) $2\{[5(b - 3) + 8] - 4\}$
35. What is the value of $2\{[5(b - 3) + 8] - 4\}$ when $b = 3.5$?
36. Simplify $2[3(x + 2y) - 6y] - 6x + 4$.
37. For what value of m does $2[7 + 3(m - 5)] = 44$?
38. For what value of x does $2[3 - (1 - x)] = 18$?
39. For what value of x does $\dfrac{x}{6} - \dfrac{x}{8} = 10$?
40. You tell a person to pick a number, double it, add 3, and then multiply this result by 5. What number did the person pick if the answer was 105?

EXERCISE SET 4 (Form B)

Section 4.1

1. What number is named by $7 + \{[3 + (5 + 2)] + (3 + 6)\}$?
2. When $n = 12$, what is the value of
    (a) $(n - 7) - 3$ (b) $n - (7 - 3)$
3. Do the expressions $(n - 7) - 3$ and $n - (7 - 3)$ have the same meaning?
4. For what value of q does
    (a) $(q + 13) + 7 = 40$ (b) $(q - 13) - 7 = 40$ (c) $q - (13 - 7) = 40$

Section 4.2

5. When $q = 20$, what is the value of
    (a) $6 + (q + 18)$ (b) $6 + (18 + q)$
6. When $q = 20$, what is the value of
    (a) $6 + (q - 18)$ (b) $6 + (18 - q)$ (c) $6 - (18 - q)$
7. For what value of t does
    (a) $11 + (t + 7) = 30$ (b) $11 + (t - 7) = 30$ (c) $11 + (7 - t) = 30$
8. What reason justifies each of the following equalities?
    (a) $8 + r + {}^-7 = 8 + (r + {}^-7)$ (b) $8 + (r + {}^-7) = 8 + ({}^-7 + r)$
    (c) $(8 + {}^-7) + r = 8 + ({}^-7 + r)$
9. Simplify $[8 + (r + {}^-7) + 2] + 3$.

Section 4.3

10. Evaluate each of the following expressions when $g = 100$.
    (a) $(g \times 10) \times 2$ (b) $g \times (10 \times 2)$
11. For what values of g does $(g \times 10) \times 2 = g \times (10 \times 2)$?

12. Evaluate each of the following expressions when $g = 100$.
    (a) $(g \div 10) \div 2$ (b) $g \div (10 \div 2)$
13. For what value of g does
    (a) $(g \times 10) \times 2 = 100$ (b) $(g \div 10) \times 2 = 100$

Section 4.4

14. In the expression $(^-3 + 5)k$, what is the coefficient of k?
15. In the expression $\frac{1}{2}(10k)$, what is the coefficient of k?
16. In the expression $(j - m)k$, what is the coefficient of k?
17. What reason justifies each of the following steps?
    (a) $(240 \times w) \div 30 = (240 \times w) \times \frac{1}{30}$
    (b) $(240 \times w) \times \frac{1}{30} = 240 \times \left(w \times \frac{1}{30}\right)$
    (c) $240 \times \left(w \times \frac{1}{30}\right) = 240 \times \left(\frac{1}{30} \times w\right)$
    (d) $240 \times \left(\frac{1}{30} \times w\right) = \left(240 \times \frac{1}{30}\right) \times w$
18. Evaluate $(240 \times w) \div 30$ when $w = 2\frac{1}{4}$.
19. For what value of w does $(240 \times w) \div 30 = 200$?

Section 4.5

20. What is the value of $-(-7.5)$?
21. If $d + {}^-d = 0$, what is the value of ${}^-d + d + 7$?
22. What is the value of ${}^-c$ if $c = -8.8$?
23. What reason justifies each of the following steps?
    (a) $[p - q] + [q - p] = [p + {}^-q] + [q + {}^-p]$
    (b) $[p + {}^-q] + [q + {}^-p] = [p + ({}^-q + q)] + {}^-p$
    (c) $[p + ({}^-q + q)] + {}^-p = [p + 0] + {}^-p$
    (d) $[p + 0] + {}^-p = p + {}^-p$
24. What is the additive inverse of $p - q$? (Express the answer without grouping symbols.)

Section 4.6

25. For what value of b does the expression $b + 4$ not have a multiplicative inverse?
26. What is the multiplicative inverse of $b + 4$ when $b = -\frac{3}{4}$?
27. To solve the equation $\frac{1}{5}v = 12$, you begin by multiplying both sides of the equation by 5 to get $5\left(\frac{1}{5}v\right) = 5(12) = 60$. Justify each of the following steps.
    (a) $\left[5\left(\frac{1}{5}\right)\right]v = 60$ (b) $1v = 60$ (c) $v = 60$

Section 4.7

28. What number is named by
    (a) $(11 - 7) - (1 - 3)$ (b) $[11 - (7 - 1)] - 3$
    (c) $11 - [(7 - 1) - 3]$ (d) $11 - 7 - 1 - 3$

Exercise Set 4   (Form B)

29. Evaluate each of the following when $a = 1$, $b = {}^-2$, $c = {}^-1$, and $d = 3$.

    (a) $a(b+c)d$ (b) $a(b+cd)$ (c) $ab+cd$

30. Simplify $8(-6) \div 3(^-5 + 3)$.
31. Simplify $8(-6) \div [3(^-5 + 3)]$.
32. Simplify

    (a) $-(5^2)$ (b) $(-5)^2$ (c) -5^2

Section 4.8

33. Simplify $8 + \{3[(3-5) + 5] + {}^-3\}$.
34. Simplify

    (a) $6(c+4) - 18$ (b) $3\{[6(c+4) - 18] + 4\}$

35. What is the value of $3\{[6(c+4) - 18] + 4\}$ when $c = \dfrac{1}{9}$?
36. Simplify $^-4[2(x-2y) - 2x] - 16y - 7$.
37. For what value of q does $3[2 + 4(q-5)] = -30$?
38. For what value of y does $^-3[2 - (y+3)] = 0$?
39. For what value of x does $\dfrac{x}{5} - \dfrac{x}{6} = \dfrac{1}{15}$?
40. A person picks a number, adds 3, doubles this result, and subtracts 6. What number did the person pick if the final answer was 40?

Lesson 5 An Introduction to Linearity

Overview   A woman was watching a professional football game on television with her five-year-old son. The announcer was giving the lineups and the weights of the players. "Jones — 280; Smith — 265; McCoy — 255; . . ." The son looked at his mother in amazement and said, "Gee, Mom, no wonder those guys are so good — look how old they are!"

The son made a mistake that lots of older people make too. He had assumed that if you're twice as old, you must be twice as good. Of course, this need not be true. For example, if you number 100 people according to height, with 1 standing for the shortest and 100 for the tallest, it does not follow that number 30 is twice as tall as number 15.

But there *are* times when doubling one variable in a formula also doubles the other. If pencils cost 5 cents each and there is no discount for buying a greater number, it costs twice as much to buy 30 pencils as it does to buy 15 pencils.

When one variable increases at the same rate as the other, their relationship is usually referred to as linear. (Linear sounds as if it comes from the word "line"; it does, and we'll explore this in Lesson 7.)

Although many relationships are not linear, the fact is that linear relationships play a very important part in all phases of mathematics, from elementary to advanced. For this reason, linear relationships are the major topic of the next five lessons in this book. As it happens, linear equations also give us an excellent topic for applying the material in Lesson 4, when we discussed the structure of arithmetic.

Section 5.1
Linear Expression

Suppose you see an advertisement for foreign cheese. The advertised price is $3 per pound plus a $2 charge to cover the cost of shipping. You can then calculate the price for any number of pounds of cheese you wanted to buy.

Example 1

According to the above conditions, what does it cost you if you buy five pounds of the cheese?

It costs you $17. You pay $3 per pound, so five pounds costs $3 × 5 or $15. In addition, there is the $2 charge to cover shipping cost. So the total charge is $15 + $2, or $17.

Of course, it is possible that you wanted to purchase other than a whole number of pounds. You still compute the price the same way.

Example 2

Under the same conditions as in Example 1, what would it cost to buy 6.7 pounds of cheese?

It would cost $22.10. You still multiply the number of pounds by $3 and add the $2 shipping charge. In this case you get

$$(6.7 \times \$3) + \$2$$
or $20.10 + $2 = $22.10

The point is that we have a formula for finding the cost of the cheese. Namely

1. Multiply the number of pounds of cheese by the price per pound ($3).
2. Then add $2 to this product.

Example 3

Using the same conditions as in Examples 1 and 2, let p be the number of pounds of cheese you are going to buy and let c be the cost of the cheese in dollars. Write the formula that expresses c in terms of p.

The formula is $c = 3p + 2$.
As before, you multiply the number of pounds (p) by 3 to get $3p$. Then you add 2 (dollars) to get $3p + 2$ as the cost of p pounds of cheese.
Because c denotes the cost of p pounds of cheese in dollars, you obtain the result

$$c = 3p + 2 \qquad [1]$$

Notice that in the expression $3p + 2$, the variable p is being multiplied by a constant (3), and that a constant (2) is then added to the result.

> **DEFINITION**
> When we multiply a variable by a constant and then add a constant to the result, the expression we get is said to be **linear** in that variable.

OBJECTIVE
To recognize as linear all expressions of the form

$$mx + b$$

where m and b are constants and x is a variable.

Notice that the $2 shipping charge is a one-time charge per order, regardless of the number of pounds you buy.

REVIEW
$3p + 2$ means that first we multiply p by 3 and then we add 2. That is,

$3p + 2$ means $(3p) + 2$

In other words, the expression

$$mp + b$$

is linear in p if m and b are constants. For $3p + 2$, we have $m = 3$ and $b = 2$.

Example 4

Is $5p + 4$ linear in p?

Yes. That is, p is multiplied by a constant (in this case, 5) and a constant (in this case, 4) is added to the result.

That is, we have the form
$$mp + b$$
with $m = 5$ and $b = 4$.

Example 5

Is $-5p + 4$ linear in p?

Yes. In this case, you have $mp + b$, where $m = -5$ and $b = 4$. That is, the expression is

$$(-5)p + 4$$
$$\uparrow \qquad \uparrow$$
$$m \qquad b$$

Although -5 is negative, it is still a constant.

Example 6

Is $5p - 4$ linear in p?

Yes. You can write $5p - 4$ as

$$5p + (-4)$$
$$\uparrow \qquad \uparrow$$
$$m \qquad b$$

In other words, in the expression $mx + b$, b need not be positive. We may have $5p + (-4)$, in which case we abbreviate it as $5p - 4$. But what we call b in the expression $mx + b$ must follow a *plus* sign.

Example 7

If y is a variable, is $5y + 3$ linear in y?

Yes. You still have a variable (y) multiplied by a constant (5) to which is added a constant (3). That is,

$$my + b$$
$$\uparrow \quad \uparrow$$
$$5 \quad 3$$

Example 7 indicates that the idea of a linear expression does not depend on the name given to the variable. Rather, a linear expression in a variable means an expression in which that variable is multiplied by a constant and a constant is added to the product.

Sometimes the $mp + b$ form is not so obvious.

To avoid any possible misinterpretation, we usually say that $5p + 3$ is linear *in p*. In a similar way, we say that $5y + 3$ is linear *in y*.

But when it's clear, for example, that the variable is y, we often say that $5y + 3$ is linear, rather than linear *in y*. For example, in Example 10 we say that $4p^2 + 3$ is not linear. We mean that it is not linear in p — but it is linear in p^2.

Example 8

Is the expression $5p$ linear in p?

Yes. Remember that adding 0 to a number doesn't change the value of the number. Therefore, $5p$ is the same as

$$5p + 0$$
$$\uparrow \quad \uparrow$$
$$m \quad b$$

Remember that 0 is a constant.

Section 5.1   Linear Expression

Example 9

Is the expression $p + 3$ linear in p?

Yes. Here you may assume that any number is being multiplied by 1. That is, $p + 3$ is the same as

$$\underset{\underset{m}{\uparrow}}{1p} + \underset{\underset{b}{\uparrow}}{3}$$

In the last lesson, we referred to this as the multiplicative identity. That is, for any number p, $1p = p$.

Examples 4 through 9 may make it seem that most expressions are linear. This is not true. To be linear, the variable must be multiplied by a *constant*.

Example 10

Is the expression $4p^2 + 3$ linear in p?

No. In this case you have

$$4p(p) + 3$$

The variable p is being multiplied by a variable (p itself), as well as by the constant 4.

REVIEW

p^2 means the product of p and itself; that is, $p \times p$. $4p^2 + 3$ is linear in p^2. That is, if we treat p^2 as the variable, we have

$$\underset{\underset{m}{\uparrow}}{4(p^2)} + \underset{\underset{b}{\uparrow}}{3}$$

Sometimes linear expressions are disguised.

Example 11

Show that the expression

$$3(x + 5) - 8$$

can be written in the form $mx + b$, where m and b are constants.

$3(x + 5) - 8$ is equal to $3x + 7$.
  By the distributive property, you know that $3(x + 5) = 3x + 15$. Hence

$$3(x + 5) - 8 = 3x + 15 - 8 = 3x + 7$$

which has the form $mx + b$ with $m = 3$ and $b = 7$.
  The "disguise" is the fact that $x + 5$, and not just the x, is multiplied by 3.

> **DEFINITION**
> A linear expression in x is said to be in **standard form** if we write it as $mx + b$.

We can treat the entire expression $x + 5$ as a variable. In this case, we can say that $3(x + 5) - 8$ is linear in $x + 5$. But to be linear in x means that x is multiplied by a constant, and a constant is added to that product.

$$\underset{\underset{m}{\uparrow}}{3x} + \underset{\underset{b}{\uparrow}}{7} \text{ is in standard form.}$$

$3(x + 5) - 8$ is not in standard form even though it may be simplified as $3x + 7$, which is in standard form.

Example 12

Rewrite the expression

$$5(2x + 3) + 7$$

so that it is in standard form.

In standard form, $5(2x + 3) + 7 = 10x + 22$. That is,

$$\begin{aligned} 5(2x + 3) + 7 &= 5(2x) + 5(3) + 7 \\ &= 10x + 15 + 7 \\ &= 10x + 22 \end{aligned}$$

In other words, $5(2x + 3) + 7 = mx + b$, where $m = 10$ and $b = 22$.

For each given value of x, $5(2x + 3) + 7$ and $10x + 22$ have the same value. But $10x + 22$ is said to be in standard form, while $5(2x + 3) + 7$ isn't.

Many important formulas involve linear expressions. For example the relationship between Celsius (C) and Farenheit (F) temperature readings is given by

$$C = \frac{5}{9}(F - 32)$$
$$= \frac{5}{9}F - \frac{5}{9}(32)$$
$$= \underset{m}{\frac{5}{9}F} - \underset{b}{\frac{160}{9}}$$

Remember, m and b are constants. They do not have to be whole numbers.

Linear expressions have a rather special property. They increase at a constant rate. If an expression is linear in p, then the expression increases by the same amount every time p increases by 1.

Example 13

Evaluate the expression $3p + 2$ when $p = 1$, $p = 2$, $p = 3$, and $p = 4$.

You have

p	$3p$	$3p + 2$
1	3	5
2	6	8
3	9	11
4	12	14

This shows that each time p increases by 1, $3p + 2$ increases by 3.

The fact that $3p + 2$ increases by 3 every time p increases by 1 is correct even if p is not a whole number.

Example 14

Evaluate the expression $3p + 2$ when $p = \frac{4}{3}$, $p = \frac{7}{3}$, and $p = \frac{10}{3}$.

You have

p	$3p$	$3p + 2$
$\frac{4}{3}$	4	6
$\frac{7}{3}$	7	9
$\frac{10}{3}$	10	12

So $3p + 2$ still increases by 3 every time p increases by 1.

Of course, linear expressions in p do not always increase by 3 whenever p increases by 1.

Section 5.1 Linear Expression

Example 15

Evaluate the linear expression $5p + 4$ for $p = 1$, $p = 2$, $p = 3$, and $p = 4$.

You have

p	$5p$	$5p + 4$
1	5	9
2	10	14
3	15	19
4	20	24

(with brackets showing: p increases by 1 each step, $5p$ increases by 5, and $5p+4$ increases by 5)

Note that in this case $5p + 4$ increases by 5 every time p increases by 1.

This doesn't happen in expressions that aren't linear.

To see why this happens, replace p by $p + 1$ in $5p + 4$. We get $5(p + 1) + 4$. By the distributive property, this is $5p + 5 + 4$, which is 5 more than $5p + 4$.

Example 16

Evaluate the expression $p^2 + 3$ when $p = 1$, $p = 2$, and $p = 3$.

You have

p	p^2	$p^2 + 3$
1	1	4
2	4	7
3	9	12

When p increases by 1 from 1 to 2, $p^2 + 3$ increases by 3, from 4 to 7. But when p increases by 1 from 2 to 3, $p^2 + 3$ increases by 5, from 7 to 12. In other words, $p^2 + 3$ is not linear in p — because it doesn't change by the same amount every time p increases by 1.

PRACTICE DRILL

1. What expression is named by $mx + b$ if
   (a) $m = 3$ and $b = 7$
   (b) $m = 3$ and $b = -7$
   (c) $m = 1$ and $b = 2$
   (d) $m = 2$ and $b = 0$

2. If you can order fish for $4 per pound plus a handling charge of $3, what is the cost of p pounds of fish (in dollars)?

3. Write each of the following linear expressions in standard form.
   (a) $3x + 4x + 9$
   (b) $2(x + 4) - 7$

Answers

1. (a) $3x + 7$ (b) $3x - 7$, or $3x + (-7)$ (c) $x + 2$, or $1x + 2$ (d) $2x$ (that is, $2x + 0 = 2x$)
2. $4p + 3$
3. (a) $7x + 9$ (b) $2x + 1$

CHECK THE MAIN IDEAS

1. In order for the expression $mx + b$ to be linear in x, both m and b must be _____.

2. For example, $3x + 4$ is said to be linear in _____ because both 3 and 4 are constants.

Answers

1. constants 2. x 3. y 4. $3(x + 2)$ 5. 2 (The important thing is that $2x + 5$ increases by *the same amount* each time x increases by 1.)

3. On the other hand, $3y + 4$ is linear in _____ .
4. $3(x + 2)$ is equal to $3x + 6$; and because $3x + 6$ is linear in x, _____ is also linear in x.
5. Another reason for calling an expression such as $2x + 5$ linear in x is that every time x increases by 1, $2x + 5$ increases by _____ .

Section 5.2
Equations of the Form $mx + b = c$

OBJECTIVE
Given an equation of the form
$$mx + b = c$$
to be able to solve for x in terms of m, b, and c.

In the last section, we talked about the example of buying cheese at $3 a pound plus a $2 shipping charge. We saw that the cost c (in dollars) for p pounds of cheese was given by

$$c = 3p + 2 \qquad [1]$$

Formula [1] requires only the use of arithmetic if we are given the value of p and asked to find the value of c.

Example 1

What is the value of c if $c = 3p + 2$, and $p = 6$?

When $p = 6$, $c = 20$. Replace p by 6 in the expression $3p + 2$ to obtain $3(6) + 2 = 18 + 2 = 20$.

Sometimes we are given the value of c and asked to find the value of p. For example, we may know how much money we have to spend, and we want to know how much cheese we can buy for that amount.

Example 2

In the formula $c = 3p + 2$, what equation do you have to solve if you want to find the value of p for which $c = 35$?

You have to solve the equation $35 = 3p + 2$. All you do is replace c by 35 in the formula

$$c = 3p + 2$$

to get the equation

$$35 = 3p + 2$$

Equation [2] is called a **linear equation.**

> **DEFINITION**
> An equation is said to be linear (in x) if one side of the equation is linear (in x) and the other side is either linear (in x) or a constant.

Example 3

Is $3p + 7 = 4p + 8$ linear in p?

Yes. Both sides of the equation are linear in p.

Section 5.2 Equations of the Form $mx + b = c$

Example 4

Is $3p + 7 = 4p^2 + 8$ linear in p?

No. The left side ($3p + 7$) is linear in p. But the right side is neither linear in p nor a constant.

> Following the definition in the previous section, $4p^2 + 8$ is linear in p^2, but not in p.

Example 5

Is $3(2p - 3) = 6$ linear in p?

Yes. Since the right side is 6, you know that the right side is a constant. By the distributive property, the left side of the equation is equal to $3(2p) - 3(3)$, or $6p - 9$, which is linear. If one side is linear in p and the other side is a constant, we say that the equation is linear in p.

> REVIEW
>
> $3(2p - 3)$ is linear in p but not in standard form. When written in standard form, it becomes $6p + {}^-9$, or more commonly, $6p - 9$.
> We often say "linear equation" rather than "linear equation in p."

Our goal in this lesson is to learn to solve linear equations. We shall divide the problem into two parts. In this section we shall treat the case where one side of the equation is a linear expression and the other side is a constant. In the next section, we shall treat the case where both sides of the equation are (non-constant) linear expressions.

> Technically speaking, a constant has the form
>
> $$mx + b$$
>
> where $m = 0$. For example,
>
> $$3 = 0x + 3$$
>
> However, the usual agreement is that when we say that $mx + b$ is linear, we mean that m is not zero.

Example 6

For what value of x does

$$2x = 8 \qquad [3]$$

$2x = 8$ only when $x = 4$. That is, $2(4) = 8$ is a true statement.
   This example is easy enough for you to solve by trial and error. Aside from trial and error, there are at least two other ways to solve the problem.
   One method is the one you used in Lesson 2, where you "undid" various operations. For example, you see that on the left side of [3] x is multiplied by 2. To "undo" multiplying x by 2, divide by 2. But to keep the equation balanced, if you divide one side by 2, you must also divide the other side by 2. This leads to

$$\frac{2x}{2} = \frac{8}{2} \qquad \text{or} \qquad x = 4$$

The other method is to use the properties of arithmetic shown in Lesson 4.

STEP 1: Multiply both sides of [3] by $\frac{1}{2}$ (which is the multiplicative inverse of 2).

> $\frac{1}{2}(2x) = \frac{1}{2}(8) = 4$

STEP 2: Use the associative property of multiplication on the left side of the equation.

> $\left[\frac{1}{2}(2)\right]x = 4$

STEP 3: By the definition of multiplicative inverse, $2\left(\frac{1}{2}\right) = 1$.

> $1x = 4$

STEP 4: 1 is the multiplicative identity, so $1x = x$.

> $x = 4$

Which of the last two methods is the better one to use? Well, they both say about the same thing. The difference is that one says it informally and the other says it more formally.
   As for trial and error, the fact is that in more complicated problems it may be difficult to guess the exact answer.

> When we're sure that what we're doing is right, we usually use the informal method. When we have doubts, we try to stay as close as possible to the "rules."

Example 7

For what value of x does $2.7x = 57$?

$2.7x = 57$ when $x = 21\frac{1}{9}$.

Just as you did in the last example, look at the left side of the equation to see that x is being multiplied by 2.7. To undo multiplying by 2.7, divide both sides of the equation by 2.7 to get

$$\frac{2.7x}{2.7} = \frac{57}{2.7} = \frac{570}{27} = 21\frac{1}{9}$$

If you divide 57 by 2.7 using the hand calculator, the answer is $x = 21.11111$. Remember, the calculator writes all quotients as decimals, and "chops off" repeating decimals.

The numbers are a bit more complicated in this example than in the previous one, but the idea is the same. In general, if

$$mx = c \quad \text{and} \quad m \text{ is not } 0$$

we can divide both sides of the equation by m to obtain

$$\frac{mx}{m} = \frac{c}{m} \quad \text{or} \quad x = \frac{c}{m}$$

In summary, if m is not equal to 0, $mx = c$ is true if and only if

$$x = \frac{c}{m}$$

Example 8

For what value of x does

$$2x + 3 = 11 \qquad [4]$$

$2x + 3 = 11$ when $x = 4$.

Again, you can check this by replacing x by 4 in $2x + 3 = 11$ to get

$$2(4) + 3 = 11$$

which is a true statement ($11 = 11$).

If you prefer to use the technique of Lesson 2, you will notice that to make the left side of [4] equal to $2x$, you must undo adding 3. So you subtract 3 from both sides of [4] to obtain

$$\begin{array}{rcl} 2x + 3 &=& 11 \\ -3 & & -3 \\ \hline 2x &=& 8 \end{array} \qquad [5]$$

REVIEW

Notice that you subtract 3 rather than divide by 2. The idea is that $2x + 3$ means $(2x) + 3$, and you treat everything in the parentheses as a single number. In effect, you think of $(2x) + 3$ as being

$$() + 3$$

and you subtract 3 to get (). That is, we isolate the expression in parentheses.

But equation [5] is the same as equation [3], which you solved in Example 6.

In other words, by subtracting 3 from both sides of $2x + 3 = 11$, you convert the new equation [4] to an equation you have already learned to solve.

We could also solve equation [4] by the method of Lesson 4.

STEP 1: Add (-3) to both sides of [4].

$$(2x + 3) + (-3) = 11 + (-3)$$

STEP 2: Apply the associative property of addition.

$$2x + [3 + (-3)] = 11 + (-3)$$

STEP 3: By definition, $3 + (-3) = 0$.

$$2x + 0 = 11 + (-3)$$

STEP 4: 0 is the additive identity, so $2x = 8$.

$$2x = 8$$

Section 5.2   Equations of the Form $mx+b=c$ 111

Example 9

For what value of x does $5x + 9 = 89$?

$5x + 9 = 89$ when $x = 16$. The idea is to "isolate" x by itself on one side of the equation. You begin by subtracting 9 from both sides of the equation. This gives you

$$\begin{aligned} 5x + 9 &= 89 \\ -9 \quad & -9 \\ \hline 5x &= 80 \end{aligned}$$

> Because 9 is being added to $5x$, we "undo" it by subtracting 9. Then to keep the equation balanced, we also subtract 9 from the other side.

Then because x is multiplied by 5, you undo this by dividing both sides of the new equation by 5 to get

$$\frac{5x}{5} = \frac{80}{5} \quad \text{or} \quad x = 16$$

> We may also use a horizontal form. That is, we can say that $5x + 9 - 9 = 89 - 9$. Use whichever form you prefer.

Example 10

For what value of x does $5x - 9 = 91$?

$5x - 9 = 91$ when $x = 20$.
  To "undo" subtracting 9, you add 9. This gives you

$$\begin{aligned} 5x - 9 &= 91 \\ +9 \quad & +9 \\ \hline 5x &= 100 \end{aligned}$$

> In terms of signed numbers, $5x - 9$ may also be thought of as addition. That is, $5x - 9 = 5x + {}^-9$.

Then you divide both sides by 5 to get

$$\frac{5x}{5} = \frac{100}{5} \quad \text{or} \quad x = 20$$

Before ending this section, we should notice that it is possible to have a linear equation that is not in the standard form but can be converted into the standard form. Let's look at a few examples.

Example 11

For what value of m does

$$3(m + 2) = 18 \qquad [6]$$

$3(m + 2) = 18$ when $m = 4$. First apply the distributive property to the left side of [6] to obtain

$$3(m) + 3(2) = 18$$
or $\qquad 3m + 6 = 18 \qquad [7]$

Equation [7] is now in the standard form that you have already learned how to solve in this section.

$$\begin{aligned} 3m + 6 &= 18 \\ -6 \quad & -6 \\ \hline 3m &= 12, \quad \frac{3m}{3} = \frac{12}{3}, \quad \text{or} \quad m = 4 \end{aligned}$$

> If we want to solve $3m + 6 = 18$, we do not divide both sides by 3 unless we make sure that the entire left side is divided by 3. That is, we write
>
> $$\frac{3m + 6}{3} = \frac{18}{3} = 6$$
>
> The distributive rule tells us $3(m + 2) = 3m + 6$, so we have
>
> $$\frac{3(m + 2)}{3} = 6$$
>
> which is the same as $m + 2 = 6$.

ANOTHER METHOD

Remember that you treat everything within grouping symbols as a single term. Hence 3 is a factor of the left side of [6]. That is, 3 is multiplying $m + 2$. You can undo multiplying by 3 if you divide by 3. Dividing both sides of [6] by 3 gives

$$\frac{3(m+2)}{3} = \frac{18}{3} \quad \text{or} \quad (m+2) = 6$$

You can then solve $m + 2 = 6$ by subtracting 2 from both sides. This is another correct way for concluding that $m = 4$.

Example 11 gives us an idea of why the structure of arithmetic that we talked about in Lesson 4 is so important. Sometimes we have to rewrite algebraic expressions in order to convert them into a form that we have already studied. This is illustrated further in the next example.

Example 12

Find the value of x for which

$$3(2x + 5) + 4x = 175 \qquad [8]$$

Equation [8] is true when $x = 16$.

By the distributive property, you have that $3(2x + 5) = 3(2x) + 3(5)$, or $6x + 15$. Hence you may replace [8] by

$$6x + 15 + 4x = 175 \qquad [9]$$

Then because $6x + 4x = 10x$, you may rewrite [9] as

$$10x + 15 = 175 \qquad [10]$$

Equation [10] is in standard form, and you can solve it by the method already studied in this section.

$$\begin{array}{rl} 10x + 15 = & 175 \\ -15 & -15 \\ \hline 10x = & 160, \end{array} \qquad \frac{10x}{10} = \frac{160}{10}, \quad \text{or} \quad x = 16$$

NOTE

If you aren't sure that your answer is correct, you can always check it in the original equation. In this example, replace x by 16 in [8] to obtain

$$3(2[16] + 5) + 4(16)$$
$$= 3(32 + 5) + 64$$
$$= 3(37) + 64$$
$$= 111 + 64 = 175$$

If we want to work more formally with the left side of [9] using the results of Lesson 4, we have

$$\begin{array}{ll} (6x + 15) + 4x & \\ = 6x + (15 + 4x) & \text{(assoc.)} \\ = 6x + (4x + 15) & \text{(comm.)} \\ = (6x + 4x) + 15 & \text{(assoc.)} \\ = ([6 + 4]x) + 15 & \text{(distrib.)} \\ = (10x) + 15 & \\ = 10x + 15 & \end{array}$$

See how important Lesson 4 is here? Without using the various properties of numbers, we could replace $3(2x + 5) + 4x$ by the simpler expression $10x + 15$. When we rewrite $3(2x + 5) + 4x$ as $10x + 15$, we say that we have *simplified* $3(2x + 5) + 4x$.

PRACTICE DRILL

Solve each of the following linear equations.
1. $3x = 21$ 2. $3x + 6 = 21$ 3. $3x - 6 = 21$ 4. $3(x - 6) = 21$

Answers
1. $x = 7$ 2. $x = 5$
3. $x = 9$ 4. $x = 13$

CHECK THE MAIN IDEAS

1. Suppose you want to solve the _____ equation:

$$4(p + 2) = 28 \qquad [a]$$

2. You may first use the _____ property to rewrite [a] as:

$$4p + 8 = 28 \qquad [b]$$

3. To solve [b], you subtract _____ from both sides to get:

$$4p = 20 \qquad [c]$$

4. The solution to equation [a] is $p = $ _____.

Answers
1. linear   2. distributive
3. 8   4. 5

Section 5.3
Solving Equations of the Form $mx + b = cx + d$

It often happens that the variable appears on both sides of a linear equation. In this section, we want to learn how to solve such equations.

First let's see how such equations may arise. At the beginning of this lesson, we talked about the advertisement for cheese. The cheese was $3 per pound plus a $2 shipping charge. We saw that

$$c = 3p + 2 \qquad [1]$$

where p is the number of pounds of cheese we bought and c is the cost of the cheese in dollars.

Suppose that a second cheese company advertises the same cheese at a different price. The second company charges $5 for shipping costs but only $2 per pound for the cheese.

OBJECTIVE
Given the equation
$$mx + b = cx + d$$
to be able to solve for x in terms of the constants m, b, c, and d.

Example 1
How much would 6 pounds of cheese cost if you buy it from the second company?

6 pounds of cheese would cost $17.
You multiply the number of pounds of cheese (6) by the price per pound ($2) to get $12. Then you add the $5 shipping charge to this cost to get $17.

If you buy 6 pounds of cheese, it is better to use the second company. If you used the first company, the shipping charge would be only $2 but the cheese would cost $18 ($3 times 6 pounds) for a total cost of $20.

Example 2
How much would $1\frac{1}{2}$ pounds of cheese cost if you bought it from the second company?

$1\frac{1}{2}$ pounds would cost $8. You multiply the number of pounds of cheese by the price per pound (in this case, $1\frac{1}{2} \times 2$) to get $3 as the price of the cheese. Then you add the $5 shipping charge.

If you bought the cheese from the first company, it would cost $6.50. At $3 per pound, you would pay $4.50 for the cheese plus the $2 charge. This time it is a better buy to use the first company.

We can generalize the results of Example 1 and 2 by expressing the cost c of p pounds of cheese in the following way. We multiply the number of pounds p of cheese by the price per pound ($2) to find that p pounds of cheese costs $2p$ dollars. We then add the $5 shipping charge to obtain

$$c = 2p + 5 \qquad [2]$$

If we know how to read the formulas $c = 3p + 2$ and $c = 2p + 5$, we can use arithmetic to compare the costs of p pounds of cheese offered by the two companies

Number of Pounds (p)	Cost Using First Company ($3p + 2$)	Cost Using Second Company ($2p + 5$)	
1	$3(1) + 2 = \$5$	$2(1) + 5 = \$7$	
2	$3(2) + 2 = \$8$	$2(2) + 5 = \$9$	
3	$3(3) + 2 = \$11$	$2(3) + 5 = \$11$	←
4	$3(4) + 2 = \$14$	$2(4) + 5 = \$13$	
5	$3(5) + 2 = \$16$	$2(5) + 5 = \$15$	

Notice that the cost is the same with each company for 3 pounds. For less than 3 pounds, the first company offers the better buy; and for more than 3 pounds, the second company offers the better buy.

In this case we don't need algebra to determine that when we buy 3 pounds of cheese, the cost is the same regardless of which company we use.

If we want to, we can use algebra to find the answer by solving a linear equation. At the first company, the cost of p pounds of cheese is $3p + 2$ dollars. At the second company, the same p pounds of cheese costs $2p + 5$ dollars. We want to know for what number of pounds (p) the cost at the first company ($3p + 2$) is equal to the cost at the second company ($2p + 5$). Using an equation, we are asking for the value of p for which

$$3p + 2 = 2p + 5 \qquad [3]$$

Although we're using p instead of x, this is exactly the type of equation we're trying to solve in this section. We have the form

$$mp + b = cp + d$$
$$\uparrow \quad \uparrow \quad \uparrow \quad \uparrow$$
$$3 \quad 2 \quad 2 \quad 5$$

Equation [3] differs from the type of linear equation we solved in the previous section only because the variable now appears on both sides of the equation.

How does p appear on the right side of [3]? We see that $2p$ is being added to 5. To "undo" adding $2p$, we must subtract $2p$. If we subtract $2p$ from both sides of [3], we get

$$\begin{array}{rl} 3p + 2 = & 2p + 5 \\ -2p & -2p \\ \hline p + 2 = & 5 \end{array} \qquad [4]$$

Actually, $3p - 2p = 1p$, but we know that $1p = p$.

Equation [4] has the form

$$mp + b = c$$
$$\uparrow \quad \uparrow \quad \uparrow$$
$$1 \quad 2 \quad 5$$

which we learned to solve in the previous section. To solve [4], we add ⁻2 to (or subtract 2 from) both sides of the equation to get

$$\begin{array}{rl} p + 2 = & 5 \\ -2 & -2 \\ \hline p = & 3 \end{array}$$

As a check, if we replace p by 3 we find that

$$3p + 2 = 3(3) + 2 = 9 + 2 = 11$$
and
$$2p + 5 = 2(3) + 5 = 6 + 5 = 11$$

This checks with our earlier trial-and-error method. Let's try a few more problems by this method.

We've really shown that an equation of the form

$$mx + b = cx + d$$

can always be converted to an equation of the form we studied in the last section, if we subtract cx from both sides. That is,

$$\begin{array}{rl} mx + b = & cx + d \\ -cx & -cx \\ \hline mx - cx + b = & d \end{array}$$
$$(m - c)x + b = d$$

Example 3

Solve the equation

$$6m + 7 = 2m + 75 \qquad [5]$$

The solution of [5] is $m = 17$.

As $2m$ is being added to 75 on the right side of [5], you can undo this by subtracting $2m$ from both sides of [5] to obtain

$$\begin{array}{rl} 6m + 7 = & 2m + 75 \\ -2m & -2m \\ \hline 4m + 7 = & 75 \end{array} \qquad [6]$$

There is no one right way to solve an equation. For example, we might have tried to solve [5] by first subtracting 7 from each side to get

$$\begin{array}{rl} 6m + 7 = & 2m + 75 \\ -7 & -7 \\ \hline 6m = & 2m + 68 \\ -2m & -2m \\ \hline 4m = & 68 \end{array}$$

Equation [6] is in the form you studied in the previous section. The technique is to subtract 7 from both sides of [6] to get

$$\begin{array}{rl} 4m + 7 = & 75 \\ -7 & -7 \\ \hline 4m = & 68 \end{array} \qquad [7]$$

which is the same equation as equation [7].

Then you find m by dividing both sides of equation [7] by 4. That is,

$$\frac{4m}{4} = \frac{68}{4} \quad \text{or} \quad m = 17$$

Section 5.3   Solving Equations of the Form $mx+b=cx+d$

Equations [5] and [7] are quite different. That is, the equations $6m + 7 = 2m + 75$ and $4m = 68$ do not look alike. Yet one is derived from the other by using the accepted rules (properties) of arithmetic. That means that although the two equations look different, they have the same solution.

> **DEFINITION**
> Equations that have the same solution are said to be **equivalent** equations.

For example, we say that the two equations
$$6m + 7 = 2m + 75$$
and
$$4m = 68$$
are equivalent. Any solution of one will also be a solution of the other. In this case, it is easier to solve $4m = 68$ than to solve $6m + 7 = 2m + 75$. So we use the fact that the two equations are equivalent and solve the easier one to find a solution of the harder one.

Example 4

Solve the equation
$$3n + 7 = 8n - 63 \qquad [8]$$

The solution of [8] is $n = 14$.

To get the terms involving n on one side of the equation, subtract $3n$ from both sides to get

$$\begin{array}{r} 3n + 7 = 8n - 63 \\ -3n \quad\quad -3n \\ \hline 7 = 5n - 63 \end{array} \qquad [9]$$

You can then undo subtracting 63 from $5n$ by adding 63 to both sides of [9] to get

$$\begin{array}{r} 7 = 5n - 63 \\ +63 \quad\quad +63 \\ \hline 70 = 5n \end{array} \qquad [10]$$

Equation [10] is equivalent to equation [8], and you can solve [10] by dividing both sides by 5 to obtain

$$\frac{70}{5} = \frac{5n}{5} \quad \text{or} \quad 14 = n$$

NOTE

No law says that the variable must be on the left side of an equation. We could solve [8] by first subtracting $8n$ from both sides of the equation. Then we would deal with a negative coefficient for n. This isn't the worse thing in the world, but usually we prefer positive coefficients.

Of course, there are times when no matter how we try to avoid negative numbers, we have to use them.

For example:

$$\begin{array}{r} 3n + 7 = 8n - 63 \\ -8n \quad\quad -8n \\ \hline -5n + 7 = -63 \\ -7 \quad\quad -7 \\ \hline -5n = -70 \end{array}$$

$$n = \frac{-70}{-5} = 14$$

Do you see that it is less confusing to work with positive coefficients?

Example 5

For what value of n does
$$2n + 6 = 5n + 36 \qquad [11]$$

[11] is true when $n = {}^-10$. To keep the coefficient of n positive, you begin by subtracting $2n$ from both sides of equation [11] to get

$$\begin{array}{r} 2n + 6 = 5n + 36 \\ -2n \quad\quad -2n \\ \hline 6 = 3n + 36 \end{array} \qquad [12]$$

CHECK

When $n = {}^-10$,
$$2n + 6 = 2({}^-10) + 6$$
$$= {}^-20 + 6 = {}^-14$$
$$5n + 36 = 5({}^-10) + 36$$
$$= {}^-50 + 36 = {}^-14$$

Then you subtract 36 from both sides of [12] to get

$$6 = 3n + 36$$
$$-36 \quad -36$$
$$\overline{-30 = 3n} \qquad [13]$$

Divide both sides of [13] by 3 to obtain $\frac{-30}{3} = \frac{3n}{3}$, or $^-10 = n$.

We could try to avoid negative numbers by subtracting 6 from both sides of [12], but we want to get the variables on one side and the constants on the other. If we subtract 6 from both sides of [12], we get $0 = 3n + 30$, and n still has to equal $^-10$.

Example 6
Solve the equation

$$3(2m + 7) - 14 = 5(m + 15) - 3m \qquad [14]$$

The solution of this equation is $m = 17$.
  By the distributive property

$$3(2m + 7) = 6m + 21 \quad \text{and} \quad 5(m + 15) = 5m + 75$$

Therefore, equation [14] becomes

$$6m + 21 - 14 = 5m + 75 - 3m$$
or $\qquad 6m + 7 = 2m + 75$

This is the same equation you solved in Example 3.

PRACTICE DRILL
Solve each of the following linear equations.
1. $3x + 5 = 2x + 7$
2. $3x + 7 = 2x + 5$
3. $3x - 7 = 2x + 5$
4. $3(x + 7) = 2x + 5$
5. $4x + 9 = 2x + 14$

Answers
1. $x = 2$ 2. $x = -2$
3. $x = 12$ 4. $x = -16$
5. $x = 2\frac{1}{2}$

CHECK THE MAIN IDEAS
Suppose you want to solve the linear equation

$$4x + 5 = 3x + 7 \qquad [a]$$

1. To get x to appear only on the left side of [a], you add _____ to both sides to get

$$x + 5 = 7 \qquad [b]$$

2. To solve [b], you add _____ to both sides to get $x = 2$.
3. To check that $x = 2$ is a solution of [a], you replace x by 2 in [a] and get the true statement that _____ = _____.
4. You have shown that if x is not equal to _____, then $4x + 5$ cannot be equal to $3x + 7$.

Answers
1. $-3x$ (that is, you subtract $3x$)
2. -5 (that is, you subtract 5)
3. 13 [that is, $4(2) + 5 = 3(2) + 7$]
4. 2

Section 5.4
Inconsistent Equations and Identities

  Start with any number and add 1 to it. The new number can't possibly equal the original number. In other words, no number can equal 1 more than itself.
  If we let m stand for any number, then the equation

$$m = m + 1 \qquad [1]$$

should have no number that is a solution.

OBJECTIVE
To show that there are cases in which a linear equation

$$mx + b = cx + d$$

has no numbers for which the equation is true, and other cases in which the equation is true for all numbers.

Section 5.4 Inconsistent Equations and Identities

Suppose we didn't know this, and we tried to use the techniques from the previous sections. Something interesting happens.

We may begin by subtracting m from both sides of [1]. We get

$$\begin{aligned} m &= m + 1 \\ -m & -m \\ \hline 0 &= 1 \end{aligned} \qquad [2]$$

[2] happens to be a false statement.

The false statement that $0 = 1$ came from the *assumption* that $m = m + 1$, so it means that there is no number m for which $m = m + 1$.

Example 1

Show that the equation

$$3(m + 2) = 6m + 9 \qquad [3]$$

has no number that makes it true.

We start with a linear equation, just as in the previous sections, and we use the same techniques. But now we find that no number is a solution of the equation.

You begin by using the distributive property to rewrite $3(m + 2)$ as $3m + 6$. Then [3] becomes

$$3m + 6 = 3m + 9 \qquad [4]$$

If you subtract $3m$ from both sides of [4], you get $6 = 9$, which is a false statement.

$$\begin{aligned} 3m + 6 &= 3m + 9 \\ -3m & -3m \\ \hline 6 &= 9 \end{aligned}$$

This leads to:

> **DEFINITION**
> A linear equation is said to be **inconsistent** if there is no number for which it is true.

We saw in Example 1 and our earlier discussion that there is a practical way to decide whether a linear equation is inconsistent or not. We solve it the same way as we did in the previous sections. If we get a false statement (such as $0 = 1$ or $6 = 9$), then the equation is inconsistent.

We shall study inconsistent equations in more detail in Lesson 8.

Let's look at yet a different situation. By the commutative property of addition, we know that any number plus 1 is the same as 1 plus that number. Using an equation, if we let m stand for any number, this fact becomes

$$m + 1 = 1 + m \qquad [5]$$

We know that [5] is true for every number m, but suppose we don't know this. We can still try to solve [5] with the methods from the previous sections. For example, we may elect to subtract m from each side of [5], in which case we get

$$\begin{aligned} m + 1 &= 1 + m \\ -m & -m \\ \hline 1 &= 1 \end{aligned} \qquad [6]$$

[6] must be equivalent to [5], but [6] is simply a true statement. Hence [5] must also always be a true statement.

This shows us that it is possible for a linear equation to have more than one number as a solution.

REVIEW

If an equation is true for all values of the variable, we call the equation an identity. For example, [5] is an identity.

Example 2

Show that

$$2(3p + 4) + 5p = 11p + 8 \qquad [7]$$

is an identity; that is, that [7] is true for every number p.

This is just another way of saying that $11p + 8$ and $2(3p + 4) + 5p$ are equivalent. In algebra, it means that we may always replace $2(3p + 4) + 5p$ by the simpler expression $11p + 8$.

There are several ways to show this. We may begin rewriting $2(3p + 4)$ as $6p + 8$ by use of the distributive property. In this way, [7] becomes

$$6p + 8 + 5p = 11p + 8$$
or
$$11p + 8 = 11p + 8 \quad [8]$$

[8] is clearly an identity. Because it is equivalent to [7], then [7] is also an identity.

If we weren't content with [8], we could subtract $11p$ from both sides to get the true statement that $8 = 8$; and if we still weren't content, we could subtract 8 from both sides to get $0 = 0$.

If we combine the results of this section with those of the previous sections, we have the following general idea for solving linear equations.

For example,

$$3(2x + 5) - 2(x - 1) = 2x + 7$$
$$6x + 15 - 2x + 2 = 2x + 7$$
$$4x + 17 = 2x + 7$$

STEP 1: Rewrite the equation in a form in which all grouping symbols have been removed.

STEP 2: On each side of the equation, combine like terms. This leaves you with just two terms on each side of the equation — one that involves the unknown and one that doesn't. In symbolic form, the equation will look like

$$mx + b = cx + d \quad [9]$$

STEP 3: If m and c are unequal, you may solve equation [9] as described in the previous section. In this case, you find that [9] has one and only one solution. If m and c are equal, then

(a) Equation [9] has no solutions if b and d are unequal.

For example, $2x + 1 = 2x + 4$, which implies that $1 = 4$.

(b) Equation [9] has a great many solutions (it is an identity) if $b = d$.

For example, $2x + 1 = 2x + 1$, which says that $1 = 1$, and so on.

The main point is that the identity and the inconsistent equation are special cases of the general linear equation $mx + b = cx + d$. The difference is that the inconsistent case means that we get a false statement when we try to solve the equation; the identity means that we get a true statement when we try to solve the equation.

Whichever case we get, we still use the same method to solve the equation

$$mx + b = cx + d$$

PRACTICE DRILL

What value(s) of x, if any, satisfy each of the following equations?
1. $2x + 3 = 2(x + 3)$
2. $2x + 2 = 2(x + 1)$
3. $2x + 2 = x + 2$
4. $2x + 4 = 4 + 2x$
5. $2x - 4 = 4 - 2x$

Answers
1. No values of x
2. All values of x
3. $x = 0$ (This is a value of x. It is not the same as saying no values of x.)
4. All values of x
5. $x = 2$

CHECK THE MAIN IDEAS

1. An equation such as $x + 3 = 3 + x$ is called an _____.
2. $x + 1 = 1 + x$ is called an identity because for each given value of x, the expressions $x + 1$ and $1 + x$ are _____.
3. This is not the same as $2x = x + 1$, because the expressions $2x$ and $x + 1$ are not always _____. In fact, $2x = x + 1$ if and only if $x = 1$.
4. The equation $x + 1 = x + 2$ is the "opposite" of an identity. That is, no matter what number you replace x by, $x + 1$ and $x + 2$ can never be _____.
5. Unless a linear equation is an identity, it cannot have more than _____ solution.
6. If it is an identity, then any value of x is a _____ of the equation.

Answers
1. identity
2. equal
3. equal
4. equal
5. one (but it can have none, such as for $x + 1 = x + 2$)
6. solution

Section 5.5
Inverting Linear Relationships

In the first section of this lesson, we looked at equations such as
$$c = 3p + 2 \qquad [1]$$

In Section 5.2, we solved [1] for p, given a value for c. Let' review.

OBJECTIVE
Given that one variable is expressed in terms of a second variable, to learn how to express the second variable in terms of the first.

Example 1

For what value of p will $c = 62$ if $c = 3p + 2$?

$c = 62$ if $p = 20$. Replace c by 62 in [1] to get
$$62 = 3p + 2 \qquad [2]$$

Then subtract 2 from both sides of [2] to get
$$60 = 3p \qquad [3]$$

Then divide both sides of [3] by 3 to get $20 = p$.

There was nothing special about picking c to be 62 in [1]. We could pick any value for c, and still use the same idea to find what value p has to have.

Sometimes we leave c in [1] and *pretend* it is a number. We then solve for p just as we did before, except that now our answer comes out in terms of c.

Let's try a problem.

1. We start with
$$c = 3p + 2 \qquad [1]$$

    Our aim is to "isolate" p. That is, we want to get p by itself on one side of the formula [1].

2. Because 2 is being added to $3p$, we can undo the 2 by subtracting it from both sides of [1] to obtain
$$\begin{array}{r} c = 3p + 2 \\ -2 \quad -2 \\ \hline c - 2 = 3p \end{array} \qquad [1a]$$

3. Now p is being multiplied by 3, so we can undo this by dividing both sides of [1a] by 3. This gives us
$$\frac{c-2}{3} = p \qquad [1b]$$

[1] and [1b] are the same relationship but with different emphasis. For example, [1] is the more convenient formula if we are given p and want to find the value of c. On the other hand, if we are given c and want to find the value of p, then the more convenient formula is [1b].

It is often the custom to write the equation so that the isolated variable is on the left side. In this case we write
$$p = \frac{c-2}{3}$$

Example 2

Use equation [1b] to find the value of p when $c = 62$.

When $c = 62$, $p = 20$.

You saw this in Example 1. Now, however, you're trying to show that you can get the answer quickly from [1b]. That is, you replace c by 62 in [1b] and get
$$\frac{62-2}{3} = p \quad \text{or} \quad \frac{60}{3} = p$$

From this we conclude that $p = 20$.

The main idea is that there are times when we want to know how p is expressed in terms of c rather than to know how we can find p for a certain value of c.

Example 3

Rewrite the formula

$$m = 6n + 7 \qquad [4]$$

in a way that expresses n in terms of m.

$m = 6n + 7$ means the same as $n = \dfrac{m-7}{6}$.

To isolate n in [4], notice that 7 is being added to $6n$. So you begin by subtracting 7 from both sides of [4] to obtain

$$\begin{array}{r} m = 6n + 7 \\ -7 \quad\quad -7 \\ \hline m - 7 = 6n \end{array} \qquad [5]$$

Complete your isolation of n by dividing both sides of [5] by 6 to get

$$\frac{m-7}{6} = \frac{6n}{6} = n \qquad [6]$$

As we go from [4] to [6], the important thing to notice is that we proceed the same way as we would if m were replaced by a number. For example, if we replace m by 61 in [4], we get the equation $61 = 6n + 7$. We solve for n by first subtracting 7 and then dividing by 6. That is,

$$\begin{array}{r} 61 = 6n + 7 \\ -7 \quad\quad -7 \\ \hline 54 = 6n \end{array}$$

so that $\dfrac{54}{6} = \dfrac{6n}{6}$ and $n = 9$. But [6] tells us how to find n for any value of m — not just for $m = 61$.

The arithmetic can become more complicated if the relationship is more complicated, but the idea is always the same.

SUMMARY
[4] tells us that starting with n, we first multiply by 6 and then add 7 to get m. [6] tells us that if we start with m, we first subtract 7 and then divide by 6 to find n. But [4] and [6] are two different ways of stating the same relationship.

Example 4

Celsius temperature (C) is related to Fahrenheit temperature (F) by the formula

$$C = \frac{5}{9}(F - 32) \qquad [7]$$

Rewrite this formula so that F is expressed in terms of C.

$C = \dfrac{5}{9}(F - 32)$ means the same as $F = \dfrac{9}{5}C + 32$.

Because $\dfrac{5}{9}$ is a factor of the right side of [7], you can undo the formula by multiplying both sides of [7] by $\dfrac{9}{5}$ (which is the same as dividing by $\dfrac{5}{9}$). This gives

$$\frac{9}{5}(C) = \frac{9}{5}\left[\frac{5}{9}(F - 32)\right] = \left[\frac{9}{5}\left(\frac{5}{9}\right)\right](F - 32)$$

or

$$\frac{9}{5}C = F - 32 \qquad [8]$$

You then add 32 to both sides of [8] to obtain

$$\frac{9}{5}C + 32 = F - 32 + 32 = F$$

or

$$F = \frac{9}{5}C + 32 \qquad [9]$$

In other words, to find C we subtract 32 from F and then multiply the answer by $\dfrac{5}{9}$. But to find F from C, we first multiply C by $\dfrac{9}{5}$ and then add 32 to this result.

Again, the point is that we would use [7] if we were converting Fahrenheit to Celsius. But we would use [8] to convert Celsius to Fahrenheit.

Section 5.5 Inverting Linear Relationships

Now, if we know that the temperature is 68°F, we can replace F by 68 in [7] to see that

$$C = \frac{5}{9}(68 - 32) = \frac{5}{9}(36) = 20$$

This tells us that 68°F is equivalent to 20°C. But if we know that the temperature is 20°C, we use [9] to conclude that

$$F = \frac{9}{5}(20) + 32 = 36 + 32 = 68$$

That is, [9] tells us that 20°C is the same temperature as 68°F.
In summary, [7] and [9] give us the same information with a different emphasis.

We also could replace F by 68 in [9] to obtain

$$68 = \frac{9}{5}C + 32$$

But we would have to use the undoing method if we want to find C.

NOTE

The process of going from $C = \frac{5}{9}(F - 32)$ to $F = \frac{9}{5}C + 32$ is called **inverting** the relationship.

In general, when one variable (y) is expressed in terms of another (x), and we use the properties of arithmetic to express x in terms of y, we say that we've inverted the formula (relationship).

Example 5

Given the relationship

$$p = 3(q - 4) + 2q \qquad [10]$$

solve for q in terms of p.

You have that

$$p = 3q - 12 + 2q \qquad \text{or} \qquad p = 5q - 12$$

Hence

$$\begin{array}{r} p = 5q - 12 \\ +12 \qquad +12 \\ \hline p + 12 = 5q \end{array} \qquad [11]$$

If you divide both sides of [11] by 5, you get

$$q = \frac{p + 12}{5} \qquad [12]$$

Example 6

Use the result of Example 5 to solve the equation

$$63 = 3(q - 4) + 2q \qquad [13]$$

This is just [10] with $p = 63$. According to [12], you first add 12 to p (in this case, $p = 63$) to get 75, and then you divide 75 by 5 to get 15.
 As a check, note that

$$\begin{aligned} 63 &= 3(15 - 4) + 2(15) \\ &= 3(11) + 2(15) \\ &= 33 + 30 \\ &= 63 \end{aligned}$$

We could solve [13] by the method of the previous sections. However, if we later change the 63 to a different number, we would have to do all the work again. If we use [12], we know that for *any* value of p, we first add 12 and then divide by 5 to find the value of q in [10].

PRACTICE DRILL

Rewrite each of the following in such a way that x is expressed in terms of y.
1. $y = 2x$
2. $y = 2x + 3$
3. $y = 2(x + 3)$
4. $y = \frac{1}{2}(x + 3)$

Answers
1. $x = \frac{1}{2}y$ or $\frac{y}{2}$
2. $x = \frac{1}{2}(y - 3)$ or $\frac{y - 3}{2}$
3. $x = \frac{y}{2} - 3$ or $\frac{y - 6}{2}$
4. $x = 2y - 3$

CHECK THE MAIN IDEAS

1. The formula $c = 2p + 3$ expresses c in terms of p. If you want to express p in terms of c, you must _____ the formula.
2. To invert $c = 2p + 3$, you can first subtract _____ from both sides to get $c - 3 = 2p$.
3. Then you can _____ both sides by 2 to get $p = \frac{c - 3}{2}$.
4. This tells you that if you are given c and want to find p, you first (a) _____ 3 from c and then divide this answer by (b) _____.

Answers
1. invert
2. 3
3. divide
4. (a) subtract (b) 2

EXERCISE SET 5 (Form A)

Section 5.1
1. Is $2.34 - 7.5p$ linear in p?
2. Is $3p^2 + 4$ linear in p?
3. Write the following in the form $mn + b$, where m and b are constants.
   (a) $2[3(4n + 1)] + 4$ (b) $3(2n + 1) - 6n + 5$ (c) $3\{[4n - (3n - 1)] + n - 2\}$
4. Find the value of $3\{[4n - (3n - 1)] + n - 2\}$ when $n = 2.34$.

Section 5.2
Solve each of the following equations.
5. $2[3(4n + 1)] + 4 = 106$
6. $3\{[4n - (3n - 1)] + n - 2\} = 60$
7. $2(5n - 7) - 2(3 - 4n) = 16$
8. $2(5n - 7) - 2(3 - 4n) = 1$
9. $-5(n - 2) + 6(n + 4) = 7$

Section 5.3
Solve each of the following equations.
10. $7.5n + 13.6 = 1.5n + 12.7$
11. $7(n + 2) = 3(n + 2) + 5n$
12. $2[8 - (3 - n)] + 4n = 2n + 3$

Section 5.4
13. Simplify $2\{[3n - (2n + 4)] + 5 - n\}$.
14. For what value(s) of n will $2\{[3n - (2n + 4)] + 5 - n\} = 5$?
15. For what value(s) of n will $2\{[3n - (2n + 4)] + 5 - n\} = 2$?
16. For what value(s) of n does $3[2(n + 2) + 1] - 21 = 6(n - 1)$?
17. For what value(s) of n does $4[3(n - 1) + 2] + 5 = 12(n + 2)$?

Section 5.5
18. If $c = \frac{1}{2}(p - 8) + 13$, express p in terms of c.
19. For what value of p does $\frac{1}{2}(p - 8) + 13$ equal
    (a) 20 (b) 5 (c) 13.7

Answers: Exercise Set 5
Form A
1. yes
2. no
3. (a) $24n + 10$ (b) $0n + 8$ (or 8) (c) $6n - 3$ (or $6n + {}^-3$)
4. 11.04
5. $n = 4$
6. $n = 10.5$ (or $10\frac{1}{2}$)
7. $n = 2$
8. $n = \frac{7}{6}$ (or $1\frac{1}{6}$ or $1.166\ldots$)
9. $n = -27$
10. $n = -0.15$ (or $-\frac{.9}{6}$ or $-\frac{.3}{2}$)
11. $n = 8$
12. $n = -1\frac{3}{4}$ (or $-\frac{7}{4}$ or -1.75)
13. 2
14. no values of n
15. all values of n
16. all values of n
17. no values of n
18. $2(c - 13) + 8$, or $2c - 18$
19. (a) 22 (b) -8 (c) 9.4
20. $\frac{Q + 3}{6}$
21. $b = y - mx$ (or $^-mx + y$, or $y + {}^-mx$)
22. $x = \frac{y - b}{m}$

20. If $Q = 3\{[4n - (3n - 1)] + n - 2\}$, express n in terms of Q.
21. In the formula $y = mx + b$, express b in terms of x, m, and y.
22. In the formula $y = mx + b$, express x in terms of b, y, and $m(m \neq 0)$.

EXERCISE SET 5 (Form B)

Section 5.1

1. Is $\frac{2}{3}r - \frac{4}{7}$ linear in r?
2. Is $5r^2 - 4$ linear in r?
3. Write each of the following in the form $cp + d$, where c and d are constants:
   (a) $-3[4(2p + 1)] + 8$ (b) $-2(5p + 4) + 5(2p - 3)$
   (c) $2\{[5p - (4p + 3)] - 2p + 6\}$
4. What is the value of $2\{[5p - (4p + 3)] - 2p + 6\}$ when $p = 2.37$?

Section 5.2
Solve each of the following equations
5. $-3[4(2p + 1)] + 8 = 44$
6. $2\{[5p - (4p + 3)] - 2p + 6\} = -16$
7. $4(3p + 5) - 3(2 - 4p) = 110$
8. $4(3p + 5) - 3(2 - 4p) = 11$
9. $-7(p - 2) + 8(p + 5) = 30$

Section 5.3
Solve each of the following equations
10. $8.7p + 22.56 = 4.7p + 15.32$
11. $12(p + 3) = 5(p + 3) + 8p$
12. $3[11 - (5 - p)] + 2p = 6p + 11$

Section 5.4
13. Simplify $3\{[4p - (5p - 3)] + p + 2\}$.
14. For what value(s) of p does $3\{[4p - (5p - 3)] + p + 2\} = 8$?
15. For what value(s) of p does $3\{[4p - (5p - 3)] + p + 2\} = 15$?
16. For what value(s) of p does $4[2(p - 3) + 4] + 16 = 8(p + 1)$?
17. For what value(s) of p does $2[3(p - 7 + 4] + 12 = 6(p + 4)$?

Section 5.5
18. If $c = 5(p + 4) - 22$, express p in terms of c.
19. For what value of p does $5(p + 4) - 22$ equal:
    (a) 28   (b) -17 (c) 100
20. If $Q = \frac{1}{2}\{[4n - 3(n - 1)] + 4\}$, express n in terms of Q.
21. In the formula $s = vt + p$, express p in terms of s, v, and t.
22. In the formula $s = vt + p$, express v in terms of s, p, and t $(t \neq 0)$.

An Introduction to Word Problems

Overview There is an old story about a man who walked into a tailor shop with a button and said to the tailor, "Please sew a suit on!"

It would seem much more natural to come in with a suit and ask to have a button sewed on!

From a practical point of view, that's the point we have reached in our course. Put simply, most people who study algebra want to use the material to solve problems in the real world. They are not particularly interested in the distributive property or the associative property or the fact that one is the multiplicative identity — unless these facts can be used to solve problems.

Up to now, we have been preparing the way to introduce the study of problem-solving. First we reviewed arithmetic so that we were sure of the facts we may have to know. Then we went over the structure of arithmetic, so that we could, if necessary, undo certain equations.

In this lesson, we introduce some ways in which the material of the first five lessons can be applied toward solving real-world problems.

Because our emphasis has been on linear equations, the emphasis in this lesson will be on those "word" problems whose solutions use the format of linear equations.

Our approach emphasizes how we must learn to translate from English into the language of mathematics — how we can look at a given situation to determine what equation must be solved. There is no sure way of learning this idea, but we can note certain types of situations. That's what we do in this lesson.

It is important to realize that in many cases where we may need to use algebra, there is an excellent chance that someone will give us the proper formula to use. For this reason, we begin our study with the situation in which a formula is given. Later we explore typical situations in which we may be expected to make up our own formula.

Section 6.1
Formulas Leading to Linear Equations

One way to get linear equations is from formulas. Very often we are shown a formula that applies to a given situation. In Lesson 1, we saw how the area of a rectangle is related to the length and width of the rectangle. Or we may know how the distance a freely falling body falls is related to the time that the body falls. No matter what subject we study, if measurements are involved, there is usually a known relationship between the variables.

Let's look at a few examples in this section.

OBJECTIVE
To learn how linear equations arise from certain formulas.

Example 1

The area (A) of a rectangle is related to the length (L) and the width (W) by the formula

$$A = LW \qquad [1]$$

What is the area of a rectangle whose width is 5 feet and whose length is 8 feet?

The area of the rectangle is 40 square feet.
In terms of the symbols in [1], you are given that L is 8 and W is 5. So you replace L by 8 and W by 5 in [1] to obtain

$$A = 8(5) = 40 \qquad [2]$$

Notice that to solve Example 1, we have to know only how to read the formula $A = LW$. We do not have to know where it comes from. Once we accept [1], we solve the problem just by arithmetic. That is, we multiply 8 and 5.

Example 2

Using the same formula as in Example 1, find the value of L if $A = 400$ and $W = 16$.

In this case $L = 25$. You replace A by 400 and W by 16 in [1] to obtain

$$400 = L(16) = 16L \qquad [3]$$

[3] is a linear equation that you solve by dividing both sides by 16 to get

$$\frac{400}{16} = \frac{16L}{16} \quad \text{or} \quad 25 = L$$

This problem would arise if we know, for example, that the area of the rectangle is 400 square feet and that the width is 16 feet.

Equations [2] and [3] are, in fact, linear equations that are derived from [1]. But we can solve [2] by arithmetic, while we need algebra to solve [3].

Example 3

The cost in dollars (c) of p pounds of cheese is given by the formula

$$c = 4p + 7 \qquad [4]$$

What is the cost of 24 pounds of cheese?

24 pounds of cheese costs $103
All you do is replace p by 24 (because p is the number of pounds of cheese you're buying) in [4] to obtain

$$c = 4(24) + 7 = 96 + 7 = 103$$

Formula [4] would arise if the cheese cost $4 per pound and there was a $7 handling charge. But once [4] is given, we don't have to know this. All we have to do is know how to read the formula.

Example 4

Using the same formula as in Example 3, find out how many pounds of cheese you can buy for $367.

You can buy 90 pounds of cheese for $367.

In this case, 367 is the cost of the cheese in dollars, and you are told that this is represented by c in formula [4]. So you replace c by 367 in [4] to obtain

$$367 = 4p + 7 \qquad [5]$$

[5] is a linear equation that you can solve by the method of the previous lesson.

$$\begin{array}{r} 367 = 4p + 7 \\ -7 -7 \\ \hline 360 = 4p \end{array}$$

or $\qquad \dfrac{360}{4} = p$

Hence $p = 90$ (pounds of cheese).

Example 5

An object moves a distance of d feet in t seconds according to the formula

$$d = 70t + At^2 \qquad [6]$$

Find the value of A if $d = 160$ when $t = 2$.

The value of A is 5. You replace d by 160 and t by 2 in [6] to get

$$160 = 70(2) + A(2)^2$$
or $\qquad 160 = 140 + 4A \qquad [7]$

[7] is linear in A, and you can solve it by

$$\begin{array}{r} 160 = 140 + 4A \\ -140 -140 \\ \hline 20 = 4A \end{array}$$

$$A = \dfrac{20}{4} = 5$$

IMPORTANT NOTE

[6] is linear in A but not in t (t is squared in [6]). This is one reason why we say an equation is linear in A; that is, the equation can be linear in one variable but not in another.

Example 6

Given the formula

$$I = prt \qquad [8]$$

find the value of I when $p = 600$, $r = 0.08$, and $t = 5$.

In this case, $I = 240$. All you do is replace p by 600, r by 0.08, and t by 5 in [8] to obtain

$$I = 600(0.08)(5) = 48(5) = 240$$

Example 7

Using the same formula as in Example 6, find the value of r if $I = 1{,}000$, $p = 5{,}000$, and $t = 4$.

In this case, $r = 0.05$. You replace I by 1,000, p by 5,000, and t by 4 in [8] to obtain

$$1{,}000 = 5{,}000(r)4 = 20{,}000r$$

which is a linear equation in r. Divide both sides of this equation by 20,000 to conclude that

$$r = \dfrac{1{,}000}{20{,}000} = \dfrac{1}{20} = 0.05$$

Notice that we did this example without even trying to explain what [8] means. In fact, [8] tells us the amount of interest we would earn if we invest p dollars at $r\%$ simple annual interest for t years. In Example 6, we have shown that $600 invested for 5 years at 8% simple interest would earn $240.

Section 6.2 Translating Words into Algebra

It is impossible to list all the situations in which we get linear equations from formulas. The important point is that whenever this happens, we solve the equation according to the methods discussed in Lesson 5.

There are times, however, in which the formula is not given and it is our job to figure out what the formula is. This is not always easy. But in the remainder of this lesson, we'll try to explain a few general cases in which this occurs.

PRACTICE DRILL

1. Given the formula $I = PRT$
    (a) Find I if $P = 2{,}000$, $R = 0.05$, and $T = 8$
    (b) Find T if $I = 800$, $P = 5{,}000$, and $R = 0.04$
    (c) Find R if $I = 840$, $P = 6{,}000$, and $T = 2$

2. Given the formula $d = At + Bt^2$
    (a) Find d if $A = 5$, $B = 6$, and $t = 4$
    (b) Find A if $d = 100$, $t = 5$, and $B = 2$

Answers
1. (a) 800 (b) 4 (c) 0.07
2. (a) 116 (b) 10

CHECK THE MAIN IDEAS

1. A relationship such as $A = LW$ is often called a _____.
2. A formula tells you the _____ between the various quantities (variables).
3. Sometimes you are given the values of all but _____ variable.
4. In that case, you replace each of those variables in the formula by their given _____.
5. Then you solve the resulting _____.
6. In the formula $A = LW$, if you are told that $A = 50$ and $L = 10$, you find the value of W by solving the equation _____.

Answers
1. formula   2. relationship
3. one   4. value
5. equation   6. $50 = 10W$

Section 6.2
Translating Words into Algebra

OBJECTIVE
To learn how verbal statements are converted into algebraic equations.

In many ways mathematics is just like any other language. We must often translate from English into mathematics (or, in some cases, from mathematics into English).

When the translations involve concrete numbers, we usually have much less trouble than when we also have to work with "letters." But if we understand what happens when we deal with numbers, we get a better idea of what is happening when we work with letters.

Let's look at a few examples.

Example 1

John has 50 marbles. Bill has 5 more marbles than John. How many marbles does Bill have?

Bill must have 55 marbles, because 55 is 5 more than 50.

This problem is probably very easy for you. Perhaps it is so easy that you are not aware of the fact that what you actually did was the arithmetic problem $50 + 5$. That is, you know that 5 more than 50 means $50 + 5$, but it is easier just to say 55.

Suppose that we had the following, instead:

Example 2

John has J marbles. Bill has 5 more marbles than John. How many marbles does Bill have?

Bill has $J + 5$ marbles.
    It is still true that to have 5 more than John means that you add 5 to John's number to get Bill's number. In this case, you add 5 to J. But J is a variable, so you don't know its exact numerical value. Therefore, you cannot further simplify $J + 5$. You must leave the answer as $J + 5$.

    As with any language, we must listen carefully when we translate. A little change in a word can change the meaning of a whole statement or paragraph.

> In Example 1, J is 50. In that case $J + 5$ is 55. In general, the value of $J + 5$ depends on the value of J. In Example 1, we are saying that the value of the *expression* $J + 5$ is 55 when J is 50.

Example 3

John has 50 marbles. Bill has 5 times as many marbles as John. How many marbles does Bill have?

Bill has 250 marbles.
    The phrase "5 times as many" means to multiply by 5. $5 \times 50 = 250$, which gives you the answer.
    Did you notice that Examples 1 and 3 look almost alike? In one case, the phrase is "5 more," which tells you to add 5. In the other case, the phrase is "5 times as many," which tells you to multiply by 5.

    Once we know how to translate, the idea stays the same whether we use a variable or a constant.

Example 4

John has J marbles. How many marbles does Bill have if Bill has 5 times as many marbles as John?

Bill has $5J$ marbles. You are told that to find the number of marbles Bill has, you multiply the number John has by 5. When you multiply J by 5, you write the product as $5J$.

> If this seems difficult, it may help to think in terms of concrete numbers. For example, we know that 5 times 50 is 250, $5 \times 20 = 100$, and so on. If we look at enough examples, we notice that we multiply the given number by 5. If the given number happens to be called J, we multiply J by 5 and we get $5J$.

Example 5

John has 50 marbles. Bill has half the number of marbles John has. How many marbles does Bill have?

Bill has 25 marbles. You just take half of 50.
    Using common fractions, you translate "half" as $\frac{1}{2}$. Then you note that "of" means "times". Hence you translate "half of 50" as $\frac{1}{2} \times 50$. Performing the multiplication gives you the correct answer.

> In decimal fractions "half" translates as 0.5, and we get the answer by calculating 0.5×50. The reasoning is the same in either case, but the translation is different because we are using different numerals.

Example 6

John has J marbles. Bill has half of the number of marbles that John has. How many marbles does Bill have?

Section 6.2   Translating Words into Algebra

Bill has $\frac{1}{2}J$ marbles. You know the number Bill has is equal to half the number John has. So you find the answer by taking half of John's number, or half of J.

$\frac{1}{2}J$, which means $\frac{1}{2} \times J$, may also be written as $\frac{J}{2}$. One way to see this is to write $\frac{1}{2} \times J$ as $\frac{1}{2} \times \frac{J}{1} = \frac{1 \times J}{2 \times 1} = \frac{J}{2}$.

Sometimes we are required to translate more complicated sentences that require two or more operations.

Example 7

John has 12 marbles. Bill has 5 more than 3 times as many marbles as John. How many marbles does Bill have?

Bill has 41 marbles.
If Bill had 3 times as many marbles as John he would have 3×12 or 36 marbles. But you are told that he has 5 more than this number — that is, 5 more than 36, and 5 more than 36 is 41. Step by step, you have

$12 =$ number of marbles John has
$36 = (3 \times 12) = 3$ times as many marbles as John
$36 + 5 = 5$ more than 3 times as many (36) as John $=$ number of marbles Bill has

Since $36 + 5 = 41$, 41 is the answer to this example.

Example 8

John has J marbles. Bill has 5 more than 3 times as many marbles as John. How many marbles does Bill have?

Bill has $3J + 5$ marbles. That is, since John has J marbles, $3J$ names 3 times this number. To get 5 more than this amount, you must add 5 to it, which gives you $3J + 5$.
As a check, let $J = 50$ and see whether the value of $3J + 5$ agrees with your answer to the example.

$$3(50) + 5 = 150 + 5 = 155$$

Once we know how to translate English expressions into algebra expressions, we must also know how to interpret verbal relations.

We use J as the variable to denote the number of marbles that John has because J suggests John. But we may use any symbol we wish to name the number of marbles that John has.

Notice the importance of grouping symbols. $3J + 5$ means that first we multiply J by 3 and then we add 5 to the result.
If we want first to add 5 and then multiply this sum by 3, we would write $3(J + 5)$. $3(J + 5)$ means 3 times 5 more than the number John has. In this case, if John had 50 marbles, 5 more would be 55 and 3 times this would be 165, not 155.

Example 9

John and Bill together have 75 marbles. Bill has 5 more marbles than John. How many marbles does each have?

Bill has 40 marbles and John has 35. Again, let J stand for the number of marbles John has. Then, $J + 5$ stands for the number of marbles Bill has.
The key point is to understand that the verbal relationship is

$$\begin{pmatrix} \text{number of marbles} \\ \text{John has} \end{pmatrix} + \begin{pmatrix} \text{number of marbles} \\ \text{Bill has} \end{pmatrix} = \begin{pmatrix} \text{total number} \\ \text{of marbles} \end{pmatrix}$$

We find the total number of marbles by adding the number each of the boys has.

Then you translate the verbal equation into an algebraic equation by replacing the number of marbles John has by J, the number of marbles Bill has by $J + 5$, and the total number of marbles by 75. This leads to the equations

$$J + (J + 5) = 75$$
$$(J + J) + 5 = 75$$
$$2J + 5 = 75$$

and this gives us

$$2J + 5 = 75$$
$$\underline{-5 \quad -5}$$
$$2J \quad = 70 \quad \text{or} \quad J = \frac{70}{2} = 35$$

Then $J + 5 = 35 + 5 = 40 =$ number Bill has.

As usual, there is more than one way to translate a statement correctly. For example, we can solve Example 9 another way. If J is the number of marbles John has, and John and Bill together have 75 marbles, then Bill must have the difference between 75 and the number John has. That is, Bill's number of marbles may be written as $75 - J$.

But Bill has 5 more marbles than John. This says that if we add 5 to John's amount, it will equal Bill's amount. That is,

(John's amount) + 5 = Bill's amount

If we replace John's amount by J and Bill's amount by $75 - J$, we get

$$J + 5 = 75 - J$$

That is, $J + B = 75$ and if we subtract J from both sides, we have $B = 75 - J$.

This equation can also be solved by the methods discussed in the last lesson.

$$J + 5 = 75 - J$$
$$\underline{+ J \qquad +J}$$
$$2J + 5 = 75$$
$$\underline{-5 \quad -5}$$
$$2J \quad = 70 \quad \text{hence} \quad J = \frac{70}{2} = 35$$

We are showing that there is no one right way to solve a problem. We translate the problem from English into an algebraic equation, which can often be done in more than one way. Then we solve the resulting equation.

We may look at Example 9 as a special case of a more general result. If we let T stand for the total number of marbles John and Bill have, then if Bill has 5 more marbles than John and John has J marbles, we have the *formula*

$$T = J + (J + 5)$$
$$T = (J + J) + 5$$
$$T = 2J + 5 \qquad [1]$$

Then Example 9 is the special case of finding the value of J in [1] for which $T = 75$.

In Section 6.1, we mentioned that we are often given the required formula. Now we see that even in the case where the formula isn't given, it is usually our job to deduce what the formula should be. In summary, instead of being given the wording of Example 9, we could be given formula [1] in order to find the value of J for a given value of T (in this case, 75).

Sometimes a given verbal relationship may be too hard for us to translate into the proper algebraic equation. When this happens, we can always take a guess and see what happens.

For example, suppose we couldn't get the correct equation in Example 9. We could take a guess as to how many marbles John had. If we guessed 50, then Bill would have 55 (because he has 5 more than John). But 50 + 55 is 110, which is more than the given total of marbles (75). This tells us that our guess that John had 50 marbles was too large.

Next we may guess that John has 30 marbles. Then, because we know Bill has 5 more than John, Bill must have 35 marbles. But 30 + 35 is 65, which is less than the correct number (75). This means that our guess that John has 30 marbles was on the low side.

If 50 is too high and 30 too low, we conclude that John has between 30 and 50 marbles. We continue guessing in this way until we get the exact answer or a sufficiently close estimate.

Section 6.2 Translating Words into Algebra

Let's review some of these ideas with a few more problems.

Example 10

John's age is J years. Express Bill's age if Bill is 4 years more than twice John's age.

Bill's age is $2J + 4$. Twice John's age means 2 times John's age, and this is written as $2J$. 4 more than this number $(2J)$ is written as $2J + 4$.

Example 11

Using the information in Example 10, let S denote the sum of the ages of John and Bill. Express S in terms of J.

In this case, $S = 3J + 4$. That is, to find the sum, you add the two ages. This gives you $J + (2J + 4) = (J + 2J) + 4 = 3J + 4$.

Example 12

Using the information in Example 10, find John's age if the sum of the two ages is 70.

John is 22.
From Example 11, you already know that the sum of their ages (S) in terms of John's age (J) is given by

$$S = 3J + 4$$

Replace S by 70 (the given sum) to get

$$70 = 3J + 4$$
$$\underline{-4 \quad\quad -4}$$
$$66 = 3J \quad\quad \text{so } J = \frac{66}{3} = 22$$

CHECK
If John is 22, Then Bill is 48. That is, 4 more than twice 22 is $4 + 2(22) = 4 + 44 = 48$. In this case, the sum of their ages is $22 + 48$, which is 70.

Example 13

By guessing, show that John's age (following Example 12) is between 20 and 25.

If John is 20, then twice his age is 40 and 4 more than that is 44, which is Bill's age. But in this case, the sum of their ages is $20 + 44$ or 64, which is less than 70.
But if $J = 25$, then 4 more than twice J is $4 + 2(25) = 4 + 50 = 54$, which is Bill's age. In this case, the sum of their ages is $25 + 54$ or 79, which is more than 70. So J must be between 20 and 25.

That is, $J = 20$ is too small.

That is, $J = 25$ is too big.

PRACTICE DRILL

1. Let M denote Mary's age and B, Bill's age. Express B in terms of M if
    (a) Bill's age is three times Mary's age
    (b) Bill's age is 3 years more than Mary's age
    (c) Bill's age is 3 years less than 4 times Mary's age
    (d) The sum of their ages is 40
2. The sum of Bill's age and Mary's age is 40 years. Find the age of each if Bill is three times as old as Mary.
3. The sum of Bill's age and Mary's age is 40 years. Bill's age is 4 years more than twice Mary's age. How old is each?

Answers
1. (a) $B = 3M$ (b) $B = M + 3$
   (c) $B = 4M - 3$ (d) $B = 40 - M$
   (don't write $B + M = 40$, because this doesn't express B in terms of M)
2. Bill is 30 years old, and Mary is 10 years old. 3. Mary is 12 years old, and Bill is 28 years old.

CHECK THE MAIN IDEAS

1. Just as with other foreign languages, you often have to translate from English into _____.
2. Suppose you are given the phrase "Bill has twice as many marbles as John." You may let B denote the number of marbles Bill has, and let J denote _____.
3. $B = 2J$ stands for the fact that Bill has _____ as many marbles as John.
4. You can also indicate that Bill has twice as many marbles as John by writing $J =$ _____.
5. If you want to say that Bill has 2 more marbles than John, you would not write $B = 2J$ but rather $B =$ _____.

Answers
1. algebra or mathematics
2. the number of marbles John has
3. twice
4. $\frac{1}{2}B$ (that is, invert $B = 2J$ by dividing both sides by 2)
5. $2 + J$

Section 6.3
Mixture Problems

There are many situations in which we form a mixture of two or more items. A company may add water to alcohol to sell a mixture of water and alcohol called antifreeze. Perhaps a grocer is mixing two or more brands of coffee to make a special blend. A bank customer may have two or more sums of money invested at different rates of interest, in which case the "mixture" determines the amount of interest she earns. A theater may sell two or more differently priced tickets for a performance, in which case the "mixture" determines how much money the show makes. Someone may have a mixture of coins and want to find the total value of the coins.

OBJECTIVE
To be able to formulate and solve linear equations that occur in problems where two or more substances are combined to form a mixture.

> The basic thought behind solving mixture problems is that the total value of the mixture is the sum of the values of the parts that make up the mixture.

Example 1

Tickets to a certain event cost 75¢ for children. What is the cost of C children's tickets?

The cost of C children's tickets is $75C$ cents or $0.75C$ dollars.
  You find the cost of the tickets by multiplying the price per ticket by the number of tickets. So you have

$$75 \frac{\text{cents}}{\text{ticket}} \times C \text{ tickets} = 75C \text{ cents}$$

or

$$\$0.75 \frac{\text{dollars}}{\text{ticket}} \times C \text{ tickets} = 0.75C \text{ dollars}$$

Example 2

Referring to Example 1, three times as many adult tickets were sold as children's tickets. If an adult ticket costs $2, how much money was made on adult tickets?

The value of the sales for adult tickets was $6C$ dollars (or $600C$ cents).
  In Example 1, you saw that C children's tickets were sold. If three times as many adult tickets were sold, $3C$ adult tickets must be sold.

Section 6.3   Mixture Problems

Each adult ticket costs $2 dollars. The cost of any number of adult tickets, in dollars, is found by multiplying the number of tickets by 2. Hence $3C$ tickets cost $2(3C)$ dollars. But

$$2(3C) = [2(3)]C = 6C$$

A dollar is worth 100 cents, so $6C$ dollars is worth $100 \times 6C$ or $600C$ cents.

> Notice how we use the structural properties discussed in Lesson 4. They help us to write expressions in the simplest form.

Example 3

If T denotes the value of the total number of tickets sold in Examples 1 and 2, how is T related to C?

You have that T (in cents) $= 75C + 600C = 675C$. That is, the total value is equal to the sum of the value of the children's tickets (which you saw in Example 1 was $75C$ cents) plus the total value of the adult tickets (which you saw in Example 2 was $600C$ cents). That is,

$$T = 75C + 600C$$
or
$$T = 675C \qquad [1]$$

Example 4

Tickets to a certain event cost $0.75 for children and $2.00 for adults. The total ticket sales are $2,430. If there were 3 times as many adult tickets sold as there were children's tickets, how many children's tickets were sold?

360 children's tickets were sold.
Formula [1], from the solution to Example 3, gives the relationship between T in cents and the number of children's tickets (C) sold. T is given as $2,430, so you rewrite it as 243,000 (in cents). Then [1] becomes

$$243{,}000 = 675C$$
$$C = \frac{243{,}000}{675} = 360$$

If you want to replace T by 2,430, you cannot use formula [1] because that formula requires that T be in cents. Rather, you would have to use the formula

$$T = 0.75C + 6.00C \qquad \text{or} \qquad T = 6.75C$$

In that way, you are correct in expressing T in dollars. The point is that you must make sure that your formulas are expressed in the proper units.

If we had trouble, we could have tried guessing. For example, if 300 children's tickets had been sold, then 900 adult tickets were sold (3 times the number of children's tickets). At $0.75 each, 300 children's tickets are worth $225. At $2 each, 900 adult tickets are worth $1,800. The total value would be $225 + $1,800 or $2,025, which is less than the $2,430 we are given as the total value. This means that the correct number of children's tickets is more than 300. We can also show that if we thought 400 children's tickets were sold, the total sales would be greater than $2,430, which tells us that fewer than 400 children's tickets were sold.

The ideas we have used so far are the only ones we really have to know in order to do any mixture problem. Let's practice some more problems.

> You would not want to express one value in cents and the other in dollars. So you would write either $T = 675C$ or $T = 6.75C$. You would not write $T = 75C + 6C$ because then the first price is in cents and the second is in dollars.

> We still have to know how to do arithmetic, but we must also know what arithmetic must be done.

> 400 tickets at $0.75 are $300 and 1,200 at $2 are $2400; $2400 + $300 is $2700.

Example 5

A piggy bank contains 23 coins, consisting of dimes and quarters. The total value of the coins is $3.35. How many coins of each kind are there?

There are 16 dimes and 7 quarters.

The key is that the total value ($3.35) is the sum of the values of the dimes and the quarters. If you let d stand for the number of dimes, the value of d dimes is $10d$ cents or $0.1d$ dollars (that is, you multiply the number of dimes by 10 to find the value of the dimes in cents).

Similarly, the value of the quarters is 25 cents times the number of quarters. There are 23 coins altogether, of which d are dimes. The rest are quarters. Therefore, you find the number of quarters by subtracting d from 23 to get $23 - d$. The value of the quarters (in cents) is then given by $25(23 - d)$.

In cents, the total value of the dimes is $10d$, the total value of the quarters is $25(23 - d)$, and the total value of all the coins is 335 cents. So the equation you want is

$$\underset{\substack{\text{value} \\ \text{of} \\ \text{dimes}}}{10d} + \underset{\substack{\text{value} \\ \text{of} \\ \text{quarters}}}{25(23 - d)} = \underset{\substack{\text{total} \\ \text{value} \\ \text{of} \\ \text{coins}}}{335} \qquad [2]$$

CHECK

16 dimes at 10¢ each are worth $1.60, and 7 quarters are worth $1.75. The total value of the coins is

$1.60 + $1.75 or $3.35

You could start with

$$d + q = 23$$

where q is the number of quarters. Then subtract d from both sides.

Notice the use of the grouping symbols. $23 - d$ is *the* number of quarters; and it is that number which we must multiply by 25. So we place $23 - d$ in parentheses and then multiply by 25.

Equation [2] is linear. To solve it you have

$$10d + 25(23) - 25d = 335$$
$$-15d + 575 = 335$$
$$-15d = 335 - 575 = -240$$
$$d = \frac{-240}{-15} = 16$$

There is an advantage in letting d stand for dimes. As soon as we get $d = 16$, we know that it means 16 dimes.

Instead of saying that the coins were worth $3.35 or 335 cents, we could have said that the coins were worth C cents. In this case, instead of [2] we would get

$$C = 10d + 25(23 - d) = 10d + 575 - 25d$$
$$\text{or } C = 575 - 15d \qquad [3]$$

[3] tells us how the value is related to the number of dimes. As a check, we see that if none of the 23 coins is a dime ($d = 0$), [3] becomes $C = 575$, and the value of 23 quarters is 575 cents. At the other extreme, if all 23 coins are dimes, (so that $d = 23$), [3] becomes $C = 575 - 15(23)$, or, $C = 575 - 345 = 230$, which is the value of 23 dimes.

Again, we can use trial and error to guess the right answer. But even when we use trial and error, there are quicker ways than others. For example, if all the coins were dimes, the value would be $2.30. Every time we change a dime to a quarter, we make up $0.15 (that is, 25¢ − 10¢ = 15¢). To get from $2.30 to $3.35, we need $1.05, which is $0.15 seven times. That is, we have to exchange 7 dimes for quarters, leaving us with 16 dimes and 7 quarters.

In other words, we can use common sense to estimate the answer. If each coin were a dime, the total value of the 23 coins would be $2.30; and if each coin were a quarter, the total value of the 23 coins would be $5.75. So regardless of what the right answer is, we know that it is between $2.30 and $5.75.

This may not be the best way for you to do the problem. But if you know enough arithmetic and understand the problem, you can get the answer in a lot of ways without having to know algebra.

However, if you know algebra, it is often less complicated to solve the problem using algebra. In other words, algebra is meant to supplement your common sense, not to replace it.

Example 6

You have 24 pounds of candy worth 75¢ per pound. You have another kind of candy that is worth $1.30 a pound. How many pounds of the $1.30 candy should you mix with the other 24 pounds to get a mixture that is worth $1 per pound?

You need 20 pounds of the $1.30 candy.

Let p be the number of pounds of $1.30 candy needed to bring the value of the mixture to $1 per pound. To keep track of things, you can make up the following chart:

Section 6.3 Mixture Problems

Item	Unit Value	Number of Units	Value
Candy	75 cents	24 pounds	$75(24) = 1800$ cents
Candy	130 cents	p pounds	$130p$ cents
Mixture	100 cents	$p + 24$	$100(p + 24)$ cents

From one point of view, the total value is $1800 + 130p$. From another point of view, it is $100(p + 24)$.

So the equation is

$$1800 + 130p = 100(p + 24)$$
$$1800 + 130p = 100p + 2400$$

That is, we add the values of each kind of candy to find the total value.

Therefore

$$\begin{array}{rcl} 1800 + 130p &=& 100p + 2400 \\ -100p & & -100p \\ \hline 1800 + 30p &=& 2400 \\ -1800 & & -1800 \\ \hline 30p &=& 600 \end{array} \quad \text{or} \quad p = \frac{600}{30} = 20$$

CHECK
You have 24 pounds worth $18 and 20 pounds worth $26 ($20 \times \$1.30$). In all, you have 44 ($20 + 24$) pounds worth $44 ($18 + $26), or $1 per pound.

Of course, there are many other versions of mixture problems that we have left for the exercise set, but now you have a general idea of how mixture problems lead to linear equations, which you can solve by the methods of the previous lesson.

PRACTICE DRILL

1. If steak costs $2.30 a pound, find the cost of
   (a) 10 pounds of steak (b) 100 pounds of steak
   (c) 53 pounds of steak (d) 8 ounces of steak (16 ounces = 1 pound).

2. (a) At $2.30 a pound, 40 pounds of steak is worth how many dollars?
   (b) At $2.80 a pound, 60 pounds of steak is worth how many dollars?
   (c) A mixture of 40 pounds of ground steak at $2.30 a pound and 60 pounds of ground steak at $2.80 at $2.80 a pound has a total value of how many dollars?
   (d) What is the value of each pound of the mixture?

Answers
1. (a) $23 (b) $230 (c) $121.90
   (d) $1.15   2. (a) $92
   (b) $168 (c) $260 (d) $2.60

CHECK THE MAIN IDEAS

1. In studying mixtures, the basic idea is that if you know the unit price of an item, you find the total price of that item by multiplying the number of items by the _____.
2. The total value of 5 cakes worth $2 each is $_____.
3. The total value of 5 boxes of candy worth $_____ each is $15.
4. The total value of the cakes and candy is $_____.

Answers
1. unit price of that item
2. 10   3. 3   4. 25

Section 6.4
Application to Constant Speed

In most situations involving travel, we are unlikely to move at the same speed all the time. On a long trip, we may go for a long stretch on a highway at 55 miles per hour. In the city, our speed may vary from being stopped in a traffic jam to about 25 miles per hour.

Nevertheless, in many mathematical applications we assume that motion occurs at constant speed. We may talk about a car that travels at 40 miles per hour. In this case, we say the car has *constant* speed. The speed of the car remains the same throughout the problem.

> The main idea in problems of constant speed is to find the distance the object travels by multiplying the (*constant*) speed by the time.
>
> In more mathematical form, if a car (or any other object) travels at a rate of r miles per hour for t hours, then it travels a distance of d miles according to the formula:
>
> $$d = rt \qquad [1]$$

OBJECTIVE

To be able to formulate and solve linear equations that occur in the study of problems involving constant speed.

Sometimes we use *average* speed. We obtain the average speed by dividing the distance by the time it took. For example, if a car goes 200 miles in 5 hours, we say that its average speed is 40 miles per hour ($200 \div 5$). This means that if the car had moved at the constant speed of 40 miles per hour, it would have covered the 200 miles in 5 hours.

We can check this by looking at "labels." That is, r is in miles per hour, and since "per" means "divided by" we have

$$rt = \frac{\text{miles}}{\cancel{\text{hours}}} \times \cancel{\text{hours}} = \text{miles}$$

That is the label of rt is distance, as d should be.

Example 1

A car goes at a rate of 40 miles per hour for 6 hours. How many miles does it travel during this time?

It travels 240 miles.

You multiply 40 by 6. That is, each hour you go 40 miles, so in 6 hours you go 240 miles. Using [1], replace r by 40 and t by 6 to get

$$d = 40(6) \qquad \text{or} \qquad d = 240$$

Example 2

A car travels at a speed of 40 miles per hour for t hours. How many miles does it travel in this time?

The car travels a distance of $40t$ miles.

You still multiply r by t, but in this case t is given in literal form rather than as a number. So all you can do is symbolically multiply 40 by t, and this is written as $40t$.

The difference between Examples 1 and 2 is that in Example 2 the distance is an expression involving t. The answer depends on the value of t. To evaluate the distance, you must pick a value for t.

From another point of view, the answer to Example 1 is a special case of the answer to Example 2. That is, you get the answer to Example 1 by replacing t by 6 in the expression $40t$.

Sometimes we must be careful to see that we are using the correct units.

Example 3

A car travels at a speed of 40 miles per hour for 45 minutes. How far does the car travel during this time?

The car travels 30 miles during this time.

You do *not* multiply 40 by 45. In the formula $d = rt$, if r is miles per hour

Section 6.4   Application to Constant Speed

then t is also in hours. 45 minutes is $\frac{45}{60}$ or $\frac{3}{4}$ of an hour, so you replace t by $\frac{3}{4}$ to get
$$d = 40 \times \frac{3}{4} = 30$$

If you wanted the time to be in minutes, you would have to replace t by $\frac{t}{60}$. That is, in this case the formula would be
$$d = r\left(\frac{t}{60}\right)$$
where t is the time in minutes (not hours).

HINT
Because 45 minutes is less than an hour, we should expect that at 40 miles per hour, the car must travel less than 40 miles in 45 minutes. This should warn us not to multiply 40 by 45 to get the distance.

$\frac{\text{miles}}{\text{hours}} \times$ minutes is not miles.

Sometimes our motion problems involve two vehicles.

Example 4

Two cars start from the same place but go in opposite directions. The first car travels at a constant speed of 40 miles per hour, and the second car travels at a constant rate of 55 miles per hour. How far apart are the two cars after 2 hours?

The two cars are 190 miles apart after 2 hours.

Because the cars are moving in opposite directions, you add their speeds to see how far apart they are after one hour. $40 + 55 = 95$, so you see that the cars move apart at a rate of 95 miles per hour. Hence in two hours, they are twice 95 miles apart, or 190 miles apart.

You did not use algebraic equations here. But even if you didn't say it, you were using a formula. In words, the formula says

In other words, after one hour the first car is 40 miles to the right of the starting point while the second car is 55 miles to the left. That is,

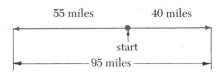

Distance travelled by the first car in 2 hours	+	Distance travelled by the second car in 2 hours	=	Distance between the two cars after 2 hours
↓		↓		↓
40(2)	+	55(2)	=	
80	+	110	=	190

If the cars travel for t hours instead of 2 hours, the equation for the distance between the two cars after t hours is
$$40(t) + 55(t) \quad \text{or} \quad 95t$$

In other words, under the conditions of Example 4, after t hours the distance (d) between the two cars is given by
$$d = 95t \qquad [2]$$

Equation [2] is general enough to let us solve many variations of Example 4.

Example 5

Two cars leave the same point moving in opposite directions. The first car goes at a constant speed of 40 miles per hour, and the second car goes at the constant speed of 55 miles per hour. After how long will the two cars be 380 miles apart?

The two cars will be 380 miles apart in 4 hours. In Example 4, you already saw that
$$d = 95t \qquad [2]$$

You want the value of t in [2] for which $d = 380$. This leads to the linear equation
$$380 = 95t \qquad [3]$$
which you solve by dividing both sides of [3] by 95. You find that $t = 4$.

If you had not done Example 4, you could have solved this example in a way similar to what you did in Example 4.

FORMULA

Distance the first car goes in t hours	+	Distance the second car goes in t hours	=	Distance between the two car after t hours
$40t$	+	$55t$	=	380

SKETCH

```
       55t        40t
   <────────┬────────>
          start
   <──────380──────>
```

Hence from the diagram
$$55t + 40t = 380$$
$$\text{or } 95t = 380$$

Sometimes we get a different kind of motion problem. In the last examples, we thought about two cars leaving the same place at the same time but moving in opposite directions. Sometimes, we are interested in a case where the two cars leave the same place at different times and move in the same direction. Such problems are often called pursuit problems because the car that started later is trying to catch up to the car that started earlier.

Example 6

At 7 A.M. a car leaves a particular place and travels at a constant speed of 30 miles per hour. At 9 A.M. the same day, a second car leaves from the same place and travels with a constant speed of 45 miles per hour in the same direction as the first car. How far apart are the two cars at 10 A.M.?

At 10 A.M. the two cars are 45 miles apart.

At 10 A.M. the first car has been going for 3 hours. So at 30 mph it has gone 90 miles. But at 10 A.M. the second car has been travelling for only 1 hour, so it has moved 45 miles. That leaves the first car still 45 miles ahead of the second car.

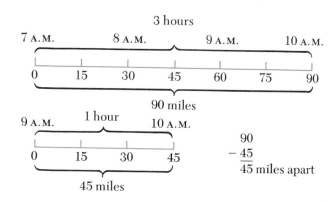

$$\begin{array}{r} 90 \\ -45 \\ \hline 45 \end{array} \text{ miles apart}$$

Remember, the second car travels for only 1 hour because it doesn't start until 9 A.M.

In terms of a formula

Distance travelled by first car	−	Distance travelled by second car	=	Distance between the two cars
rate × time	−	rate × time		
30 × 3	−	45 × 1		
90	−	45	=	45 (miles)

Section 6.4 Application to Constant Speed

Notice that when the second car started, the first was 60 miles ahead of it because it had been moving for 2 hours at 30 mph. One hour later, the second car was only 45 miles behind the first. The next natural question might be to find out where the second car overtook the first.

Perhaps the problem would be more intriguing if we didn't already know how fast the second car was going. So let's change the emphasis a little bit and try another version of this kind of problem.

Even here we may not really need algebra if we understand the problem. Since 45 mph is 15 mph more than 30 mph, every hour the second car goes, it gains 15 miles on the first. But the first car has a 60 mile "head start." Therefore, the second car overtakes the first after 4 hours.

Example 7

At 7 A.M. a car leaves a certain place at a constant speed of 30 mph. At 9 A.M. a second car leaves the same place in the same direction with a constant speed. The second car catches up with the first car at noon. What is the speed of the second car?

The second car has a speed of 50 mph.
Suppose you let s stand for the speed of the second car. At noon the second car has been travelling for 3 hours (it started at 9 A.M.), and the first car has been travelling for 5 hours (it started at 7 A.M.).
The first car has travelled a total of 150 miles (5×30) and the second car has gone $3s$ miles (you are trying to find the value of s). The distance between the two cars is given by

$$\begin{array}{ccc} \text{Distance travelled} & - & \text{Distance travelled} \\ \text{by the first} & & \text{by the second} \\ 150 & - & 3s \end{array}$$

At noon the distance between the two cars is 0 (because that's when the second caught up to the first). Therefore, the equation is

$$150 - 3s = 0$$

from which it quickly follows that $s = 50$. Because s is in mph, the answer is 50 mph.

We shall see several problems of this type in the exercise set. In this section, our goal is to get used to seeing another situation that gives rise to the study of linear equations.

CHECK
At 50 mph, the second car gains 20 miles per hour over the first car (that is, 50 is 20 more than 30). To make up the 60 miles head start, the second car must travel for 3 hours. It starts at 9 A.M., so 3 hours later is noon.

NOTE
If you have trouble with the method and can't think of another algebraic way, use trial and error or arithmetic. You know that at noon the first car has gone 150 miles and the second car has travelled for 3 hours. This means that the second car has to go the same 150 miles, but in only 3 hours. This tells you that at constant speed the second car *must* move at 50 mph. So you can use either algebra or arithmetic, depending on which makes you feel more comfortable. This is just another example which tries to show that algebra is used to supplement common sense, not to replace it.

PRACTICE DRILL

1. Find the distance a car travels if it goes
    (a) 40 mph for 3 hours   (b) 40 mph for 3 minutes
    (c) 40 mph for $3\frac{1}{2}$ hours

2. A car starts at 7 A.M. and travels at 40 mph until 1 P.M. (the same day). How far does the car travel during this time?

3. One car is 100 miles ahead of a second car and travels at 40 mph. If the second car travels in the same direction as the first car and moves at a speed of 60 mph, how long will it take the second car to overtake the first?

Answers
1. (a) 120 miles (b) 2 miles
(c) 140 miles   2. (a) 240 miles
(b) 5 hours

CHECK THE MAIN IDEAS

1. This section studies motion in which objects move with _____ speed.
2. When the speed is constant, you can find the distance an object travels by _____ the speed of the object and the time it travels at this speed.
3. Using symbols, if d is miles, r is the speed, and t is the time the object travels, the formula is given by $d = $ _____.
4. If a car travels at a speed of 40 mph for 5 hours, it travels a distance of _____ miles.
5. If the car, moving at constant speed, had covered the 200 miles in 4 hours instead of 5 hours, its speed would have been _____ mph.
6. Even if the car didn't travel at constant speed, if it still covered 200 miles in 4 hours you could say that its _____ speed was 50 mph.

Answers
1. constant 2. multiplying
3. rt 4. 200 5. 50 mph
6. average

Section 6.5
Applications to Percents

If there is any one part of arithmetic that seems to come up over and over again in our lives, it is probably the study of percents. Wherever we turn, it seems that we are involved with percentages. Income tax, Social Security tax, sales taxes, meal taxes, investments — everything seems to be based on percentages.

OBJECTIVE
To be able to formulate and solve linear equations that occur in various kinds of problems involving percents.

Example 1

Your weekly income, before any deductions, is $150. Your take-home pay, however, is only $114 a week. What percent of your gross pay is being deducted?

24% of your gross pay is being deducted.

If you are taking home only $114 of the $150 you are being paid, then $36 of your gross pay is being deducted. That is, 36 of your 150 dollars is being deducted. This is $\frac{36}{150}$ of your pay. In lowest terms, $\frac{36}{150}$ is the same as $\frac{6}{25}$. To convert this to a percent, you multiply by 100, which gives

$$\frac{6}{25} = \left(\frac{6}{25} \times 100\right)\% = 24\%$$

We may not have realized it, but we used a formula to solve Example 1. The formula is

$$\frac{\text{amount deducted}}{\text{gross pay}} \times 100 = \text{percent of gross pay deducted}$$

We can write this as an algebraic formula by letting d = amount of money deducted, g = gross pay, and p = percentage of gross pay deducted. Then we have

$$\frac{d}{g}(100) = p \qquad [1]$$

Let's see how we can use [1] in a problem.

VOCABULARY
gross pay: The money you earn before any deductions are made.
net pay: The amount of money you actually receive in your paycheck, after deductions.

Remember the 100% refers to the whole amount. So $\frac{6}{25}$ means $\frac{6}{25}$ of the whole amount, or $\frac{6}{25}$ of 100%; and since $\frac{6}{25}$ of 100 is 24, the answer is 24%.

CAUTION
In [1], g appears as a denominator. From our study of fractions, we know that the denominator can't be 0. We must remember that g is not 0. However, since g denotes gross pay, it is reasonable to assume g is not 0.

Example 2

Suppose you know that $36 was deducted from your gross pay and that this amount is 24% of your gross pay. How much is your gross pay?

Section 6.5 Applications to Percents

Your gross pay is $150.

To use formula [1], you replace d by 36 and p by 24 to obtain

$$\frac{36}{g}(100) = 24 \qquad [2]$$

$$\text{or } \frac{3600}{g} = 24 \qquad [3]$$

In its present form, [3] is not linear in g because g appears in the denominator. But you can convert [3] into a linear equation by multiplying both sides of [3] by g. You get

$$3600 = 24g \qquad [4]$$

You can solve the linear equation [4] by dividing both sides by 24. This gives us

$$150 = g$$

which means your gross pay is $150.

Notice that Example 2 is just another version of Example 1. This emphasizes how the same situation can be presented in different ways.

REVIEW
[3] and [4] are called equivalent equations because they have the same solutions, provided that g is not equal to 0.

NOTE

We do not have to use algebra or [1] to solve this problem if we understand percents well enough. For example, the fact that 24% of the gross pay is $36 tells us that 1% of the gross pay is $1.50 (that is, $\frac{1}{24}$ of $36). The entire gross pay is 100% of itself, and 1% of it is $1.50, so we see that the entire gross pay is $150 ($1.50 × 100).

Sometimes we need to find the sales tax on an item we have purchased.

Example 3

What is the tax on a radio if the radio costs $75 and the tax is 6% of the selling price?

The tax is $4.50. The selling price is $75 and the tax is 6% of the selling price, so the tax is

$$6\% \times \$75 = .06 \times \$75 = \$4.50$$

or

$$\frac{6}{100} \times \$75 = \$4.50$$

Remember, "of" means "times" and "percent" means "divided by 100."

In problems such as Example 3, we usually want to find the total price rather than the tax. In that case, we use the fact that the total cost is the selling price plus the tax.

Example 4

What is the price of the radio in Example 3, including the tax?

Including tax, the radio costs $79.50.

You already know that the selling price before the tax is $75. You saw in Example 3 that the tax is an additional $4.50. So altogether, the total cost is $75 + $4.50, or $79.50.

Algebra can be helpful if the problem is stated in such a way that the tax is already included.

Example 5

A radio costs $110. The price includes a 10% tax on the selling price. What is the price of the radio before the tax?

Note! Don't subtract 10% of $110. The $110 already includes the tax.

The radio costs $100 before the tax.
 If you let c be the cost of the radio without the tax, then the tax, which is 10% of c, is $0.1c$. The total price is still cost plus tax, which is $c + 0.1c$, or $1.1c$. The total cost is also $110, so you have

$$1.1c = 110 \quad \text{or} \quad c = \frac{110}{1.1} = 100$$

which is the cost in (dollars) without tax.

To summarize:
$$c + (10\% \text{ of } c)$$
$$= c + 0.1c$$
$$= 1c + 0.1c$$
$$= (1 + 0.1)c$$
$$= 1.1c$$

To check:
$$10\% \text{ of } \$100 = \$10$$
$$\$100 + \$10 = \$110$$

Example 6

The price of a ring is $70.20. The price already includes an 8% tax. What is the price of the ring, without the tax?

Without the tax, the ring costs $65.
 If you let r denote the price of the ring without the tax, you know that the tax is 8% of r. In decimal form, this is the same as $0.08r$.
 The total cost is the cost without tax (r) plus the tax ($0.08r$). You also know that the total cost is $70.20. Putting this together tells you that

$$r + 0.08r = \$70.20$$
$$1r + 0.08r = \$70.20$$
$$(1 + 0.08)r = \$70.20$$
$$1.08r = \$70.20$$
$$r = \frac{\$70.20}{1.08} = \$65$$

CHECK
$$8\% \text{ of } \$65 = .08 \times 65 = \$5.20$$
$$\$65 + \$5.20 = \$70.20$$

NOTE

Again the problem can be done by arithmetic. The ring costs 100% of itself, but the 8% sales tax means that you are paying for 108% of the ring. If 108% of the ring is worth $70.20, 1% is worth $70.20 ÷ 108, or $0.65. Therefore, 100% of the ring is worth $65.

Did you notice that a hand calculator may be helpful in dividing $70.20 by 1.08? But the hand calculator can't help if you don't know that you're supposed to divide $70.20 by 1.08.

The same concept is involved when we compute the interest our money earns in the bank. For example, if the bank pays us 5% interest annually, we earn 5% of the value of our account. If we have $1,000 in the bank, we shall earn 5% of $1,000, or $50, in interest. This is the same idea as saying that if there is a 5% tax, we pay a $50 tax on a $1,000 item. The formula to follow is that the interest we earn annually is the interest rate times the money we have in the bank.
 Algebra is useful when we know the interest rate and the amount of the interest, and we want to find how much we have in the bank.

Example 7

A bank pays an annual interest of 6%. If your account earns $180 in interest, how much money is in the account?

You have $3,000 in the account.
 If you let A denote the money in the account, then the interest is 6% of A, or $0.06A$. But you are told that the interest is $180. So

$$0.06A = \$180$$

[5]

Section 6.5 Applications to Percents

[5] is a linear equation that you solve by dividing both sides by 0.06 to obtain

$$A = \frac{\$180}{0.06} = \$3{,}000$$

As the problems get more and more complicated, it becomes more and more difficult to keep track of things without using algebra to help. Let's look at a more complicated problem.

If you want to use arithmetic, you know that 6% of the account is worth $180. So 1% is worth one-sixth of $180, or $30. Hence 100% of the account is $100 \times \$30$ or $3,000.

Example 8

A couple invests a total of $10,000 in two types of accounts. One account pays an annual interest rate of 4%, and the other, 5%. The two investments earn a total of $440 in a year. How much money did they invest at each rate?

They invested $4,000 at 5% and the rest at 4%.

Let x be the amount invested at 5%. Then the rest (which is $10,000 - x$) is invested at 4%. They earn 5% of $\$x$ and 4% of $\$(10{,}000 - x)$. This is a total of $[0.05x + 0.04(10{,}000 - x)]$ dollars. But the total interest is given as $40. Therefore the equation is

$$0.05x + 0.04(10{,}000 - x) = 440 \qquad [6]$$

[6] is a linear equation which you can solve by the methods of the previous lesson.

$$0.05x + 0.04(10{,}000) - 0.04x = 440$$
$$0.01x + 400 = 440$$
$$0.01x = 440 - 400 \quad \text{or} \quad 0.01x = 40$$

Hence,

$$x = \frac{40}{.01} = 4{,}000 \text{ (dollars at 5\%)}$$

and

$$\$10{,}000 - x = 10{,}000 - 4{,}000$$
$$= 6{,}000 \text{ (dollars at 4\%)}$$

The mathematics doesn't tell us why they didn't invest more at 5%. What's important is that the math gives us the numerical answers, but we still have to make our own decisions.

Once again, there are several nonalgebraic ways of solving this problem. For example, we may notice that if the entire amount were invested at 4%, the yield would have been $400, which is less than $440. On the other hand, if the entire amount were invested at 5%, the yield would have been $500, which is more than $440.

Every time we take $1,000 and transfer it from the 4% to the 5% account, we make an extra $10 (that is, 4% of $1,000 is $40 while 5% of $1,000 is $50). Therefore to get from $400 to $440 we must transfer $4,000 from the 4% account to the 5% account. This is the same as saying that we must invest $4,000 at 5% and the rest at 4%.

More important, this approach tells us the best and the worst that can happen. We must make at least $400, but we can't make more than $500.

There are other ways we could have tried to find the answer, but it is obviously worthwhile to know the algebra. The formula and the techniques for solving the resulting equations stay the same even if the problems become more complicated.

PRACTICE DRILL

1. (a) What is 15% of 60? (b) What percent of 60 is 15?
   (c) 15% of what number is 60?

2. What is your take-home pay if you earn $200 and the deductions are 30% of what you earn?

Answers
1. (a) 9 (b) 25 (c) 400
2. $140   3. $242
4. $200

3. A clock costs $220. In addition, there is a 10% sales tax. What is the price of the clock, including the tax?
4. If the clock costs $220 including the 10% sales tax, what is the cost of the clock without the tax?

CHECK THE MAIN IDEAS

1. If we make 6% interest on an investment, it means that for each $100 you invest, you earn $_____ interest.
2. If you invest $500, you would earn $6, (a) _____ times, or (b) $_____.
3. If the sales tax is 6% of the price of the object, you pay $0.06 for each $_____ the object costs.
4. If the object costs $50, the sales tax is $_____.
5. If the object costs $212 including the 6% sales tax, then 106% of the cost of the object is $_____.
6. If 106% of the cost is $212, then 1% of the cost is $_____.
7. The price of the object is 1%, 100 times, or $_____.

Answers
1. 6 2. (a) five (5) (b) 30
3. 1 4. 3 5. 212
6. 2 7. 200

EXERCISE SET 6 (Form A)

Section 6.1

1. In the formula $S = \frac{n}{2}(a + L)$, find the value of L if $S = 600$, $n = 40$, and $a = 20$.
2. In the formula $D = A(n - 1)$, find the value of n if $D = 32$ when $A = 5$.
3. In a simple electric circuit (such as a flashlight), the voltage produced by the batteries is related to the current and the resistance in the circuit by *Ohm's Law*, which states

$$V = IR$$

 where:

 V = voltage (volts)
 I = current (amperes)
 R = resistance (ohms)

 If a battery puts out 3 volts and a current of 0.2 amperes flows through the circuit, what is the resistance in the circuit?
4. How much current will flow through a flashlight circuit if the battery puts out 7.5 volts and the resistance is 30 ohms?
5. How much current will flow through a flashlight circuit if the circuit has a five-cell battery putting out 1.5 volts per cell, and the resistance is 30 ohms?

Section 6.2

6. The perimeter of a rectangle is defined as twice the sum of the length and the width of the rectangle. If L denotes the length of the rectangle and W the width (in feet), express L in terms of W if the length is 4 feet more than 3 times the width.
7. Express the perimeter of a rectangle in terms of its width (W) if the length is 4 feet more than 3 times the width.
8. The length of a rectangle is 4 feet more than 3 times the width of the rectangle. What is the width of the rectangle if the perimeter of the rectangle is 240 feet?

Answers: Exercise Set 6
(Form A)
1. 10   2. 7.4   3. 15 ohms
4. 0.25 amps   5. 0.25 amps
6. $3W + 4$ 7. $8W + 8$
8. 29 feet   9. second $= f + 4$
third $= 2f + 20$ 10. 39°
11. $x, x+1, x+2, x+3$
12. $4x + 6$ 13. 24, 25, 26, 27
14. 53, 54, 55, 56   15. there are none   16. any four
17. (a) $5n$ (b) $200 - 10n$
(c) 12 nickels, 8 dimes
18. 7 quarters, 10 dimes
19. mechanic works 10 hrs, assistant works 8 hrs
20. (a) $200 + g$ (b) $12,000 + 60g$
21. 100   22. $31\frac{1}{4}$ lb at 12¢; $18\frac{3}{4}$ lb at 20¢
23. (a) $3s$ (b) $3s + 240$
24. 325 mph   25. (a) $3A$
(b) $2A - 20$ 26. 40 mph
27. 1,200 miles   28. (a) $0.06s$
(b) $0.105s$ 29. $5,000 at 6%
10,000 at 5.25%   30. (a) $0.4s$
(b) $0.26s$ 31. $1,850
32. (a) 3 lbs   (b) 3 lbs
(c) $50 - x$ (d) 20 lbs

Exercise Set 6 (Form A)

9. The sum of the measures of the angles in a triangle is 180 degrees (180°). Let f denote the measure of one of the angles. Express the measures of the other two angles in terms of f if the second angle is 4° more than the first angle and the third angle is 12° more than twice the second angle.
10. Find the value of f in the previous problem.
11. If n denotes any integer, the next consecutive integer is defined as $n + 1$. Write four consecutive integers if x denotes the first of the four.
12. If x is the first of four consecutive integers, express the sum of the four in terms of x.
13. What are the four consecutive integers whose sum is 102?
14. What are the four consecutive integers if the sum of the first three exceeds twice the fourth by 50?
15. Find four consecutive integers for which the sum of the first and the third equals the sum of the second and the fourth.
16. What four consecutive integers have the property that the sum of the first and fourth is the same as the sum of the second and the third?

Section 6.3

17. (a) What is the value in cents of n nickels?
    (b) What is the value in cents of $(20 - n)$ dimes?
    (c) You have twenty coins, all of which are nickels or dimes. How many of each do you have if the total value of the coins is $1.40?

18. In a group of coins consisting only of quarters and dimes, there are three more dimes than quarters. How many of each type of coin are there if the total value of the coins is $2.75?
19. A mechanic charges $6 an hour for labor, and his assistant charges $4 an hour. On a repair job, the total charge for labor by the two men is $92. If the mechanic worked two hours more than his assistant, how many hours did each person work?
20. Suppose g gallons of gasoline worth 50¢ a gallon is mixed with 200 gallons of gas worth 65¢ a gallon. Express in terms of g:

    (a) The total number of gallons of gasoline in the mixture
    (b) The total value of the mixture (in cents) if it is worth 60¢ a gallon?

21. How many gallons of gasoline worth 50¢ a gallon must be mixed with 200 gallons of gas worth 65¢ a gallon to produce a mixture that's worth 60¢ a gallon?
22. A farmer mixes seed worth 12¢ a pound with seed worth 20¢ a pound to produce a mixture of 50 pounds that's worth 15¢ a pound. How many pounds of each type of seed did he use in the mixture?

Section 6.4

23. Two planes leave an airport at the same time and travel in opposite directions. One plane travels 80 mph faster than the other. Let s denote the speed of the slower plane. Express in terms of s:

    (a) The distance in miles that the slower plane travels in 3 hours
    (b) The distance that the faster plane travels in 3 hours

24. Two airplanes leave an airport at the same time and travel in opposite directions. One plane travels 80 mph faster than the other. What is the speed of the slower plane if the two planes are 2,190 miles apart at the end of three hours?
25. Ann lives 230 miles from her father. They decide to meet. Ann leaves for her father's house at 9 A.M. Her father leaves the house at 10 A.M. and drives along the same road toward Ann. Her father drives at an average

speed that is 10 mph less than Ann's average speed. Let A stand for Ann's average speed. Express in terms of A:

(a) The distance (in miles) Ann travels by noon that day
(b) The distance her father travels by noon that day

26. According to the previous problem, what was the average speed of Ann's father if they meet at noon that day?
27. An airplane travels from one airport to another and back in 5 hours. Going one way, the pilot averages 600 mph; going the other way, the pilot averages 400 mph. What is the distance between the two airports?

Section 6.5

28. A man invests some money at an annual interest rate of 6%. He invests twice that amount at 5.25% annual interest. If s denotes the amount of money (in dollars) invested at 6%, express in terms of s:

    (a) The amount of money (in dollars) earned in one year by the 6% investment
    (b) The amount of money earned in one year by the 5.25% investment

29. A man invests some money at 6% annual interest and twice that amount at 5.25% annual interest. How much does he have invested at each rate if in one year he earns $825?

30. In constructing a sewer, the county pays 40% as much as the state, and the city pays 65% as much as the county. If s denotes the amount (in dollars) paid by the state, express in the terms of s:

    (a) The amount spent by the county   (b) The amount spent by the city

31. In the previous problem, what is the cost to the state if the entire cost is $3,071?
32. (a) How many pounds of salt are there in a fifty-pound mixture of salt and water if the mixture is 6% salt?
    (b) How many pounds of salt are left if x pounds of water evaporate?
    (c) How many pounds of mixture are left when x pounds of water evaporate?
    (d) How many pounds of water must evaporate in order for the mixture to become 10% salt?

EXERCISE SET 6 (Form B)

Section 6.1

1. In the formula $S = \frac{n}{2}(a + L)$, find the value of n if $S = 1,800$, $a = 60$, and $L = 30$.
2. In the formula $D = A(n - 1)$, find the value of n if $D = 85$ and $A = 4$.
3. In a simple electric circuit (such as a flashlight), the voltage produced by the batteries is related to the current and the resistance in the circuit by *Ohm's Law*, which states

$$V = IR$$

where:

$$V = \text{voltage (volts)}$$
$$I = \text{current (amperes)}$$
$$R = \text{resistance (ohms)}$$

If a battery puts out 6 volts and a current of 0.15 amperes flows through the circuit, what is the resistance in the circuit?

Exercise Set 6   (Form B)

4. How much current will flow through a flashlight circuit if the battery puts out 10 volts and the resistance is 50 ohms?
5. How much current will flow through a flashlight circuit if the circuit has a four-cell battery putting out 2.5 volts per cell and the resistance is 50 ohms?

Section 6.2

6. The perimeter of a rectangle is defined as twice the sum of the length and the width of the rectangle. If L denotes the length and W the width of the rectangle (in meters), express L in terms of W if the length is 3 meters less than five times the width.
7. Express the perimeter of the rectangle in terms of its width (W) if the length of the rectangle is 3 meters less than five times the width.
8. The length of a rectangle is 3 meters less than five times the width of the rectangle. What is the width of the rectangle if the perimeter of the rectangle is 234 meters?
9. The sum of the measures of the angles in a triangle is 180°. Let f denote the measure of the first angle. Express, in terms of f, the measures of the other two angles if the second is 6° more than the measure of the first and the third is 6° more than three times the measure of the second.
10. Find the value of f in the previous problem.
11. If n denotes any integer, the next consecutive integer is defined as $n + 1$. Write five consecutive integers if x denotes the first of the five.
12. If x denotes the first of five consecutive integers, express the sum of the five consecutive integers in terms of x.
13. What are the five consecutive integers whose sum is 300?
14. What are the five consecutive integers if the sum of the first four exceeds 3 times the fifth by 30?
15. Find five consecutive integers for which the sum of the first four equals the fifth.
16. What five consecutive integers have the property that the sum of the second and fifth equals the sum of the third and fourth?

Section 6.3

17. (a) What is the value in cents of d dimes?
    (b) What is the value in cents of $(30 - d)$ quarters?
    (c) You have thirty coins, all of which are dimes or quarters. How many of each kind do you have if the total value of the coins is $5.10?

18. In a group of coins consisting only of nickels and dimes, there are nine fewer nickels than dimes. How many coins of each type are there if the total value of the coins is $3.60?
19. A painter charges $8 an hour, and his helper charges $5 an hour. Together, they charge $115 for a paint job. If the painter worked for 3 hours more than his helper, how many hours did the helper work?
20. Suppose g gallons of gasoline worth 70¢ a gallon is mixed with 250 gallons of gasoline worth 80¢ a gallon. Express in terms of g:

    (a) The total number of gallons of gasoline in the mixture
    (b) The total value of the mixture (in cents), if it's worth 74¢ a gallon

21. How many gallons of gasoline worth 70¢ a gallon must be mixed with 250 gallons of gasoline worth 80¢ a gallon to produce a mixture that's worth 74¢ a gallon?
22. A farmer mixes seed worth 15¢ per pound with seed worth 20¢ per pound to produce a mixture of 50 pounds of seed worth 18¢ per pound. How many pounds of each type of seed did she use in the mixture?

Section 6.4

23. Two planes leave an airport at the same time and travel in opposite directions. One plane travels 100 mph faster than the other. Let s denote the speed of the slower plane. Express in terms of s:

    (a) The distance in miles that the slower plane travels in 4 hours
    (b) The distance the faster plane travels in 4 hours

24. Two airplanes leave an airport at the same time and travel in opposite directions. One plane travels 100 mph faster than the other. What is the speed of the slower plane if the two planes are 6,000 miles apart at the end of 4 hours?

25. Bill lives 330 miles from his father. They decide to meet. Bill leaves for his father's house at 8 A.M. His father leaves the house at 10 A.M. and drives along the same road toward Bill. His father drives at an average speed that's 15 mph less than Bill's average speed. Let B stand for Bill's average speed. Express in terms of B:

    (a) The distance in miles Bill travels by noon that day
    (b) The distance his father travels by noon that day

26. According to the previous problem, what was the average speed of Bill's father if they met at noon that day?

27. A car makes a round trip between two cities in 11 hours. Going one way, the driver averaged 60 mph; going the other way, the driver averaged 50 mph. What is the distance between the two cities?

Section 6.5

28. A woman invests some money at an annual interest rate of 7%. She invests three times that amount at 6.75% annual interest. If s denotes the amount of money (in dollars) that she's invested at 7%, express in terms of s:

    (a) The amount (in dollars) earned in one year by the 7% investment
    (b) The amount (in dollars) earned in one year by the 6.75% investment

29. A person invests some money at 7% annual interest and three times that amount at 6.75% annual interest. How much is invested at each rate if the total amount earned by the two investments for the year is $1,090?

30. In constructing a playground, the county pays 30% of what the state pays, and the city pays 70% of what the county pays. If s denotes (in dollars) the amount the state must pay, express in terms of s:

    (a) The amount (in dollars) the county must pay
    (b) The amount (in dollars) the city must pay

31. According to the previous problems, what is the state's share of the cost if the entire cost of the project is $45,300?

32. (a) How many pounds of salt are there in a 75-pound mixture of salt and water if the mixture is 8% salt?
    (b) How many pounds of salt are left if you evaporate x pounds of water?
    (c) How many pounds of mixture are left if you evaporate x pounds of water?
    (d) How many pounds of water must be evaporated from the original mixture to get a mixture that's 15% salt?

Additional Topics in Linearity
unit 3

An Introduction to Graphs of Linear Equations

Overview

Girl: Do you believe television can replace the newspaper?
Boy: No — how can you wrap fish in a television set?

This rather silly joke suggests the advantage of *seeing* over *reading*. The story implies that television is less useful than a newspaper only in rather far-fetched situations. We often say this more directly in the maxim, "A picture is worth a thousand words."

Surprisingly, this idea is also true in mathematics. We often say that the temperature *rose*; yet temperature neither rises nor falls. It *increases* or *decreases*. What rises or falls is the level of mercury in a tube. In short, a mercury thermometer gives us a visual way of measuring temperature. Even today, when we see digital thermometers on buildings or billboards, we still talk about the temperature's rising and falling.

In economics, we talk about profits that *rise* or *fall*; yet profits *increase* or *decrease*. When we say that profits rise, we are probably thinking of a *graph*, in which increasing is indicated in the upward direction and decreasing in the downward direction.

In this lesson, we study the nature of a graph. In effect, a graph is a pictorial way of viewing a mathematical situation. From a very elementary point of view, an ordinary ruler is a perfect example of a graph. A ruler consists of a line (a geometric concept) on which we mark off points (also a geometric concept) at regular intervals. We then name these points by numerals (an arithmetical concept). That is how a ruler relates numbers to points on a line.

At the end of this lesson, we shall know how to visualize linear equations. In fact, we shall understand the relationship between the word "linear" and the word "line." In this sense, we shall learn to use the picture of a straight line to solve an algebraic linear equation.

Section 7.1
The Number Line

The **number line** is a geometric way of thinking about numbers. It is an extension of what a ruler is. We can make a number line as follows.

1. We draw a straight line.

2. We place arrowheads at each end of the line to indicate that the line may be extended as far as we wish in either direction.

3. Somewhere on the line (perhaps near the middle of the length) we pick a point and label it 0.

4. We pick a convenient length. Then we start at 0 and mark this length off successively on both sides of 0.

5. We label the points we've marked off to the right of 0 as 1, 2, 3, and so on. We label the points to the left −1, −2, and so on.

OBJECTIVE
To see how a picture can help us better understand arithmetic and algebra.

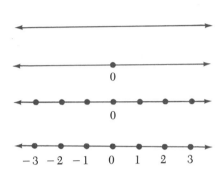

The number line relates arithmetic to geometry because it takes points on a line and names them by numbers.

Perhaps the number line looks familiar to you. In a way, we examined it when we discussed signed numbers in Lesson 3. The magnitude of the signed number tells us how far from 0 the point named by the signed number is. The sign of the number tells us whether the point is to the left (negative) or the right (positive) of 0. Let's practice locating numbers on the number line.

VOCABULARY REVIEW
Given a signed number such as −6, 6 is called the magnitude, and the negative sign is called the sign or signature.

Example 1

On the line shown below, find the point P that represents the number 4.

The number 4 is represented below by the point P.

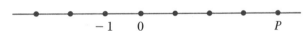

Because the spacing between the integers is always the same, you can use the distance between −1 and 0 as a guide. Remember that 4 means +4, which means that you start at 0 and go 4 units to the right. That is,

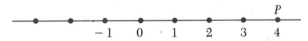

Example 2

On the number line shown below, find the point that represents the number −4.

The number −4 is represented below by Q.

Use the distance between 0 and 1 as a "standard unit." Then remember that −4 means that you want to go 4 units, but this time to the *left* of 0. So begin at 0

and mark off 4 units to the left. This gives you the point Q, which represents the number -4. That is,

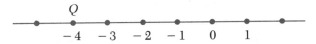

This idea is not limited to the integers.

Example 3

On the number line shown below, find the point that represents 2.25.

2.25 is represented by the point R below.

If you look at the line, you see that the points divide the distance between 0 and 1 into four equal parts. Therefore, the distance between two consecutive points on the line is now $\frac{1}{4}$ or 0.25. This means that the first point after 2 represents 2.25.

Until now we have started with the signed number and located the point on the number line that represented the number. We can also do the problem in reverse. We can start with the point on the number line and decide what number it represents.

Example 4

What number is represented by the point P on the number line shown below?

P represents the number 6 ($+6$).

The labeling that appears on the number line tells you that the distance between two consecutive points is 1 standard unit. By counting, you see that P is 6 units to the right of 0. Hence the magnitude of the number represented by P is 6, and the sign is positive because positive means to the right of 0. So P represents the number $+6$.

Example 5

What number is represented by the point Q on the number line shown below?

Q represents the point -3. Using the distance between 0 and 1 as a guide, you see that Q is 3 units to the *left* of 0. So the number represented by Q has 3 as its magnitude, and its signature is negative. This means that Q represents -3.

Example 6

What number is represented by the point R on the number line shown below?

REVIEW
The integers are those signed numbers whose magnitudes are whole numbers. The whole numbers are 0, 1, 2, 3, 4, and so on.

It is really important to understand that each point on the number line represents a number and that every number can be represented by a point on the number line. Then it is clear why the number line is a good way to picture numbers.

In Lesson 3, we mentioned that any number whose square is nonnegative is called a real number. On the number line, all points on the line represent real numbers. The nonreal numbers are represented by the points that aren't on the number line. In the course, when we say "number," we mean "real number."

R represents the number $3\frac{1}{2}$, or 3.5. Looking at the number line, you see that the distance between 0 and 1 is divided into two parts of equal length. That is, the distance between a pair of consecutive points is just a half unit. If you label the point to the right and to the left of R, you see that

R is midway between the points that represent 3 and 4. That is, R is 3.5 units to the *right* of 0. Therefore, R represents the number $+3.5$, or more simply, 3.5.

The number line is more than just a way to draw pictures of numbers. One of the reasons we use positive and negative rather than such phrases as "to the left of 0" and "to the right of 0" is that there is a very strong connection between finding the distance between points on the number line and being able to subtract signed numbers. Let's see what this means in some problems.

Example 7

How far must you go, and in what direction, to get from the point 1 to the point 5 on the number line?

By drawing the line and locating the points 1 and 5, you see that you must go 4 units to the right to get from 1 to 5.
   You don't have to draw the number line to get this answer. All you have to do is subtract 1 from 5. You are already 1 unit to the right of 0, and you want to be 5 units to the right of 0. So you just go 4 more units to the right of 0. Visually:

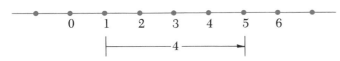

What happens if the motion is from right to left? What happens if negative numbers are involved?

> Notice how we use point and number interchangeably? We use the number line to identify points and numbers.

Example 8

How far must you go, and in what direction, to get from 5 to 1?

Now you go 4 units toward the left. That is, you are starting 5 units to the right of 0 and you want to finish 1 unit to the right of 0, so you must return 4 units to the left.

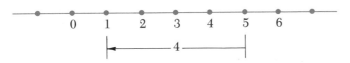

Suppose you subtract the starting point (in this case, 5) from the ending point (in this case, 1). You get $1 - 5$, or -4. If you interpret the negative signature to mean that you move from right to left, the answer tells you that you move 4 units from right to left.

So far, it appears that to find the directed (signed) distance in going from P to Q, we need only subtract P (the starting point) from Q (the finishing point). Let's see if the method continues to work.

> In either Example we must go 4 units, but the directions are different.
>
> Since positive means motion from left to right and negative means motion from right to left, we often draw the number line as
>
> $\longrightarrow +$
>
> to indicate that the positive direction is to the right.

Section 7.1   The Number Line

Example 9

Find the directed distance (that is, how far and in what direction) from 3 to −4.

To get from 3 to −4, you must go 7 units to the left. You can see this at once by looking at the number line.

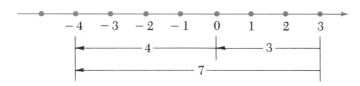

Without using a picture, see what happens when you subtract the starting point, 3, from the ending point, −4? You get

$$-4 - 3 = -4 + (-3) = -7$$

This tells you that you must go 7 units toward the left.

Recall that subtraction is not commutative. If we subtracted −4 from 3, we would get

$$3 - (-4) = 3 + 4 = 7$$

This would be the directed distance from −4 to 3. In summary, if we subtract in the wrong order, we get the right distance but the wrong direction.

Example 10

Find the directed distance from −7 to −4.

To get from −7 to −4, you must go 3 units toward the right.

In a diagram, you have

If you subtract the starting point, −7, from the finishing point, −4, you get

$$(-4) - (-7) = (-4) + (+7) = +3$$

which means that you go 3 units toward the right.

REVIEW OF SUBTRACTION
To subtract signed numbers, we leave the first alone, change the operation to addition, and change the sign of the second number.

NOTE

The rule for subtracting signed numbers as given in Lesson 3 is exactly the rule we need to be able to compute directed distances by subtraction.

This section shows only one of many ways in which arithmetic and geometry are related. These ideas are used elsewhere in the lesson.

PRACTICE DRILL

1. On the number line below, locate the points

   (a) 3   (b) −2   (c) 0   (d) $1\frac{1}{2}$

2. Without drawing the number line, find the directed distance from

   (a) 12 to 15   (b) −12 to −15   (c) 12 to −15
   (d) −12 to 15   (e) 15 to −12

Answers

1.

2. (a) +3, or 3 to the right   (b) −3, or 3 to the left
   (c) −27, or 27 to the left
   (d) 27, or 27 to the right
   (e) −27, or 27 to the left

CHECK THE MAIN IDEAS

1. The _____ is a geometric way of representing numbers.
2. The reference point on the number line from which you locate all other numbers is represented by the number _____.
3. The sign of any number to the right of 0 is _____.
4. In fact, the left-to-right direction is called the _____ direction.
5. The sign of a number is negative if the number is to the _____ of 0.
6. For example, −3 means the number that is (a) _____ units to the (b) _____ of 0.

Answers
1. number line 2. 0
3. positive 4. positive
5. left 6. (a) 3 (b) left

Section 7.2
Cartesian Coordinates

OBJECTIVE
To show how points in the plane can be identified with ordered pairs of numbers.

Have you ever tried to look up a city on a road map, or have you ever played the game "Battleship"? If you have, chances are that you have seen a way in which **coordinates** are used to locate positions.

For example, suppose you are looking up a street in the index of a street map and you see the symbol $C2$ following the name. If you look at the map, you will see a series of horizontal and vertical lines. One set of lines will be marked by numbers and the other by letters. So you may see

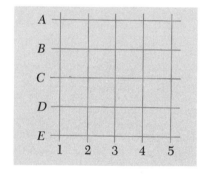

The label $C2$ tells you to look at or near the point where the lines marked C and 2 meet. That is,

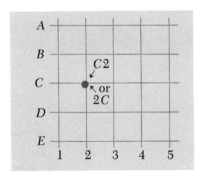

Let's practice locating a few points this way.

Section 7.2 Cartesian Coordinates

Example 1

Where would you look for the point A4?

The point A4 is where the line marked A and the line marked 4 meet. For example,

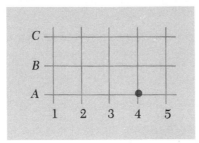

In the previous map, we had A on the top rather than on the bottom. But as long as the lines are marked, they can be in any order.

However, it is much easier to locate things if the lines are marked in some kind of regular order. For example, it would be confusing to mark the lines in the order 3, 6, 4, 2, 1, 7.

Example 2

Where would you look for the point 4A?

The point 4A is located in the same place as the point A4. As long as the lines are marked A and 4, it makes no difference which you name first.

However, if the horizontal lines and the vertical lines are both marked by numbers or both by letters, then you have to indicate which is the horizontal line and which is the vertical.

When both sets of lines are named by numerals, we talk about **ordered pairs** of numerals. For example, it makes a difference whether the horizontal line is 1 and the vertical line 2 or the horizontal line is 2 and the vertical line is 1. That is,

Example 3

In the diagram below, how should the point P be labeled?

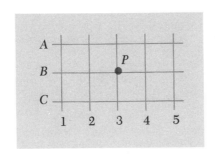

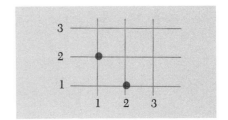

P is the point at which the two lines B and 3 meet, so you should label P either as B3 or as 3B.

In the seventeenth century, the French philosopher, mathematician, and theologian René Descartes used a variation of the "map" idea to carry out the idea of naming points other than those on the number line.

Descartes started with the number line, which he named the *x*-axis. The *x*-axis is usually the number line that is drawn horizontally and for which the positive direction is from left to right.

Then he drew another number line through 0 that was perpendicular to the *x*-axis. This second number line was considered positive in the direction of down to up. This number line was given the name *y*-axis.

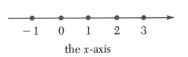

the *x*-axis

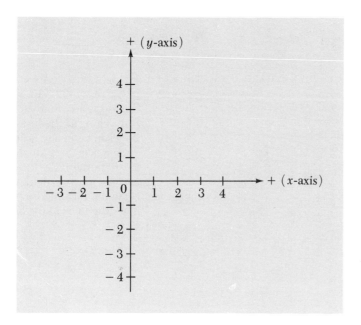

Every point in a picture can now be located by giving a pair of numbers, much as we did on the road map. As we noted in Example 2, the only difference is that both lines are labeled with numbers. So we must be careful about how we name our points.

Let's look at a few more examples.

Example 4

What is the location of the point P in the diagram below?

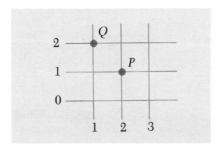

P is where the horizontal line 1 and the vertical line 2 meet. But it would be confusing and misleading to say that P is where the lines marked 1 and 2 meet. There are two lines marked 1 and two lines marked 2. In fact, the point Q above is also on lines marked 1 and 2. The difference is that Q is on the vertical line 1 and the horizontal line 2.

It is customary to locate the point P relative to 0. So we say that P is 2 units to the right of 0 and 1 unit above 0.

We give special names to the numbers that locate the point. The number that tells us how far the point is to the right or left of 0 is called the *x*-coordinate of the point. The number that tells us how far the point is above or below 0 is called the *y*-coordinate of the point.

Section 7.2 Cartesian Coordinates

Example 5

In the diagram below, what is the *x*-coordinate of the point *P*?

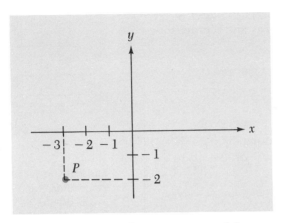

P is 3 units to the left of 0, so its *x*-coordinate is -3, and this is true no matter how far above or below *P* is from the *x*-axis.

In summary, if *P* is to the right of 0, its *x*-coordinate is positive. If it is to the left of 0, its *x*-coordinate is negative.

Example 6

With *P* located as in Example 5, what is the *y*-coordinate of *P*?

P is 2 units below the *y*-axis, so its *y*-coordinate is -2.

It is awkward to keep saying "the *x*-coordinate is . . ." and "the *y*-coordinate is. . . ." So as shorthand, we agree to write the two coordinates as a pair of numbers, in which the first member of the pair stands for the *x*-coordinate and the second member of the pair stands for the *y*-coordinate.

If we say the point $(3,4)$, we mean the point whose *x*-coordinate is 3 and whose *y*-coordinate is 4. That is, $(3,4)$ names the point which is 3 units to the right of 0 and 4 units above 0. If we want the point that is 4 units to the right of 0 and 3 units above 0, we would write this point as $(4,3)$.

Let's try to locate a few points.

If *P* is above 0 (that is, above the *x*-axis), its *y*-coordinate is positive. If it is below the *x*-axis, its *y*-coordinate is negative.

There is no natural reason for listing the *x*-coordinate first. But one of the two had to come first. The accepted order tells us very compactly which is the *x*-coordinate and which is the *y*-coordinate.

Example 7

Describe the point *P* if *P* is $(5,1)$.

By the agreement, the first member of the pair names the *x*-coordinate and the second member names the *y*-coordinate. So $(5,1)$ is the point whose *x*-coordinate is 5 and whose *y*-coordinate is 1. Therefore, *P* is the point that lies 5 units to the right of 0 and 1 unit above 0.

Rather than write that *P* is the point $(5,1)$, we usually just write $P(5,1)$. In terms of the axes, $(5,1)$ is 5 units to the right of the *y*-axis and 1 unit above the *x*-axis.

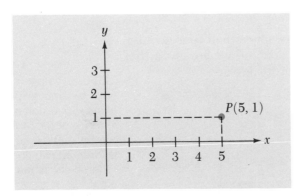

Example 8

What point is named by $Q(-2,3)$?

This is the point whose x-coordinate is -2 and whose y-coordinate is 3. Therefore, it is the point that lies 2 units to the left of the y-axis and 3 units above the x-axis.

Example 9

What point is named by $R(3,-2)$?

R is the point whose x-coordinate is 3 and whose y-coordinate is -2. This means that R is located 3 units to the right of 0 and 2 units below 0; or, 3 units to the right of the y-axis and 2 units below the x-axis.

Example 10

What point is named by $T(-2,-3)$?

The x-coordinate of T is -2 and the y-coordinate is -3. Therefore, T is located 2 units to the left of the y-axis and 3 units below the x-axis.

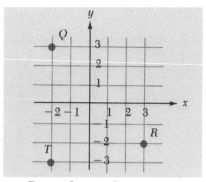

Diagram for Examples 8, 9, and 10

It is easy to confuse the points Q, R, and T if you aren't sure which is the x-coordinate and which is the y-coordinate.

Example 11

What point is named by $(0,3)$?

It is the point whose x-coordinate is 0 and whose y-coordinate is 3. The x-coordinate tells us how far to the right or left of 0 the point is. The fact that the x-coordinate is 0 means that the point is neither to the left nor right of 0.

This means that the point is on the y-axis. That is, $(0,3)$ is on the y-axis, 3 units above the x-axis.

More generally, any point whose x-coordinate is 0 must lie on the y-axis.

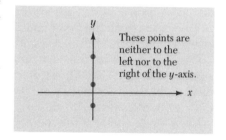

These points are neither to the left nor to the right of the y-axis.

Example 12

What point is named by $(3,0)$?

The point is 3 units to the right of 0, and no units above or below 0. That is, the point is on the x-axis, 3 units to the right of 0.

Any point whose y-coordinate is 0 must lie on the x-axis. Visually:

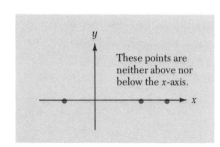

These points are neither above nor below the x-axis.

So far we have located the position of various points, given their coordinates. It is important to understand that we can do the reverse. If we are given the location of the point in the plane, we can list its coordinates.

This is very important. For it now gives us a way to match points in the plane with ordered pairs of numbers and vice versa. In other words, we can always think of a point as being an ordered pair of numbers; and an ordered pair of numbers as naming a point.

Section 7.2 Cartesian Coordinates

Example 13

In the diagram below, what ordered pair of numbers names the point *P*?

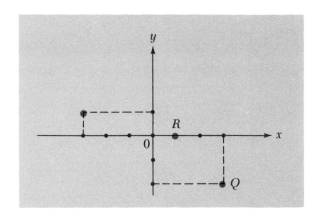

OPTIONAL NOTE

In more advanced courses we talk about **complex numbers** as well as real numbers. In terms of points or ordered pairs of numbers, any point names a complex number. The real numbers are those special points that lie on the *x*-axis; that is, those points whose *y*-coordinate is zero.

P is named by the ordered pair $(-3,1)$. That is, *P* is located 3 units to the left of 0. This makes its *x*-coordinate -3. It is also 1 unit above 0, so its *y*-coordinate is 1. Writing first the *x*-coordinate and then the *y*-coordinate, you have that *P* is named by $(-3,1)$.

Whenever we say "above (or below) 0," we really mean "above (or below) the *x*-axis." When we say "to the right (or left) of 0," we really mean "to the right (or left) of the *y*-axis."

Example 14

Using the diagram for Example 13, what ordered pair of numbers names *Q*?

Q is named by $(3,-2)$. That is, *Q* is 3 units to the right of 0 but 2 units below 0.

Example 15

Using the diagram for Example 13, what ordered pair of numbers names *R*?

R is named by $(1,0)$. That is, *R*, is 1 unit to the right of 0 and on the *y*-axis. Because it is neither above nor below the *y*-axis, the *y*-coordinate of *R* is 0.

Example 16

What ordered pair of numbers names the point 0 in the diagram for Example 13?

0 is the point $(0,0)$. It is on both the *x*-axis and the *y*-axis.

PRACTICE DRILL

All problems refer to the diagram on this page. In the diagram, the distance between consecutive parallel lines is 1 unit.

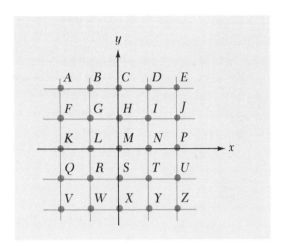

1. Name each of the following points in Cartesian coordinate form.

   (a) A   (b) E   (c) P   (d) C   (e) M   (f) U   (g) Q
   (h) B   (i) W   (j) S   (k) L

2. Which point is named by

   (a) (1,2)   (b) (2,1)   (c) (1,−2)   (d) (−1,2)   (e) (−1,−2)
   (f) (0,−2) (g) (−2,0) (h) (0,0)

Answers
1. (a) (−2,2) (b) (2,2) (c) (2,0) (d) (0,2) (e) (0,0) (f) (2,−1) (g) (−2,−1) (h) (−1,2) (i) (−1,−2) (j) (0,−1) (k) (−1,0)
2. (a) D (b) J (c) Y (d) B (e) W (f) X (g) K (h) M

CHECK THE MAIN IDEAS

1. To develop Cartesian coordinates, you begin by taking two _____ lines and placing them at right angles so that the two lines meet at 0.
2. The two number lines are usually chosen so that one is horizontal and the other is vertical. The horizontal line is called the _____-axis.
3. The vertical line is called the _____-axis.
4. A point can now be described by two numbers, called the x- and y-_____.
5. The x-coordinate tells you how far the point is to the right or left of the (a) _____ and is written first. The (b) _____ tells you how far above or below the x-axis the point is.
6. On the y-axis, the upward direction is called (a) _____ and the (b) _____ direction is called negative.
7. For example, (3,−4) means the point that is 3 units to the right of the y-axis and 4 units below the x-axis. That is, starting from (0,0), you move (a) _____ units to the right and 4 units (b) _____.
8. If a point is on the x-axis, it is neither above nor below the y-axis. Hence the _____-coordinate of such a point is 0.
9. So (3,0) is on the _____-axis because its y-coordinate is 0.

Answers
1. number   2. x 3. y
4. coordinates   5. (a) y-axis (b) y-coordinate
6. (a) positive (b) downward
7. (a) 3 (b) down   8. y
9. x

Section 7.3
Graphing Linear Relationships

Now that we have introduced Cartesian coordinates and assigned an ordered pair of numbers (x,y) to name each point in the plane, we can give a new interpretation to formulas.

For example, consider the formula
$$y = 2x + 3 \qquad [1]$$

This is like the formula in Lesson 5 for computing the cost in dollars (y) of x pounds of cheese if the cheese cost \$2 per pound and there is a \$3 shipping charge.

But we can also use [1] to describe how the x- and y-coordinates of a point are related.

OBJECTIVE
To be able to interpret the linear relationship
$$y = mx + b$$
in terms of a straight line.

In Lesson 5, we used p rather than x for the number of pounds of cheese, and c rather than y for the cost of the cheese. But by this time it should be clear that the letters we use to represent the amounts aren't what's important. The important thing is the relationship between the amounts — not the symbols that name the amounts.

Example 1

Where is the point P if its x-coordinate is 1 and the y-coordinate is related to the x-coordinate by the rule $y = 2x + 3$?

The point is $(1,5)$. That is, the point is 1 unit to the right of 0 and 5 units above 0.
The formula that relates the x- and the y-coordinates is $y = 2x + 3$. This says that given x, you double it $(2x)$ and then add 3 to get y. In this case, x is 1, so $2x$ is 2 and $2x + 3$ is 5. This tells you that if the x-coordinate is 1, the y-coordinate is 5. You write this as $(1,5)$.

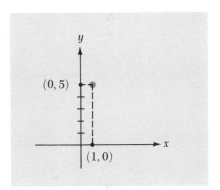

Example 2

Where is the point R if its x-coordinate is -1 and the y-coordinate is given by $y = 2x + 3$?

R is the point $(-1,1)$. It is 1 unit to left of 0 and 1 unit above 0. The rule is the same as in Example 1. You double the x-coordinate and add 3 to find the y-coordinate. In this case, x is -1. So $2x = 2(-1)$ or -2, and $2x + 3$ is $-2 + 3$, or 1. If x is -1 and y is 1, you write $(-1,1)$.

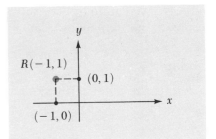

Example 3

Where is the point S if the x-coordinate of S is 0 and the y-coordinate is given by $y = 2x + 3$?

S is the point $(0,3)$. It is on the y-axis, 3 units above the x-axis.
You find y by doubling x and adding 3. In this case, x is 0, so twice x is also 0. Thus, $2x + 3$ is 3. If the x-coordinate of S is 0 and the y-coordinate is 3, then S is the point $(0,3)$.

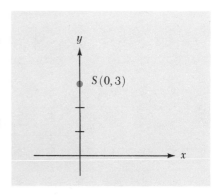

Let's take the points P, R, and S, which we located in Examples 1, 2, and 3, and place them in the same diagram. We have

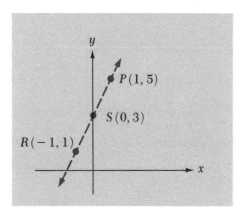

These three points lie on the same straight line. For these three points, every time the x-coordinate increases by 1, the y-coordinate increases by 2. That is,

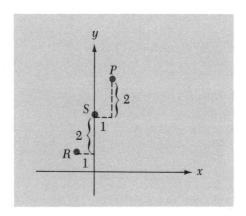

This is not a coincidence. Look at $y = 2x + 3$. If x is increased by 1 it becomes $(x + 1)$. If we double $(x + 1)$ we double both x and 1. That is, $2(x + 1) = 2x + 2$ by the distributive principle.

Before we continue with this section, it is important to emphasize that our results do not depend on the coordinates being whole numbers.

Example 4

Where is the point T if its x-coordinate is 1.5 and its y-coordinate is given by $y = 2x + 3$?

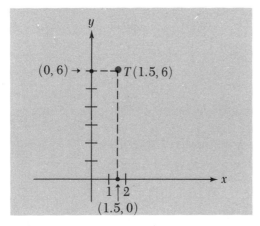

T is the point $(1.5, 6)$. That is, the point is 1.5 units to the right of 0 and 6 units above 0.

Just as in the previous examples, the rule for finding the y-coordinate of the point when you have the x-coordinate is to double the x-coordinate and add 3. In this case

$$y = 2(1.5) + 3 = 3 + 3 = 6$$

So the x-coordinate of T is 1.5 and the y-coordinate is 6. You write T as $T(1.5, 6)$.

If you locate T in the same diagram as P, R, and S, you see that these points all lie on the same straight line.

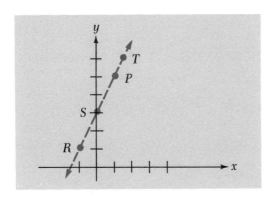

To summarize the results, *all* points (x,y) whose coordinates satisfy the equation

$$y = 2x + 3 \qquad [1]$$

lie on the same straight line.

There is some basic terminology associated with equation [1].

Of course, Examples 1 through 4 don't give us all the points on the line. All we showed is that the four points $P(1,5)$, $R(-1,1)$, $S(0,3)$, and $T(1.5,6)$ are on the line.
While every point (x,y) for which $y = 2x + 3$ is on this line, we haven't proven this yet.

DEFINITION

1. The **set** (collection) of points (x,y) whose coordinates satisfy an equation is called the **graph** of the equation. For example, in Examples 1 through 4, the points we found all belonged to the graph of the equation $y = 2x + 3$, because the coordinates of these points (x,y) satisfied the equation $y = 2x + 3$.
2. If the points that form the graph of an equation are "connected," we call the graph a **curve**.
3. A point at which a curve (and in this lesson, all curves are straight lines) crosses the y-axis is called a **y-intercept** of the curve.
    We find the y-intercept of the graph of an equation by replacing x by 0 in the equation and solving for y. That is, every point on the y-axis has its x-coordinate equal to 0. For example, in Example 3 we showed that the y-intercept of the graph of the equation $y = 2x + 3$ is $(0,3)$.
4. A point at which the curve crosses the x-axis is called an **x-intercept** of the curve. We find the x-intercept of the graph of an equation by replacing y by 0 and solving for x.

The best way to understand the new vocabulary is to use it in a few examples.

This does not say that the graph of $y = 2x + 3$ contains only these four points. As we saw, the graph is a straight line, and these are only four of the many points on the line.

This may seem strange, since we think of a curve as "bent," but by this definition a straight line is a curve, because it is a connected set of points.

REVIEW

Any point on the y-axis lies neither to the left nor to the right of the y-axis. Because the x-coordinate measures the distance of the point to the left or right of the y-axis, then a point is on the y-axis if and only if its x-coordinate is 0.

Example 5

What is the y-intercept of the line

$$y = 3x + 1 \quad [2]$$

The y-intercept is the point (0,1).
Just replace x by 0 in [2] to get

$$y = 3(0) + 1 = 1$$

The y-coordinate of the point is 1 and the x-coordinate is 0, so the point is (0,1).

NOTE

There is a tendency to confuse an equation with a graph. For example, we often call $y = 2x + 3$ a line. It is actually the graph of $y = 2x + 3$ which is the line.
However, if we agree to think of $y = 2x + 3$ as meaning the set of all points (x,y) for which $y = 2x + 3$, then it is correct to think of $y = 2x + 3$ as a line.

Example 6

What is the x-intercept of the line

$$y = 3x + 1 \quad [2]$$

The x-intercept is $\left(-\frac{1}{3}, 0\right)$.

All you have to do is replace y by 0 in [2] to get the equation

$$0 = 3x + 1 \quad [3]$$

Equation [3] is linear, and you solve it by the method of the previous lessons to get

$$\begin{array}{r} 0 = 3x + 1 \\ -1 \quad\quad -1 \\ \hline -1 = 3x \end{array}$$

so you know that $\frac{-1}{3} = x$.

We may use the results of Examples 5 and 6 to draw the line $y = 3x + 1$. We need to know only two points to determine a line, and we have already found out that (0,1) and $\left(-\frac{1}{3}, 0\right)$ are on the line.

We can write this in the form of an example.

Example 7

Draw the line $y = 3x + 1$.

The line is shown in the sketch on the page. You draw it by making it pass through the two points (0,1) and $\left(-\frac{1}{3}, 0\right)$. These happen to be the y- and x-intercepts. It is common to draw a line by locating the x- and y-intercepts and then passing the line through these two points.

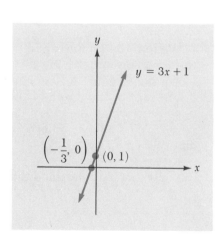

Of course, there are many other points that are on the line $y = 3x + 1$. To find any such point, all we have to do is pick a value of x at random and solve for y, or vice versa.

Section 7.3   Graphing Linear Relationships

Example 8

What point on the line $y = 3x + 1$ has an x-coordinate equal to 2?

The point $(2,7)$ is the only point on the line $y = 3x + 1$ whose x-coordinate is 2. Again, all you do is replace x by 2 in $y = 3x + 1$. This gives

$$y = 3(2) + 1 = 6 + 1 = 7$$

In other words, for $(2,y)$ to be on the line, y must be 7. So $(2,7)$ is also on the line $y = 3x + 1$, and this is shown in the illustration.

We can also start with an arbitrary value of y and then solve the equation to find the corresponding value of x.

Remember that for (x,y) to be on the line $y = 3x + 1$, the x and y coordinates of the point must be related by $y = 3x + 1$.

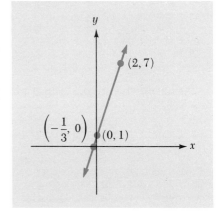

Example 9

What point on the line $y = 3x + 1$ has a y-coordinate equal to -2?

The point is $(-1,-2)$. In this case, you are told that $y = -2$, so you replace y by -2 in the equation $y = 3x + 1$ to get

$$\begin{array}{r} -2 = 3x + 1 \\ \underline{-1 \quad\quad -1} \\ -3 = 3x \quad\text{or}\quad x = -1 \end{array}$$

That is, when the y-coordinate is -2, the x-coordinate must be -1 if the point is to be on the line $y = 3x + 1$, so the point is $(-1,-2)$. This is also shown in an illustration.

Do you begin to see a pattern to the points? Every time the x-coordinate of a point on the line increases by 1, the y-coordinate must increase by 3 in order for the point to remain on the line. This is not self-evident, and the next example may help you to understand.

See how natural negative numbers are in terms of graphs? Example 9 asks us simply to find the point on the line whose y-coordinate is 2 units below the x-axis.

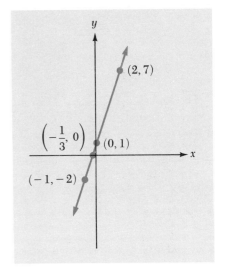

Example 10

Let

$$y = 3x + 1 \qquad\qquad [2]$$

Use the distributive property to show that whenever x increases by 1, y increases by 3.

If you replace x by $x + 1$ in [2], you get

$$\begin{aligned} y &= 3(x + 1) + 1 \\ &= 3x + 3 + 1 \\ &= 3x + 4 \end{aligned}$$

and $3x + 4$ is 3 more than $3x + 1$. As a check, you have

x	3x	3x + 1	=	y	(x,y)
−2	−6	−5		−5	(−2,−5)
−1	−1	−2		−2	(−1,−2)
0	0	1		1	(0,1)
1	3	4		4	(1,4)
2	6	7		7	(2,7)

(With each increase of 1 in x, y increases by 3.)

We didn't have to pick these particular values for x.

If you locate these points in a plane, you have

It is important to see that every time x increases by 1, y increases by 3. The 3 is not a coincidence. It comes from the 3 in the equation $y = 3x + 1$.

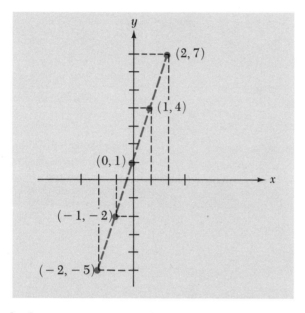

The fact that the line $y = 3x + 1$ rises by 3 units every time x increases by 1 is given a special name. We say that the slope of the line is 3.

> **DEFINITION**
> The **slope** of a line is the amount the line rises (or falls) every time we move one unit to the right.

That is, the slope is the increase (or decrease) in y per unit increase in x.

Example 11

What is the slope of the line

$$y = 2x - 1 \qquad [4]$$

The slope of this line is 2.

The 2 comes from the coefficient of x in the equation $y = 2x - 1$.

VOCABULARY REVIEW
By the **coefficient** of x we mean the number that multiplies x.

There are many ways to show this. One way is to replace x by x + 1 in [4] to obtain

$$y = 2(x + 1) - 1$$
$$= 2x + 2 - 1$$
$$= 2x + 1$$
$$= (2x - 1) + 2$$

This shows you that when x increases by 1, y increases by 2.

Section 7.3   Graphing Linear Relationships

Sometimes it's easier to see this by picking a few points on the line whose x-coordinates differ by 1 and see what happens to y. For example,

x	$2x$	$2x-1\ (=y)$	(x,y)
-2	-4	-5	$(-2,-5)$
-1	-2	-3	$(-1,-3)$
0	0	-1	$(0,-1)\ \leftarrow y$-intercept
1	2	1	$(1,1)$
2	4	3	$(2,3)$

(successive y-values differ by 2)

When the equation of the line is given in the form $y = mx + b$, it is easy to find the slope and y-intercept. In this case, m is always the slope and $(0,b)$ is always the y-intercept.

Do you see why? The y-intercept is found by letting $x = 0$. In this case, $y = mx + b$ becomes $y = 0x + b$, or $y = b$. Hence the y-intercept is $(0,b)$.

If we replace x by $x+1$, we get

$$y = m(x+1) + b$$
$$= mx + m + b$$
$$= mx + b + m$$
$$= (mx + b) + m$$

Example 12

What are the slope and y-intercept of the line $y = 4x + 3$?

The slope is 4 and the y-intercept is 3. That is,

$$y = \underset{\underset{m}{\uparrow}}{4}x + \underset{\underset{b}{\uparrow}}{3}$$

As a check, you have

x	$4x$	$4x+3\ (=y)$	(x,y)
0	0	3	$(0,3) \leftarrow y$-intercept
1	4	7	$(1,7)$
2	8	11	$(2,11)$
3	12	15	$(3,15)$

(successive y-values differ by 4)

We do not need the equation in the form $y = mx + b$ to find the slope and y-intercept.

Example 13

What is the y-intercept of the line $3x + 4y = 12$?

The y-intercept is $(0,3)$. Remember, to find the y-intercept, you always replace x by 0 in the equation. In this case, you get

$$3(0) + 4y = 12$$
$$4y = 12$$

Hence $y = 3$ when $x = 0$. So the y intercept is $(0,3)$.

Regardless of the form of the equation, we find the y-intercept by letting $x = 0$ and then solving the resulting equation for y.

Remember that we can have a negative slope. All this means is that as we move to the right, the line seems to be falling. That is, as x increases, y decreases. This situation is shown in the illustration.

In terms of the form $y = mx + b$, we have a negative slope whenever m is negative. This is shown in Example 14.

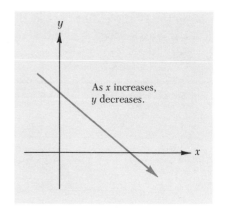

As x increases, y decreases.

Example 14

What are the slope and y-intercept of the line $y = -3x - 4$?

The slope of this line is -3, and it crosses the y-axis at the point $(0, -4)$. You have

$$y = -3x - 4 = \underset{m}{(-3)}x + \underset{b}{(-4)}$$

In this form, m gives you the slope and b gives you the y-intercept.

Knowing the slope and y-intercept is enough information to draw the line. The y-intercept tells us where the line crosses the y-axis. We can then use the slope to locate, for example, the point whose x-coordinate is 1. All we have to do is increase or decrease the y-intercept by the magnitude of the slope.

We do not have to know the slope and the y-intercept to draw the line. All we need is any two points. We can pick x at random and use $y = -3x - 4$ to find the corresponding y-coordinate. For example,

x	$-3x$	$-3x - 4$	$(= y)$	(x, y)
-1	3	$3 - 4$	(-1)	$(-1, -1)$
-2	6	$6 - 4$	(2)	$(-2, 2)$
0	0	$0 - 4$	(-4)	$(0, -4)$

This tells us the line is

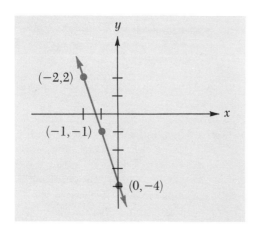

We needed only two points to be able to draw the line, but by locating three points, we have a check against locating one of the points incorrectly. However,

Remember, "x increases" means x goes from left to right.

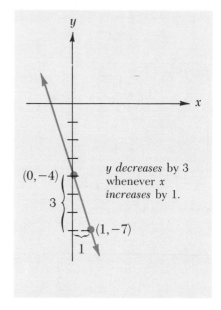

y decreases by 3 whenever x increases by 1.

For example, in Example 14,

1. The line meets the y-axis at $(0, -4)$.
2. The slope is -3, which means that the line drops 3 units for every unit we move to the right.

Given two points, we can always pass a straight line through them. But given three points, they need not lie on the same straight line.

Section 7.3   Graphing Linear Relationships

we did not have to pick the three points we did. Any points on the line would work as well.

Now we can use algebra to solve problems involving lines.

Example 15

What point on the line $y = 4x - 3$ has its x-coordinate equal to 5?

The point is $(5, 17)$. Since $y = 4x - 3$, you find y by replacing x by 5. This gives
$$y = 4(5) - 3 = 20 - 3 = 17$$
So the point is $(5, 17)$.

Example 16

What point on the line $y = 15x + 93$ has its x-coordinate equal to -6?

The point is $(-6, 3)$. You replace x by -6 in the equation $y = 15x + 93$ to obtain
$$y = 15(-6) + 93 = -90 + 93 = 3$$
So the point is $(-6, 3)$.

This seems easier than having to draw the line $y = 15x + 93$ on a sheet of paper.

We don't have to draw a picture. If the picture is "messy," we can use the equation instead. If the equation is complicated, then we can use the picture instead.

Example 17

What point on the line $y = 3x - 5$ has its y-coordinate equal to 16?

The point is $(7, 16)$.

Here you must use algebra to solve a linear equation. That is, you are given the y-coordinate and want to find the x-coordinate. The coordinates are related by the equation
$$y = 3x - 5 \qquad [5]$$
If you replace y by 16 in [5], you get
$$16 = 3x - 5 \qquad [6]$$
You solve [6] as a regular linear equation.
$$\begin{array}{r} 16 = 3x - 5 \\ +5 \qquad +5 \\ \hline 21 = 3x \end{array}$$
so $x = \dfrac{21}{3} = 7$.

PRACTICE DRILL

1. What are the coordinates of P if
   (a) P is on the line $y = 3x + 2$ and its x-coordinate is 0
   (b) P is on the line $y = 3x + 2$ and its x-coordinate is 1
   (c) P is on the line $y = 3x + 2$ and its x-coordinate is -1
   (d) P is on the line $y = 3x + 2$ and its x-coordinate is 200
   (e) P is on the line $y = 3x + 2$ and its y-coordinate is 200

2. What is the slope of each of the following lines?
   (a) $y = 3x + 2$ (b) $y = 3x - 2$ (c) $y = -3x + 2$

3. At what point does each of the following lines cross the y-axis?
   (a) $y = 3x + 2$ (b) $y = 3x - 2$ (c) $y = -3x + 2$

Answers
1. (a) (0,2)   (b) (1,5)
   (c) (−1,−1)   (d) (200, 602)
   (e) (66, 200)   2. (a) 3   (b) 3
   (c) −3   3. (a) (0,2)
   (b) (0,−2)   (c) (0,2)

CHECK THE MAIN IDEAS

1. The set of all points (x,y) for which $y = 3x + 2$ is called the _____ of $y = 3x + 2$.
2. To find a point in the graph of $y = 3x + 2$, you may pick any value for the x-coordinate. You then multiply this number by 3 and add 2 to get the _____ of the point.
3. $3(4) + 2 = 14$, so you know that the point $(4,\underline{})$ belongs to the graph of $y = 3x + 2$.
4. If you locate a few points in this way, you see that the graph of $y = 3x + 2$ is a _____. It is for this reason that $y = 3x + 2$ is called linear.
5. People usually talk about the line $y = 3x + 2$ instead of saying the _____ of $y = 3x + 2$ is a line.
6. The slope of the line $y = 3x + 2$ is (a) _____. This means that if the x-coordinate increases by 1, the y-coordinate of the point on the line increases by (b) _____ units.
7. If the slope is -3, when the x-coordinate increases by 1 unit, the y-coordinate _____ by 3 units.
8. To see if a point belongs to the line $y = 3x + 2$, you replace x and y by the coordinates of the point and check that $y = 3x + 2$ becomes a _____ statement.
9. If you replace x by 0, you see that y must equal 2 in order for $y = 3x + 2$ to be true. This means that $y = 3x + 2$ crosses the y-axis at _____.

Answers
1. graph   2. y-coordinate
3. 14   4. line   5. graph
6. (a) 3   (b) 3   7. decreases
8. true   9. (0,2)

Section 7.4
The General Straight-Line Equation

In the last section, we studied straight lines and the points where they crossed the y-axis, but there is one type of line that does not cross the y-axis.

OBJECTIVE
To see how every straight line can be represented by an equation of the form
$$ax + by = c$$
where a, b, and c are constants.

Example 1

Describe the line that passes through the points $(2,1)$ and $(2,3)$.

This is the line that is parallel to the y-axis and lies 2 units to the right of it. If you look at the line that joins the two given points, you see that it must be a vertical line and that it passes through every point whose x-coordinate is 2.

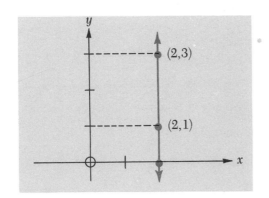

The equation of the line described in Example 1 is written quite simply as
$$x = 2 \qquad [1]$$

This is shorthand for saying, "The set of all points (x,y) whose x-coordinates are 2." In other words, the equation $x = 2$ tells us to consider all points whose x-coordinates are 2 regardless of the value of the y-coordinate.

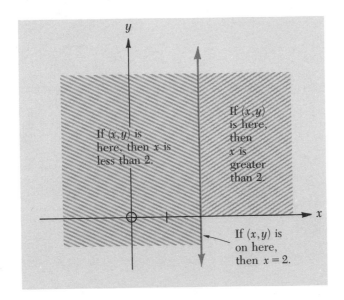

Example 2

What line is described by the equation $x = -1$?

This is the line parallel to the y-axis, and lying 1 unit to the left of it. $x = -1$ means that you want all points whose x-coordinates are -1, including such points as $(-1, -1)$, $(-1, 0)$, and $(-1, 2)$. It is not hard to see that this is the line that lies 1 unit to the left of the y-axis.

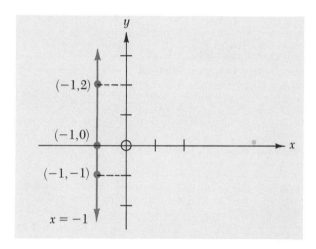

Example 3

What line is described by the equation $x = 0$?

$x = 0$ describes the y-axis.

$x = 0$ stands for the set of all points whose x-coordinates are 0. If the x-coordinate is 0, the point is neither to the left nor to the right of the y-axis; hence, it is on the y-axis.

These first three examples tell us that every line parallel to the y-axis has an equation of the form

$$x = c \qquad [2]$$

where c is a constant.

If a line is not parallel to the y-axis, it has an equation of the form $y = mx + b$, as discussed in the previous section.

$x = 0$ describes the y-axis, not the x-axis. The x-axis is defined by $y = 0$. That is, the x-axis is made up of these points which are neither above nor below the x-axis.

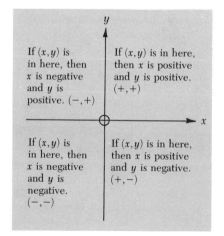

Section 7.4 **The General Straight-Line Equation**

In summary, if L denotes any straight line, then the equation of L has the form

$$y = mx + b \qquad [3]$$

if L is not parallel to the y-axis; and

$$x = c \qquad [2]$$

if L is parallel (or equal) to the y-axis. Equations [2] and [3] are included in the more general form

$$Ax + By = C \qquad [4]$$

where A, B, and C are constants.

We often refer to [4] as the standard linear equation. That is, every straight line has an equation of the form given in [4].

NOTE

$y = c$ is the equation of a line parallel to the x-axis. We do not study this case separately, because $y = c$ is a form of equation [3]. That is, $y = c$ means

$$y = 0x + c$$

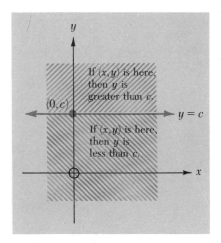

Example 4

Describe the straight line whose equation is [5]

$$-4x + 2y = 8$$

by telling where the line crosses each axis.

It is the straight line that meets the y-axis at $(0,4)$ and the x-axis at $(-2,0)$.
First, observe that [5] is a special case of [4]. That is,

$$\underset{A}{(-4)}x + \underset{B}{2}y = \underset{C}{8}$$

To draw a straight line, you need to know only two points on the line. It is usually easy to find where the line crosses either axis. To find where a line crosses the x-axis, let $y = 0$; and to find where it crosses the y-axis, let $x = 0$.
If you let $x = 0$ in [5], you get

$$-4(0) + 2y = 8 \quad \text{or} \quad 2y = 8$$

When $x = 0$, $y = 4$; this tells you that $(0,4)$ belongs to the line.
If you let $y = 0$ in [5], you get

$$-4x + 2(0) = 8 \quad \text{or} \quad -4x = 8$$

This tells you that when $y = 0$, $x = -2$. Therefore, $(-2,0)$ belongs to the line.

We could pick x or y completely at random and still find a point on the line. For example, if we let $x = 3$, [5] becomes

$$-4(3) + 2y = 8 \quad \text{or} \quad -15 + 2y = 8$$

Hence $2y = 8 + 15 = 23$, and $y = \dfrac{23}{2} = 11.5$. Therefore $(3, 11.5)$ is also a point of the graph of $-4x + 2y = 8$.

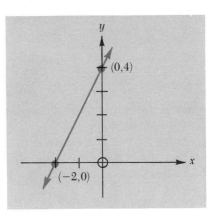

The Graph of $-4x + 2y = 8$

All points found in this way will lie on the same straight line. For example, $(-2,0)$, $(0,4)$, and $(3,11.5)$ are on the same line.

We could also add $4x$ to both sides of [5] to get

$$-4x + 2y = 8$$
$$+\ 4x \qquad\quad + 4x$$
$$\overline{\qquad\qquad\qquad\qquad}$$
$$2y = 4x + 8 \qquad \text{or} \quad y = \frac{4x + 8}{2}$$

That is,

$$y = \frac{2(2x + 4)}{2} = 2x + 4$$

This tells us that we have the line whose slope is 2 and whose y-intercept is 4. The important point is that whenever the line is not parallel to the y-axis, [5] can always be converted to the form $y = mx + b$.

Notice that the same line can be represented by several different equations.

Example 5

What point on the line $2x + 3y = 6$ has its x-coordinate equal to 0?

The point $(0,2)$ has its x-coordinate equal to 0.
  All you have to do is replace x by 0 in the equation $2x + 3y = 6$. This gives you

$$2(0) + 3y = 6 \quad \text{or} \quad 3y = 6 \quad \text{or} \quad y = 2$$

Hence $(0,2)$ is a point on the line. In fact, $(0,2)$ is the y-intercept of the line $2x + 3y = 6$.

Remember, to be on the line $2x + 3y = 6$, the point (x,y) must satisfy the equation $2x + 3y = 6$.

Example 6

What is the x-intercept of the line $2x + 3y = 6$?

The line crosses the x-axis at $(3,0)$.
  The x-intercept is the point at which the line crosses the x-axis. This point has its y-coordinate equal to 0. To find this point, replace y by 0 in the equation $2x + 3y = 6$.
  This leads to $2x + 3(0) = 6$, or $2x = 6$. Hence $(3,0)$ is the point at which the line crosses the x-axis.

Example 7

Use the results of Example 5 and 6 to describe the line $2x + 3y = 6$. That is, name the points at which the line crosses each axis.

It is the line that passes through the two points $(3,0)$ and $(0,2)$. We know that these two points are on the line, and two points are all we need in order to draw the line.

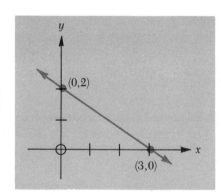

Section 7.4 The General Straight-Line Equation

Example 8

What point on $2x + 3y = 6$ has its y-coordinate equal to 4?

The point $(-3,4)$ has its y-coordinate equal to 4. All you have to do is replace y by 4 in the equation $2x + 3y = 6$. You get

$$2x + 3(4) = 6 \quad \text{or} \quad \begin{array}{r} 2x + 12 = 6 \\ -12 = -12 \\ \hline 2x = -6 \end{array}$$

So $x = -3$. The equation $2x + 3y = 6$ is satisfied when $x = -3$ and $y = 4$, so $(-3,4)$ is a point on the line $2x + 3y = 6$.

Let's try a few more examples illustrating the fact that $Ax + By = C$ is the standard form of a linear equation.

It is less work to find the x- or y-intercept, because in that case either x or y is 0. But in Example 8, there is more work because neither x nor y is 0.

Example 9

Describe the graph of $3x + 4y = 12$ in terms of the x- and y-intercepts.

This is the line that passes through the two points $(4,0)$ and $(0,3)$.
To find the y-intercept, let $x = 0$. This leads to

$$3(0) + 4y = 12 \quad \text{or } 4y = 12 \quad \text{or } y = 3$$

So $(0,3)$ is the point where the line crosses the y-axis.
Letting $y = 0$, you get $3x + 4(0) = 12$, or $3x = 12$. This tells you that $(4,0)$ is the point where the line crosses the x-axis.

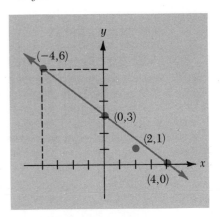

Graphical Interpretation of Examples 9, 10, and 11

Example 10

What point on the line $3x + 4y = 12$ has its x-coordinate equal to -4?

The point $(-4,6)$ is the correct answer.
Replace x by -4 in $3x + 4y = 12$ to obtain

$$3(-4) + 4y = 12$$
$$-12 + 4y = 12$$

Hence, $4y = 12 + 12 = 24$, or $y = 6$.
This is satisfied when $x = -4$ and $y = 6$; therefore $(-4,6)$ is on the line $3x + 4y = 12$.

Example 11

Is the point $(2,1)$ on the line $3x + 4y = 12$?

No. You don't have to draw the line to see this. All you have to do is replace x by 2 and y by 1 in $3x + 4y = 12$ to see if you get a true statement.
Replacing x by 2 and y by 1, you get

$$3(2) + 4(1) = 12 \quad \text{or } 6 + 4 = 12$$

which is a false statement. Therefore $(2,1)$ is not a point on the line $3x + 4y = 12$.

The graph of $Ax + By = C$ is quite special. We should not believe that all graphs are straight lines. For example, the graph of $y = \frac{1}{x}$ is not a straight line. To see this, let's locate a few points that belong to the graph and see what happens.

We call $Ax + By = C$ a linear equation to emphasize that its graph is a straight line.

x	$\frac{1}{x} (= y)$	(x,y)
1	$\frac{1}{1} = 1$	$(1,1)$
2	$\frac{1}{2}$	$\left(2, -\frac{1}{2}\right)$
3	$\frac{1}{3}$	$\left(3, \frac{1}{3}\right)$
4	$\frac{1}{4}$	$\left(4, \frac{1}{4}\right)$

The expression $\frac{1}{x}$ is not defined when $x = 0$ (that is, we cannot divide by 0). Accordingly, no point on $y = \frac{1}{x}$ can have its x-coordinate equal to 0. This means that the curve $y = \frac{1}{x}$ never meets the y-axis.

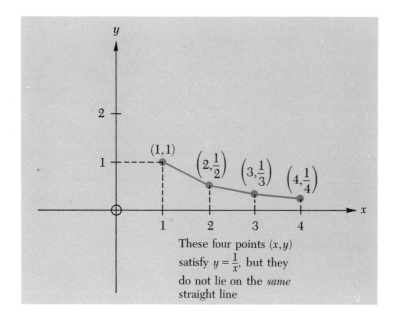

These four points (x,y) satisfy $y = \frac{1}{x}$, but they do not lie on the *same* straight line

The lines we drew to connect the points are not the true graph. We drew the lines just to show that the graph was not a straight line.

Section 7.4 The General Straight-Line Equation   179

PRACTICE DRILL

1. What point on the line $3x + 2y = 6$ has

    (a) its x-coordinate equal to 0
    (b) its y-coordinate equal to 0
    (c) its x-coordinate equal to 4
    (d) its y-coordinate equal to 6

2. What point on the line $x = 4$ has

    (a) its y-coordinate equal to 3
    (b) its x-coordinate equal to 3
    (c) its x-coordinate equal to 4

3. In the diagram below, draw the line

    (a) L, where the equation of L is $3x + 2y = 6$
    (b) M, where the equation of M is $3x - 2y = 6$
    (c) N, where the equation of N is $x = -2$
    (d) P, where the equation of P is $y = -2$

Answers

1. (a) (0,3) (b) (2,0) (c) (4,−3)
   (d) (−2,6)   2. (a) (4,3)
   (b) no such point (c) every point on the line
3.

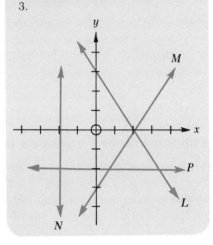

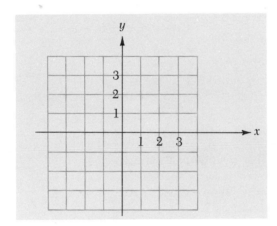

CHECK THE MAIN IDEAS

1. The equation $y = mx + b$ can never describe a line that is parallel to _____.

2. Consider the line that is parallel to the y-axis and passes through the point (4,0). Every point on this line has its x-coordinate equal to (a) _____, regardless of its y-coordinate. In other words, there is no relationship between the two coordinates, only the fact that the (b) _____-coordinate of the point is 4.

3. Therefore, the equation of this line is _____ = 4.

4. The general equation $Ax + By = C$ includes the equation $x = 4$. Namely, let $A = 1$, $B =$ _____ , and $C = 4$ to get $1x + 0y = 4$, or $x = 4$.

5. The equation $y = 3$ is the line that is parallel to the (a) _____-axis and lies 3 units (b) _____ it.

6. The equation $y = 3$ is included in the equation $y = mx + b$. You let m equal _____ and b equal 3.

7. The equation of any _____ can be written in the form $Ax + By = C$.

Answers

1. y-axis   2. (a) 4 (b) x
3. x 4. 0   5. (a) x
(b) above   6. 0   7. line (in the plane)   8. x 9. y
10. (a) axis (b) y (c) x

8. If the line is parallel to the y-axis, its equation has the form _____ = c.
9. If the line is parallel to the x-axis, its equation has the form _____ = c.
10. If the line is not parallel to either axis, you draw the line by finding where the line crosses each (a) _____. To find where it crosses the x-axis, let (b) _____ = 0; and to find where it crosses the y-axis, let (c) _____ = 0. Once you know where the line crosses each axis, you draw the line through these two points.

Section 7.5
Solving Linear Equations by Graphs

OBJECTIVE
To be able to use graphs to solve linear equations in one unknown.

One use of graphs is to help in solving equations. We have already seen that there is a system for solving linear equations by using the structure of arithmetic. In this section, we shall see how we may use graphs to solve these equations. This technique makes it easier to solve more complicated equations.

Suppose we are given an equation such as

$$7 = 2x + 3 \qquad [1]$$

By the methods of Lesson 5, it is not hard to see that the solution of [1] is $x = 2$.

We can also solve [1] by the use of graphs. We replace 7 by y to get the equation

$$y = 2x + 3 \qquad [2]$$

The graph of equation [2] is a straight line. The line has a slope of 2, and its y-intercept is the point (0,3).

The relationship between [1] and [2] is that [1] is the equation we must solve if we want to find the point on $y = 2x + 3$ whose y-coordinate is 7.

This is shown in more detail in Example 1.

Example 1

What point on the line $y = 2x + 3$ has its y-coordinate equal to 7?

The point (2,7) is the point on $y = 2x + 3$ whose y-coordinate is 7.

You are told that y is 7, so you replace y by 7 and obtain an equation that you can solve for x. If you replace y by 7, you get the equation

$$7 = 2x + 3$$

which is precisely equation [1].

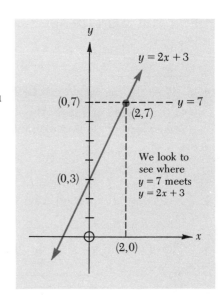

Section 7.5 Solving Linear Equations by Graphs

Example 2

Use a graph to solve the equation

$$-5 = 2x + 3 \qquad [3]$$

Draw the graph of $y = 2x + 3$ and look for the point on the graph whose y-coordinate is -5. That is, if you replace y by -5 in the equation $y = 2x + 3$, you are finding the x-coordinate of the point on the line whose y-coordinate is -5.

We get:

1. Draw $y = 2x + 3$.
2. See where it meets $y = -5$.

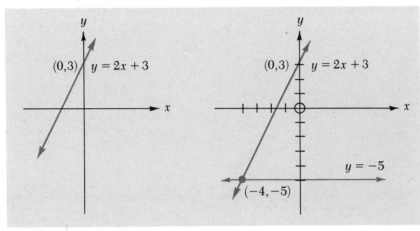

As a check, we have

$$-5 = 2x + 3$$
$$-3 = -3$$
$$-8 = 2x,$$

so $x = -4$.

Example 3

Use a graph to solve the equation $3 = 4x - 1$.

You graph $y = 4x - 1$ and look for the point whose y-coordinate is 3. That is,

1. Draw $y = 4x - 1$.
2. Then see where it meets $y = 3$.

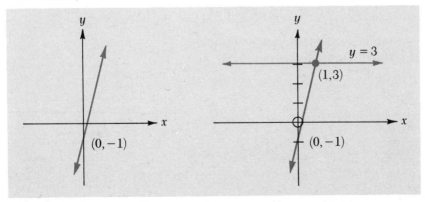

To solve Examples 1 and 2, we used the line $y = 2x + 3$. To solve Example 3, we used the line $y = 4x - 1$. We can solve more complicated problems by combining these two results.

Example 4

Use graphs to solve the equation $4x - 1 = 2x + 3$.

You draw the lines $y = 4x - 1$ and $y = 2x + 3$ in the same diagram. The point at which the two lines meet is on each line, therefore, its coordinates satisfy both equations.

This means that you can compute the y-coordinate of the point at which the two lines meet in two ways. Using one equation, y is $4x - 1$, and using the other, y is $2x + 3$. Because you have only one point, the y-coordinate is the same no matter which you use.

Therefore, $4x - 1 = 2x + 3$ must tell you the x-coordinate of the point at which the two lines meet. Drawing the lines, you have

By algebra, we have

$$\begin{array}{rr} 4x - 1 = & 2x + 3 \\ -2x & -2x \\ \hline 2x - 1 = & 3 \\ +1 = & +1 \\ \hline 2x = & 4 \end{array}$$

Therefore, $x = 2$.

We're not claiming that it is easier to use graphs in this example. Rather, we are showing that we can use graphs as a general method that works for much more complicated equations. In other words, although it is easy to solve linear equations without the concept of a graph, it often happens that we need graphs to handle nonlinear equations.

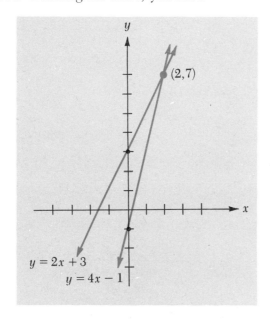

Example 5

Use graphs to find the solution of the equation

$$3x - 4 = 2x + 1 \qquad [4]$$

Draw the lines $y = 3x - 4$ and $y = 2x + 1$. The solution of [4] is the x-coordinate of the point at which these two lines meet. Since the lines meet at $(5, 11)$, the required x-coordinate is 5.

For example,

x	$2x$	$3x$	$3x - 4$	$2x + 1$
1	2	3	-1	3
2	4	6	2	5

So $(1, 3)$ and $(2, 5)$ are on the line $y = 2x + 1$; and $(1, -1)$ and $(2, 2)$ are on line $y = 3x - 4$.

As a check,

$$\begin{array}{rr} 3x - 4 = & 2x + 1 \\ -2x & -2x \\ \hline x - 4 = & 1 \\ +4 & +4 \\ \hline x = & 5 \end{array}$$

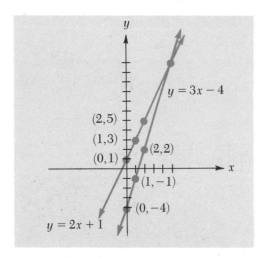

Section 7.5 Solving Linear Equations by Graphs

In closing this lesson, we should notice that there is no reason why the axes have to be called the x- and y-axes. For example, if we have the formula

$$c = 2p + 3 \qquad [5]$$

we can label the horizontal axis the p-axis and the vertical axis the c-axis. If we did this, the graph of [5] would be

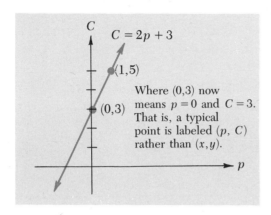

This line looks exactly like the line in Examples 1 and 2. That is, aside from the name of the variables, $y = 2x + 3$ and $c = 2p + 3$ are the same equation. In either case, we find one variable by taking 3 more than twice the other.

Now we can see how a graph allows us to visualize more readily what an algebraic relationship actually looks like. This idea is particularly important when the relationship is not linear, because nonlinear equations are very often difficult to solve by algebraic methods.

PRACTICE DRILL

1. Sketch the line $y = 2x + 3$.
2. Sketch the line $y = x + 5$.
3. Use the graph to estimate where the two lines meet.

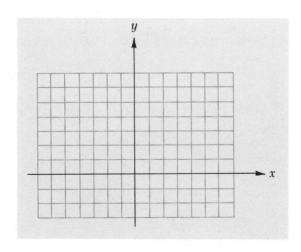

Answers

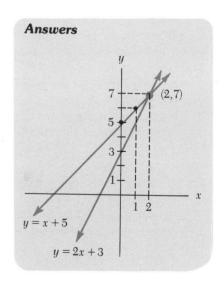

CHECK THE MAIN IDEAS

1. You use the equation $y = 2(4) + 3$ to find the y-coordinate of the point on the line $y = 2x + 3$ whose x-coordinate is _____.
2. You use the equation $16 = 2x + 3$ to find the x-coordinate of the point on the line $y = 2x + 3$ whose _____-coordinate is 16.
3. If x is the x-coordinate of any point on the line $y = 2x + 3$, the y-coordinate of that point is _____.
4. If x represents the x-coordinate of any point on the line $y = x + 5$, the y-coordinate of that point is _____.
5. If these two lines meet at (x,y), then y must equal $2x + 3$, because the point is on the line $y = 2x + 3$. But the point is also on $y = x + 5$, so y must also equal _____.
6. That is, you can find the x-coordinate of the point at which the two lines meet by solving the equation $2x + 3 = $ _____.
7. Because the solution of $2x + 3 = x + 5$ is $x = 2$, and because when $x = 2$ both $2x + 3$ and $x + 5$ equal 7, you may conclude that the two lines $y = 2x + 3$ and $y = x + 5$ meet at the point _____. This checks with the geometric solution given in the practice drill.

Answers
1. 4 2. y 3. $2x + 3$
4. $x + 5$ 5. $x + 5$
6. $x + 5$ 7. $(2,7)$

EXERCISE SET 7 (Form A)

Section 7.1

1. How far and in what direction do you have to move along the number line to get from $^-3.6$ to $^-7.4$?
2. How far and in what direction do you have to move along the number line to get from $^-287$ to $2,713$?
3. What is your change in altitude if you go from 287 feet below sea level to 2,713 feet above sea level?
4. Suppose after moving 386 units in the positive direction along the number line, you arrive at the point 250. Form what point on the number line did you start?
5. If after increasing your altitude by 386 feet, you are 250 feet above sea level, what was your original altitude?

Section 7.2

6. A corner of a geometric figure is called a vertex. The plural of vertex is vertices. If three vertices of a square are at the points $(0,0)$, $(0,4)$, and $(4,0)$, at what point is the fourth vertex of the square?
7. A rectangle has three of its vertices at the points $(-3,4)$, $(3,4)$, and $(3,6)$.

   (a) At what point is the fourth vertex of the rectangle?
   (b) What is the perimeter of the rectangle?

Answers: Exercise Set 7
(Form A)
1. 3.8 units to the left (negative)
2. 3,000 units to the right (positive) 3. a rise of 3,000 feet
4. -136 5. 136 feet below sea level   6. $(4,4)$
7. (a) $(-3,6)$ (b) 16 units

Exercise Set 7 (Form A)

8. In the space below, draw the triangle whose vertices are at the points $A(-2,-1)$, $B(-1,2)$, and $C(3,1)$.

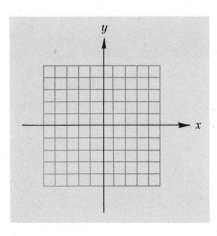

8. see graph

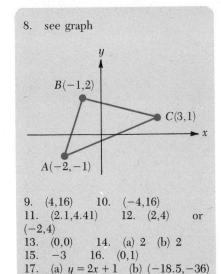

9. (4,16)   10. (−4,16)
11. (2.1,4.41)   12. (2,4)   or (−2,4)
13. (0,0)   14. (a) 2 (b) 2
15. −3   16. (0,1)
17. (a) $y = 2x + 1$ (b) (−18.5,−36)
 (c) (33,67)   18. (a) 2 (b) (0,3)
19. $\frac{5}{9}$

Section 7.3

Suppose the x- and y-coordinates of a point are related by $y = x^2$.

9. What must the point be if the x-coordinate is 4?
10. What must the point be if the x-coordinate is −4?
11. What must the point be if the x-coordinate is 2.1?
12. What must the point be if the y-coordinate is 4?
13. What must the point be if the y-coordinate is 0?
14. What is the slope of the line that passes through the points

    (a) (2,5) and (8,17)   (b) (8,17) and (2,5)

15. What is the slope of the line that passes through the points (5,1) and (3,7)?
16. Suppose L is the line that passes through (2,5) and (8,17). What is the point at which L crosses the y-axis?
17. Suppose L is the line which passes through (2,5) and (8,17).

    (a) How must y be related to x if (x,y) is on L?
    (b) What point on L has its x-coordinate equal to −18.5?
    (c) What point on L has its y-coordinate equal to 67?

18. Let L be the line whose equation is $y - x - 2 = x + 1$.

    (a) What is the slope of L?
    (b) At what point does L cross the y-axis?

19. The relationship between Celsius (C) and Fahrenheit (F) is linear. Suppose you graph this relationship using C as the vertical axis and F as the horizontal axis. What is the slope of the line? (Use the facts that when $C = 0$, $F = 32$, and when $C = 100$, $F = 212$.)

20. In the space below, draw each of the lines $y = mx + 1$ if

(a) $m = 2$ (b) $m = -2$ (c) $m = 0$ (d) $m = \dfrac{1}{2}$ (e) $m = \dfrac{5}{3}$

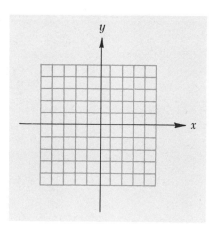

21. In the space below, draw the line $y = -2x + b$ if

(a) $b = -2$ (b) $b = 2$ (c) $b = 0$

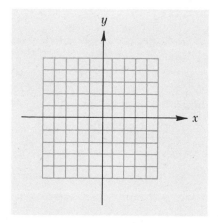

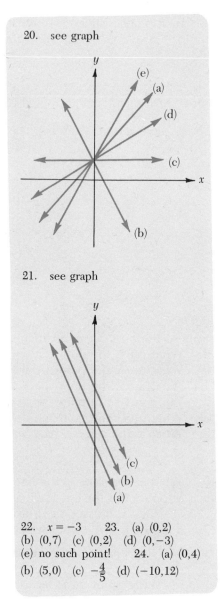

20. see graph

21. see graph

22. $x = -3$ 23. (a) (0,2) (b) (0,7) (c) (0,2) (d) (0,−3) (e) no such point!   24. (a) (0,4) (b) (5,0) (c) $-\dfrac{4}{5}$ (d) (−10,12)

Section 7.4
22. What is the equation of the vertical line that passes through (−3,5)?
23. What is the point at which each of the following lines crosses the y-axis?

(a) $x + 2y = 4$ (b) $y = 7$ (c) $2x + 3y = 6$
(d) $3x - 5y = 15$ (e) $x = 5$

24. Suppose L is the line whose equation is $4x + 5y = 20$.

(a) At what point does L cross the y-axis?
(b) At what point does L cross the x-axis?
(c) What is the slope of L?
(d) What point on L has its y-coordinate equal to 12?

Exercise Set 7 (Form A)

Section 7.5

25. In the space below, draw the lines $y - 2x = 3$ and $y = 3x + 5$. From the graphs, guess the point at which the two lines meet. Check the accuracy by seeing if the coordinates of the point of intersection satisfy the equations of the two lines.

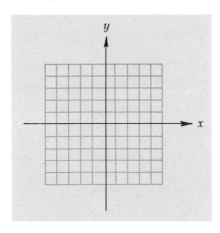

26. Use the algebraic method to determine the point at which the lines $y = 2x + 3$ and $y = 5x + 4$ meet. Then draw both lines in the space below and use the graphs to see if you can determine the point of intersection exactly.

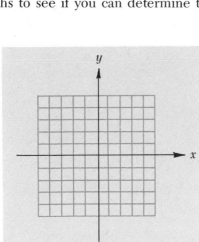

27. At what point do the lines $y = 2x + 5$ and $y = 3x + 32$ meet?

---

25. see graph

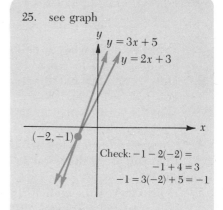

Check: $-1 - 2(-2) =$
$-1 + 4 = 3$
$-1 = 3(-2) + 5 = -1$

26. $\left(-\frac{1}{3}, \frac{7}{3}\right)$; see graph

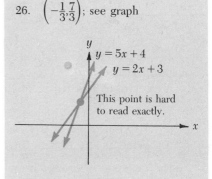

This point is hard to read exactly.

27. $(-27, -49)$

EXERCISE SET 7 (Form B)

Section 7.1

1. How far and in what direction do you have to move along the number line to get from −4.7 to −3.2?
2. How far and in what direction do you have to move along the number line to get from −117 to 1,883?
3. What is your change in altitude if you go from 117 feet below sea level to 1,883 feet above sea level?
4. Suppose after moving 275 units in the positive direction along the number line, you are at the point 250. From what point on the number line did you start?
5. If after increasing your altitude by 275 feet, you are 250 feet above sea level, what was your original altitude?

Section 7.2

6. If three vertices of a square are at the points (0,0), (−3,0), and (0,3), at what point is the fourth vertex of the square?
7. A rectangle has three of its vertices at the points (−4,−2), (2,−2), and (−4,3).

    (a) At what point is the fourth vertex of the rectangle?
    (b) What is the perimeter of the rectangle?

8. In the space below, draw the triangle whose vertices are at the points $A(3,-2)$, $B(2,4)$, and $C(-3,2)$.

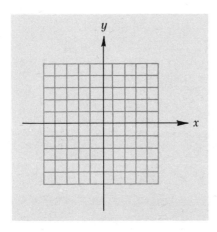

Section 7.3

Suppose the x and y coordinates of a point are related by $y = x^2$.

9. What must the point be if the x-coordinate is 3?
10. What must the point be if the x-coordinate is −3?
11. What must the point be if the x-coordinate is 1.2?
12. What must the point be if the y-coordinate is 16?
13. What must the x-coordinate be if the y-coordinate is 0?
14. What is the slope of the line that passes through the points

    (a) (−2,5) and (8,25)   (b) (8,25) and (−2,5)

15. What is the slope of the line that passes through the points (8,3) and (6,7)?
16. Suppose L is the line that passes through (−2,5) and (8,25). At what point does L cross the y-axis?

Exercise Set 7 (Form B)   189

17. Suppose L is the line that passes through $(-2,5)$ and $(8,25)$.

   (a) How must y be related to x if (x,y) is on L?
   (b) What point on L has its x-coordinate equal to -4.3?
   (c) What point on L has its y-coordinate equal to 29?

18. Let L be the line whose equation is $y + x - 2 = 3x + 4$.

   (a) What is the slope of L?
   (b) At what point does L cross the y-axis?

19. The relationship between Celsius and Fahrenheit is linear. Suppose you graph this relationship using C as the horizontal axis and F as the vertical axis. What is the slope of this line? (Use the facts that when $C = 0$, $F = 32$, and when $C = 100$, $F = 212$.)

20. In the space below, draw each of the lines $y = mx - 2$ if

   (a) $m = 2$ (b) $m = -2$ (c) $m = 0$ (d) $m = \dfrac{1}{3}$ (e) $m = \dfrac{4}{3}$

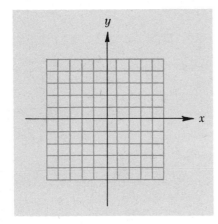

21. In the space below, draw the line $y = -x + b$ if

   (a) $b = -3$ (b) $b = 3$ (c) $b = 0$

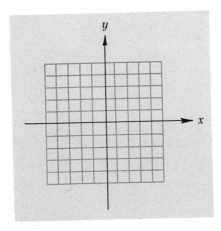

Section 7.4

22. What is the equation of the vertical line that passes through $(-7,7)$?

23. At what point does each of the following lines cross the y-axis?

   (a) $2x + 3y = 6$ (b) $y = -4$ (c) $4x + 3y = 6$
   (d) $4x - 5y = 15$ (e) $x = 7$

24. Suppose L is the line whose equation is $6x + 5y = 30$.

    (a) At what point does L cross the y-axis?
    (b) At what point does L cross the x-axis?
    (c) What is the slope of L?
    (d) What point on L has its y-coordinate equal to 12?

Section 7.5

25. In the space below, draw the lines $y + x = 5$ and $y = 2x - 4$. From the graphs, guess the point at which the two lines meet. Check the accuracy of your answer by seeing if the coordinates of the point of intersection of the two lines satisfy the equation of each of the two lines.

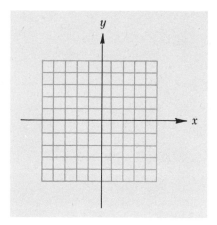

26. Use the algebraic method to find where the lines $y = 5 - 3x$ and $y = 4x + 3$ meet. Then draw both lines in the space below and use the picture to see if you can find the exact point of intersection of the lines.

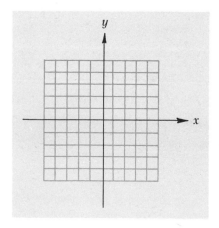

27. At what point do the lines $y = 3x + 40$ and $y = 2x - 7$ meet?

Lesson 8

Simultaneous Linear Equations

Overview In a math class in a high school, the teacher posed the following problem. A bottle and a cork together cost $1.10. The bottle costs $1 more than the cork. How much does each cost?

Almost the whole class blurted out that the bottle cost $1 and the cork cost a dime. Of course, this is the wrong answer: the bottle would cost only 90 cents more than the cork. For some reason most people give the same incorrect answer the first time they hear the question.

The correct answer is that the bottle costs $1.05 and the cork costs $0.05. The teacher used this example to introduce the topic of this lesson: simultaneous equations.

If we let B stand for the cost of the bottle and C for the cost of the cork, we know that $B + C = \$1.10$ and, at the same time, $B - C = \$1.00$. That is, we want:

$$B + C = \$1.10 \quad [1]$$
$$\text{and} \quad B - C = \$1.00 \quad [2]$$

to be true simultaneously (at the same time).

If we add equations [1] and [2], we see that the left side becomes $2B$ (the C and $-C$ add up to 0), and the right side becomes $2.10. In other words, if we add [1] and [2], we get the rather simple equation:

$$2B = \$2.10$$

from which it follows that $B = \$1.05$.

The idea of finding the answer by a method no more complicated than adding two equations is both simple and appealing, so it has wide application. We shall explore the application of this idea in this lesson.

Our main goal in this lesson is to develop a general way, to find the solution if we are given two linear equations in two variables that must be satisfied for the same values of the variables. We call this solving simultaneous linear equations.

Section 8.1
Introduction to Simultaneous Equations

In the last set of exercises, it was sometimes difficult to pinpoint the exact place that two lines intersect. The closer the lines are to being parallel, the more the point of intersection can be affected by a carelessly drawn line. Let's see what this means as we use this section to review.

OBJECTIVE
To understand what is meant by simultaneous equations.

Example 1

Use the x- and y-intercepts to graph the equation $2x + 3y = 6$.

The line meets the x-axis at (3,0) because at that point the y-coordinate is 0. Replacing y by 0 in the equation $2x + 3y = 6$, you get $2x + 3(0) = 6$. When $y = 0$, $x = 3$.
To find the y-intercept, let $x = 0$ in the equation $2x + 3y = 6$. This gives you $2(0) + 3y = 6$, or $3y = 6$. Therefore, when $x = 0$, $y = 2$; and the line crosses the y-axis at (0,2).

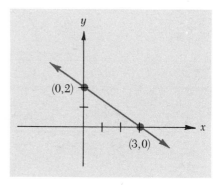

Graph of $2x + 3y = 6$

Example 2

Use the x- and y-intercepts to draw the line $3x + 4y = 12$.

Letting $x = 0$, you have that $3(0) + 4y = 12$, so that $y = 3$. This means that (0,3) is a point on the line. Letting $y = 0$, you get that $3x + 4(0) = 12$, so that $x = 4$. This means that (4,0) is also a point on the line.

Suppose we want to find the point at which the lines in Examples 1 and 2 meet. One way is to draw both lines in the same diagram. We would then see

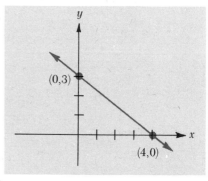

Graph of $3x + 4y = 12$

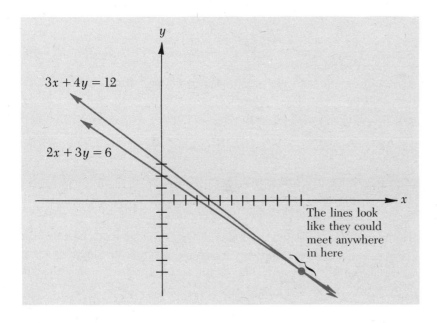

The lines look like they could meet anywhere in here

Even when it's easy to locate the point at which the two lines meet, it is often hard to express the coordinates of the point exactly. At other times, after we draw the two lines, we find that the point at which they meet is off the paper.

Notice how difficult it is to pick out the point at which the two lines meet. As the lines start to meet, there seems to be an almost solid section at which they come together. In the last lesson, we saw how a formula could be represented in terms of a graph. In effect, we learned how we can use pictures to solve algebraic equations. In this lesson, we shall see how we can use algebraic equations to help us locate the point at which a pair of lines meet.

Let's begin our study with a simple case. Suppose we want to find the point at which the lines $y = 3x + 5$ and $y = 2x + 7$ meet. Without drawing either line, we can reason as follows.

One equation tells us that the y value is given by $3x + 5$. The other equation tells us that it is also given by $2x + 7$. This means that the same number (in this case, the y-coordinate of the point at which the two lines meet) is named by both $3x + 5$ and $2x + 7$.

You can see this by picking two lines that seem to pass through the dots that represent $(0,3)$ and $(4,0)$. Notice how the point of intersection seems to change depending on which line you use.

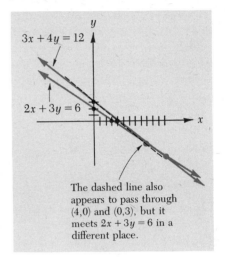

The dashed line also appears to pass through $(4,0)$ and $(0,3)$, but it meets $2x + 3y = 6$ in a different place.

In the language of algebra, this says that
$$3x + 5 = 2x + 7 \qquad [1]$$

Equation [1] is quite familiar by now, and we can solve it at once.

$$\begin{array}{rr} 3x + 5 = & 2x + 7 \\ -2x & -2x \\ \hline x + 5 = & 7 \\ -5 & -5 \\ \hline x = & 2 \end{array}$$

With $x = 2$, we may use either $y = 3x + 5$ or $y = 2x + 7$ to find y. For example, letting $x = 2$ in $y = 3x + 5$, we get
$$y = 3(2) + 5 = 6 + 5 = 11$$

Letting $x = 2$ in $y = 2x + 7$, we get
$$y = 2(2) + 7 = 4 + 7 = 11$$

In either case, when $x = 2$, $y = 11$. This tells us that the lines meet at the point $(2, 11)$.

We sometimes use an arithmetically simpler technique to find where $y = 3x + 5$ and $y = 2x + 7$ meet. We line up the two equations so that the y's are in a vertical line.

$$y = 3x + 5 \qquad [2]$$
$$y = 2x + 7 \qquad [3]$$

In the last lesson, we started with equation [1] and tried to solve the equation by finding where the graphs met. In this lesson, we started with the two lines and decided to find where they met by solving equation [1]. In other words, previously we replaced an algebra problem by a geometry problem. Now we are replacing a geometry problem by an algebra problem.

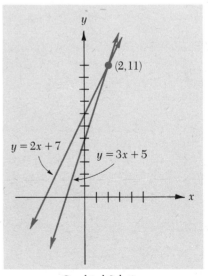

Graphical Solution

Then we subtract [3] from [2] to eliminate y.

$$\begin{array}{r} y = 3x + 5 \\ -y = 2x + 7 \\ \hline 0 = x - 2 \end{array}$$ [4]

We see at once from [4] that $x = 2$. We then find that $y = 11$ just as we did before.

In fact, [1] says that $3x + 5 = 2x + 7$, even if we aren't thinking in terms of lines. For example, we would get equation [1] if we had asked for the value of x that made the expressions $3x + 5$ and $2x + 7$ equal. In this sense, we showed that when $x = 2$, both expressions equal 11.

This shows the complete relationship between algebra and geometry. By the use of Cartesian coordinates, any algebraic equation can be expressed in terms of a geometric relationship; conversely, any geometric relationship can be stated in algebraic terms.

That is, we subtract y from y to get 0; we subtract $2x$ from $3x$ to get $1x$ or x; and we subtract 7 from 5 to get -2.

The point is that although we had to let $x = 2$ in either [2] or [3] only, by doing it in both, we can check our results. That is, if the value we get for y in [2] is not the same as the value we get for y in [3], we know that we have made a mistake: the values for x and y must satisfy both equations.

Example 3

For what value of x are the expressions $3x + 5$ and $x + 11$ equal?

They are equal when $x = 3$.

Since you want $3x + 5$ and $x + 11$ to be equal, the equation is simply

$$3x + 5 = x + 11 \qquad [5]$$

This is solved by the method of Lesson 5.

$$\begin{array}{rr} 3x + 5 = & x + 11 \\ -x & -x \\ \hline 2x + 5 = & 11 \\ -5 = & -5 \\ \hline 2x = & 6 \end{array}$$

so $x = 3$.

CHECK
When $x = 3$,

$$\begin{aligned} 3x + 5 &= 3(3) + 5 \\ &= 14 \\ x + 11 &= 3 + 11 \\ &= 14 \end{aligned}$$

Example 4

At what point do the lines $y = 3x + 5$ and $y = x + 11$ meet?

These two lines meet at the point (3,14).

Because the y-coordinate of the point of intersection is given by both $3x + 5$ and $x + 11$, you can find the x-coordinate of the point of intersection by solving the equation

$$3x + 5 = x + 11$$

This is equation [5] of Example 3.

In other words, by Example 3 you know that $x = 3$. By letting $x = 3$ in either $y = 3x + 5$ or $y = x + 11$, you see that when $x = 3$, $y = 14$. Therefore, the point of intersection is (3, 14).

You can also solve this problem by writing

$$\begin{array}{r} y = 3x + 5 \\ (-)y = x + 11 \\ \hline 0 = 2x - 6 \end{array}$$

So $2x = 6$ and $x = 3$, just as before.

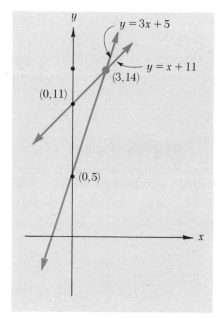

That is, we subtracted y from y to get 0; x from $3x$ to get $2x$; and 11 from 5 to get -6.

Section 8.1   Introduction to Simultaneous Equations

The point is that Examples 3 and 4 are very different conceptually. In Example 3, we are dealing primarily with algebra. In Example 4, we are asking a geometric type of question. But in either case, the resulting algebraic equation is the same.

Let's try just a few more examples.

Example 5

At what point do the lines $y = 4x + 7$ and $y = 3x + 2$ meet?

They meet at the point $(-5, -13)$. You have

$$y = 4x + 7 \qquad [6]$$
and $$y = 3x + 2 \qquad [7]$$

You subtract [7] from [6] to obtain

$$0 = x + 5 \quad \text{or} \quad x = -5$$

But if $x = -5$, then $3x + 2 = 3(-5) + 2 = -15 + 2 = -13$; and $4x + 7 = 4(-5) + 7 = -20 + 7 = -13$.

In either case, when $x = -5$, $y = -13$. So the point at which the two lines meet is $(-5, -13)$.

When we subtract [7] from [6], we are finding the x-coordinate of the point of intersection. Then we use this value of x in either [6] or [7] to find the y-coordinate of that point.

We omit the graph just to emphasize the fact that we do not need the picture once we agree to use the algebraic equations of the lines.

Example 6

At what point do the lines $y = 2x + 7$ and $y = 4x + 8$ meet?

They meet at the point $\left(-\frac{1}{2}, 6\right)$. You have

$$y = 4x + 8 \qquad [8]$$
and $$y = 2x + 7 \qquad [9]$$

If you subtract [9] from [8], you get

$$0 = 2x + 1 \quad \text{or} \quad 2x = -1 \quad \text{or} \quad x = -\frac{1}{2}$$

You then replace x by $-\frac{1}{2}$ in either $y = 4x + 8$ or $y = 2x + 7$ to find that $y = 6$.

We wrote $y = 4x + 8$ first because when we subtract $2x$ from $4x$, we get $2x$. The other way, we would subtract $4x$ from $2x$ to get $-2x$. In general, it is easier to work with positive coefficients, but we don't have to.

Example 7

For what value of x are the expressions $4x + 8$ and $2x + 7$ equal?

The two expressions are equal when $x = -\frac{1}{2}$, in which case each expression equals 6.

In effect, you did this problem when you did Example 6. Asking when $4x + 8 = 2x + 7$ is the same thing as asking to find the point at which the lines $y = 4x + 8$ and $y = 2x + 7$ meet.

Solving $4x + 8 = 2x + 7$ gives the x-coordinate of the point at which the lines meet.

With these examples as background, we are now ready to define the concept of **simultaneous equations.** There are many points (x,y) that belong to the line $y = 2x + 7$. To find such a point, all we have to do is pick a value for x at random and then let $y = 2x + 7$. For example, when $x = 1$, $2x + 7 = 2(1) + 7 = 9$. So $(1,9)$ belongs to the line.

There are also many points that belong to the line $y = 4x - 1$. For example, if we again pick x to be 1, we see that $4x - 1$ is $4(1) - 1$, so that y is 3. This means that the point $(1,3)$ belongs to the line.

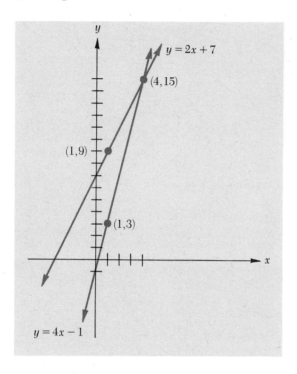

$(1,3)$ satisfies the equation $y = 4x - 1$ but not the equation $y = 2x + 7$. So $(1,3)$ belongs to the line $y = 4x - 1$ but not to the line $y = 2x + 7$. On the other hand, $(1,9)$ is on the line $y = 2x + 7$ but not on the line $y = 4x - 1$.

When we find the point at which $y = 2x + 7$ and $y = 4x - 1$ meet, then we are finding a point (x,y) whose coordinates satisfy both equations.

In the examples we have been studying in this section, we have been looking for *the* point that belongs to both lines. That is, the point at which two lines intersect *belongs to both lines at the same time.* This means that the point is on both lines *simultaneously.*

In other words, when we want to solve the *simultaneous* equations

$$y = 2x + 7 \qquad [10]$$
and
$$y = 4x - 1 \qquad [11]$$

we want to find *the* value of x and *the* of y for which [10] and [11] are simultaneously true. Geometrically, we want to find the point at which the two lines represented by equations [10] and [11] meet.

To indicate that we want the simultaneous solution of both equations, we usually couple the equations with a brace. That is, we write

$$\left. \begin{array}{l} y = 2x + 7 \\ y = 4x - 1 \end{array} \right\}$$

We can then use the method of this section to show that the solution is $x = 4$ and $y = 15$. The braces are interpreted as: "Find all (x,y) for which $y = 2x + 7$ *and* (at the same time) $y = 4x + 1$."

Section 8.2 A Special Case

PRACTICE DRILL

1. At what point do the lines $y = 4x + 3$ and $y = 2x + 9$ meet?
2. At what point do the lines $y = 4x + 3$ and $y = -2x + 9$ meet?
3. For what value of x are the expressions $x + 3$ and $4x - 15$ equal, and what is the value of each expression when they are equal?

Answers
1. (3,15) 2. (1,7)
3. The expressions equal 9 when x equals 6.

CHECK THE MAIN IDEAS

1. There are many points on the line $y = 4x - 15$. One such point is $(0,\underline{})$.
2. But $(0, -15)$ is not on the line $y = x + 3$. In fact, $y = x + 3$ crosses the y-axis at $(0,\underline{})$. This point is not on the line $y = 4x - 15$.
3. There are points on $y = 4x - 15$ that don't belong to the line $y = x + 3$, and there are points on $y = x + 3$ that are not on $y = 4x - 15$. In fact, there is only \underline{} which is on both lines.
4. Geometrically, this is the point at which the \underline{} intersect (cross).
5. Algebraically, you find this point by solving the equation $x + 3 = \underline{}$.
6. This tells you that the \underline{}-coordinate of the point of intersection is 6.
7. When x is 6, both $x + 3$ and $4x - 15$ equal 9. Therefore, the point at which the two lines meet is \underline{}.
8. You refer to this point as the \underline{} solution of the equations $y = x + 3$ and $y = 4x - 15$.

Answers
1. -15 2. 3 3. one point 4. (two) lines
5. $4x - 15$ 6. x 7. (6,9)
8. simultaneous

Section 8.2
A Special Case

In dealing with simultaneous equations of the form

$$\left.\begin{array}{l} y = 4x + 7 \\ y = 2x - 3 \end{array}\right\}$$

all we had to do was equate the two values of y to get the linear equation

$$4x + 7 = 2x - 3 \quad [1]$$

We also could have subtracted the bottom equation from the top to obtain

$$\begin{array}{r} y = 4x + 7 \\ (-)y = 2x - 3 \\ \hline 0 = 2x + 10 \end{array} \quad [2]$$

Supposing the equations are given in the more general form

$$\left.\begin{array}{l} 2x + 3y = 6 \\ 3x + 4y = 12 \end{array}\right\} \quad [3]$$

In the form the equations have in [3], we cannot get rid of either of the variables by adding or subtracting the two equations.

We can rewrite each equation in [3] in the form $y = mx + b$ and then use the method of the last section. But, as the next two examples show, this creates quite a bit of extra algebra. We have to solve each equation and express y in terms of x.

OBJECTIVE
To learn how to solve simultaneous equations of the form

$$\left.\begin{array}{l} ax + by = c \\ dx - by = f \end{array}\right\}$$

NOTE
Sometimes we get confused when we subtract signed numbers. So we often multiply both sides of the bottom (or top) equation by -1. This replaces y by $-y$. For example, if we multiply the bottom equation by -1, we get

$$\left.\begin{array}{l} y = 4x + 7 \\ -y = -2x + 3 \end{array}\right\}$$

Now we can add to get

$$0 = 2x + 10$$

Addition seems easier for most people to handle. Recall that by the distributive property

$$-1(2x - 3) = -1(2x) - 1(-3)$$
$$= -2x + 3$$

Example 1

Rewrite the equation

$$2x + 3y = 6 \qquad [4]$$

so that y is expressed in terms of x.

$2x + 3y = 6$ is the same as

$$y = -\frac{2}{3}x + 2 \qquad [5]$$

Add $-2x$ to both sides of [4] to obtain

$$-2x + (2x + 3y) = -2x + 6$$
$$0 + 3y = -2x + 6$$
$$3y = -2x + 6$$

You can now multiply both sides of this equation by $\frac{1}{3}$ (that is, divide both sides by 3) to obtain

$$y = \frac{1}{3}(-2x + 6) = \frac{1}{3}(-2x) + \frac{1}{3}(6)$$

So the answer is

$$y = -\frac{2}{3}x + 2$$

Equation [4] is more convenient if we want to describe the line in terms of its intercepts. By letting $x = 0$, we see that $3y = 6$, or $y = 2$. And by letting $y = 0$, we get $2x = 6$ so that $x = 3$. This tells us that the line passes through $(0,2)$ and $(3,0)$.

Equation [5] defines the same line but stresses that its slope is $-\frac{2}{3}$ and its y-intercept is 2.

Example 2

Use the equation

$$3x + 4y = 12 \qquad [6]$$

to express y in terms of x.

$3x + 4y = 12$ is the same as $y = -\frac{3}{4}x + 3$. You add $-3x$ to both sides of [6] to obtain

$$4y = -3x + 12$$

Then you divide both sides of this equation by 4 to obtain

$$y = \frac{-3x + 12}{4} = \frac{(-3x)}{4} + \frac{12}{4}$$

Examples 1 and 2 illustrate that there is a lot of work to do if we want to convert the simultaneous equations

$$\left.\begin{array}{r} 2x + 3y = 6 \\ 3x + 4y = 12 \end{array}\right\} \qquad [3]$$

into the form of the last section.

We shall learn how to solve [3] another way in the next section. In this section, we shall treat an easier special case.

Suppose we are given the simultaneous equations

$$\left.\begin{array}{r} 4x + 3y = 15 \\ 2x - 3y = 3 \end{array}\right\} \qquad [7]$$

Nothing says that we have to express y in terms of x (except that we were asked to). We could solve for x in terms of y. That is,

$$\begin{array}{r} 3x + 4y = 12 \\ -4y -4y \\ \hline 3x = 12 - 4y \\ x = \dfrac{12}{3} - \dfrac{4y}{3} \end{array}$$

We chose to solve for y because $y = mx + b$ is a more common way to express the equation of a straight line. It exhibits the slope and y-intercept. That is,

$$y = mx + b$$
$$\downarrow \downarrow$$
slope $(0,b)$ is the
y-intercept

Section 8.2 A Special Case

Notice that in [7], the coefficients of y are opposites, or additive inverses; that is, they have the same magnitude but different signs. If we add the two equations, the terms involving y will cancel, as $3y + (-3y) = 0$. So we add the two equations that make up [7] to obtain

$$4x + 3y = 15$$
$$2x - 3y = 3$$
$$\overline{6x = 18} \quad \text{or} \quad x = 3$$

Once we know that $x = 3$, we may replace x by 3 in either of the two equations in [7] and solve for y. For example, if we replace x by 3 in $2x - 3y = 3$, we get

$$2(3) - 3y = 3 \quad \text{or } 6 - 3y = 3$$

So $-3y = 3 - 6$, or $-3y = -3$. This tells us that when $x = 3$, $y = 1$.

In summary, the simultaneous solution of [7] is $x = 3$ and $y = 1$. In terms of graphs, we have shown that the two lines $4x + 3y = 15$ and $2x - 3y = 3$ intersect at the point $(3,1)$.

Let's try a similar example.

We can easily make a mistake in solving the equation $2x - 3y = 3$ when $x = 3$. To guard against this, we also let $x = 3$ in the other equation and check that we get the same value for y. In this case, if $x = 3$, $4x + 3y = 15$ becomes

$$4(3) + 3y = 15$$
$$12 \phantom{{}+{}} + 3y = 15$$
$$3y = 3$$
$$y = 1$$

Example 3

For what value of x and y are $2x - 4y = 18$ and $3x + 4y = 7$ both true?

They are both true when $x = 5$ and $y = -2$.

You just add the two equations to get

$$2x - 4y = 18$$
$$3x + 4y = 7$$
$$\overline{5x = 25}$$

so $x = 5$.

If you let $x = 5$ in $3x + 4y = 7$, you get

$$3(5) + 4y = 7 \quad \text{or} \quad 15 + 4y = 7$$

Therefore, $4y = 7 - 15 = -8$, or $y = -2$.

In terms of graphs, you are saying that the lines $2x - 4y = 18$ and $3x + 4y = 7$ meet at the point $(5,-2)$.

Remember, the addition works because the y terms drop out. They drop out because the coefficients are equal in magnitude and opposite in sign.

> **A SPECIAL CASE**
> If in a pair of simultaneous equations, the coefficients of y are opposites, we may cancel y by adding the two equations.

Of course, a similar result holds if the coefficients of x are opposites. In this case, if we add the two equations, we cancel the x terms.

As a check, we also let $x = 5$ in $2x - 4y = 18$. We get

$$2(5) - 4y = 18$$
$$10 - 4y = 18$$
$$-4y = 18 - 10 = 8$$
$$y = \frac{8}{-4} = -2$$

Or we can just replace x by 5 and y by -2 in $2x - 4y = 18$ to see if we get a true statement.

$$2(5) - 4(-2)$$
$$= 10 + 8 = 18$$

Example 4

At what point do the lines $3x + 2y = 10$ and $-3x + 4y = 2$ meet?

These two lines meet at the point $(2,2)$.

You add the two equations to eliminate x.

$$3x + 2y = 10$$
$$-3x + 4y = 2$$
$$\overline{6y = 12}$$

or $y = 2$. Now replace x by 2 in $3x + 2y = 10$ to get $3x + 2(2) = 10$, or $3x + 4 = 10$, or $3x = 6$, or $x = 2$. So when $x = 2$, $y = 2$. Thus, the lines meet at $(2,2)$.

As a check, we replace x by 2 and y by 2 in $-3x + 4y = 2$. We get the true statement $-3(2) + 4(2) = 2$.

If we don't mind subtracting one equation from another, the coefficients of one of the variables need not be opposites. They can be equal.

For example, suppose we want to solve the simultaneous equations

$$3x + 4y = 19$$
$$3x - 2y = -5$$ [8]

If we add the two equations in [8], we get

$$6x + 2y = 14$$ [8a]

There is nothing wrong with [8a], but it doesn't simplify anything. We still haven't found the value of either x or y.

However, if we subtract the bottom equation in [8] from the top, we do eliminate the x terms. We get

$$3x + 4y = 19$$
$$3x - 2y = -5$$
$$6y = 24 \quad \text{or} \quad y = 4$$

When we know that $y = 4$, we may replace y by 4 in either of the two equations in [8]. For example, if we replace y by 4 in $3x + 4y = 19$, we get $3x + 16 = 19$, or $3x = 3$, or $x = 1$. This indicates that the solution is given by $x = 1$ and $y = 4$. As a check, we can replace x by 1 and y by 4 in the other equation of [8] to see if the answer works. We get

$$3(1) - 2(4) = 3 - 8 = -5$$

which checks with $3x - 2y = -5$.

Let's try one other example of this type.

Were you tempted to say $2y$ instead of $6y$? Remember, you subtract by changing signs and adding.

$$3x + 4y - (3x - 2y)$$
$$= 3x + 4y + (-3)x + 2y = 6y$$

and

$$19 - (-5) = 19 + 5 = 24$$

As an additional check, we may return to [8a] and replace x by 1 and y by 4. If we do, we get

$$6x + 2y = 6(1) + 2(4)$$
$$= 6 + 8 = 14$$

which checks.

Example 5

Solve the simultaneous equations

$$3x + 4y = 11$$
$$5x + 4y = 13$$ [9]

The solution is given by $x = 1$ and $y = 2$. In other words, the lines $3x + 4y = 11$ and $5x + 4y = 13$ meet at the point (1,2).

If you subtract the bottom equation of [9] from the top, you get

$$-2x = -2 \quad \text{or} \quad x = 1$$

Knowing that x must be 1, you may replace x by 1 in $3x + 4y = 11$ to get $3(1) + 4y = 11$ or $3 + 4y = 11$. This means that $4y = 8$ or $y = 2$.

As a check, replace x by 1 and y by 2 in $5x + 4y = 13$ to obtain $5(1) + 4(2) = 13$ or $5 + 8 = 13$, which is a true statement.

Some people make more careless errors when they subtract than when they add. For this reason, they prefer to have the coefficients of one of the variables opposite rather than equal.

If the variables are equal, all we have to do is multiply one of the equations by -1. This will make the coefficients opposite. For example, if we multiply the top equation in [9] by -1, it becomes

$$-1(3x + 4y) = -1(11)$$
or
$$-3x - 4y = -11$$

We can also subtract the top from the bottom. This helps us avoid negative coefficients, since $5x - 3x$ is $2x$ and $13 - 11$ is 2.

REVIEW

If we perform the same operation on each side of an equation, we get an equivalent equation. That is, we get different equations but they have the same solution. We often say this as "equals added to equals are equal" or "equals subtracted from equals are equal."

Section 8.2   A Special Case

If we now replace the top equation [9] by the new one, we get

$$\left.\begin{array}{r}-3x - 4y = -11\\5x + 4y = 13\end{array}\right\} \quad [10]$$

If we now add the two equations in [10], we get

$$2x = 2$$

Let's try an example in which we multiply one of the equations by -1.

[10] Of course, we can multiply the bottom equation by -1 instead. This leads to

$$\begin{array}{r}3x + 4y = 11\\-5x - 4y = -13\\\hline-2x = -2\end{array} \quad \text{or} \quad x = 1$$

But the first way gives us fewer minus signs to worry about.

Example 6

Solve the simultaneous equation

$$\left.\begin{array}{r}3x + 4y = 29\\7x + 4y = 41\end{array}\right\} \quad [11]$$

The solution is $x = 3$ and $y = 5$.

You may multiply the top equation by -1 to get

$$-3x - 4y = -29$$

If you combine this with the bottom equation of [11], you get the new simultaneous equations

$$\left.\begin{array}{r}-3x - 4y = -29\\7x + 4y = 41\end{array}\right\}$$

Adding these two equations, you get

$$4x = 12 \quad \text{or} \quad x = 3$$

Knowing that $x = 3$, you may find from either $3x + 4y = 29$ or $7x + 4y = 41$ that $y = 5$.

[11] Remember, we can also solve [11] by subtracting the two equations. If, for example, we subtract the top from the bottom, we get $4x = 12$, or $x = 3$.

A MORE GENERAL RESULT

If in a pair of simultaneous equations the coefficients of y (or x) are equal, we may eliminate y (or x) by subtracting one equation from the other.

If we prefer, we may multiply one of the equations by -1 and then add this equation to the second to eliminate y (or x).

In the next section, we shall see how we may convert simultaneous equations in which the coefficients of neither variable are equal or opposite into the type of equation we have studied in this section.

$$\begin{array}{r}4x + 5y = 25\\2x + 5y = 15\\\hline 2x = 10,\end{array} \quad x = 5$$

$$\begin{array}{r}4x + 5y = 25\\-2x - 5y = -15\\\hline 2x = 10\end{array}$$

or

$$\begin{array}{r}-4x - 5y = -25\\2x + 5y = 15\\\hline -2x = -10,\end{array} \quad x = 5$$

PRACTICE DRILL

Solve each of the following pairs of simultaneous equations.

1. $\left.\begin{array}{r}2x + 3y = 13\\2x - 3y = 7\end{array}\right\}$ 2. $\left.\begin{array}{r}2x + 3y = 5\\x + 3y = 1\end{array}\right\}$ 3. $\left.\begin{array}{r}2x + 3y = 11\\-2x + 2y = 14\end{array}\right\}$

Answers
1. $x = 5$ and $y = 1$
2. $x = 4$ and $y = -1$
3. $x = -2$ and $y = 5$

CHECK THE MAIN IDEAS

1. Suppose you are given two equations of the form $Ax + By = C$. Then unless one or both of the equations represents a line parallel to the _____ _____, you can write each equation in the form $y = mx + b$ and use the method of the previous section to solve the simultaneous equations.

Answers
1. y-axis   2. opposite
3. adding   4. 8   5. y
6. subtracting   7. 6
8. equal

2. You can eliminate y from both equations without having to rewrite the equations if the coefficients of y in the two equations are either equal or _____.

3. If the coefficients of y are opposite, you eliminate y by _____ the two equations.

4. As an illustration, if $\left.\begin{array}{l}3x + 4y = 7 \\ 5x - 4y = 1\end{array}\right\}$ the coefficients of y are opposites, and you can add the two equations to get $8x = $ _____.

5. If the coefficients of y are neither equal nor opposite, you can still add or subtract the equations, but this process will not eliminate _____ in this case.

6. There is no reason why you have to eliminate y. For example, if the coefficients of x are equal in the two equations, you can eliminate x by _____ one equation from the other.

7. If you are given:
$$\left.\begin{array}{l}3x + 7y = 12 \\ 3x + 5y = 6\end{array}\right\}$$
you can subtract the bottom equation from the top to get $2y = $ _____.

8. The principle behind what you did here is usually stated as "equals added to (or, subtracted from) equals are _____."

Section 8.3
The More General Case

Suppose we want to solve the pair of equations

$$\left.\begin{array}{l}2x + 3y = 6 \\ 3x + 4y = 12\end{array}\right\} \qquad [1]$$

Now the coefficients of y (and x) in the two equations are neither equal nor opposite. If we add the two equations, neither variable will be eliminated. In fact, if we add the two equations in [1], we get

$$5x + 7y = 18 \qquad [2]$$

If we subtract the top equation in [1] from the bottom, we get

$$x + y = 6 \qquad [3]$$

We'd like to be able to convert the equations in [1] into equations in which the coefficients of y (or x) are equal. Then we can use the method of the previous section.

> The technique we use is based on the fact that *if we multiply both sides of an equation by the same non-zero number*, the new equation has the *same solutions* as the original equation.

Example 1

What pair of simultaneous equations do you get if you multiply the top equation in [1] by 4 and the bottom equation by 3?

You get the pair of equations

$$\left.\begin{array}{l}8x + 12y = 24 \\ 9x + 12y = 36\end{array}\right\} \qquad [4]$$

OBJECTIVE
To solve the pair of simultaneous equations

$$Ax + By = C$$
$$Dx + Ey = F$$

even if the coefficients of y (or x) are neither equal nor opposite.

Neither [2] nor [3] is incorrect, but the equations do not help us isolate x or y.

For example, if we multiply both sides of $x + y = 1$ by 3, we get $3(x + y) = 3(1)$ or, $3x + 3y = 3$.
Though $x + y = 1$ and $3x + 3y = 3$ are different equations, they have the same solutions. In terms of graphs, both equations describe the line that passes through $(0,1)$ and $(1,0)$. In terms of algebra, $x + y = 1$ and $3(x + y) = 3$ are equivalent equations, because they have the same solutions.

Section 8.3 The More General Case

That is, $4(2x + 3y) = 4(6)$ or $8x + 12y = 24$; and $3(3x + 4y) = 3(12)$ or $9x + 12y = 36$.

Equations [4] and [1] have the same solution, but in [4], the coefficients of y are equal. So, as seen in the previous section, we can eliminate y by subtracting one equation from the other in [4].

What made us multiply the top equation in [1] by 4 and the bottom equation by 3? We did the multiplication this way to make sure that the coefficients of y would be the same in both equations. In other words, the coefficients are 3 and 4, and 3×4 and 4×3 are both 12.

If we wanted to eliminate x, we would notice that the coefficients of x are 2 and 3; therefore, we would multiply the 2 by 3 and the 3 by 2. This way, the coefficient of x, would be 6 in both of the new equations.

Example 2

Rewrite equations [1] so that the new equations have the same coefficient of x.

The new equations could be

$$\left.\begin{array}{r}6x + 9y = 18 \\ 6x + 8y = 24\end{array}\right\} \qquad [5]$$

That is, you start with

$$\left.\begin{array}{r}2x + 3y = 6 \\ 3x + 4y = 12\end{array}\right\}$$

You multiply the top equation by 3 to get

$$3(2x + 3y) = 3(6)$$
or $\qquad 6x + 9y = 18$

You multiply the bottom equation by 2 to get

$$2(3x + 4y) = 2(12)$$
or $\qquad 6x + 8y = 24$

Remember that by the distributive property, to multiply $2x + 3y$ by 3 we must multiply both $2x$ and $3y$ by 3.

Notice that if we subtract the top equation from the bottom in [4], we get $x = 12$. If we subtract the bottom equation from the top in [5], we get $y = -6$. Since [4] and [5] have the same solution as [1], the solution of the simultaneous equations

$$\left.\begin{array}{r}2x + 3y = 6 \\ 3x + 4y = 12\end{array}\right\} \qquad [1]$$

is given by $x = 12$ and $y = -6$.

As a check, we need only replace x by 12 and y by -6.

$$2x + 3y = 2(12) + 3(-6) = 24 - 18 = 6$$
$$3x + 4y = 3(12) + 4(-6) = 36 - 24 = 12$$

Let's go through another example step by step.
Suppose we want to solve the system

$$\left.\begin{array}{r}4x + 5y = 7 \\ 5x + 6y = 9\end{array}\right\} \qquad [6]$$

STEP 1: We eliminate y first. Since the coefficient of y is 5 in the first equation and 6 in the second equation, we multiply the first equation by 6 and the second by 5 to obtain

We could choose first to eliminate x. The variables may be eliminated in any order we want.

$$\left.\begin{array}{r}6(4x + 5y) = 6(7) \\ 5(5x + 6y) = 5(9)\end{array}\right. \quad \text{or} \quad \left.\begin{array}{r}24x + 30y = 42 \\ 25x + 30y = 45\end{array}\right\} \qquad [7]$$

STEP 2: If we subtract the top equation from the bottom in [7], we obtain that $x = 3$.

We could choose to subtract the bottom from the top, but we would get a negative coefficient. The important thing is to eliminate y by subtracting either equation in [7] from the other.

STEP 3: To begin eliminating x, we notice that the coefficient of x in the top equation of [6] is 4, and the coefficient of x in the bottom equation is 5. Therefore, we multiply both sides of the top equation by 5, and both sides of the bottom equation by 4, to obtain

$$\left.\begin{array}{l}5(4x + 5y) = 5(7) \\ 4(5x + 6y) = 4(9)\end{array}\right\} \quad \text{or} \quad \left.\begin{array}{l}20x + 25y = 35 \\ 20x + 24y = 36\end{array}\right\} \qquad [8]$$

STEP 4: We now subtract the bottom equation of [8] from the top equation to get $y = -1$, so that the solution to [6] is $x = 3$ and $y = -1$. We check by replacing x by 3 and y by -1.

$$4x + 5y = 4(3) + 5(-1) = 12 - 5 = 7$$
$$5x + 6y = 5(3) + 6(-1) = 15 - 6 = 9$$

From a geometric point of view, we have shown that the lines defined in equations [6] meet at the point $(3, -1)$.

Let's try another example.

Example 3

Solve the simultaneous equations

$$\left.\begin{array}{l}3x + 5y = 6 \\ 4x + 7y = 13\end{array}\right\} \qquad [9]$$

The solution is $x = -23$ and $y = 15$.

If you want to find x first, then you eliminate y. To do this, multiply the top equation by 7 and the bottom equation by 5 to get

$$\left.\begin{array}{l}21x + 35y = 42 \\ 20x + 35y = 65\end{array}\right\} \qquad [10]$$

Now subtract the bottom equation from the top equation in [10] to get

$$x = 42 - 65 \quad \text{or} \quad -23$$

To find the value of y, eliminate x. This is done by multiplying the top equation in [9] by 4 and the bottom equation by 3. You get

$$\left.\begin{array}{l}12x + 20y = 24 \\ 12x + 21y = 39\end{array}\right\} \qquad [11]$$

If you subtract the top equation from the bottom in [11] you get

$$y = 15$$

Check:

$$3x + 5y = 3(-23) + 5(15) = -69 + 75 = 6$$
$$4x + 7y = 4(-23) + 7(15) = -92 + 105 = 13$$

We could let $x = -23$ in either of the equations in [9]. We would get

$$3(-23) + 5y = 6$$

or else

$$4(-23) + 7y = 13$$

Either of these equations has $y = 15$ as the solution.

However, the method we are using allows us to follow the same principle whether we're solving for x or for y. In effect, we have a simple "recipe" that helps us find rather quickly that the solution of

$$\left.\begin{array}{l}3x + 5y = 6 \\ 4x + 7y = 13\end{array}\right\}$$

is $x = -23$ and $y = 15$.

Geometrically, we've shown the lines $3x + 5y = 6$ and $4x + 7y = 13$ meet at the point $(-23, 15)$.

If we prefer addition to subtraction, we can rewrite the equations so that the coefficients are opposites rather than equal. For example, we can multiply the top equation in [9] by 7 and the bottom equation by -5 to get

$$\left.\begin{array}{r}21x + 35y = 42 \\ -20x - 35y = -65\end{array}\right\} \qquad [12]$$

Now we can add the two equations in [12] to find that $x = -23$.

Section 8.3   The More General Case

Example 4

Solve the simultaneous equations

$$\left.\begin{array}{r}2x - 3y = 11 \\ 3x + 4y = 8\end{array}\right\} \quad [13]$$

The solution is given by $x = 4$ and $y = -1$.

You can eliminate y first by multiplying the top equation by 4 and the bottom equation by 3 to obtain

$$\left.\begin{array}{r}8x - 12y = 44 \\ 9x + 12y = 24\end{array}\right\} \quad [14]$$

But since the coefficients of y in [14] are opposites, you eliminate y by adding the two equations in [14].

$$17x = 68 \quad \text{or} \quad x = 4$$

Without substituting this value for x and then solving for y, you can now eliminate x by multiplying the top equation of [13] by -3 and the bottom equation by 2 to obtain

$$\begin{array}{r}-6x + 9y = -33 \\ 6x + 8y = 16\end{array} \quad [15]$$

Adding the two equations in [15], you get

$$17y = -17 \quad \text{or} \quad y = -1$$

Once we know $x = 4$, we can replace x by 4 in either of the two equations in [13] or [14] and solve for y. For example, if we replace x by 4 in the top equation of [13], we get

$$2(4) - 3y = 11$$
$$8 - 3y = 11$$
$$-3y = 11 - 8 = 3$$

So $y = -\dfrac{3}{3}$, or -1.

To check your results:

$$2x - 3y = 2(4) - 3(-1) = 8 + 3 = 11$$
$$3x + 4y = 3(4) + 4(-1) = 12 - 4 = 8$$

There are cases in which a pair of simultaneous equations has either no solutions or more than one solution. From a geometric point of view, every linear equation represents a line. As we noticed earlier, two lines meet at one point, or are parallel and never meet, or the two equations name the same line. Let's look at a few more examples.

Example 5

Find all solutions of the simultaneous equations

$$\left.\begin{array}{r}x + 2y = 3 \\ 3x + 6y = 9\end{array}\right\} \quad [16]$$

Every solution of $x + 2y = 3$ is also a solution of $3x + 6y = 9$. In fact, the bottom equation is obtained by multiplying both sides of the top equation by 3.

If you didn't notice this, you could proceed as in the other problems in this section. For example, you could multiply both sides of the top equation in [16] by -3 to obtain

$$\left.\begin{array}{r}-3x - 6y = -9 \\ 3x + 6y = 9\end{array}\right\} \quad [17]$$

If you add the two equations in [17], you get

$$0 = 0$$

which indicates an identity. This tells you that the equations in [16] are equivalent.

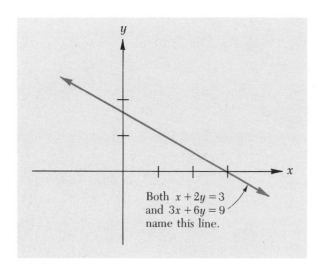

Both $x + 2y = 3$ and $3x + 6y = 9$ name this line.

Example 6

Find all solutions of the simultaneous equations

$$\left. \begin{array}{r} x + 2y = 4 \\ 2x + 4y = 9 \end{array} \right\} \qquad [18]$$

In this case, there are no solutions. For any pair of numbers, x and y, it is *impossible* for both equations in [18] to be true statements.

If you use the method of this section, you can multiply the top equation in [18] by -2 to get

$$\left. \begin{array}{r} -2x - 4y = -8 \\ 2x + 4y = 9 \end{array} \right\} \qquad [19]$$

Now if you add the two equations in [19], you get

$$0 = 1$$

which is impossible. This means that the two equations in [18] are *inconsistent*. Another way to see what happened is to multiply the top equation of [18] by 2. We get

$$\left. \begin{array}{r} 2x + 4y = 8 \\ 2x + 4y = 9 \end{array} \right\}$$

This would make $8 = 9$, which is not possible.

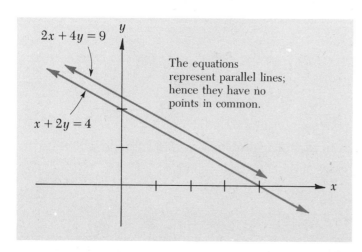

The equations represent parallel lines; hence they have no points in common.

Section 8.3 The More General Case

Let's summarize these possibilities. Given a pair of simultaneous equations, one of three possibilities *must* occur.

1. The equations have one only one solution (that is, one x value and one y value that satisfy both equations simultaneously). For example,

$$\left.\begin{array}{r}x+y=9\\x-y=5\end{array}\right\}$$

have only the solution $x=7$ and $y=2$.

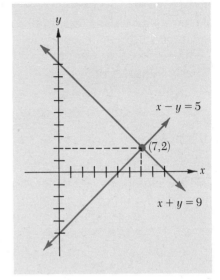

$x+y=9$ and $x-y=5$ are two nonparallel lines, and they meet at $(7,2)$.

2. The equations are inconsistent and have no simultaneous solution. For example,

$$\left.\begin{array}{r}2x+2y=7\\x+y=3\end{array}\right\}$$

is the same as:

$$\begin{array}{r}2x+2y=7\\-2x-2y=-6\end{array}$$

If we add these two equations, we get $0=1$.

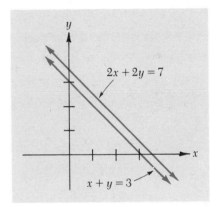

$2x+2y=7$ and $x+y=3$ represent a pair of separate parallel lines. Such lines have no common points.

3. The equations are dependent. That is, one of the equations is "extra," so that every solution of one of the equations is automatically a solution of the other. For example,

$$\left.\begin{array}{r}2x+2y=8\\x+y=4\end{array}\right\}$$

has as a solution every solution of the equation $x+y=4$. If we multiply the bottom equation by -2, we get

$$\begin{array}{r}2x+2y=8\\-2x-2y=-8\end{array}$$

If we subtract these two equations, we get the identity $0=0$.

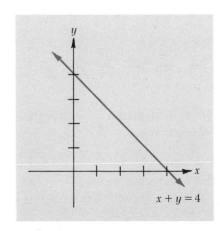

$x+y=4$ and $2x+2y=8$ are two different equations that name the same line. Algebraically, the equations have the same solution because the top equation is obtained by multiplying the bottom equation by 2.

AN OPTIONAL NOTE

The method we have just studied can be applied to three (or more) simultaneous equations in three (or more) variables.

Suppose that we want to find the values of x, y, and z that solve the following simultaneous equations:

$$\left.\begin{array}{r} x + y + 2z = 3 \\ 2x + y + 3z = 4 \\ 3x + 2y + 4z = 8 \end{array}\right\} \quad [1]$$

We can multiply each equation by a number and try to get a new set of equations in which, for example, the coefficients of x have the same magnitude.

For example, the coefficients of x in [1] are 1, 2, and 3. All of these are divisors of 6. So we can multiply the top equation of [1] by 6, the second equation by -3, and the bottom equation by -2. We obtain

$$\left.\begin{array}{r} 6x + 6y + 12z = 18 \\ -6x - 3y - 9z = -12 \\ -6x - 4y - 8z = -16 \end{array}\right\} \quad [2]$$

We want to get equations in which the magnitude of each coefficient of x is 6. The reason for getting -6 in the bottom two equations is that we can eliminate variables by adding rather than by subtracting.

If we replace the second equation in [2] by the sum of the second and the first, and if we replace the third equation by the sum of the third and the first, we can now eliminate x from all but the first equation.

We then get

$$\left.\begin{array}{r} 6x + 6y + 12z = 18 \\ 3y + 3z = 6 \\ 2y + 4z = 2 \end{array}\right\} \quad [3]$$

For example,

$$\begin{array}{r} 6x + 6y + 12z = 18 \\ \underline{-6x - 3y - 9z = -12} \\ 3y + 3z = 6 \end{array}$$

As a check, we get the last equation in [3] by adding the third and the first equation in [2]. That is,

$$\begin{array}{r} 6x + 6y + 12z = 18 \\ \underline{-6x - 4y - 8z = -16} \\ 2y + 4z = 2 \end{array}$$

Though [3] is still three simultaneous equations in three unknowns, if we look at the last two equations, we see that we have two linear equations in the two unknowns, y and z. We can solve the last two equations to find the value of y and the value of z. Then we can use these values of y and z in the first equation of [3] to solve the resulting equation for x.

We can solve [3] in an even simpler way. First, notice that we can divide the first equation in [3] by 6, the second by 3, and the third by 2 to obtain

$$\left.\begin{array}{r} x + y + 2z = 3 \\ y + z = 2 \\ y + 2z = 1 \end{array}\right\} \quad [4]$$

That is, we get

$$6x + 6y + 12z = 18$$

by multiplying both sides of

$$x + y + 2z = 3$$

by 6. But we already know that we don't change the solutions of an equation if we multiply both sides of that equation by the same non-zero number.

Using the technique of the previous section, we can eliminate y from the last equation in [4] by replacing the last (third) equation by the third equation *minus* the second equation. This gives us

$$\left.\begin{array}{r} x + y + 2z = 3 \\ y + z = 2 \\ z = -1 \end{array}\right\} \quad [5]$$

If we rewrite the order of the second and third equation and subtract, we get

$$\begin{array}{r} y + 2z = 1 \\ \underline{y + z = 2} \\ z = 1 - 2 = -1 \end{array}$$

The third equation in [5] tells us that $z = -1$. If we replace z by -1 in the second equation of [5], we get

$$y + (-1) = 2$$

This tells us that $y = 3$. Knowing that $y = 3$ (and that $z = -1$), we can use this information in the first equation in [5] to get

$$x + 3 + 2(-1) = 3 \quad \text{or} \quad x + 1 = 3$$

This tells us that $x = 2$.

Because [1] and [5] have the same solution, the fact that $x = 2$, $y = 3$, and $z = -1$ is the solution of [5] means that it is also the solution of [1].

Section 8.3 The More General Case

As a check, we can replace x by 2, y by 3, and z by -1 to obtain

$$x + y + 2z = 2 + 3 + 2(-1) = 2 + 3 - 2 = 3$$
$$2x + y + 3z = 2(2) + 3 + 3(-1) = 4 + 3 - 3 = 4$$
$$3x + 2y + 4z = 3(2) + 2(3) + 4(-1) = 6 + 6 - 4 = 8$$

This agrees with [1].

In a way, this material goes beyond a beginners' algebra course. But the idea is simple and relatively easy to master.

PRACTICE DRILL

Find the solution of each pair of simultaneous equations.

1. $\left.\begin{array}{l} x + y = 100 \\ x - y = 4 \end{array}\right\}$
2. $\left.\begin{array}{l} x + 3y = 17 \\ x + 2y = 12 \end{array}\right\}$
3. $\left.\begin{array}{l} 2x + 3y = 12 \\ 3x + 4y = 15 \end{array}\right\}$
4. $\left.\begin{array}{l} 2x - 3y = 1 \\ 4x + 2y = 26 \end{array}\right\}$
5. $\left.\begin{array}{l} 2x - 3y = 1 \\ 4x - 6y = 3 \end{array}\right\}$

Answers
1. $x = 52$ and $y = 48$
2. $x = 2$ and $y = 5$
3. $x = -3$ and $y = 6$
4. $x = 5$ and $y = 3$
5. No simultaneous solutions (The equations represent parallel lines.)

CHECK THE MAIN IDEAS

1. This particular section involves pairs of simultaneous equations of the form $Ax + By = C$, where the coefficients of y (and of x) are neither _____ nor opposite.
2. In such a case, you use the fact that if you multiply both sides of an equation by the same non-zero number, the new equation and the original equation have the same _____.
3. Look at the specific example $\left.\begin{array}{l} 2x + 3y = 12 \\ 3x + 4y = 15 \end{array}\right\}$. If you add these two equations, you get _____.
4. You do not eliminate y this way because the coefficients of y in the two equations are not _____.
5. You can get a different pair of equations with the same solutions as the first pair by multiplying the top equation by -4 and the bottom equation by _____.
6. This gives you the equations $\left.\begin{array}{l} -8x - 12y = \underline{\hspace{1cm}} \\ 9x + 12y = 45 \end{array}\right\}$.
7. If you add these two equations, you get $x = $ _____.
8. If you now want to find y, you can use the same idea, but you multiply the top equation by 3 and the bottom equation by _____, and subtract.

Answers
1. equal 2. solutions
3. $5x + 7y = 27$ 4. opposite
5. 3 6. -48 7. -3
8. -2

Section 8.4
Applications to Word Problems

Sometimes it is easier to use two variables than one in solving a problem. Let's begin by reviewing a problem that we could solve in Lesson 6 without the idea of simultaneous equations.

OBJECTIVE
To apply the ideas of this lesson toward the solution of word problems.

Example 1

The sum of two numbers is 15. When the lesser is subtracted from the greater, the difference is 3. What are the two numbers?

The two numbers are 6 and 9.
There are several ways to solve this problem. One way is to let s denote the smaller number. Then, since the greater number is 3 more than the smaller number, you denote the greater number by $s + 3$.
You are told that the sum of the two numbers is 15. That is,

$$s + (s + 3) = 15 \qquad [1]$$

Putting [1] into simpler form, you get

$$2s + 3 = 15 \qquad [2]$$

or

$$\begin{aligned} 2s + 3 &= 15 \\ -3 & \;\; -3 \\ \hline s &= 12 \end{aligned}$$

or $s = 6$; hence $s = 6$ and $s + 3 = 9$.

CHECK
$$9 + 6 = 15$$
$$9 - 6 = 3$$

If we want to use two variables, we can let s stand for the smaller number and let b stand for the bigger number. Then

$$b + s = 15 \quad \text{(that is, the sum of } b \text{ and } s \text{ is 15)}$$
$$b - s = 3 \quad \text{(the difference is 3)}$$

If we add these two equations, we get

$$2b = 18 \quad \text{or} \quad b = 9$$

It's easy to say that $b - s = 3$. Then if we want, we can add s to each side to get $b = s + 3$.

Often it seems easier to use two variables than to try to express the second unknown in terms of the first unknown.

Example 2

The sum of two numbers is 1,004, and their difference is 6. What is the greater number?

The greater number is 505.
If you let b denote the greater and s, the lesser, you have

$$\left. \begin{aligned} b + s &= 1004 \\ b - s &= 6 \end{aligned} \right\} \qquad [3]$$

Adding the two equations in [3], you get

$$2b = 1010$$

so $b = 505$.

Section 8.4   Applications to Word Problems

If you subtract the bottom equation from the top in [3], we get
$$2s = 998 \quad \text{or} \quad s = 499$$
As a check, $505 + 499 = 1{,}004$; and $505 - 499 = 6$.

> Remember, when we subtract $-s$ from s we get $2s$. That is:
> $$s - (-s) = s + s = 2s$$
> When we add s and $-s$, we get 0.

There are many types of problems in which we can use two unknowns. The basic idea is still to translate a problem from words into mathematics. To get an idea of how we use two unknowns, let's look at the same type of problem we attempted in Lesson 6 with one unknown.

Example 3

John's age is 3 less than twice Bill's age. How old is each if the sum of their ages is 60?

Bill is 21 and John is 39.

Let J stand for John's age and B, for Bill's age. Since the sum of their ages is 60, one equation will be $J + B = 60$.

The other equation comes from translating

"John's age is 3 less than twice Bill's"
$$J = -3 + 2B$$

That means that our simultaneous equations are
$$\left. \begin{array}{l} J + B = 60 \\ J = 2B - 3 \end{array} \right\} \qquad [4]$$

In the top equation of [4], it is usually more convenient to replace J by $2B - 3$. This is called **substitution**. That is, we substitute for J in the top equation, its value from the bottom equation. We get
$$(2B - 3) + B = 60 \qquad [5]$$

> Perhaps this is still difficult. Try comparing J and $2B$. Which is more and by how much? Where do you add the 3?
> You are told that J is 3 less than $2B$. To make J and $2B$ equal, you must add 3 to J. This gives you $J + 3 = 2B$, and you can now subtract 3 from both sides (or add $^-3$).

Equation [5] is exactly what you would have gotten using the method of Lesson 6, when you did equations using one unknown. Using two unknowns seems to make it easier to keep track of the relationship between the two ages.

> The method of substitution is used when one of the variables in an equation is already expressed in terms of the other. This method can always be used regardless of the form of the equation, because in a linear equation we can always express one variable in terms of the other.

We can solve [5] by writing
$$3B - 3 = 60, \quad 3B = 63, \quad B = 21$$
$$J = 2B - 3 = 2(21) - 3 = 42 - 3 = 39$$

If you want to, you can still eliminate J by subtracting the bottom equation of [4] from the top. That is,
$$\begin{array}{r} J + B = 60 + (0B) \\ \underline{J = -3 + 2B} \\ B = 63 - 2B \end{array}$$
or
$$3B = 63 \qquad [6]$$

> We write $0B$ to remember that we are subtracting $2B$ from 0, and this is $0 - 2B$, or $-2B$.

Example 4

Tickets to a certain event cost \$1 for children and \$2 for adults. Find the number of each type of ticket sold if the total sales were \$140 and three times as many adult tickets were sold as children's.

20 children's and 60 adult's tickets were sold.

You can let A stand for the number of adult tickets sold, and C for the number of children's tickets. Then

$$\left. \begin{array}{r} 2A + C = 140 \\ A = 3C \end{array} \right\}$$

[7] That is, each adult ticket is \$2, so A adult tickets are $2A$ dollars. Similarly, at \$1 each, C children's tickets cost C dollars.

You can replace A by $3C$ in the top equation to get

$$2(3C) + C = 140$$
$$6C + C = 140$$

or
$$7C = 140$$

Then,
$$C = 20 \quad (= \text{number of children's tickets})$$
$$A = 3C = 3(20) = 60 \quad (= \text{number of adult tickets})$$

Example 5

A piggy bank contains 23 coins, consisting only of dimes and quarters. The total value of the coins is \$3.35. How many coins of each kind are there?

This is exactly the same problem as Example 5 in Section 6.3. If you compare these two examples, you'll see how alike the two methods are, and also how they differ.

There are 16 dimes and 7 quarters.
Let D denote the number of dimes, and Q, the number of quarters. Since there are 23 coins altogether,

$$Q + D = 23$$

A dime is worth 10¢ and a quarter 25¢, so D dimes and Q quarters are worth $10D + 25Q$. Since \$3.35 is 335¢, you have

$$25Q + 10D = 335$$

The simultaneous equations are

$$\left. \begin{array}{r} 25Q + 10D = 335 \\ Q + D = 23 \end{array} \right\}$$

[8] The fact that $Q + D = 23$ makes it easy to see that $Q = 23 - D$ or that $D = 23 - Q$.

You can solve [8] by multiplying the bottom equation by -10.

$$\begin{array}{r} 25Q + 10D = 335 \\ -10Q - 10D = -230 \\ \hline 15Q = 105 \end{array}$$

so $Q = 7$.
If you multiply the bottom equation of [8] by -25, you find that

$$\begin{array}{r} 25Q + 10D = 335 \\ -25Q - 25D = -575 \\ \hline -15D = -240 \end{array}$$

So $D = \dfrac{-240}{-15} = 16$.

We also could choose to multiply the top equation in [8] by -1, and the bottom equation by 25, to get

$$\begin{array}{r} ^-25Q + {^-}10D = {^-}335 \\ 25Q + 25D = 575 \\ \hline 15D = 240 \end{array}$$

This way, we avoid a negative coefficient.

We do not use two unknowns merely as another way of solving a problem we can solve using just one unknown. There are times when it is decidedly harder to think in terms of one unknown. For example, it is easier to handle motion problems that deal with moving against the current or moving with the current in two unknowns. The idea is that if someone travels at 30 mph and there is a current of 5 mph, then if the person travels with the current, his or her speed is 35 mph (30 + 5); but if someone travels against the current, the speed is only 25 mph (30 − 5).

Section 8.4 **Applications to Word Problems**

Example 6

A jet traveling with the wind flies 2325 miles in 3 hours. Flying against the wind, the same jet takes 4 hours to go 2900 miles. What is the speed of the wind and the speed of the plane in still air?

The speed of the jet is 750 mph, and the speed of the wind is 25 mph.

Let J stand for the speed of the jet, and W, for the speed of the wind. When the jet moves with the wind, its speed is $(J + W)$ mph. When it moves against the wind, its speed is $(J - W)$ mph.

Since distance traveled equals the speed times the time, the two equations are

$$\left.\begin{array}{l}3(J + W) = 2325 \\ 4(J - W) = 2900\end{array}\right\} \quad [9]$$

You can simplify [9] if you divide the top equation by 3 and the bottom equation by 4.

$$\left.\begin{array}{l}J + W = 775 \\ J - W = 725\end{array}\right\} \quad [10]$$

Adding the two equations in [10] gives

$$2J = 1500 \quad \text{or } J = 750$$

Subtracting the same two equations gives

$$2W = 50 \quad \text{or } W = 25$$

PRACTICE DRILL

1. The sum of two numbers is 25 and the difference between the greater and the lesser is 7. What are the two numbers?
2. Bill's age is 1 year less than 3 times Joan's age. What is the age of each if the sum of their ages is 47?

Answers
1. 16 and 9   2. Joan is 12 and Bill is 35.

CHECK THE MAIN IDEAS

1. If you let B stand for Bill's age, and J, for Joan's, the fact that Bill's age is one year less than three times Joan's age means that $B = $ _____.
2. If the sum of their ages is 47, the equation is $B + J = $ _____.
3. If the sum of their ages is 47, and, at the same time, Bill's age is 1 year less than 3 times Joan's age, then you must solve the simultaneous equations:

$$\left.\begin{array}{l}B + J = 47 \\ \underline{}\end{array}\right\}$$

4. One way to solve this system is to eliminate B by _____ the bottom equation from the top.
5. If you do this, the left-hand side becomes $(B + J) - B$ or (a) _____, while the right side becomes $47 - $ (b) (_____) or $47 - 3J + 1$.
6. In other words you find Joan's age from the equation: _____ $= 48 - 3J$.
7. If you solve this equation, you find that $J = 12$ and therefore that $3J - 1 = $ _____.
8. Since $3J - 1$ stands for _____'s age, and since J stands for Joan's age, you see that Joan is 12 and Bill is 35.

Answers
1. $3J - 1$ 2. 47
3. $B = 3J - 1$ 4. subtracting
5. (a) J (b) $3J - 1$ 6. J
7. 35   8. Bill

EXERCISE SET 8 (Form A)

Section 8.1
1. At what point do the lines $y = 0.08x$ and $y = 0.05x + 51$ meet?
2. Mary is offered a sales position by a company that is willing to pay her $51 a week guaranteed salary plus a commission of 5% on all sales she makes. Express her weekly salary (y) in terms of her weekly sales (x).
3. A second company makes Mary a different offer. They guarantee her nothing but offer a straight 8% commission on all sales she makes. At what amount of weekly sales does the second company's offer become as good as the first company's offer?
4. At what point do the lines $y = \frac{9}{5}x + 32$ and $y = x$ meet?
5. If you express Fahrenheit temperature readings (F) in terms of Celsius temperature readings (C), the relationship is given by

$$F = \frac{9}{5}C + 32$$

   At what temperature are the Celsius and Fahrenheit readings the same?

Section 8.2
Solve each of the following systems of equations.

6. $\begin{aligned} 3x + 8y &= 12 \\ 5x - 8y &= 20 \end{aligned}$

7. $\begin{aligned} 2.5x + 9.4y &= 12.2 \\ -3.5x - 9.4y &= -15.2 \end{aligned}$

8. $\begin{aligned} 2.5x + 9.4y &= 33.8 \\ -3.5x + 9.4y &= -2.2 \end{aligned}$

9. Solve the system $\begin{aligned} 9 &= 4m + b \\ 5 &= 2m + b \end{aligned}$ for m and b.

10. (a) Use the results of the previous problem to find the values of m and b if the line $y = mx + b$ passes through the points (4,9) and (2,5).
    (b) What is the equation of the line?

Section 8.3
At what point(s) do each of the following pairs of lines meet?
11. $4x + 5y = 15$ and $7x + 9y = 7$
12. $4x + 5y = 15$ and $y = 2x + 3$
13. $4x + 5y = 15$ and $8x + 10y = 30$
14. $4x + 5y = 15$ and $12x + 15y = 40$
15. Find the value of x and y for which

$$\begin{aligned} 4x + 5y &= 118 \\ 7x + 9y &= 210 \end{aligned}$$

16. Suppose that 4 pears and 5 oranges cost $1.18 and that 7 pears and 9 oranges cost $2.10. What is the price of (a) each pear? (b) each orange?

Section 8.4
17. (a) At what point do the lines $y = 30,000 - 550x$ and $y = 18,000 + 1,450x$ meet?
    (b) Town A has a population of 30,000, but the population is decreasing at a rate of 550 persons a year. Town B has a population of 18,000, but the population is increasing at a rate of 1,450 people per year. In how many years will the two towns have the same population, and what will the population of each town be at that time?

Answers: Exercise Set 8

Form A
1. (1700, 136)
2. $y = 0.05x + 51$ 3. $1,700
4. (-40, -40)   5. -40°
6. $x = 4$ and $y = 0$ 7. $x = 3$ and $y = \frac{1}{2}$ 8. $x = 6$ and $y = 2$
9. $m = 2$ and $b = 1$
10. (a) $m = 2$ and $b = 1$
(b) $y = 2x + 1$ 11. (100, -77)
12. (0, 3)   13. all points on either line   14. the lines never meet   15. $x = 12$ and $y = 14$
16. (a) 12¢ (b) 14¢
17. (a) (6, 26,700) (b) In 6 years, both towns have 26,700 people.   18. (a) 12 (b) 6
19. (a) 12¢ (b) 14¢
20. (a) 5 (b) 7
21. (a) 40 mph (b) 800 mph
22. (a) $\frac{1}{2}(h + w) = 93$ or $h + w = 186$ (b) 26 mph
23. $x = -2$, $y = 1$, and $z = 2$
24. $x = 10$, $y = 12$, and $z = 14$
25. (a) 14¢ (b) 12¢ (c) 10¢
26. none — the equations are inconsistent

18. You buy twice as many oranges as pears. If oranges cost 15¢ each, pears cost 18¢ each, and the total cost is $2.88,

    (a) How many oranges did you buy?
    (b) How many pears did you buy?

19. If 11 oranges and 13 pears cost $3.14 but 5 oranges and 6 pears cost $1.44,

    (a) How much does each orange cost?
    (b) How much does each pear cost?

20. A trucking company has two different size trucks. Three full loads of the smaller truck and 4 full loads of the larger truck hold a total of 43 tons, but 3 full loads of the larger truck and 4 full loads of the smaller truck hold a total of 41 tons.

    (a) How many tons is a full load in the smaller truck?
    (b) How many tons is a full load in the larger truck?

21. An airplane traveling with the wind goes 4,200 miles in 5 hours. When it travels against the wind, the plane goes 6,080 miles in 8 hours.

    (a) What is the speed of the wind?
    (b) What is the speed of the plane in still air?

22. Cruising at half-speed against the wind, a traffic helicopter flies 36 miles in 40 minutes. Flying at full speed with the same wind, the helicopter flies 93 miles in 30 minutes. Let h stand for the full speed of the helicopter and let w stand for the speed of the wind, both in mph.

    (a) Write the equation (in terms of h and w) that says the helicopter, flying at full speed with the wind, went 93 miles in 30 minutes.
    (b) What was the speed of the wind?

Optional Exercises for Systems of Three Linear Equations

23. For what values of x, y, and z does

    $$\left. \begin{array}{r} 3x + 4y + 2z = 2 \\ x + y + 3z = 5 \\ 2x + 3y + 4z = 7 \end{array} \right\}$$

24. Find the values of x, y, and z if

    $$\left. \begin{array}{r} 3x + 4y + 2z = 106 \\ 4x + 5y + 3z = 142 \\ 6x + 8y + 5z = 226 \end{array} \right\}$$

25. Three oranges, 4 apples, and 2 pears cost a total of $1.06. Four oranges, 5 apples, and 3 pears cost a total of $1.42. Six oranges, 8 apples, and 5 pears cost $2.26. What is the cost of (a) each pear? (b) each apple? (c) each orange?

26. For what values of x, y, and z does

    $$\left. \begin{array}{r} 3x + 4y + 2z = 106 \\ 4x + 5y + 3z = 142 \\ 6x + 7y + 5z = 226 \end{array} \right\}$$

EXERCISE SET 8 (Form B)

Section 8.1
1. At what point do the lines $y = 0.07x$ and $y = 0.04x + 63$ meet?
2. John is offered a sales position that consists of $63 a week in guaranteed salary plus a commission of 4% on all sales he makes. Express his weekly salary (y) in terms of his weekly sales (x).
3. John is offered another sales position in which he gets no guaranteed salary but does get a 7% commission on all sales he makes. At what amount of weekly sales are the two offers the same?
4. At what point do the lines $y = \frac{2}{3}x + 7$ and $y = x$ meet?
5. There are 7 more than two-thirds as many men as women in a group. How many of each are there if the group has the same number of men as women?

Section 8.2
Solve each of the following systems of equations.

6. $\left. \begin{array}{l} 4x + 11y = 12 \\ 6x - 11y = 18 \end{array} \right\}$

7. $\left. \begin{array}{l} 4.5x + 7.2y = 21.6 \\ 0.5x - 7.2y = -1.6 \end{array} \right\}$

8. $\left. \begin{array}{l} 4.5x + 7.2y = 72 \\ 0.5x + 7.2y = 40 \end{array} \right\}$

9. Solve the system $\left. \begin{array}{l} 11 = 5m + b \\ 12 = 4m + b \end{array} \right\}$ for m and b.

10. (a) Use the previous problem to find m and b if the line $y = mx + b$ passes through the two points $(5, 11)$ and $(4, 12)$.
    (b) Write the equation of this line.

Section 8.3
At what point(s) do each of the following lines meet?
11. $3x + 2y = 15$ and $4x + 3y = 12$
12. $3x + 2y = 15$ and $y = 2x + 4$
13. $3x + 2y = 15$ and $9x + 6y = 45$
14. $3x + 2y = 15$ and $6x + 4y = 20$
15. Find the value of x and y for which

$$\left. \begin{array}{l} 3x + 2y = 42 \\ 4x + 3y = 59 \end{array} \right\}$$

16. Three onions and two peppers cost 42¢. Four onions and three peppers cost 59¢.

    (a) What is the price of each onion?
    (b) What is the price of each pepper?

Section 8.4
17. (a) At what point do the lines $y = 40{,}000 - 600x$ and $y = 12{,}000 + 400x$ meet?
    (b) One town has a population of 40,000, but the population is decreasing at a rate of 600 persons per year. A second town has a population of 12,000, but the population is increasing at a rate of 400 persons per year. In how many years will the two towns have the same population, and what will their population be at that time?

Exercise Set 8 (Form B)

18. You buy three times as many apples as bananas. Apples cost 8¢ each and bananas cost 7¢ each. If the total cost of the fruit is $3.10,

    (a) How many apples did you buy?
    (b) How many bananas did you buy?

19. If 11 oranges and 8 pears cost $1.47, but 4 oranges and 3 pears cost $0.54,

    (a) How much does each orange cost?
    (b) How much does each pear cost?

20. Two full loads of the big truck and 3 full loads of the small truck hold a total of 25 tons. But 3 full loads of the big truck and 4 full loads of the small truck hold a total of 36 tons.

    (a) How many tons is a full load of the smaller truck?
    (b) How many tons is a full load of the larger truck?

21. An airplane traveling with the wind goes 2,190 miles in 3 hours. When it travels against the wind, the same plane goes 2,680 miles in 4 hours.

    (a) What is the speed of the wind?
    (b) What is the speed of the plane in still air?

22. Cruising at two-thirds speed with the wind, a helicopter goes 62 miles in 30 minutes. Flying at full speed against the same wind, the helicopter goes 42 miles in 20 minutes.

    (a) Write the equation that says that at two-thirds speed with the wind, the helicopter goes 62 miles in 30 minutes; use h for the speed of the helicopter and w for the speed of the wind.
    (b) What is the speed of the wind?

Optional Exercises for Systems of Three Linear Equations

23. For what values of x, y, and z does

    $$\left. \begin{array}{r} 2x + y + 3z = 9 \\ x + y + z = 2 \\ 4x - y + 2z = 12 \end{array} \right\}$$

24. Find the values of x, y, and z if

    $$\left. \begin{array}{r} 2x + 3y + 2z = 51 \\ 3x + 2y + 3z = 59 \\ 4x + 4y + 2z = 70 \end{array} \right\}$$

25. Two oranges, 3 apples, and 2 pears cost 51¢. Three oranges, 2 apples, and 3 pears cost 59¢. Four oranges, 4 apples, and 2 pears cost 70¢. What is the cost of (a) each orange? (b) each apple? (c) each pear?

26. Find the values of x, y, and z for which

    $$\left. \begin{array}{r} 2x + 3y + 2z = 51 \\ 3x + 2y + 3z = 59 \\ x + 4y + z = 70 \end{array} \right\}$$

Lesson 9

An Introduction to Inequalities

Overview In George Orwell's book *Animal Farm*, the animals write their own constitution. After their society has existed for a while, the statement that all animals are created equal is replaced by "All animals are equal, but some animals are more equal than others."

The analogy may be farfetched, but in the world of real measurements, the ideas of equality and exactness are hard to define. Can a piece of string be exactly six inches long? Perhaps, but we don't know. To find out, we have to use a ruler or some other measuring device, but the markings on a ruler have thickness. What part of the six-inch marking do we use when we measure a length?

This kind of question has philosophical value, but in the real world, we settle for measurements that are "close enough." We may say that we want the exact length to be 6 inches — within a hundredth of an inch. That is, we might be satisfied to know that the length is *greater than* 5.99 inches but, *less than* 6.01 inches.

Inequalities involve the concepts of "greater than" and "less than". We shall study these statements in this lesson, applying the ideas to relationships in both algebra and graphs. Equalities are a very important topic to study and understand, but it is at least as important to extend the study of relationships from equalities to inequalities.

Section 9.1
An Introduction to Inequalities

OBJECTIVE
To understand the importance of inequalities and to learn the meaning of the symbols < and >.

In real-life situations we often talk about "more than" and "less than" instead of "equal to." How often do we talk about making a salary of *exactly* $10,000? Generally, we say that we want a salary of *at least* $10,000.

Yet until now this course has been devoted to the study of equality rather than to the more general study of **inequality** ("more than" and "less than"). The reason is that we often solve inequalities by knowing enough about equalities. For example, when we say that we want to make more than $10,000 a year, we can think about first making exactly $10,000 a year. Then any amount more than $10,000 is acceptable.

Example 1

A salesperson gets 10% commission on all sales (that is, the salesperson gets $10 for each $100 sale). How big a sale should the salesperson make in order to make a commission of at least $50?

The sale should be at least $500. That is, the salesperson makes $10 for each $100. For a $500 sale, the salesperson earns $50.

Any sale greater than $500 earns the salesperson a commission of more than $50. Therefore, to earn at least $50, the salesperson must make a sale of at least $500.

To solve this problem, essentially we first solve the problem of how to make a commission of exactly $50. Once we know this, the rest of the problem is relatively simple.

Let's try another problem of this type to make sure that we really understand how inequalities are related to equalities.

Example 2

A car traveling at a speed of 30 mph is 100 miles ahead of another car. How fast must the other car travel to overtake the first car in under 5 hours?

The second car can travel at *any* speed that is greater than 50 miles per hour.

The second car has to gain 100 miles on the first car to overtake it. To overtake it in exactly 5 hours, the first car has to gain 20 miles each hour (100 ÷ 5): that is, its speed has to be 20 mph greater than that of the first car. Since the first car is moving at 30 mph, the second car has to move at 50 mph (30 + 20).

To overtake the first car in under 5 hours, the second car can travel at *any* speed that *exceeds* 50 mph.

We can also solve this problem by algebra, but our goal is to stress that first we work with an equality and then we use this answer to solve the inequality.

For example, at 55 mph the other car gains 25 (55 − 30) miles each hour. In that case, it would make up the 100 miles in 4 hours — which is certainly less than 5 hours.

This example shows what we mean when we say that even though inequalities occur more often than equalities, it is important to understand equalities before we can properly study inequalities. Once we agree that inequalities are important, we can turn our attention to inventing symbols that express inequalities.

There is an interesting explanation as to why = was chosen to mean "is equal to." The equal sign consists of two parallel lines, so the space between the two lines is the same at each end. The fact that the spacing at each end is equal symbolizes that the numbers at each end are also equal.

For example, we invented the symbol = to stand for "is equal to." Rather than write

3 + 2 is equal to 5

we write

3 + 2 = 5

This fact led to a rather clever way of inventing symbols for "less than" and "greater than." Rather than keep the spacing between the two lines the same, people decided to slant the lines toward one another so that the spacing at one end would be less than the spacing at the other end. The lesser number is on the side of smaller space.

Of course, when we write quickly it may be hard to determine whether we have an inequality symbol or a "sloppy" equal sign. To avoid this problem, we close the lines completely at the end where the lines are supposed to be closer together.

For example, to indicate that 5 is less than 7, we write $5 < 7$. This places the lesser number next to the lesser (zero) space. Once we realize how this works, we can reverse the order and write $7 > 5$. That way, the lesser number is still next to the smaller space.

That is:

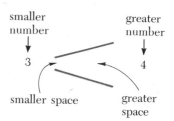

In other words, the symbol became an "arrowhead" with the closed end (lesser distance) representing the side where we write the smaller number.

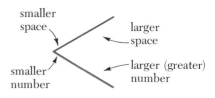

Example 3

Which inequality symbol should replace the question mark to make the following statement true?

$$8 \; ? \; 5$$

The question mark should be replaced by $>$.

Since 8 is greater than 5, you want the closed end of the arrow to be next to the 5. Therefore, you write $8 > 5$. If you want to write the 5 first, then the correct expression is $5 < 8$.

Some people prefer to remember the inequality symbol by the fact that the arrowhead always points to the lesser number.

We may also use the number line to help visualize inequalities. To draw the number line, we pick a starting point (0) and then mark off distance to either side. The direction toward the right is the positive direction and the direction toward the left is the negative direction. That is,

In terms of the number line, "greater than" means "to the right of." As we go toward the right, the numbers become greater.

Example 4

Which inequality symbol should replace the question mark to make $-8 \; ? \; -5$ a true statement?

In this case, you want the question mark replaced by $<$. That is, $-8 < -5$.

On the number line, -5 is to the right of -8, so it is greater than -8. For negative numbers, the *greater* the magnitude, the *smaller* the number.

This idea makes signed numbers a special case of inequalities. Remember, we call a number positive if it is to the right of 0. A positive number is one that is *greater than* 0. A negative number is one that is less than 0.

In symbols, if p is any positive number and n is any negative number, then

$$p > 0 \; (\text{or} \; 0 < p)$$

and

$$n < 0 \; (\text{or} \; 0 > n)$$

Section 9.2 Solving Linear Inequalities

Example 5

Which inequality symbol should replace the question mark to make 5 ? −8 a true statement?

You should replace the question mark by >: that is, 5 > −8 (5 is greater than −8).
On the number line, 5 is to the right of −8. Using signed numbers, you must add the positive number 13 to −8 to get 5, so 5 is greater than −8. In business terms, a $5 profit is more profit than an $8 loss.

Use the explanation you like best. The important point is that any positive number is greater than any negative number.

PRACTICE DRILL

With what inequality symbol should you replace the question mark to make each of the following a true statement?
1. 4 ? 5   2. 4 ? −5   3. −4 ? 5   4. −4 ? −5

Answers
1. <   2. >   3. <
4. >

CHECK THE MAIN IDEAS

1. To indicate that b is _____ than c, you write $b < c$.
2. If you want to write c first, you write c _____ b.
3. To remember whether to use < or >, you may think of an equal sign with one end pinched together. The closed end always points to the _____ number.
4. You may also think of inequalities in terms of the number line. In this context, $b < c$ means that c is to the _____ of b.
5. You write $-7 < -2$ because on the number line, -2 is to the _____ of -7.
6. You can also describe signed numbers in terms of inequalities. For example, the negative numbers are _____ than 0.

Answers
1. less (smaller)   2. >   3. lesser
4. right
5. right   6. less than

Section 9.2
Solving Linear Inequalities

We often solve problems by first translating the problem from words into an equation. In the same way, we often solve inequalities by first translating them from words to symbols.

OBJECTIVE
To be able to solve inequalities of the form

$$Ax + B < Cx + D$$

where A, B, C, D are constants.

Example 1

Write the inequality that is needed to solve the following problem: How large a sale should a saleswoman make if she wants to earn more than $90 and her commission is 15% of all sales?

Let S denote the amount of the sales. The inequality is given by $0.15S > 90$. In words, you have

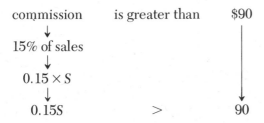

We proceed just the same as we did when we dealt with linear equalities, except that we use > instead of =.

To solve the inequality in the problem, we proceed very much as we do with equalities. We divide both sides of the inequality by 0.15 to get

$$S > \frac{90}{.15} \quad \text{or} \quad S > \frac{90000}{15} \quad \text{or} \quad S > 600$$

This tells us that in order to earn more than $90, the saleswoman must make more than $600 in sales.

Why can we divide both sides of an inequality by the same positive number without changing the sense of the inequality? If Jane has more money than Joe, and each of them triples the amount, then Jane still has more money than Joe. (See Example 3 for more details.)

Example 2

Solve the inequality

$$3x < 60 \qquad [1]$$

The solution is given by $x < 20$. Since x is being multiplied by 3, you undo the inequality by dividing both sides of [1] by 3.

So just as we talk about "equals multiplied by equals are equal," we can talk about "unequals multiplied (divided) by equals are still unequal" — but there is an important trouble spot.

Notice that $x < 20$ names more than one number. In fact, this inequality names every number that is less than 20. The solutions include 18, 7, 3.5, $\sqrt{2}$, 0, −1, −1.5, and many other numbers.

By the solution of $x < 20$, we mean the set of all values of x for which $x < 20$ is a true statement.

Example 3

What inequality do you get if you multiply both sides of $5 < 7$ by -1?

You get $-5 > -7$.

It's easy to see that when you multiply 5 by -1 you get -5, and when you multiply 7 by -1, you get -7. The surprise is that the inequality symbol *must change*. You know from the previous section that -5 is greater than -7, which means that writing $-5 < -7$ is incorrect.

If we want mathematics to give us realistic results, we must pick our rules to agree with what we believe is reality — even if this means that some of our rules don't seem "natural."

> **SUMMARY**
> In multiplying both sides of an inequality by the same non-zero constant, the inequality symbol remains the same if the constant is positive. For example, $3 < 4$ and $5(3) < 5(4)$, since $15 < 20$.
>
> The inequality symbol is changed if the constant is negative (when we change the symbol, we sometimes say that we've changed the *sense* of the inequality). For example, $3 < 4$ but $-2(3) > -2(4)$, since $-6 > -8$.
>
> If we multiply both sides of an inequality by 0, we get the equality $0 = 0$, since any number times 0 is 0.

Using symbols, if $x < y$, then

$$cx < cy \text{ if } c \text{ is positive}$$
$$cx > cy \text{ if } c \text{ is negative}$$
$$cx = cy \text{ if } c \text{ is } 0$$

Example 4

Solve the inequality

$$-3x < 60 \qquad [2]$$

The solution is given by $x > -20$.

To isolate x in [2], you must multiply both sides of the equation by $-\frac{1}{3}$ (that is, you divide by -3). This means that you must change the sense of the inequality. So any number x is a solution of [2], provided that x is greater than

Section 9.2   Solving Linear Inequalities   223

−20. Except for taking special care when you multiply both sides of an inequality by a negative constant, you handle inequalities in the same way that you handle regular equations.

Example 5

Solve the inequality

$$x + 5 > 60 \qquad [3]$$

The solution is given by $x > 55$. Any value of x that is greater than 55 makes [3] a true statement.

To isolate x in [3], you need only subtract 5 from both sides (or add −5 to both sides). This gives you

$$\begin{array}{r} x + 5 > 60 \\ -5 \quad -5 \\ \hline x \quad > 55 \end{array}$$

NOTE
If you feel that the solution should be $x < -20$, pick a value of x that is less than −20 and show that [2] is not satisfied for this choice of x. For example, if you pick x to be −30 (which is less than −20), [2] becomes

$$-3(-30) < 60$$
$$90 < 60$$

which is a false statement. That is, $-3x < 60$ is an open sentence just as $-3x = 60$ is an open sentence.

Example 6

Solve the inequality

$$x - 7 < 30 \qquad [4]$$

The solution is given by $x < 37$. You get this answer by adding 7 to both sides of [4].

To solve an inequality, sometimes we have to perform several "undoings," just as we do to solve equations.

When we add a constant to both sides of an inequality, the inequality has the same sense, regardless of the sign of the constant.

Example 7

Solve the inequality

$$2(x - 7) < 30 \qquad [5]$$

The solution of [5] is all values of x for which $x < 22$.

There are a few ways of tackling this problem. One method is to divide both sides of [5] by 2 to get

$$x - 7 < 15 \qquad [6]$$

Then, add 7 to both sides of [6] to get the answer.

You can also proceed by using the distributive property on the left side of [5] to get

$$2x - 14 < 30 \qquad [7]$$

Then you solve [7] just as you would solve an equality

$$\begin{array}{r} 2x - 14 < 30 \\ +14 \quad +14 \\ \hline 2x \quad < 44 \end{array}$$

Hence, $x < 22$.

Example 8

Solve the inequality

$$7 - x > 3 \qquad [8]$$

The solution is given by $x < 4$.

One way to get the solution is to subtract 7 from both sides of [8] to get

$$-x > 3 - 7$$
$$-x > -4 \qquad [9]$$

You can then multiply both sides of [9] by -1 (so that the left side becomes x). Remember that multiplying by a negative number changes the sense of the inequality, so you get

$$(-1)(-x) < (-1)(-4) \qquad \text{or} \qquad x < 4$$

If you prefer to avoid negative coefficients, you can add x to both sides of [8] to get

$$\begin{array}{r} 7 - x > 3 \\ +x \quad +x \\ \hline 7 \quad > 3 + x \end{array} \qquad [10]$$

Then you subtract 3 from both sides of [10] to get

$$4 > x$$

This is the same as $x < 4$.

If the variable appears on both sides of the inequality, we can still proceed the way we did when we worked with linear equations.

For example, if x is 5, we get

$$7 - 5 > 3$$

which is a false statement.

Any number less than 4 is a solution. For example, -6 is less than 4. If we replace x by -6 in [8], we get

$$7 - (-6) > 3$$
$$7 + 6 > 3$$

which is a true statement.

It's not the order in which we write x and 4 that matters, it's where the arrowhead points.

Example 9

Solve the inequality

$$3x + 5 < x + 9 \qquad [11]$$

The solution is given by $x < 2$.

You may begin by subtracting x from both sides of [11] to obtain:

$$\begin{array}{r} 3x + 5 < \quad x + 9 \\ -x \qquad -x \\ \hline 2x + 5 < \qquad 9 \\ -5 \qquad -5 \\ \hline 2x \qquad \qquad 4 \end{array}$$

So $x < 2$.

That is, [11] is a true statement for any number x provided that x is less than 2.

Example 10

Solve the inequality

$$5 - 3x < 9 - x \qquad [12]$$

The solution is given by $x > -2$.

To isolate x, you can add $3x$ to both sides of [12] to eliminate x from the left side of [12]. You also add x to both sides of [12] and eliminate x from the right side of [12].

Section 9.3 Linear Inequalities in Two Variables

Either method is correct, but if you elect to add x to both sides, you get $-2x$ on the left side of [12]. This means that when you solve for x by dividing by -2, you must remember to change the sense of the inequality.

If you prefer to avoid negative coefficients, add $3x$ to both sides of [12] to obtain

$$\begin{array}{rcl} 5-3x &<& 9-x \\ +3x && +3x \\ \hline 5 &<& 9+2x \\ -9 && -9 \\ \hline -4 &<& 2x \end{array}$$

So $-2 < x$. This is the same as $x > -2$: in either case, the wider space is next to x.

That is, we get

$$\begin{array}{rcl} 5-3x &<& 9-x \\ +x && +x \\ \hline 5-2x &<& 9 \\ -5 && -5 \\ \hline -2x &<& 4 \end{array}$$

So $x > -2$.

That is, both $x > -2$ and $-2 < x$ say that x is greater than -2. Any number greater than -2 is a solution of the inequality, regardless of which form we use.

PRACTICE DRILL

Solve each of the following inequalities.
1. $x + 4 < 5$
2. $x - 4 < 5$
3. $2x + 3 > 9$
4. $4x + 3 > 2x + 9$
5. $5 - x < 4$
6. $5x + 9 > 7x + 3$

Answers
1. $x < 1$ 2. $x < 9$
3. $x > 3$ 4. $x > 3$
5. $x > 1$ 6. $x < 3$

CHECK THE MAIN IDEAS

1. Just as some relationships lead to equalities, other relationships lead to _____.
2. In many respects, you treat inequalities just as you do equalities. For example, to solve the inequality $x + 4 < 5$, you _____ 4 from both sides of the inequality.
3. However, you must be careful when you multiply both sides of an inequality by the same _____ number.
4. For example, it is true that $-3 > -6$, but if you multiply both sides by -2, you get 6 _____ 12.
5. When you multiply both sides of an inequality by the same negative number, you change the _____ of the inequality.
6. For example, to solve $5 - x < 4$, you may first subtract 4 from both sides to get _____.
7. Then you multiply both sides of the new inequality by -1 to get x _____ 1.

Answers
1. inequalities   2. subtract
3. negative   4. <
5. sense   6. $-x < -1$
(or $-1 > -x$)   7. $x > 1$

Section 9.3
Linear Inequalities in Two Variables

Suppose you rent a scraper at \$2 per hour and a sander at \$3 per hour. If you use the scraper for x hours, the cost will be $2x$ dollars; if you use the sander for y hours, the cost will be $3y$ dollars. If C denotes the total cost, you have

$$C = 2x + 3y \qquad [1]$$

If you want the total cost of renting these two items to equal \$60, you replace C by 60 in [1] to get

$$60 = 2x + 3y \qquad [2]$$

Any values of x and y that satisfy [1] tell you how many hours you can use each machine in order to spend \$60. For example, if you let $y = 0$, then $x = 30$. This means that if you use the scraper for 30 hours but don't use the sander at all, you will spend the \$60.

OBJECTIVE
To study inequalities of the form

$$Ax + By < C$$
or
$$Ax + By > C$$

and to see what they mean in terms of graphs.

Example 1

If you use the scraper for 15 hours, for how many hours can you use the sander?

You can use the sander for 10 hours.
You replace x by 15 in [2] to get

$$60 = 30 + 3y$$

CHECK
10 hrs at \$3 = \$30
15 hrs at \$2 = \$30
\$60

So $y = 10$, which stands for the number of hours the sander may be used.

To solve [2] with a graph, you draw the line whose equation is $60 = 2x + 3y$. The coordinates of each point on this line give you a solution of the equation.

Suppose you decide to spend less than \$60 rather than \$60. To indicate that you want to spend any amount provided it is less than \$60, you replace equation [2] by the inequality

For example, (15,10) is on the line, which means that a solution of [2] is $x = 15$ and $y = 10$.

$$2x + 3y < 60 \qquad [3]$$

Let's rework Example 1 using [3] instead of [2].

The left side of [3] is the total cost and the right side is the \$60. [3] states that the total cost is less than \$60.

Example 2

Use [3] to determine how long you can use the sander if you use the scraper for 15 hours.

You can use the sander for any amount of time that is less than 10 hours.
You replace x by 15 in [3] and get

$$30 + 3y < 60 \qquad [4]$$

Using the method of the last section, you have

$$\begin{array}{rl} 30 + 3y < & 60 \\ -30 & -30 \\ \hline 3y < & 30 \end{array}$$

So $y < 10$, which means you can use the sander for any amount of time less than 10 hours.

To understand the work in this section, we must notice that [2] is an example of the form $Ax + By = C$, while [3] has the form $Ax + By < C$. We want to see how $Ax + By < C$ can be explained in terms of the equation $Ax + By = C$.

In this case, $A = 2$, $B = 3$, and $C = 60$.

To start, let's assume that the line whose equation is given by $Ax + By = C$ is not parallel to the y-axis. This means that the equation can be written in the slope and y-intercept form

$$y = mx + b$$

Using $y = mx + b$ rather than $Ax + By = C$ helps to keep the arithmetic a little simpler.

Section 9.3   Linear Inequalities in Two Variables

Example 3

Is the point (2,3) on the line $y = x + 7$?

No. For (x,y) to be on the line, its coordinates must satisfy the equation $y = x + 7$. If you replace x by 2 and y by 3, the equation becomes

$$3 = 2 + 7$$

which is false. Therefore, (2,3) is not on the line.

Example 4

What point on the line $y = x + 7$ has its x-coordinate equal to 2?

The point (2,9).
   To see this, you replace x by 2 in $y = x + 7$ and get

$$y = 2 + 7 = 9$$

This tells you that the point whose x-coordinate is 2 must have 9 as its y-coordinate if the point is to be on the line.

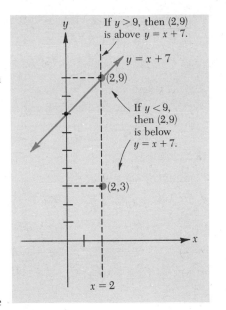

Example 5

Is the point (2,3) above or below the line $y = x + 7$?

(2,3) is below the line $y = x + 7$. From Example 3, you know (2,3) is not on the line. From Example 4, you know that (2,9) is on the line. But (2,3) is 6 units below the point (2,9). So if (2,9) is on the line, (2,3) is below the line.

Example 6

Do the coordinates of the point (2,3) satisfy the inequality $y < x + 7$?

Yes. If you replace x by 2 and y by 3 in $y < x + 7$, you get

$$3 < 2 + 7 \quad \text{or} \quad 3 < 9$$

which is a true statement.

   If we put the results of these last few examples together, we see that the point $(2,y)$ will be below the line $y = x + 7$ whenever y is less than 9. That is, if we replace x by 2 in $y < x + 7$, we get $y < 9$, which clearly is true whenever y is less than 9. However, if y is greater than 9, the point $(2,y)$ will lie above the line $y = x + 7$.
   In summary:

$(2,y)$ is on $y = x + 7$ if $y = 9$
$(2,y)$ is below $y = x + 7$ if y is less than 9
$(2,y)$ is above $y = x + 7$ if y is greater than 9

Example 7

Is the point (4,13) on, below, or above the line $y = x + 7$?

(4,13) is above the line $y = x + 7$.
   If you replace x by 4 in $y = x + 7$, you see that $y = 4 + 7 = 11$. This tells you that (4,11) is on the line. But (4,13) is 2 units above (4,11); therefore, (4,13) is above the line.

You can generalize this result to conclude

>(4,11) is on the line
>(4,y) is above the line if $y > 11$
>(4,y) is below the line if $y < 11$

Why do we have to keep working with specific x-coordinates? For any value of x, the point (x,y) will be on the line only if $y = x + 7$. Therefore, if $y < x + 7$, the point (x,y) will be below the line; and if $y > x + 7$, the point (x,y) will lie above the line. That is,

1. $y = x + 7$ is a line.
2. $y < x + 7$ describes all points (x,y) that lie below the line.
3. $y > x + 7$ describes all points (x,y) that lie above the line.

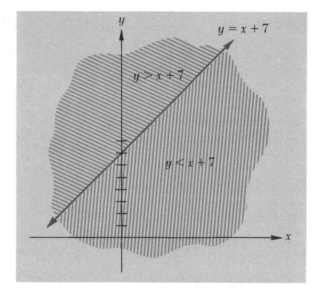

The same type of reasoning can be used for any line of the form $y = mx + b$. Given an x-coordinate, the point (x,y) will be on that line only if the y-coordinate is $mx + b$. If y is less than $mx + b$, then the point is below the line $y = mx + b$; but if y is greater than $mx + b$, the point (x,y) is above the line $y = mx + b$. In summary:

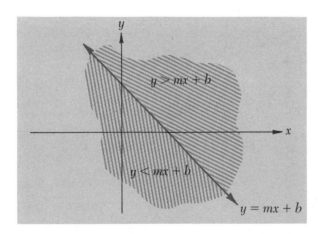

$y < mx + b$ describes the region that lies below the line $y = mx + b$.
$y > mx + b$ describes the region that lies above the line $y = mx + b$.

Section 9.3 Linear Inequalities in Two Variables

NOTE

The phrase, "the region that lies . . . ," is an abbreviation for "the set of all points (x,y) that lie. . . ." When the region consists of the points that are either above a line or below a line, the region is usually called a **half-plane**. In general, any straight line divides the plane into two half-planes — one on each side of the line.

Now let's see if we can understand what this discussion means if the inequality is in the form $Ax + By < C$.

Example 8

Describe the region that consists of those points (x,y) for which

$$x + y < 1 \qquad [5]$$

This region consists of all the points that are below the line $x + y = 1$.

You already know how to solve this problem if the line has the form $y < mx + b$. To put $x + y < 1$ into this form, subtract x from both sides to get $y < -x + 1$. In this form, you know that $y < -x + 1$ names all points that lie below the line $y = -x + 1$. This is the same line as $x + y = 1$.

Don't be fooled by the simplicity of the example. The inequality can be deceptive when it is in the form $Ax + By < C$.

Example 9

Describe the region that consists of those points (x,y) for which

$$x - y < 1 \qquad [6]$$

This region consists of all points that are *above* the line $x - y = 1$.

First rewrite $x - y < 1$ as follows.

$$x - y < 1$$
$$-y < -x + 1$$
$$y > x - 1$$

This equation represents all the points above the line $y = x - 1$ (which is the same line as $x - y = 1$).

It should be clear by now that an inequality like $2x - 3y < 6$ describes either the region below the line $2x - 3y = 6$ or the region above this line. How can we find out which?

One satisfactory way is to draw the line and pick a convenient point not on the line. Let's try another problem using this method.

The region is above the line in this example, even though we have the "less than" symbol.

Remember that multiplying both sides of an inequality by -1 changes the sense of the inequality.

In this case, we have

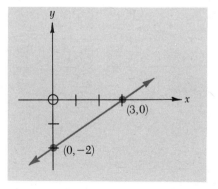

It's easy to see that $(0,0)$ is above the line.

The coordinates of $(0,0)$ are $x = 0$ and $y = 0$. So replace x by 0 and y by 0 in $2x - 3y$. We get $2(0) - 3(0)$ or 0, and $0 < 6$.

So $(0,0)$ is above the line, and its coordinates satisfy the inequality

$$2x - 3y < 6$$

This tells us that $2x - 3y < 6$ describes the region above the line $2x - 3y = 6$.

Example 10

What inequality describes the region below the line $3x - 4y = 12$?

The right answer must be either $3x - 4y < 12$ or $3x - 4y > 12$. To determine which is correct, sketch the line. You see that $(0,0)$ is above the line. If you replace x and y by 0 in $3x - 4y ? 12$, you get that $0 ? 12$, which means that ? must be replaced by $<$.

In other words, since $(0,0)$ is above the line $2x - 3y = 12$ and since the coordinates of $(0,0)$ satisfy $3x - 4y < 12$, then $3x - 4y < 12$ describes the region above the line. Therefore, the other inequality, $3x - 4y > 12$, describes the region below the line $3x - 4y = 12$.

NOTE

You need not sketch the line. Instead, find the y-intercept by letting $x = 0$ in $3x - 4y = 12$. We can see that the y-intercept is $(0,-3)$, which is below $(0,0)$. Hence $(0,0)$ is above the line $3x - 4y = 12$.

We don't have to pick $(0,0)$, but as long as $(0,0)$ is not a point on the line, it is an easy point to pick because its coordinates are $x = 0$ and $y = 0$. These values are very easy to substitute into the inequality.

PRACTICE DRILL

1. What inequality describes the x-coordinate of any point on the line $y = 3x + 2$ whose y-coordinate is less than 8?
2. What equation describes the region that is below the line $y = 2x + 1$?
3. What equation describes the region that is below the line $x - y = 7$?
4. Is the point $(1,1)$ above the line $2x + 3y = 7$?
5. Is the point $(1,1)$ above the line $5x - 2y = 10$?

Answers

1. $3x + 2 < 8$, or $x < 2$
2. $y < 2x + 1$ 3. $x - y > 7$, or $y < x - 7$ 4. No; the region below $2x + 3y = 7$ is given by $2x + 3y < 7$ 5. Yes; the region above the line $5x - 2y = 10$ is given by $5x - 2y < 10$

CHECK THE MAIN IDEAS

1. The point on the line $y = 3x + 2$ whose x-coordinate is 2 is _____.
2. Therefore, if $y < 8$, the point $(2,y)$ is _____ the line $y = 3x + 2$.
3. For example, $(2,-9)$ is _____ the line $y = 3x + 2$.
4. More generally, the point (x,y) is below the line $y = 3x + 2$ whenever its coordinates satisfy the inequality y _____ $3x + 2$.
5. The inequality $3x - y > -2$ has the same solutions as the inequality $-(3x - y)$ _____ $-(-2)$. That is, $-3x + y < 2$, or $y < 3x + 2$.
6. $y < 3x + 2$ is the region below the line $y = 3x + 2$. Therefore, $3x - y > -2$ describes the region _____ the line $y = 3x + 2$ (or $3x - y = -2$).

Answers

1. $(2,8)$ 2. below
3. below   4. $<$ 5. $<$
6. below

Section 9.4
Simultaneous Inequalities

When we studied linear equations we mentioned that there were times when we wanted the same values of x and y to satisfy two different equations. This led to the idea of simultaneous equations. It is also possible to have simultaneous inequalities.

First of all, let's review and then introduce some new symbolism.

OBJECTIVE
To solve problems involving two or more linear inequalities.

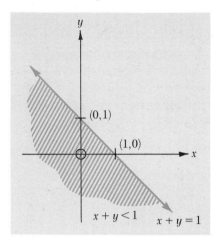

Example 1

Describe the region formed by all points (x,y) for which

$$x + y < 1 \qquad [1]$$

It is the half-plane that lies below the line $x + y = 1$. This region is shown in the illustration. Look at the line $x + y = 1$. The point $(0,0)$ is below the line. If you replace x and y by 0, you see that $x + y = 0 + 0 = 0$. Hence $(0,0)$, which lies below the line $x + y = 1$, satisfies the inequality $x + y < 1$ (since $x + y = 0$, and 0 is less than 1).

Now look again at the region we've drawn. How can we tell whether the half-plane described by $x + y < 1$ lies below the line or includes the line? From the algebra, we know that the region is below the line, but can we tell just by looking at the picture?

To avoid this problem we agree to draw the line dashed if the region is not to include the line. If the region includes the line, then we draw the line in the usual solid way. The illustration shows how we draw the region whose equation is

$$x + y < 1 \qquad [2]$$

Again, in terms of the y-intercept, $(0,1)$ is the y-intercept of $x + y = 1$; so $(0,0)$ is below the line $x + y = 1$. Look at the picture, but not at the equation. How do you know if the shaded region includes the line $x + y = 1$? Technically speaking, a line has no thickness.

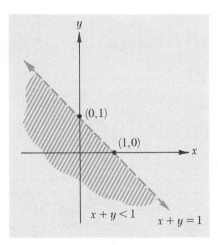

The fact that the line $x + y = 1$ is dashed (or, as we sometimes say, *broken*) means that the shaded region lies below the line and doesn't include it.

Suppose we want the line $x + y = 1$ to be included in the half-plane. We draw the line solid, as shown in the illustration; but what is the equation for this region?

We may think of the region as being divided in two parts. First there is the line $x + y = 1$ itself, and then there is the half-plane $x + y < 1$, which is below the line. To be in this region, (x,y) can satisfy

$$x + y = 1 \text{ (the point is on the line)}$$
$$\text{or } x + y < 1 \text{ (the point is below the line)}$$

We combine these two possibilities into a single inequality by writing

$$x + y \leq 1$$

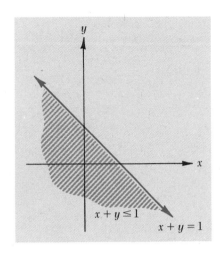

The solid line in this diagram indicates that the region is the half-plane on and below the line. The expression that tells us this is $x + y \leq 1$.

Let's review this new symbolism.

When we write $x < 3$, we mean that x is less than 3. When we write $x = 3$, we mean x is 3. When we write $x \leq 3$, we mean either that x is 3 or that x is less than 3.

It may be helpful to see how this new symbolism applies to positive and negative numbers.

A negative number is one that's less than 0. So to say that x is negative is the same as writing $x < 0$. But if we write $x \leq 0$, we're saying either that x is 0 or that x is less than 0. In other words, x is either 0 or negative. A shorter way of saying this is that x is nonpositive.

That is, $x \leq 3$ is an abbreviation for "x is 3 or less than 3". This also says that "x is no greater than 3." $\leq$ is an abbreviation for ≦, which in turn is an abbreviation for "less than or equal."

When a number is negative, it must be less than 0. But if the number is nonpositive, 0 is included. That is, while 0 is neither positive nor negative, it is both nonpositive and nonnegative.

Example 2

Does $x = 4$ satisfy the inequality

$$x < 4 \qquad [3]$$

No. If you replace x by 4 in [3], you get

$$4 < 4$$

This is a false statement, because no number is less than itself.

Using the number line, we have:

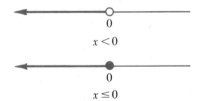

Example 3

Does $x = 4$ satisfy the inequality

$$x \leq 4 \qquad [4]$$

Yes. If you replace x by 4 in [4], you get $4 \leq 4$, which is a true statement because 4 is at least as great as itself.

To show that 0 is included, we write it as a solid point. To show that it's excluded, we write it as a hollow circle.

Sometimes it seems awkward to say that it is true that 4 is either equal to 4 or less than 4. Surely 4 isn't less than 4. The mathematically accepted meaning of "either . . . or . . ." is that at least one of the parts is true. In this case, the statement $4 = 4$ is true.

This idea occurs in grammar. If we connect two sentences by "and," then each sentence must be true for the whole statement to be true. But if we connect two sentences by "or," only one has to be true for the whole sentence to be true.

Section 9.4   Simultaneous Inequalities

Example 4

Does $x = 5$ satisfy the inequality

$$x \geq 5 \qquad [5]$$

Yes. If you replace x by 5 in [5], you get $5 \geq 5$. This means that either $5 = 5$ or 5 is greater than 5. This is an "either . . . or . . ." sentence in which the first part is true, so the whole sentence (namely, $x \geq 5$) is true.

Let's return to inequalities which have two variables.

Example 5

Describe the region formed by all points (x,y) for which

$$x - y < 1 \qquad [6]$$

It is the half-plane that lies above the line $x - y = 1$. The region is shown in the illustration.

To see why the region is above the line, notice that $(0,0)$ is above the line. If you replace x and y by 0, you get $x - y = 0 - 0 = 0$, and 0 is less than 1.

That is, $(0,0)$, which lies above the line $x - y = 1$, satisfies the equation $x - y < 1$. Therefore $x - y < 1$ represents the region above the line.

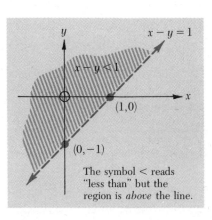

The symbol $<$ reads "less than" but the region is *above* the line.

Now that we have the necessary background, let's turn to the main point of this section. In Example 1, we learned how to find the half-plane consisting of those points (x,y) for which $x + y < 1$. In Example 5, we learned how to find the half-plane consisting of those points (x,y) for which $x - y < 1$.

We can combine the results of these two examples. That is, we may ask to find the region consisting of those points (x,y) for which *both* $x + y < 1$ and $x - y < 1$ are true. Let's try this problem.

The dashed line tells us that the line $x - y = 1$ is not included in the region.

Example 6

Describe the region that is formed by all points (x,y) for which

$$\begin{array}{ll} & x + y < 1 \qquad [1] \\ \text{and} & x - y < 1 \qquad [6] \end{array}$$

As with simultaneous equations, we use braces to denote simultaneous inequalities. So [1] and [6] are usually written as

$$\left. \begin{array}{l} x + y < 1 \\ x - y < 1 \end{array} \right\}$$

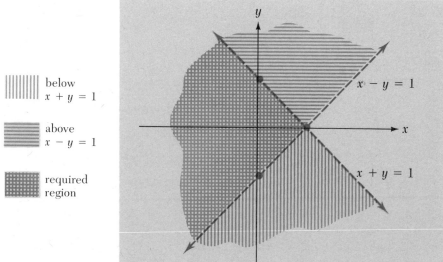

below $x + y = 1$

above $x - y = 1$

required region

233

As shown in the picture, it is the set of points (x,y) that lie below the line $x + y = 1$ but above the line $x - y = 1$.

In Example 1, you saw that to satisfy [1], the point had to be below the line $x + y = 1$. In Example 5, you saw that to satisfy [6], the point had to be above the line $x - y = 1$.

So, to satisfy [1] and [6], the point has to be below $x + y = 1$ and above $x - y = 1$.

Example 7

Describe the set of all points (x,y) for which

$$\left.\begin{array}{r} x - y \leq 4 \\ x + y > 2 \end{array}\right\} \qquad [7]$$

The region is the set of points that are above the line $x + y = 2$ and are also on or above the line $x - y = 4$. This region is shown in the illustration.

You know that the line $x - y = 4$ passes through the points $(4,0)$ and $(0,-4)$. Therefore, $(0,0)$ is above this line. But when x and y are both 0, $x - y = 0 - 0 = 0$, which is surely less than or equal to 4. Hence $x - y \leq 4$ describes the region that is on or above the line $x - y = 4$.

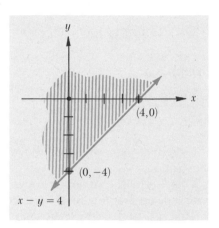

$(4,0)$ and $(0,-4)$ satisfy $x - y = 4$. The line is drawn solid to indicate that it's part of the region. It's part of the region because of the meaning of the symbol $\leq$.

You also know that the line $x + y = 2$ passes through the points $(0,2)$ and $(2,0)$. So $(0,0)$ is below this line. If x and y are both 0, $x + y < 2$ (that is, $0 + 0 < 2$). Therefore, $x + y < 2$ describes the half-plane below the line $x + y = 2$. So $x + y > 2$ must describe the half-plane that lies above the line $x + y = 2$.

Putting these two pieces of information together, you see that the region described in [7] is the set of points that are above the line $x + y = 2$ and, at the same time, on or above the line $x - y = 4$.

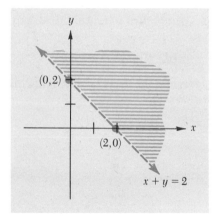

$(0,2)$ and $(2,0)$ both satisfy $x + y = 2$. The line is dashed because it's not part of the region.

The indicated region is the portion that is above $x + y = 2$ and on or below $x - y = 4$. The point $(3,-1)$ at which the lines meet is found by solving

$$\left.\begin{array}{r} x - y = 4 \\ x = y = 2 \end{array}\right\}$$

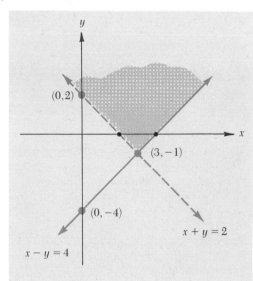

The study of linear inequalities includes many other types of situations. We shall examine a few more types before ending this section.

In the lesson on simultaneous linear equations, we did not study any case in which we had more equations than variables. The reason is simple: when we have more equations than variables, there is usually no simultaneous solution.

This may seem like a hard fact to justify, but it's fairly easy to explain using graphs. Let's use some of the results from Example 7 to see why.

The lines $x - y = 4$ and $x + y = 2$ meet only at the point $(3, -1)$. They have no other points in common. If we were to draw a third line, there is no way for all three lines to share a common point unless the third line also passes through $(3, -1)$.

Let's take any line that doesn't pass through the point $(3, -1)$. To find the equation of such a line, we just have to pick a linear equation in x and y that is not satisfied when $x = 3$ and $y = -1$. One line that doesn't pass through $(3, -1)$ is $x + 2y = 4$. There is no special reason for choosing $x + 2y = 4$. All we need is any line that doesn't pass through $(3, -1)$, and there are many such lines.

How can a point be on all three lines unless it's at least on the first two? The only point that is common to the first two is $(3, -1)$.

If we replace x by 3 and y by -1 in the equation $x + 2y = 4$, we get

$$3 + 2(-1) = 4 \quad \text{or} \quad 3 - 2 = 4$$

which is a false statement.

Example 8

Show by means of graphs that the system of simultaneous equations

$$\left.\begin{array}{r} x - y = 4 \\ x + y = 2 \\ x + 2y = 4 \end{array}\right\} \qquad [8]$$

has no solution.

To see this, look at the picture. Each pair of lines meets at one point, but there is no one point at which all three lines meet. Since a point is on a line if and only if it satisfies the equation of the line, there is no one point (x,y) whose coordinates satisfy all three of the equations in [8].

This is one important place in which inequalities differ from equalities. The basic difference comes from the fact that a linear equation in x and y represents a line, while a linear inequality in x and y represents a half-plane. Let's see why this is so important.

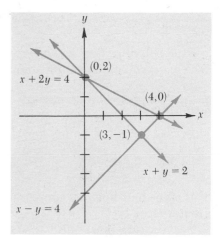

We find the points of intersection by solving each of the three *pairs* of equations simultaneously. For example, $(0,2)$ is the solution of

$$\left.\begin{array}{r} x + 2y = 4 \\ x + y = 2 \end{array}\right\}$$

But no one point belongs to all three lines.

Example 9

Describe the set of all points (x,y) that satisfy the system of inequalities

$$\left.\begin{array}{r} x - y \leq 4 \\ x + y > 2 \\ x + 2y < 4 \end{array}\right\} \qquad [9]$$

As shown in the illustration, the set of points is the region that is above $x + y = 2$, on or above the line $x - y = 4$, and, at the same time, below the line $x + 2y = 4$.

Most of this problem was already done in Example 7. All we had to do was include the half-plane defined by $x + 2y < 4$.

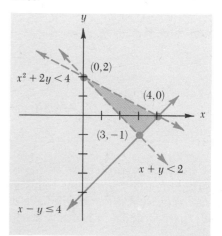

While no one point belongs to all three lines, there are many points that belong to all three half-planes. Every point in the shaded region belongs to all three half-planes.

Once we have sketched the region, we can use the region to check that we really get the solutions this way. In Example 9, we can see from the picture that $(1,0)$ is not in the shaded region. This means that at least one of the three inequalities in [9] must be false when we replace x by 1 and y by 0.

$$x - y \leq 4 \text{ becomes } 1 - 0 \leq 4 \text{ (true)}$$

This is shown by the fact that $(1,0)$ is on or above $x - y = 4$.

$$x + y > 2 \text{ becomes } 1 + 0 > 2 \text{ (false)}$$

This is shown by the fact that $(1,0)$ is not above $x + y = 2$.

$$x + 2y < 4 \text{ becomes } 1 + 2(0) < 4 \text{ (true)}$$

That is, $(1,0)$ is below the line $x + 2y = 4$.

On the other hand, it seems clear from the picture that $(3,0)$ is in the shaded region. This means that when $x = 3$ and $y = 0$, all three inequalities in [9] should become true statements. As a check, we have

$$x - y \leq 4 \text{ becomes } 3 - 0 \leq 4 \text{ (true)}$$
$$x + y > 2 \text{ becomes } 3 - 0 > 2 \text{ (true)}$$
$$x + 2y < 4 \text{ becomes } 3 < 4 \text{ (true)}$$

The picture is very helpful. We don't *have* to draw it. Instead, we can pick values for x and y and see what happens in [9]. But if we draw the region, we can see the points that are included.

In concluding this section, we should note that we can have simultaneous systems involving a combination of equalities and inequalities.

NEW TERMINOLOGY

When we talk about taking the portion of a line from a certain point on, we call that portion of a line a **half-line**.

If we pick two points on a line P and Q, and we want only that portion of the line between P and Q, we talk about the **line segment** PQ.

In everyday usage, we often say "line" when we mean "line segment," but technically a line extends indefinitely in both directions.

Example 10

Describe the set of points for which

$$\left. \begin{array}{l} y = 2x \\ y > x + 1 \end{array} \right\} \quad [10]$$

This is the **half-line** that consists of all points (x,y) on the line $y = 2x$ for which x is greater than 1.

Because (x,y) must satisfy the equation $y = 2x$, you know that the point must be on the line $y = 2x$. At the same time, the point must satisfy $y > x + 1$. Therefore, the point is above the line $y = x + 1$.

The points you want are on the line $y = 2x$ but above the line $y = x + 1$. Since $y = 2x$ and $y = x + 1$ meet at $(1,2)$, you want those points on the line $y = 2x$ for which x is greater than 1. In other words, you want the half-line that consists of those points on $y = 2x$ for which the x-coordinate is greater than 1.

The pictorial solution is also shown.

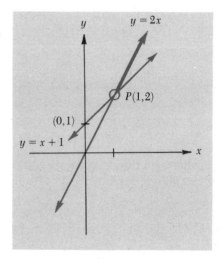

This time we don't get a region. To be sure, the point must be above the line $y = x + 1$ but it must also be *on* the line $y = 2x$.

If we want to solve this problem by algebra, we replace y by $2x$ (since $y = 2x$) in $y > x + 1$ to get

$$2x > x + 1$$

If we subtract x from both sides, we get $x > 1$.

Further examples are in the exercise set, which also includes some optional supplementary problems that show how the study of inequalities can be used in real-life situations. If you have difficulty, remember that complete solutions for all the problems in Form A are in your manual, and additional information is in the audio tapes.

Exercise Set 9 (Form A)

PRACTICE DRILL

What region of the plane consists of the points (x,y) for which

1. $2x + 3y < 1$
2. $\begin{rcases} 2x + 3y < 1 \\ 3x + 4y > 2 \end{rcases}$
3. $\begin{rcases} 2x + 3y = 1 \\ y > 2x \end{rcases}$

Answers
1. The region below the line $2x + 3y = 1$
2. The region below the line $2x + 3y = 1$ but above the line $3x + 4y = 2$
3. The region is the portion of the line $2x + 3y = 1$ that lies above the line $y = 2x$

CHECK THE MAIN IDEAS

1. $x + y > 1$ describes all points that lie _____ the line $x + y = 1$.
2. $x - y > 1$ describes all points that lie _____ the line $x - y = 1$.
3. Hence, to describe the region that consists of the points lying above $x + y = 1$ but below the line $x - y = 1$, you write the simultaneous inequalities _____.

Answers
1. above   2. below
3. $\begin{rcases} x + y > 1 \\ x - y > 1 \end{rcases}$

EXERCISE SET 9 (Form A)

Section 9.1

1. Which of the following is true and which is false?
   (a) If $n < 6$, then $6 < n$.   (b) If $n < 6$, then $6 > n$.
2. Is it true that $2 - x > 6$ if
   (a) $x = {}^-5$ (b) $x = 5$ (c) $x = {}^-4$
3. Is $2x - 3y > 6$ when
   (a) $x = 1$ and $y = {}^-3$ (b) $x = {}^-3$ and $y = 1$
   (c) $x = -4$ and $y = {}^-5$ (d) $x = 9$ and $y = 4$
4. If x is greater than b but less than c, this is written as $b < x < c$; that is, $b < x$ is the same as $x > b$, so $b < x < c$ means the same as x is less than c but (and) at the same time greater than b. List all integers n for which it is true that $3 < n < 6$.
5. If $x = 4.327$, is it true that $3 < x < 6$?
6. List all integers I for which it is true that ${}^-3 < I < 4$.
7. (a) Evaluate $\dfrac{6w}{w - x}$ when $w = {}^-8$ and $x = {}^-7$.
   (b) Evaluate $\dfrac{wx - 11}{w + x}$ when $w = {}^-8$ and $x = {}^-7$.
   (c) When $w = {}^-8$ and $x = {}^-7$, is it true that $\dfrac{6w}{w - x} - \dfrac{wx - 11}{w + x} < 50$?

Section 9.2

8. (a) For what values of x is it true that $x + 2 < 3$?
   (b) For what values of x is it true that $x + 2 > 1$?
   (c) For what values of x is it true that $1 < x + 2 < 3$?

9. For what values of x is it true that $2 - x > 6$?
10. For what values of x is it true that $5[3(x + 2) - 4] + 1 < 8(x - 3)$?
11. For what values of x is it true that $5(x + 2) - 6(3 - 2x) < -3(8 - 5x)$?

Answers: Exercise Set 9
(Form A)
1. (a) false   (b) true
2. (a) yes   (b) no   (c) no
3. (a) yes   (b) no   (c) yes
   (d) no   4. 4 and 5   5. yes
6. $-2, -1, 0, 1, 2,$ and 3
7. (a) 48   (b) -3 (c) no
8. (a) $x < 1$ (b) $x > -1$
   (c) $-1 < x < 1$ 9. $x < -4$
10. $x < -5$ 11. $x < -8$

Section 9.3

12. (a) Is the point $(0,0)$ above the line $x + 2y = 4$?
    (b) Describe the region formed by all points (x,y) in the plane for which $x + 2y < 4$.
    (c) If x and y must be whole numbers, list all points (x,y) for which $x + 2y < 4$.
13. (a) Is $(0,0)$ above the line $4x - 5y = 20$?
    (b) Describe the region consisting of those points in the plane (x,y) for which $4x - 5y > 200$.
14. A record company has a fixed expense of $300 a week before it produces even one copy of a record. Each copy of a record costs the company $1.50. What is the cost (C) in dollars for the company to produce x records a week?
15. The same company sells each copy of the record for $2. What is the company's revenue (income) R if it sells x records per week?
16. Profit is defined as $R - C$ (that is, you find the profit by subtracting the cost from the revenue). How many copies of a record does the same company have to sell in order to make a profit?
17. As dry air moves upward, it expands and cools at a rate of 5.5°F for each 1,000 feet it rises. Suppose the ground temperature is 80°F.

    (a) Write an equation that relates the temperature T (in Fahrenheit) of the air to the height h (in thousands of feet) above the ground.
    (b) What can you say about the altitude of an airplane if the temperature of the air outside the airplane is more than 14°F but less than 25°F?

Section 9.4

In the space provided, shade the region R determined by the given inequalities. In each case, label the coordinates of each point at which two lines meet.

18. $\left. \begin{array}{l} x - y \leq 3 \\ x \geq 0 \\ y \leq 0 \end{array} \right\}$

19. $\left. \begin{array}{l} x - y \leq 3 \\ x + y \leq 3 \\ x \geq 0 \end{array} \right\}$

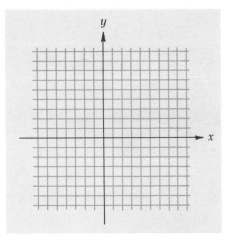

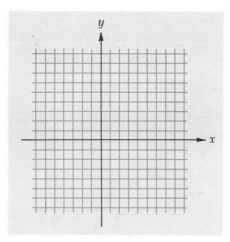

---

12. (a) no (b) all points below the line $x + 2y = 4$ (c) $(0,0)$, $(0,1)$, $(1,0)$, $(1,1)$, $(2,0)$, $(3,0)$
13. (a) yes (b) all points below the line $4x - 5y = 20$
14. $C = 300 + 1.5x$
15. $R = 2x$ 16. any number greater than 600
17. (a) $T = 80 - 5.5h$ (b) The altitude is more than 10,000 feet but less than 12,000 feet.
18. see graph

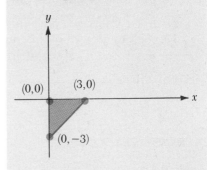

19. see graph

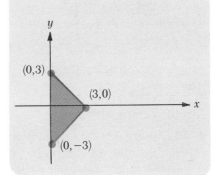

Exercise Set 9   (Form A)   239

20. $\left.\begin{array}{l} x+y>2 \\ y-x<2 \\ x\le 4 \\ y\ge 0 \end{array}\right\}$

21. $\left.\begin{array}{l} x+y\ge 2 \\ y-2x<2 \\ x+y<4 \\ y\ge 0 \end{array}\right\}$

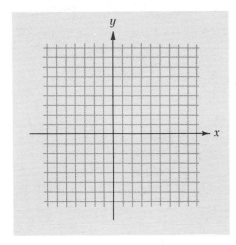
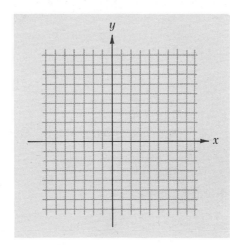

20. see graph
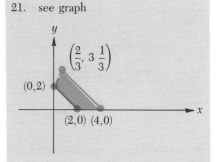

21. see graph
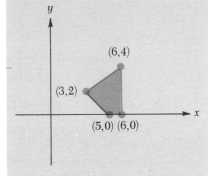

22. (Optional Problem) A family can buy books at $4 each and records at $3 each.

(a) Write the cost (C) in dollars of x books and y records.
(b) The family cannot buy a negative number of items (books or records). Write these facts as inequalities involving x and y.
(c) The family will buy at most 6 books. The family has also decided to buy no more than 6 records. Write these facts as inequalities.
(d) They also decide to buy at least 5 items. Write this fact as an inequality.
(e) In addition, the family has decided to spend at least twice as much money for books as for records. Write this fact as an inequality.
(f) In the diagram, shade the region determined by the inequalities in (b), (c), (d), and (e).
(g) The greatest value as well as the least value C can have must occur at vertex points in the shaded region. What is the least amount of money the family can spend, according to the given facts?
(h) What is the greatest amount the family can spend, according to the given facts?
(i) To spend the least money, how many records and how many books does the family buy?

22. (a) $C=4x+3y$ (b) $x\ge 0$, and $y\ge 0$ (c) $x\le 6$, and $y\le 6$ (d) $x+y\ge 5$ (e) $4x\ge 6y$, or $y\le \dfrac{2}{3}x$ (f) see graph

(g) $18   (h) $36   (i) 3 books and 2 records

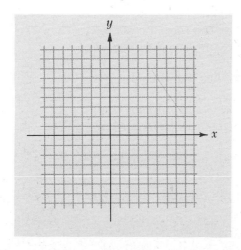

EXERCISE SET 9 (Form B)

Section 9.1

1. Which of the following is true and which is false?
   (a) If $m > -8$, then $-8 > m$.   (b) If $m > -8$, then $-8 < m$.

2. Is it true that $4 - x > 9$ if
   (a) $x = -7$ (b) $x = 11$ (c) $x = -5$

3. Is it true that $3x - 4y > 12$ if
   (a) $x = -1$ and $y = -5$ (b) $x = -5$ and $y = -1$
   (c) $x = 12$ and $y = -6$ (d) $x = 12$ and $y = 6$

4. List all integers n for which it is true that $2 < n < 7$.
5. If $n = 2.001$, is it true that $2 < n < 7$?
6. List all integers I for which it is true that $-6 < I < 3$.
7. (a) Evaluate $\dfrac{8w}{2w - x}$ when $w = 2$ and $x = -6$.

   (b) Evaluate $\dfrac{w + x}{w}$ when $w = 2$ and $x = -6$.

   (c) When $w = 2$ and $x = -6$, is it true that $\dfrac{8w}{2w - x} - \dfrac{w + x}{x} > 3$?

Section 9.2

8. (a) For what values of x is it true that $x + 4 < 8$?
   (b) For what values of x is it true that $x + 4 > 2$?
   (c) For what values of x is it true that $2 < x + 4 < 8$?

9. For what values of x is it true that $4 - x > 9$?
10. For what values of x is it true that $2[4(x - 2) + 3] + 1 > 5(2x - 3)$?
11. For what values of x is it true that $2(x - 3) - 5(4 - 3x) < -3(6 - 5x)$?

Section 9.3

12. (a) Is the point $(0,0)$ below the line $x + 3y = 6$?
    (b) Describe the region formed by all points (x,y) in the plane for which it is true that $x + 3y < 6$.
    (c) List all points (x,y) for which $x + 3y < 6$ if x and y must be whole numbers.

13. (a) Is $(0,0)$ below the line $2x - 3y = 18$?
    (b) Describe the region that consists of all points (x,y) in the plane for which $2x - 3y < 18$.

14. A record company has a fixed expense of $250 a week before it produces even one copy of a record. In addition, each copy of the record costs the company $1.75. What is the cost (C) in dollars for the company to produce r records per week?

15. The same company sells each record for $2.25. What is the company's revenue (R) if it sells r records per week?

16. Profit is $R - C$. Using the information in the two previous problems, find how many copies of the record the company must sell per week in order to make a profit.

17. As dry air moves upward, it cools at a rate of 6.5°F for each 1,000 feet it rises. Suppose the ground temperature is 73°F.

    (a) Write an equation that relates the temperature T of the dry air to the height h (in thousands of feet) above the ground.
    (b) What can you say about the altitude of an airplane if you know that the temperature is more than 44°F below 0 but less than 31°F below 0?

Exercise Set 9 (Form B)   241

Section 9.4

In the space provided, shade the region R determined by the given inequalities.

18. $\left.\begin{array}{l} x - y \leq 4 \\ x \geq 0 \\ y \leq 0 \end{array}\right\}$

19. $\left.\begin{array}{l} x - y \leq 4 \\ x + y \leq 4 \\ x \geq 0 \end{array}\right\}$

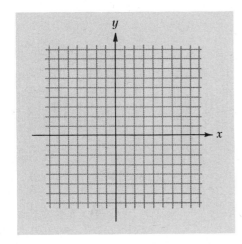
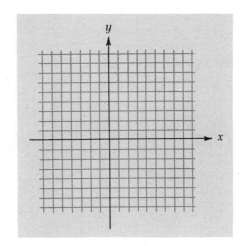

20. $\left.\begin{array}{l} x + y > 1 \\ y - x < 1 \\ x \leq 3 \\ y \geq 0 \end{array}\right\}$

21. $\left.\begin{array}{l} x + y \geq 1 \\ y - x < 1 \\ x + y < 4 \\ y \geq 0 \end{array}\right\}$

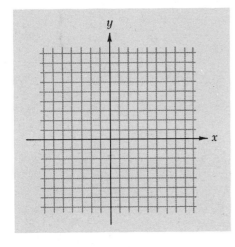
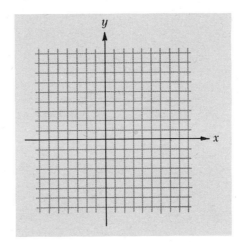

22. (Optional Problem) A family can buy old books at $2 each and records at $4 each.

   (a) Write the cost C in dollars of x books and y records.
   (b) The family members have decided to buy no more than 5 books. They have also decided to buy no more than 7 records. Write these facts as inequalities.
   (c) They will buy a combination of at least 6 books and records. Write this fact as an equality.
   (d) In addition, they have decided to spend at least twice as much for records as for books. Write this fact as an inequality.
   (e) In the diagram provided on the next page, shade the region determined by the inequalities in (b), (c), and (d).

(f) Use the fact that the greatest expense and the least expense must occur at the vertices of the region to determine the least amount of money they can spend.
(g) What is the greatest amount they can spend?
(h) When they spend the least money, how many of each item do they buy?

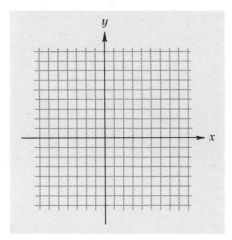

An Introduction to Nonlinear Equations
unit 4

Lesson 10

Integral Exponents

Overview Once a fourth-grader became upset trying to think of enough words to fill a five hundred-word composition. Overwhelmed, he gave up and wrote, "I have a cat. It got lost. I went out and called, 'Here kitty, kitty, kitty, kitty, kitty, kitty, kitty. . . .'"

Just as it can be monotonous to see the word "kitty" over and over again, it can be monotonous trying to read the 0's in a numeral like 1,000,000,000,000,000,000,000,000,000. Can we find an abbreviation so that we can read a number like 0.000000000000000000000000001?

We shall study questions like these in this lesson. The questions involve the idea of a product of a number multiplied by itself many times. For example, to get a number that consists of 1 followed only by 0's, we have to keep multiplying 10's. A 1 followed by seven 0's is obtained from $10 \times 10 \times 10 \times 10 \times 10 \times 10 \times 10$. Though multiplying 10's is convenient, there is no need to restrict our study to the case of 10. We may talk about products like $6 \times 6 \times 6 \times 6 \times 6 \times 6 \times 6 \times 6$.

This idea and its consequences form the topics in this lesson.

Section 10.1
Positive Integral Exponents

Tally marks become very awkward to use if we want to express large numbers. Even for not-so-large numbers, tally marks can be difficult to read. It is not easy to see quickly the difference between 23 and 24 tally marks.

To avoid this problem, we use place value. But considering the size of the numbers that we work with in modern technology, even place value can give us the same trouble that early peoples had with tally marks.

For example, look at the number

$$100,000,000,000,000,000,000,000$$

which is a 1 followed by 23 zeroes. We would like a better name for this number than "a '1' followed by twenty-three 0's." But inventing names beyond a billion or a trillion can become clumsy. What can we do?

One answer is to notice that starting with 10, we add on another 0 every time we multiply by 10. That is,

$$10 \times 10 = 100$$
$$10 \times 10 \times 10 = 1,000$$
$$10 \times 10 \times 10 \times 10 = 10,000$$

Why can't we invent a shorthand for expressing the number of times 10 appears as a factor?

For example, we can write 10 to indicate that the repeated factor is 10. Then we can write how many factors of 10 we want by placing the appropriate numeral above and to the right of the 10. In this way, the expression 10^3 means the *product* of three 10's. That is,

$$10^3 \quad \text{means} \quad 10 \times 10 \times 10$$

Example 1

What place value numeral names the same number as 10^6?

10^6 is the same as 1,000,000. That is, 10^6 means you take 10 as a factor 6 times.

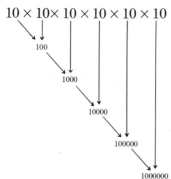

Notice that you use another 0 every time you multiply by 10. Therefore, the product of 6 factors of 10 is a 1 followed by 6 zeroes.

Example 2

What place value numeral names 10^{15}?

10^{15} means 1,000,000,000,000,000. You have the product of 15 factors of 10, which is a 1 followed by 15 zeroes.

OBJECTIVE
To understand and to use the notation

$$b^n$$

where b is any (real) number and n is any positive integer.

Notice that the commas make it a little easier to read the number of 0's, but it's still a pretty "messy" job.

If we have two factors of ten, the product is a 1 followed by two 0's. If we have three factors of ten, the product is a 1 followed by three 0's. Each additional factor of ten gives us one more 0 after the 1.

A symbol placed above a number is called a *superscript*.

Notice that the product of three 10's is 1,000 — not 30. That is, the product of three 10's means

$$10 \times 10 \times 10 \quad \text{or} \quad 1,000$$

The sum of three 10's is 30. That is,

$$10 + 10 + 10 = 30$$

It is much more compact to write 10^{15} than 1,000,000,000,000,000.

Section 10.1   Positive Integral Exponents   247

We can invert the emphasis and ask the same questions from a different point of view.

Example 3

Using the new notation, how would you write 10,000?

You would write 10,000 as 10^4. The idea is that 10,000 is the product of four 10's. That is,

$$10,000 = 10 \times 10 \times 10 \times 10$$

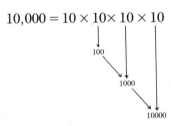

10^4 is an abbreviation for $10 \times 10 \times 10 \times 10$.
 Another way of doing the problem is to notice that 10,000 is a 1 followed by 4 zeroes; and you write this as 10^4.

Example 4

How would you write

100,000,000,000,000,000,000,000

in the new notation?

You would write this as 10^{23}. You have a 1 followed by 23 zeroes. This is the same thing as the product of twenty-three 10's. You write this as 10^{23}.

Among other things, Example 4 shows us the answer to the question we raised earlier. We can abbreviate 100,000,000,000,000,000,000,000 as 10^{23}.
 Now that we have a new shorthand, we should consider the fact that the idea goes beyond products that involve 10's. For example, we may want to express the product of three 2's or the product of twenty 2's. We may want to express the product of two 3's or of a hundred 3's. How can we do these things?
 Using the new notation, if we want the product of three 2's, we can begin by writing the 2 (to indicate what number is being used as the factor). Then we can write 3 as the superscript to tell us that the factor is to occur three times. That is,

$$2^3 \quad \text{means} \quad 2 \times 2 \times 2 \quad \text{or} \quad 8$$

Example 5

How can you abbreviate the product of twenty 2's?

You abbreviate the product of twenty 2's by writing 2^{20}. The superscript, which is 20, tells you that you want the product of twenty 2's.

As an aside, the product of twenty 2's is quite large: it is 1,048,576. Numbers increase quite quickly when we keep doubling. For example: 2; 4; 8; 16; 32; 64; 128; 256; 512; 1,024; 2,048; 4,096; 8,192; 16,384; and so on. The *sum* of twenty 2's is 40.

Example 6

How can you use the new notation to express the product of two 20's?

You write this as 20^2. That is, 20 is the factor, and it is to occur twice. The superscript is used to indicate the number of times the factor appears.
  In summary, 20^2 means 20×20, or 400.

2^{20} and 20^2 are not the same. The sum of two 20's is the same as the sum of twenty 2's (both are 40), but the product of twenty 2's is not the same as the product of two 20's. So which number is the superscript makes quite a difference.

Example 7

What number is named by 3^2?

3^2 names the number nine. That is, 3^2 means 3×3, or 9.
  Remember that since the superscript is 2, you want the product of two 3's. If you want the product of three 2's, you would have to write 2^3. In other words, $2^3 = 2 \times 2 \times 2$, or 8; while 3^2 means 3×3, or 9.
  Now we can generalize our results.

Notice the difference between 2^3 and 3^2. Order makes a difference.

> **DEFINITION**
> If b denotes any number and if n is any *positive integer*, then b^n means the product when b is used as a factor n times.

$b \times n$ stands for the *sum* of "n" b's.
b^n stands for the *product* of "n" b's.

Example 8

What does 4^5 mean?

It means the product of five 4's. That is,

$$4^5 = 4 \times 4 \times 4 \times 4 \times 4$$
$$16 \to 64 \to 256 \to 1024$$

Or, $4^5 = 1{,}024$.

Sometimes we prefer to leave the answer in the form 4^5, and other times we prefer to write 1,024. In Examples 8 and 9, we use both forms to see that 4^5 and 5^4 name different numbers. Place value numerals make this clear.

Example 9

What does 5^4 mean?
It means the product of four 5's. That is,

$$5^4 = 5 \times 5 \times 5 \times 5$$
$$25 \to 125 \to 625$$

To summarize, $4^5 = 1{,}024$ and $5^4 = 625$.

For example, in 3^5, 3 is the base and 5 is the exponent. 3^5 is read as 3 to the fifth power. It means

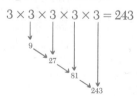

> **VOCABULARY**
> In the expression b^n
> 1. We call b the **base**.
> 2. n is called the **exponent**.
> 3. We read b^n as **b to the nth (power)**.

Section 10.1 Positive Integral Exponents 249

Example 10

How do you read 10^{30} and what number does it name?

10^{30} is read as 10 to the thirtieth power, and it is a 1 followed by 30 zeroes. In this case, 10 is the base and 30 is the exponent. This means that you want the product of thirty 10's. Each time you multiply by 10, you annex another 0. Hence you have a one followed by 30 zeroes. In other words, 10^{30} = 1,000,000,000,000,000,000,000,000,000,000.

Saying "ten to the thirtieth", is a big improvement over having to memorize names like trillions and zillions.

Example 11

How do you read 10,000,000,000,000,000,000,000,000,000?

You read this as 10 to the twenty-eighth power, and you write it as 10^{28}. In this case, 10 is the base and 28 is the exponent. Since you have a 1 followed by 28 zeroes, you have the product of 28 tens.

PRACTICE DRILL

1. What number is named by
   (a) 5^2 (b) 2^5 (c) 10^8

2. Use exponents to abbreviate each of the following.
   (a) $6 \times 6 \times 6 \times 6 \times 6$
   (b) $5 \times 5 \times 5 \times 5 \times 5 \times 5$

3. How do you read
   (a) 6^4 (b) 4^6

Answers

1. (a) 25 (b) 32
(c) 100,000,000   2. (a) 6^5
(b) 5^6 3. (a) 6 to the 4th power
(b) 4 to the 6th power

CHECK THE MAIN IDEAS

1. 3×4 is an abbreviation for the _____ of four 3's.
2. But 3^4 is the abbreviation for the _____ of four 3's.
3. 4^3 means the product of three _____'s.
4. $3^4 = 81$ but $4^3 = $ _____.
5. In the expression 3^4, 3 is called the _____.
6. 4 is called the _____.
7. The fact that 3^4 and 4^3 are not equal means that you must be careful not to interchange the base and the _____.

Answers

1. sum   2. product   3. 4
4. 64   5. base
6. exponent   7. exponent

Section 10.2
Other Integral Exponents

In the last section, we saw that 10^n names the number we usually write in place value as a 1 followed by n zeroes (as long as n is any positive integer). We used this fact to learn how exponents can give us a convenient way of representing large numbers.

If large numbers are important in today's technology, very small numbers are also important. In dealing with the size of atoms and molecules, for example, scientists commonly deal with numbers such as

$$0.00000000001 \qquad [1]$$

In this section, we want to find out if it is possible to use exponents to express numbers such as [1] more conveniently.

To begin gradually, suppose we talk about a 1 followed by 0 (no) zeroes. Presumably, this just means 1. If we define 10^5 as a 1 followed by five zeroes (100,000), let's define 10^0 as a 1 followed by no zeroes.

> **DEFINITION**
> 10^0 means 1.

Let's look at this definition from another point of view.

Let's write down 10^0 and pretend that we don't know what value it's supposed to have. Under it, let's write down the values of 10^1, 10^2, 10^3, and so on — expressions whose values we already know from Section 10.1. We get

$$10^0 = ?$$
$$10^1 = 10$$
$$10^2 = 100$$
$$10^3 = 1{,}000$$

If we look at the 10^n column and read it from bottom to top, we see that the exponent is reduced by one each time. Now we read the right-hand column from bottom to top, starting with 1,000. The next number is 100. We get from 1,000 to 100 by dividing by 10. Similarly, if we divide 100 by 10 we get the next number, 10; and if we divide 10 by 10, we get the next number, 1. In other words, if we want the right-hand column to reflect the idea that we get from one number to the one above by dividing by 10, 10^0 *must* by 1.

> **DEFINITION**
> For any non-zero number b
> $$b^0 = 1$$

The reason for this definition is the same as the reason that 10^0 is defined as 1.

$$4^0 = ?$$
$$4^1 = 4$$
$$4^2 = 16 \quad (4 \times 4)$$
$$4^3 = 64 \quad (4 \times 4 \times 4)$$

As we read this chart from bottom to top, we see that the exponent is decreased by 1 each time. In the right column, we divide by 4 to get from one number to the one above it. Therefore, to get from 4 to the number above it, we must divide 4 by 4 if the sequence is to be continued. This forces us to accept the meaning of 4^0 as 1.

OBJECTIVE
To understand the meaning of b^n when n is either 0 or a negative integer.

When we say "followed by n zeroes," we are assuming that n is a positive whole number. How can we have -5 zeroes or 3.4 zeroes?

In a way, reading this number is harder than reading

$$100{,}000{,}000{,}000$$

For most of us, not only are decimals less "natural" than whole numbers, but we don't even have the help of commas in reading the number. We could write

$$0.\ 000\ 000\ 000\ 01$$

to express [1], but that is not very helpful.

That is, the sequence of exponents is 3, 2, 1, 0.

This shouldn't be a surprise. We have already seen that to get from 10^2 to 10^3, we multiply by 10. So to reverse the process and get from 10^3 to 10^2, we divide by 10.

REVIEW
$$10^0 = 1$$
$$\uparrow$$
$$10^1 = 10$$
$$\uparrow$$
$$10^2 = 100$$
$$\uparrow$$
$$10^3 = 1{,}000$$
$$\uparrow$$
$$10^4 = 10{,}000$$

Reading the right column from bottom to top, we divide by 10 each time to get from one number to the next. Also reading from bottom to top, we decrease the exponent by 1 each time.

Section 10.2   Other Integral Exponents

There is nothing special about either 10 or 4. In general, we get from b^1 to b^0 by dividing by b. Since b divided by b is 1, we define b^0 as 1 for any *non-zero* number b.

We aren't allowed to divide by 0, so to divide by b implies that b is not 0. 0^0 is not defined. We just don't use this symbol.

Example 1

What is the value of 7^0?

7^0 is 1, because $b^0 = 1$ for all non-zero values of b.

Example 2

What is the value of $(-3)^0$?

$(-3)^0$ is 1. Remember, b^0 is defined as 1 for any non-zero value of b; and -3 is not 0.

Example 3

What is the value of $\left(\frac{1}{2}\right)^0$?

The value is 1. The value of any non-zero number raised to the 0th power is *by definition* equal to 1.

In the same way that we deduced how to define b^0, we can deduce how to define b^{-n}, where n is any positive integer. Suppose we want to find an easy way to define 4^{-1}. We can refer back to our chart for powers of 4, remembering that -1 is the integer that is 1 less than 0. Our chart now looks like this:

$$4^{-1} = ?$$
$$4^0 \phantom{^{-1}} = 1$$
$$4^1 \phantom{^{-1}} = 4$$
$$4^2 \phantom{^{-1}} = 16$$

Reading from bottom to top, we see that the exponents in the left column decrease by 1 each time we go from one entry to the one above. On the right, we see that to get from a number to the one above, we divide by 4. If we wanted this sequence to continue, the number that corresponds to 4^{-1} should come from dividing 1 by 4. Using this idea, we define

$$4^{-1} = \frac{1}{4} \quad \text{or} \quad 1 \div 4$$

That is, we divide by 4 to get from one number to the one above it.

In a similar way, we can see that

$$4^{-2} \quad \text{means} \quad 1 \div 4^2 \quad \text{or} \quad \frac{1}{16}$$

The idea is that -2 is one less than -1, and when we divide $\frac{1}{4}$ by 4, we get $\frac{1}{16}$. That is, $\frac{1}{4} \div 4 = \frac{1}{4} \times \frac{1}{4}$.

If we continue our chart a bit longer to see what happens, we realize that

$$4^{-4} = 1 \div 4^4 \text{ or } \frac{1}{4^4} = \frac{1}{256}$$
$$\uparrow \div 4$$
$$4^{-3} = 1 \div 4^3 \text{ or } \frac{1}{4^3} = \frac{1}{64}$$
$$\uparrow \div 4$$
$$4^{-2} = 1 \div 4^2 \text{ or } \frac{1}{4^2} = \frac{1}{16}$$
$$\uparrow \div 4$$
$$4^{-1} = 1 \div 4 \text{ or } \frac{1}{4}$$
$$\uparrow \div 4$$
$$4^0 = 1$$
$$\uparrow \div 4$$
$$4^1 = 4$$

The pattern is

$$4 \div 4 = 1$$
$$1 \div 4 = \frac{1}{4}$$
$$\frac{1}{4} \div 4 = \frac{1}{4^2}$$
$$\frac{1}{4^2} \div 4 = \frac{1}{4^3}$$

The exponent decreases by 1 as we read from bottom to top, and each number is obtained by dividing the one below it by 4.

Now we can develop a system for defining, say, 4^{-6} without writing the whole chart.

STEP 1: Change the sign of the exponent. ⟶ This gives us 4^6.
STEP 2: Preface the power by "$1 \div$". ⟶ This gives us $1 \div 4^6$, or $\frac{1}{4^6}$.

Example 4

What number is named by 2^{-3}?

2^{-3} means $\frac{1}{8}$. You change 2^{-3} to 2^3 and write "$1 \div$" in front of it to get

$$2^{-3} = 1 \div 2^3 = 1 \div 8 = \frac{1}{8}$$

2^{-3} is not a negative number. The negative exponent tells us only to write "$1 \div$" before the power, and to change the sign of the exponent.

If we want to use a chart:

$$2^{-3} = \frac{1}{2^3} = \frac{1}{8}$$
$$2^{-2} = \frac{1}{2^2} = \frac{1}{4}$$
$$2^{-1} = \frac{1}{2}$$
$$2^0 = 1$$

Example 5

What number is named by 3^{-2}?

3^{-2} means $\frac{1}{9}$. That is, 3^{-2} means $1 \div 3^2$, or $1 \div 9$, or $\frac{1}{9}$.

Now we can state the general definition.

> **DEFINITION**
> For any positive integer n, if $b \neq 0$ we define
> $$b^{-n}$$
> to mean
> $$1 \div b^n$$

The definition is neither natural nor self-evident. It merely guarantees that when we decrease n by 1, b^n decreases by a factor of b, regardless of whether n is positive, zero, or negative. The restriction that $b \neq 0$ agrees with our decision that we cannot divide by 0. For example, 0^{-4} would mean $\frac{1}{0^4}$, or $1 \div 0$.

Section 10.2   Other Integral Exponents

Whenever we divide a number by 10, we move the decimal point one place to the left. So, when $b = 10$, our definition of b^{-n} looks like this:

$$10^{-3} = \frac{1}{10^3} = \frac{1}{1,000} = 0.001$$

$$10^{-2} = \frac{1}{10^2} = \frac{1}{100} = 0.01$$

$$10^{-1} = \frac{1}{10} = 0.1$$

$$10^{0} = 1 = 1$$

each time we divide by 10

Suppose we want to write 10^{-7} in decimal fractions form. 10^{-7} means $1 \div 10^7$.

STEP 1: Write "1."
STEP 2: Move the decimal point 7 places to the left (because we're dividing by 10 seven times).

Example 6

What decimal fraction is named by 10^{-9}?

10^{-9} means 0.000000001.
  You start with 1 (or 1.) and then move the decimal point nine places to the left. This gives you

$$0.000000001.$$

  Now we are ready to use exponents to represent the number 0.00000000001 mentioned earlier in this section.

Example 7

How can you represent

$$0.00000000001 \qquad [2]$$

as a power of 10?

0.00000000001 is the same as 10^{-11}.
  To get from 0.00000000001 to 1, you must move the decimal point 11 places to the right. Therefore, if you start with 1., you must move the decimal point 11 places to the left to get [2]. When you move the decimal point in "1." 11 places to the left, you write 10^{-11}.

PRACTICE DRILL

1. Write each of the following without using exponents.
   (a) 6^0 (b) 6^{-2} (c) 10^{-4}

2. Write each of the following using negative exponents.
   (a) $1 \div 6^3$ (b) 0.000001

There is a pattern. When the exponent is $-n$, the value of

$$10^{-n}$$

is obtained by starting with 1 (or 1.) and moving the decimal point n places to the *left*.

REVIEW
When we use decimal fractions, we divide by 10 by moving the decimal point one place to the left; we multiply by 10 by moving the decimal point one place to the right.

$0.0000001.$

Notice that we have only 6 zeroes between the decimal point and the 1, *not* 7. (The first time we move the decimal point to the left, we go from 1 to 0.1, and there is no 0 after the decimal point.)

SUMMARY FOR INTEGRAL POWERS OF 10
Suppose that n is a positive integer. Then
1. 10^n means to start with 1 and move the decimal point n places to the right. This is the same as adding n zeroes after the 1.
2. 10^{-n} means to start with 1. and move the decimal point n places to the left. The first move brings us from 1 to .1, so we do not yet add a 0. We add one less than $n(n-1)$ zeroes to the left of the 1, and then put in the decimal point.
3. 10^0 is 1 (which we may think of as a 1 followed by no 0's).

Answers

1. (a) 1   (b) $\frac{1}{36}$ (c) $\frac{1}{10,000}$, or 0.0001   2. (a) 6^{-3} (b) 10^{-6}

CHECK THE MAIN IDEAS

1. To define b^n as the product of n factors of b, n must be a _____ integer.
2. Following this definition, whenever you decrease n by 1, b^n is decreased by a factor of _____.
3. 2^3 is 8 and 2^2 is 4. When you decrease the exponent by 1, 2^n decreases by a factor of _____ (that is, you divide 8 by _____ to get 4).
4. If it is true that b^n is decreased by a factor of b whenever n is decreased by 1, you *must* define b^0 as _____.
5. For the same reason, you must define b^{-3} as $1 \div$ _____.
6. More generally, for any positive integer n, if b is not 0, you define b^{-n} as $1 \div$ _____.

Answers

1. positive 2. b 3. 2
4. 1 5. b^3 6. b^n

Section 10.3
Multiplying Like Bases

In the first two sections of this lesson, we saw how the use of exponents helps us to express very large and very small numbers. In this section, we want to see how the use of exponents makes it easier for us to perform the operation of multiplication.

OBJECTIVE
To see how exponents allow us a convenient way for computing
$$b^m \times b^n$$

Example 1

Find the product of 10,000 and 100,000.

$10,000 \times 100,000 = 1,000,000,000$.

Every time you multiply 10,000 by 10, you annex another 0. Since 10,000 may be viewed as the product of four 10's, you may think of this problem as multiplying 100,000 by 10, four times.

You annex a 0 for each factor of 10, so you must annex four 0's to 100,000 in order to multiply it by 10, four times. Adding four 0's to the five that are already there gives you a total of nine 0's. That is, the product is a 1 followed by nine 0's, or 1,000,000,000.

$$\begin{aligned} 10{,}000 \\ 100{,}000 &= 10{,}000 \times 10 \\ 1{,}000{,}000 &= 10{,}000 \times 100 \\ 10{,}000{,}000 &= 10{,}000 \times 1{,}000 \\ 100{,}000{,}000 &= 10{,}000 \times 10{,}000 \\ 1{,}000{,}000{,}000 &= 10{,}000 \times 100{,}000 \end{aligned}$$

We didn't use exponents in Example 1, but what would happen if we did? First, since 10,000 is a one followed by 4 zeroes, it can be written as 10^4. Similarly, since 100,000 is a one followed by five zeroes, we can write it as 10^5. Therefore, $10,000 \times 100,000$ is the same as $10^4 \times 10^5$. The product of four factors of 10 and five more factors of 10 gives us a total product of nine factors of 10. That is,

$$(10 \times 10 \times 10 \times 10) \times (10 \times 10 \times 10 \times 10 \times 10)$$
$$\begin{array}{cccccccccc} 1 & 2 & 3 & 4 & (+) & 1 & 2 & 3 & 4 & 5 \\ 1 & 2 & 3 & 4 & & 5 & 6 & 7 & 8 & 9 \end{array}$$

In other words, we have:
$$10^4 \times 10^5 = 10^{4+5} = 10^9$$

Now we can use this idea much more generally.

In other words, we do not have to write out the 9 factors. All we have to do is see that four factors of 10 multiplied by 5 more factors of 10 gives us a total of 9 factors of 10.

Section 10.3   Multiplying Like Bases

Example 2

Express $3^4 \times 3^2$ as a single power of 3.

$3^4 \times 3^2 = 3^6$.

All you have to do is remember that 3^4 is the product of four 3's and 3^2 is the product of two 3's. That is,

$$3^4 \times 3^2 = \underbrace{(3 \times 3 \times 3 \times 3)}_{1 \; 2 \; 3 \; 4} \times \underbrace{(3 \times 3)}_{5 \; 6}$$

Since $4 + 2 = 6$, you see that you have the product of six 3's. That is,

$$3^4 \times 3^2 = 3^{4+2} = 3^6$$

Note that we added the exponents. We did *not* multiply them. If we multiply the two exponents, we get $4 \times 2 = 8$. But we can see that $3^4 \times 3^2$ is not equal to 3^8, because we already know that the answer is the product of six 3's, not the product of eight 3's.

Notice that we did not combine the bases. We did *not* multiply 3 by 3 and write $3^4 \times 3^2 = 9^6$. The point is that we have six factors of 3, not six factors of 9, when we multiply the product of four 3's by the product of two 3's. That is, $3^4 \times 3^2 = (3 \times 3 \times 3 \times 3) \times (3 \times 3) = 3^6$, not 9^6.

Let's try one more example.

Example 3

Express the product $5^6 \times 5^3$ as a power of 5.

$5^6 \times 5^3 = 5^9$. You have the product of six factors of 5, and you multiply this by the product of three more factors of 5. Altogether, you have the product of nine 5's. That is,

$$5^6 \times 5^3 = \underbrace{(5 \times 5 \times 5 \times 5 \times 5 \times 5)}_{1 \; 2 \; 3 \; 4 \; 5 \; 6} \times \underbrace{(5 \times 5 \times 5)}_{7 \; 8 \; 9}$$

As long as the two numbers we're multiplying have the same base, we find the product by keeping the common base and taking the sum of the exponents.

> **THE RULE FOR MULTIPLYING LIKE BASES**
> If b is any (real) number, and if m and n are any positive integers, then
> $$b^m \times b^n = b^{m+n}$$

Example 4

Use the rule to express $6^7 \times 6^5$.

$6^7 \times 6^5 = 6^{12}$.

The common base is 6 and the sum of the exponents is $7 + 5$, or 12. By the rule, you keep the common base (6) and use as your exponent the sum of the two exponents (12). So the answer is 6^{12}.

Example 5

Express $10^{15} \times 10^9$ as a single power of 10.

$10^{15} \times 10^9 = 10^{24}$.

As a check notice that

$3^4 = 3 \times 3 \times 3 \times 3 = 81$
$3^2 = 3 \times 3 = 9$
$3^6 = 3 \times 3 \times 3 \times 3 \times 3 \times 3$
$\quad = 729$

So

$3^4 \times 3^2 = 81 \times 9$
$\quad = 729 = 3^6$

It's not important whether it "seems right" to say that:

$$3^4 \times 3^2 = 3^8$$

By our *definition* of exponents, this is a false statement. The product of four 3's and two 3's is the same as the product of six 3's.

This is somewhat similar to adding fractions that have a common denominator. There, we add the numerators but keep the common denominator. Here, we keep the common base and add the exponents.

Again, avoid these errors. $5^6 \times 5^3$ is *not* the same as 5^{18}, and $5^6 \times 5^3$ is *not* the same as 25^9.

The definition of exponent says that we have the product of six 5's multiplied by the product of three 5's. This is the product of nine 5's.

Remember that the rule is just a quick way to get the same answer as we were able to get before. If we multiply the product of seven 6's by the product of five 6's, we get a product of twelve 6's.

This says that if we take the number that is a 1 followed by 15 zeroes and multiply it by a 1 followed by 9 zeroes, the product is a 1 followed by 24 zeroes.

Again, using the rule, you keep the common base (10) and use as your exponent the sum of the two exponents (15 + 9, or 24). Therefore, the answer is 10^{24}.

It is interesting that we can use the rule
$$b^m \times b^n = b^{m+n}$$
even when m and n are not positive integers. This is true because of how we defined b^n when n is either 0 or a negative integer.

Example 6

Use the rule to find the value of $2^3 \times 2^0$.

$2^3 \times 2^0 = 2^3$ (or 8). Using the rule, you have
$$2^3 \times 2^0 = 2^{3+0} = 2^3$$

As a check, notice that without the rule, you have previously defined 2^0 as 1. Therefore
$$2^3 \times 2^0 = 2^3 \times 1 = 2^3$$
You get the same answer without the rule, provided b^0 is defined as 1.

Example 7

Use the rule to find the value of $5^6 \times 5^{-4}$.

$5^6 \times 5^{-4} = 5^2$. Using the rule, you have
$$5^6 \times 5^{-4} = 5^{6+(-4)} = 5^2$$

To check this result, remember that you already know that 5^{-4} is $\frac{1}{5^4}$. Hence
$$5^6 \times 5^{-4} = 5^6 \times \frac{1}{5^4} = \frac{5^6}{5^4}$$
$$= \frac{5 \times 5 \times 5 \times 5 \times 5 \times 5}{5 \times 5 \times 5 \times 5}$$
$$= 5 \times 5 = 5^2$$

Let's try a few more examples.

Example 8

Find the value of $5^3 \times 5^{-7}$.

$5^3 \times 5^{-7} = 5^{-4}$, or $\frac{1}{5^4}$.

One method is to use the rule. This gives you $5^3 \times 5^{-7} = 5^{3+(-7)} = 5^{-4}$ or $\frac{1}{5^4}$. The other method comes from recognizing that 5^{-7} is the reciprocal of 5^7. That is,
$$5^3 \times 5^{-7} = 5^3 \times \frac{1}{5^7} = \frac{5^3}{5^7}$$

The three 5's in the numerator cancel with three of the seven 5's in the denominator, leaving you with the product of four 5's in the denominator.

Either method tells you that the answer is 5^{-4} or $\frac{1}{5^4}$.

NOTE

Some people like to use the rule as a reason for defining b^0 as 1. According to the rule,
$$b^3 \times b^0 = b^{3+0} = b^3$$
This says that when we multiply b^3 by b^0, we get b^3, so b^0 must be 1.
In other words, if we want the rule,
$$b^m \times b^n = b^{m+n}$$
to be true even if one (or both) of the exponents is 0, we *must* define b^0 as 1.
Because this rule is so convenient, we agree to define b^0 as 1.

That is, the four 5's in the denominator cancel four of the six 5's in the numerator, leaving us with $(6 - 4)$ or two 5's in the numerator.

This shows us another reason for knowing how to add signed numbers.

$$\frac{\cancel{5 \times 5 \times 5}}{\cancel{5 \times 5 \times 5} \times 5 \times 5 \times 5 \times 5}$$

Example 9

What is the value of $10^8 \times 10^{-11}$?

$10^8 \times 10^{-11} = 10^{-3} = 0.001$. In other words, $10^8 \times 10^{-11} = 10^{8+(-11)} = 10^{-3}$. In compact form, this says that

$$100{,}000{,}000 \times 0.00000000001 = 0.001$$

This should emphasize the advantage of the exponent form over place value numerals.

The method of adding exponents requires that we *multiply* like bases. When we *add* like bases we do not add exponents.

Example 10

What is the value of $2^2 + 2^3$?

$2^2 + 2^3 = 12$. Namely, $2^2 = 4$ and $2^3 = 8$. Therefore, $2^2 + 2^3 = 4 + 8 = 12$.

We do *not* say that $2^2 + 2^3 = 2^5$. 2^5 is 32, and we just saw that $2^2 + 2^3$ is 12, not 32.

PRACTICE DRILL

Write each of the following as a single power of 3.
1. $3^4 \times 3^6$ 2. $3^4 \times 3^{-6}$ 3. $3^{-4} \times 3^6$ 4. $3^{-4} \times 3^{-6}$

Answers
1. 3^{10} 2. 3^{-2} 3. 3^2
4. 3^{-10}

CHECK THE MAIN IDEAS

1. In multiplying powers that have the same base, you (a) _____ the exponents but keep the common (b) _____.
2. For example, $2^3 \times 2^4 =$ _____ ^{7}.
3. This is not the same as adding 2^3 and 2^4. Since $2^3 = 8$ and $2^4 = 16$, $2^3 + 2^4 =$ _____, which is not an integral power of 2.

Answers
1. (a) add (b) base 2. 2
3. 24

Section 10.4
Other Arithmetical Operations

In the last section, we learned how to multiply numbers having the same base. We also saw that there is no similar rule for adding such numbers.

Since subtraction is the inverse of addition, and division is the inverse of multiplication, we already know quite a bit about these operations, too.

OBJECTIVE
To study in more detail how to use exponents in arithmetic.

Example 1

Express $10^3 \div 10^4$ as a single power of 10.

$10^3 \div 10^4 = 10^{-1}$, or $\frac{1}{10}$.

One way to do this problem is to divide the product of three 10's by the product of four 10's to get $\frac{1}{10}$. But you can use the fact that division is the inverse of multiplication, and then rewrite this problem in terms of multiplication. For example,

$$10^3 \div 10^4 = 10^3 \times \frac{1}{10^4}$$

Since 10^{-4} means $\frac{1}{10^4}$, the quotient can be further rewritten as $10^3 \times 10^{-4}$.

That is,

$$\frac{10 \times 10 \times 10}{10 \times 10 \times 10 \times 10}$$

Remember, the "invert and multiply" rule still applies.

Now you can use the rule for multiplying like bases to obtain the answer.

$$10^3 \times 10^{-4} = 10^{3+(-4)} = 10^{-1} = \frac{1}{10}$$

Let's try one more this way.

Example 2

Express $4^5 \div 4^2$ as a single power of 4.

$4^5 \div 4^2 = 4^3$. You have

$$4^5 \div 4^2 = 4^5 \times \frac{1}{4^2}$$
$$= 4^5 \times 4^{-2}$$
$$= 4^{5+(-2)}$$
$$= 4^3$$

Given

$$\frac{4 \times 4 \times 4 \times 4 \times 4}{4 \times 4}$$

the two 4's in the denominator "cancel" with two of the five 4's in the numerator.

There is a quicker way to solve Example 2. Because we divided the product of five 4's by the product of two 4's, we ended up with three (that is, 5 − 2) 4's. Based on the result of Example 2, we may predict that the following rule explains how to find the quotient of two numbers having the same base.

> **RULE FOR DIVIDING LIKE BASES**
> If b is any non-zero number and m and n are any integers, then
> $$b^m \div b^n = b^{m-n}$$

The non-zero restriction can be removed unless it gives us either 0^0 or division by zero.

Notice the order $m - n$. We subtract the exponent of the divisor (denominator) from the exponent of the dividend (numerator). Remember, subtraction is not commutative.

This rule seems easy to justify when the exponents are positive integers, but our definition of other integral exponents guarantees that the rule works even if we use negative exponents. For example, if we apply this rule to the problem we solved in Example 1, we get the answer very quickly.

$$10^3 \div 10^4 = 10^{3-4} = 10^{-1} = \frac{1}{10}$$

Let's try a few examples this way.

Example 3

Express $5^{82} \div 5^{12}$ as a single power of 5.

By the rule, you have

$$5^{82} \div 5^{12} = 5^{82-12} = 5^{70}$$

We could have obtained the same result by the "invert and multiply" approach.

$$5^{82} \div 5^{12} = 5^{82} \times 5^{-12}$$
$$= 5^{82+(-12)}$$
$$= 5^{70}$$

Example 4

Express $7^{25} \div 7^{30}$ as a single power of 7.

By the rule, you have

$$7^{25} \div 7^{30} = 7^{25-30} = 7^{-5}$$

If we don't want the answer expressed with a negative exponent, it is acceptable to write $\frac{1}{7^5}$ instead of 7^{-5}.

In using the rule for division, we must remember the rules for subtracting signed numbers. In particular, we often use the fact that subtraction is the "add the opposite" rule.

Section 10.4   Other Arithmetical Operations

Example 5

Express $6^3 \div 6^{-2}$ as a single power of 6.

By the rule for division, you have
$$6^3 \div 6^{-2} = 6^{3-(-2)} = 6^{3+(+2)} = 6^5$$

As a check, remember that 6^2 is the reciprocal of 6^{-2}, so you have
$$6^3 \div 6^{-2} = 6^3 \times 6^2 = 6^{3+2} = 6^5$$

The quotient is large because 6^{-2} is $\frac{1}{36}$, and this is very much smaller than 6^3, or 216. That is, we must multiply $\frac{1}{36}$ by 6^5 to get 216.

We subtract the exponent of the divisor, but in this case the exponent is already negative. So, we are subtracting a negative number, which is the same as adding a positive number. In other words,
$$3 - {}^-2 = 3 + {}^+2 = 3 + 2 = 5$$

Another application of multiplying like bases occurs when we raise a power to a power. Suppose we want to write $(2^4)^3$ as a single power of 2. The parentheses tell us that 2^4 is being raised to the third power. That is,
$$(2^4)^3 = 2^4 \times 2^4 \times 2^4$$
$$= 2^{4+4+4}$$
$$= 2^{12}$$

We could obtain the same answer just by multiplying the two exponents (which is another way of saying that we have the product of four 2's, taken three times).

In other words,
$$2^4 \times 2^4 \times 2^4 = (2 \times 2 \times 2 \times 2)$$
$$\times (2 \times 2 \times 2 \times 2)$$
$$\times (2 \times 2 \times 2 \times 2)$$

There is a difference between $(2^4)^3$ and $2^4 \times 2^3$. In the first case, we take the product of four factors of 2, three times. In the second case, we multiply the product of four 2's and the product of three 2's.

> **RULE FOR RAISING A POWER TO A POWER**
> If b is any non-zero number, and m and n are any integers, then
> $$(b^m)^n = b^{mn}$$

Example 6

Write $(3^5)^4$ as a single power of 3.

$(3^5)^4 = 3^{20}$.
   By the rule, you need to multiply the two exponents 4 and 5 to get the correct answer.
   Without the rule, you have that $(3^5)^4$ means
$$3^5 \times 3^5 \times 3^5 \times 3^5 \quad \text{or} \quad 3^{5+5+5+5} = 3^{20}$$

Or, by the long way,
$$(3 \times 3 \times 3 \times 3 \times 3) \times (3 \times 3 \times 3 \times 3 \times 3) \times (3 \times 3 \times 3 \times 3 \times 3) \times (3 \times 3 \times 3 \times 3 \times 3)$$

Notice that this is not the same as $3^5 \times 3^4$, which is 3^9.

Example 7

Write $(3^4)^{-2}$ as a single power of 3.

$(3^4)^{-2} = 3^{-8}$.
   One way to get this answer is by the rule: $(3^4)^{-2} = 3^{4(-2)} = 3^{-8}$. That is, you can use the rule even if the exponents are negative integers.
   Do you begin to see why the definition of integral exponents is helpful? The definition allows you to use those rules that are easy to define when the exponents are whole numbers.

We know that $(3^4)^{-2}$ means $1 \div (3^4)^2$, or $1 \div 3^8$, which is 3^{-8}. This agrees with the result we get by the rule
$$(3^4)^{-2} = 3^{4(-2)} = 3^{-8}$$
Therefore, the rule still applies just as it did for whole-number exponents.

Example 8

Write $(3^{-4})^{-2}$ as a single power of 3.

$(3^{-4})^{-2} = 3^8$.

Once again, you use the rule that tells you to multiply the exponents. By the rule for multiplying signed numbers, you know that $(-4)(-2) = 8$.
One way to check this result is

$$\begin{aligned}(3^{-4})^{-2} &= 1 \div (3^{-4})^2 \\ &= 1 \div (3^{-4} \times 3^{-4}) \\ &= 1 \div 3^{-4+(-4)} \\ &= 1 \div 3^{-8} \\ &= 3^8\end{aligned}$$

This is how we may use the rule for multiplication we learned in the last section and extend it to new situations.

Before we complete this section, there is one more situation we should study. It involves problems where the factors have different bases but the same exponents.

This is the reverse of what we had in the last section, when the bases were the same but the exponents were different.

Example 9

Write $3^4 \times 2^4$ as a power of 6.

$3^4 \times 2^4 = 6^4$.

You have $(3 \times 3 \times 3 \times 3) \times (2 \times 2 \times 2 \times 2)$. Each factor of 3 combines with a factor of 2 to give as the product one factor of 6. That is, you use both a 3 and a 2 to make one 6. It may help to rearrange the factors into the form

$$\begin{array}{r}3 \times 3 \times 3 \times 3 \\ \times\; 2 \times 2 \times 2 \times 2 \\ \hline 6 \times 6 \times 6 \times 6 = 6^4 \\ \scriptstyle 1 \quad 2 \quad 3 \quad 4 \end{array}$$

Note that the answer is *not* 6^8, which would be $6 \times 6 \times 6 \times 6 \times 6 \times 6 \times 6 \times 6$. This is more than $3 \times 3 \times 3 \times 3 \times 2 \times 2 \times 2 \times 2$. You get only four factors of 6 because you used up one 3 and one 2 each time you "made" a 6. So, altogether, you can "make" only four 6's.

Example 10

Write $4^5 \times 6^5$ as a single factor of 24.

$4^5 \times 6^5 = 24^5$. That is, each 4 combines with a 6 to give you one 24. You can form five such 24's. That is,

$$\begin{array}{r}4 \times 4 \times 4 \times 4 \times 4 \\ \times\; 6 \times 6 \times 6 \times 6 \times 6 \\ \hline 24 \times 24 \times 24 \times 24 \times 24 = 24^5\end{array}$$

Be careful. Don't give 24^{10} as the answer. You add the exponents when the bases are the same.

Now we can state a general rule for the pattern.

Section 10.5   Scientific Notation

> If b and c are any real numbers and n is any integer, then
> $$b^n \times c^n = (b \times c)^n$$
> That is, when the bases are different but the exponents are the same, we multiply by multiplying the bases and keeping the common exponent.

Of course, there are other rules that we could state, but now we simply want to feel comfortable with exponents. Other examples appear in the exercise set.

No matter how complicated expressions become, we never really have to memorize rules. Once we know the basic definitions, we can always use them to get the desired result. For example, in arriving at the last rule, we simply wrote things out, found the answer, and discovered a quick way to remember how we can write the answer down without having to go through every step each time.

PRACTICE DRILL

For what value of n is each of the following a true statement?
1. $2^3 \div 2^5 = 2^n$
2. $(2^3)^5 = 2^n$
3. $(2^5)^3 = 2^n$
4. $2^5 \times 2^3 = 2^n$
5. $2^5 + 2^3 = n$
6. $3^2 \times 5^2 = 15^n$

Answers
1. -2 2. 15   3. 15
4. 8   5. 40   6. 2

CHECK THE MAIN IDEAS

1. When you know how to multiply like bases, it is not hard to find a way to divide like bases. To compute $2^3 \div 2^5$, you may first replace the division sign by a _____ sign.
2. You then invert 2^5, which gives you _____$^{-5}$.
3. In other words, $2^3 \div 2^5 = 2^3$ _____ 2^{-5}.
4. By the rule for multiplying like bases, $2^3 \times 2^{-5} =$ _____.
5. Therefore, $2^3 \div 2^5 =$ _____.
6. A quick way of getting this result by subtracting (a) _____ from (b) _____ to obtain the exponent.

Answers
1. multiplication (times)   2. 2
3. ×   4. 2^{-2} 5. 2^{-2}
6. (a) 5   (b) 3

Section 10.5
Scientific Notation

The fact that we can work with exponents is extremely important to us in our attempt to express large numbers and small numbers. By the nature of place value, which emphasizes 10, the number 10 is a particularly useful base. Let's illustrate this with two examples.

Example 1

What number is named by the product of eight 6's?

The number is 6^8, or 1,679,616.

You can leave the answer as 6^8, but do you have a feeling for what this number really is? Perhaps 1,679,616 is more meaningful, even if it is a rather complicated number. But computing 6^8 in place-value notation is a lengthy process.

OBJECTIVE
To see that every number can be written in the form
$$d \times 10^n$$
where d is between 1 and 10, and n is an integer; and to learn why this notation is helpful.

Example 2

What number is named by the product of eight 10's?

The number is 10^8, or 100,000,000. You still want to express 10^8 in place-value form, but now it's easy. As you've already learned, it is simply a one followed by 8 zeroes: 100,000,000.

> It's easy to get from 10^8 to 100,000,000; but it's not so easy to get from 6^8 to 1,679,616.

The difference between the amount of work required to solve Example 1 and the work required to solve Example 2 gives us a good reason for the fact that scientists and engineers like to express numbers in terms of powers of 10.

Very few numbers are integral powers of 10. In fact, the only whole numbers are those that consist of a one followed by only zeroes; the decimal fractions are those that have only zeroes and a one following the decimal point. However, we can group together all the numbers that lie between two consecutive integral powers of 10. We group the numbers that are between 10^0 and 10^1; those that are between 10^1 and 10^2; those that are between 10^2 and 10^3; and so on. We also group those that are between 10^{-1} and 10^0; between 10^{-2} and 10^{-1}; and so forth.

> $0.01 = 10^{-2} < 0.0239 < 10^{-1} = 0.1$
> $0.1 = 10^{-1} < 0.239 < 10^0\phantom{^{-}} = 1$
> $1 = 10^0\phantom{^{-}} < 2.39 < 10^1\phantom{^{-}} = 10$
> $10 = 10^1\phantom{^{-}} < 23.9 < 10^2\phantom{^{-}} = 100$
> $100 = 10^2\phantom{^{-}} < 239 < 10^3\phantom{^{-}} = 1,000$

In this way, every number can be written as

(a number between 1 and 10) × an integral power of 10

Let's look at an example. Suppose we start with 239, which is between 10^2 and 10^3. 239 is not a number between 1 and 10. To be between 1 and 10, 239 needs a decimal point between the 2 and the 3. This means that we must move the decimal point two places to the left (or, divide by 100). To make up for this, we must multiply by 100, which is 10^2. In other words,

$$239 \;=\; \underset{\substack{\uparrow \\ \text{This is} \\ \text{between 1} \\ \text{and 10.}}}{2.39} \;\times\; \underset{\substack{\uparrow \\ \text{This is an} \\ \text{integral power} \\ \text{of 10.}}}{10^2}$$

> That is, 2.39 is between 1 and 10.

> That is,
> $239 = (239 \div 100) \times 100$
> $ = 2.39 \times 100$
> $ = 2.39 \times 10^2$

Let's try some problems.

Example 3

Write 93,456 as the product of an integral power of 10 and a number between 1 and 10.

$93,456 = 9.3456 \times 10^4$.

Notice that the decimal point should fall between the 9 and the 3 if you want a number between 1 and 10. This means that you must move the decimal point 4 places to the left, which is the same as dividing by 10,000. To make up for this, you must now multiply by 10,000 (which is 10^4).

As a check, recall that multiplying by 10^4 tells you to move the decimal point 4 places to the right. That is,

$$9.3456 \times 10^4 = 9\underset{1\;2\;3\;4}{.3456\curvearrowright}$$

> That is, 9.3456 is between 1 and 10.

> In other words, we multiply and divide by 10,000 to get
> $(93,456 \div 10,000) \times 10,000$
> $= 9.3456 \;\times\; 10^4$

NOTE

It is also true that $93,456 = 93.456 \times 10^3$. But this is not the correct answer to Example 3, because 93.456 is not a number between 1 and 10.

Since 10^4 is the same as 1×10^4, and since 10^5 is the same as 10×10^4, we have

$$10^4 < 9.3456 \times 10^4 < 10 \times 10^4 \;(= 10^5)$$

> That is,
> 93.456×10^3
> $= 93,456.$
> $\downarrow\downarrow\downarrow$
> $1\;2\;3$

Section 10.5   Scientific Notation

This helps us to estimate the size of 9.3456×10^4 by glancing at it. It is more than 10,000 but less than 100,000.

In other words, in the form 9.3456×10^4, we recognize at once that the number is between the consecutive integral powers 10^4 and 10^5.

Example 4

Write 45,678,000 as the product of an integral power of 10 and a number between 1 and 10.

$45{,}678{,}000 = 4.5678 \times 10^7.$

You move the decimal point 7 places to the left to get 4.5678000. You make up for this by moving the decimal point back 7 places to the right. This is done by multiplying by 10^7. The answer tells us that 45,678,000 is between 10^7 and 10^8.

4.5678000 is the same number as 4.5678.

The same principle applies when the number is less than 1.

Example 5

Express 0.00321 as the product of an integral power of 10 and a number between 1 and 10.

$0.00321 = 3.21 \times 10^{-3}.$

To convert 0.00321 into a number between 1 and 10, you must place the decimal point between the 3 and the 2. This means that you must move it 3 places to the right. To make up for this, you move the decimal point back 3 places to the left. That is, you must multiply by 10^{-3}, which gives you 3.21×10^{-3} as the answer.

This method of writing numbers is so common in science and technology that it has a special name.

That is,

$(0.00321 \times 10^3) \times 10^{-3}$

$= 3.21 \times 10^{-3}$

> **DEFINITION**
> A number is said to be written in **scientific notation** if it has the form
>
> $$K \times 10^n$$
>
> where K is a number that is at least 1 but less than 10 and n is an integer (either positive, negative, or 0).

That is, we shall write 10,000 as 1×10^4 in scientific notation, but not as 10×10^3. It is not wrong to write 10×10^3, but 10×10^3 is not scientific notation.

Scientific notation can be helpful when we want to multiply (or divide) very large or very small numbers. Let's look at a typical problem.

Example 6

Write 60,000,000 in scientific notation.

$60{,}000{,}000 = 6 \times 10^7.$ That is,

$60{,}000{,}000 = (60{,}000{,}000 \times 10^{-7}) \times 10^7$

$= 6.0000000 \times 10^7$

${\scriptstyle 7\,6\,5\,4\,3\,2\,1}$

$= 6 \times 10^7$

CAUTION

Sometimes, through carelessness, we may write 6^7 rather than 6×10^7. Notice that these two numbers are very different.

6×10^7 means a 6 followed by 7 zeroes (which is what we want in this example). But 6^7 means $6 \times 6 \times 6 \times 6 \times 6 \times 6 \times 6$, or 279,936 — which is quite different from 60,000,000.

Example 7

Write 0.0000123 in scientific notation.

$0.0000123 = 1.23 \times 10^{-5}$. That is,

$$0.0000123 = 0.\underset{1\,2\,3\,4\,5}{0000123} \times 10^5 \times 10^{-5}$$
$$= 1.23 \times 10^{-5}$$

Example 8

Use scientific notation to compute the product $60{,}000{,}000 \times 0.0000123$.

The product is 7.38×10^2 (or 738).
From Examples 6 and 7, you already know that $60{,}000{,}000 = 6 \times 10^7$ and that $0.0000123 = 1.23 \times 10^{-5}$. Therefore

$$60{,}000{,}000 \times 0.0000123 = 6 \times 10^7 \times 1.23 \times 10^{-5}$$
$$= (6 \times 1.23) \times (10^7 \times 10^{-5})$$
$$= 7.38 \times 10^{7+(-5)}$$
$$= 7.38 \times 10^2$$

It is often more useful to write the final answer as 738.

> We are assuming that multiplication is both commutative and associative when we change the order of factors as well as the grouping in getting the answer to these examples.

> Notice how scientific notation makes it easier to keep track of the zeroes so that there is less chance for a careless mistake?

Scientific notation has many advantages, and in the next section we shall study a very important use of this notation for exact measurement. However, in closing this section, it is important for us to realize that scientific notation is only one way of using exponential notation. There are times when we prefer not to use it.
For example, how comfortable is it to talk about 2.5×10^2 dollars? Most of us find nothing difficult about saying $250. In fact, if we were comparing amounts of money, it would be easier to compare the size of $195 and $76 directly than to compare 1.95×10^2 and 7.6×10^1.
If we want to add two expressions like 1.95×10^2 and 7.6×10^1, we should rewrite one of the numbers so that the two addends have the same power of 10. We may choose to rewrite 1.95×10^2 as 19.5×10^1. We want to do this because *addition of like bases is convenient only if the powers are the same.* This follows from the distributive property, as shown in the next problem.

> What I want you to understand here is the "free spirit" of mathematics. We use scientific notation when it is helpful to us to do so. But in other situations when another form may be more helpful, we don't stay "locked in" to scientific notation.

> Remember that in Example 10 of Section 10.3, we made the point that addition of like bases is not easy, especially if the powers are different.

Example 9

Express $19.5m + 7.6m$ as a single term.

$19.5m + 7.6m = 27.1m$. The key is that $(19.5 + 7.6)m = 19.5m + 7.6m$ by the distributive property. Then all you have to do is replace $19.5 + 7.6$ by 27.1.

Example 10

Express the sum $19.5 \times 10^4 + 7.6 \times 10^4$ as a single term.

The sum is 27.1×10^4.
This is the same problem as Example 9, but we have replaced m by 10^4. That is,

$$19.5 \times 10^4 + 7.6 \times 10^4 = (19.5 \times 10^4) + (7.6 \times 10^4)$$
$$= (19.5 + 7.6) \times 10^4$$
$$= \underbrace{27.1} \times 10^4$$

> If we want to, we can use scientific notation and rewrite the sum as 2.71×10^5. However, if we prefer to deal with whole numbers, it is perfectly acceptable to write the sum as 271×10^3. Either form is correct: both name the number 271,000 (which most of us read as 271 thousand).

Section 10.5 Scientific Notation   265

Now let's return to our earlier discussion.

Example 11

Express $1.95 \times 10^2 + 7.6 \times 10^1$ as a single term.

The sum is 271 and may be written as 2.71×10^2 or as 27.1×10.
   From the result of Example 9, you should decide to rewrite the two addends so that they use the same power of 10. You could decide to rewrite the terms so that both involve 10^1 or so that both involve 10^2.
   If you write 1.95×10^2 as 19.5×10^1, the problem becomes $19.5 \times 10^1 + 7.6 \times 10^1$. Since you are adding like powers, you need only add 19.5 and 7.6 to get the answer in the form 27.1×10^1.

Notice that we do *not* add 1.95 and 7.6.

Actually, we can rewrite both in terms of 10^0 and get the much more familiar $195 + 76 = 217$.

Example 12

Express $3 \times 10^4 + 2 \times 10^2$ as a single term.

The sum is 3.02×10^4 or 302×10^2.
   To make the powers equal, you rewrite 3×10^4 as 300×10^2. The sum is then

$$300 \times 10^2 + 2 \times 10^2 = (300 + 2) \times 10^2.$$

You can also rewrite 2×10^2 as 0.02×10^4 to obtain
can also rewrite 2×10^2 as 0.02×10^4 to obtain

$$3 \times 10^4 + 0.02 \times 10^4 = (3 + .02) \times 10^4$$

Of course, we know that 3×10^4 is 30,000 and 2×10^2 is 200. Therefore, the sum is $30,000 + 200$, or $30,200$. But this method is more suited to the general case in which the exponent may be very large — large enough to discourage us from using place-value notation.

PRACTICE DRILL

1. Write each of the following in scientific notation.
   (a) 18,000,000   (b) 0.000018
2. Write each of the following in place-value notation.
   (a) 2.3×10^7 (b) 2.3×10^{-5}

Answers

1. (a) 1.8×10^7 (b) 1.8×10^{-5}
2. (a) 23,000,000 (b) 0.000023

CHECK THE MAIN IDEAS

1. In scientific notation, a number is written as a number between 1 and 10 multiplied by an integral power of _____.
2. For example, 40×10^{15} is not in scientific notation because _____ is not between 1 and 10.
3. If a number is not between 1 and 10, you can always move the _____.
4. To write 18,000,000 in scientific notation, you want the decimal point between _____.
5. This means that you must move the decimal point 7 places to the _____.
6. To offset this, you then move the decimal point back _____ places to the right.
7. Moving the decimal point 7 places to the right is the same as multiplying by _____.
8. Therefore, $18,000,000 = 1.8 \times$ _____.

Answers

1. 10   2. 40   3. decimal point   4. the 1 and the 8   5. left   6. 7   7. 10^7
8. 10^7

Section 10.6
Significant Figures

OBJECTIVE
To understand how we measure accurately and how we use numerals to indicate the degree of accuracy to which a number is measured.

Have you ever found yourself using an expression like, "This piece of string is exactly two inches long"? People often use phrases that contain the word "exactly."

Sometimes we can measure exactly, and sometimes we can't. For example, in the first illustration, there are *exactly two* circles. We're sure because we can count them.

In the next illustration, we are measuring a piece of string. According to our diagram, it looks as if it is 2 inches long. But how can we be sure? After all, the mark that we call 2 on the ruler has thickness. If we enlarged it, it might look like the last illustration.

What part of the marking is "exactly 2?" Although we can make the measurement more and more accurate if we are careful enough, the fact is that as long as the marking has some thickness, we can never be sure.

Usually in technical situations, we guarantee that the measurement is as accurate as it needs to be. There is a clever way of doing this using decimal fractions.

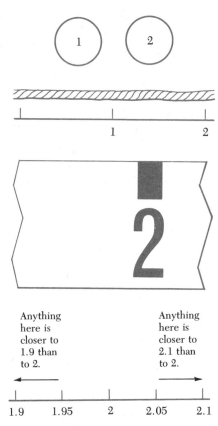

If the marking had no thickness, the problem would be even worse. We couldn't even see the marking.

Basically, we can use inequalities to pin down the exact length between two definite amounts. Suppose we are accurate enough to be sure that the exact length is more than 1.95 inches but less than 2.05 inches.

If we look at the diagram, we see that the length is closer to 2 inches than to either 1.9 inches or 2.1 inches. In other words, to the *nearest tenth*, the length of the string is 2.0 inches.

In "pure" mathematics 2, 2.0, 2.00, 2.000, and so on are all different decimals that name the number 2. But in the real world of measurement, each means something different. The zeroes are significant because they tell us the degree of accuracy our measurements have.

In other words, when we write 2.0, we mean that the length is closer to 2 than to either 1.9 or 2.1. In this sense, the 0 in 2.0 is *significant* because it is telling us that the measurement is correct to the nearest tenth of an inch.

Example 1

If a piece of string is 2.00 inches long, what does the measurement tell you about the length of the string?

It means that the length is more nearly equal to 2 inches than to either 1.99 or 2.01 inches.

In other words, the length is 2 inches measured to the nearest hundredth (of an inch). Using inequalities, if L denotes the exact length of the string, you are guaranteeing that

$$1.995 < L < 2.005$$

Perhaps the solution was too brief. Let's see, one step at a time, why a length of 2.00 inches tells us that the true length is between 1.995 and 2.005 inches.

This is what we mean by *significant*. It is significant to keep the two 0's because they tell us that the measurement is accurate to the nearest hundredth. If, for example, we write 2.0, all we are guaranteeing is that the measurement is accurate to the nearest tenth.

Section 10.6   Significant Figures   **267**

STEP 1: Write down the given number.
STEP 2: Because 2.00 indicates that you are measuring to the nearest hundredth, write the number that is 0.01 greater than 2.00 and the one that is 0.01 less than 2.00.

        2.00
        2.01
        2.00
        1.99

If this seems hard, leave out the decimal point and think of 199, 200, and 201.

STEP 3: Annex 0's to each numeral (this makes it easier to find the halfway point).

        2.010
        2.000
        1.990

STEP 4: Replace the upper and lower numbers by the midpoints.

2.010
      2.005   any number in here is closer to 2
2.000
      1.995   than to 1.99 or 2.01
1.990

Example 2

What can you say about **the exact** value of L if the measured value of L is given as 3.000 centimeters?

You can say that to the nearest thousandth of a centimeter, L is 3 centimeters. In terms of inequalities, $2.9995 < L < 3.0005$. In other words, L is closer to 3 cm than to either 2.999 cm or 3.001 cm.

Use Steps 1 through 4 as described above.

Again if the decimals are hard for you, leave them out and pretend that the numbers were, say 2,010 and 2,000. It's easier to see that 2,005 is midway between 2,000 and 2,010. Then put the decimal points back in.

STEP 1: 3.000

STEP 2: 3.001
         3.000
         2.999

STEP 3: 3.0010
         3.0000
         2.9990

STEP 4: 3.0010
                3.0005
         3.0000 L is between these two values
                2.9995
         2.9990

That is why it is important to write 3.000 cm rather than 3.0 cm. 3.0 cm tells us that the length is between 2.95 and 3.05 cm. So the zeroes in 3.000 are important. We need them to tell us how accurate our measurement is. For this reason, we say that 3.000 has four significant figures: we need the 3 and all three 0's.

        3.10
               3.05
        3.00
               2.95
        2.90

Here we have combined Steps 1 through 4.

Example 3

You try to measure the length of a piece of string as accurately as you can. When you are finished, you are sure that to the nearest millionth of a meter, the length is 4 meters. How would you write this length?

You would write it as 4.000000 meters.

You use six 0's because that puts the last 0 in the millionths place. This reflects the accuracy of your measurement. You say that 4.000000 has seven significant figures because the 4 and the six 0's are important.

This means that the exact length is between 3.9999995 and 4.0000005 meters, and there are many such lengths. But in most cases, we couldn't measure this very small difference unless we had extremely sensitive measuring instruments. Most likely, we would say that the length is "exactly 4" meters rather than "4.000000 meters."

Example 4

The length L of a rectangle is 4.0 feet, and the width is 2.0 feet. What can you say about the area of the rectangle?

You can say that the area is greater than 7.7075 square feet but less than 8.3025 square feet. Unless more information is given, any "more accurate" answer is a guess.

The point is that 4.0 tells us that the length is between 3.95 and 4.05 feet (that is, 4.0 means that the measurement is accurate to the nearest tenth of a foot). 2.0 means that the width is between 1.95 and 2.05 feet.

Remember the four steps.

```
4.10              2.10
        4.05              2.05
4.00              2.00
        3.95              1.95
3.90              1.90
```

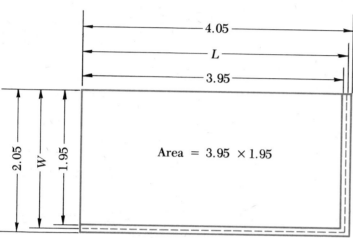

In terms of the illustration, you can see how you can find upper and lower bounds for the exact area of the rectangle. In other words,

$$\begin{array}{r} 3.95 < L < 4.05 \\ \times\, 1.95 < W < 2.05 \\ \hline 7.7025 \qquad\quad 8.3025 \end{array}$$

Unless more accuracy is guaranteed in your measurements, you can say nothing more. In fact, if you look at the bounds on the possible right answers, you see that to the nearest whole number, the area is 8 square feet.

In other words, even though each of your measurements had two significant figures, the product had only one significant figure.

Example 5

The length L of a rectangle is 4.123 feet and the width W is 2.6 feet. What can you say about the area of this rectangle?

Its area must be greater than 10.512375 square feet but less than 10.927275 square feet. That is,

$$\begin{array}{r} 4.1225 < L < 4.1235 \\ \times\, 2.55 \quad < W < 2.65 \\ \hline 10.512375 \qquad 10.927275 \end{array}$$

If you multiplied 4.123 by 2.6, you would get 10.7198 square feet, which is in the correct range of values. But notice that if you rounded off 4.123 to 4.1, the product would be 4.1×2.6, or 10.66, which is also in the required range.

All you can say from the given information is that the exact answer is between 10.512375 and 10.927275 square feet. So you would be no less accurate if you used 4.1×2.6 to get the answer instead of 4.123×2.6.

This means that in multiplying significant figures, we never get more accuracy than the least number of significant figures. For example, in this problem, the least accurate number is 2.6, so we may just as well use 4.1 as 4.123.

Section 10.6 Significant Figures

What does all this have to do with scientific notation?

The answer is that we can use the part of the number that multiplies the power of 10 to tell us the number of significant figures. For example, suppose we measure a length and say it is 1,000 feet. How do we know if the measurement is accurate to the nearest 10 feet or the nearest foot or the nearest thousandth of a foot?

If we want to show that all four digits in 1,000 are significant, we write

$$1.000 \times 10^3$$

The fact that the multiplier of 10^3 is written with four significant figures means that we have four significant figures. That is, the answer is 1,000 feet, correct to the nearest foot (since the place of the last significant figure is the ones place).

Example 6

Use scientific notation to indicate that the measurement 1,000 feet is correct to the nearest hundred feet.

You write 1.0×10^3 feet.

The idea is that the hundred place is held by the 0 just to the right of the 1. Therefore, all you definitely have are two significant figures (the 1 and the first 0). By writing 1.0×10^3, you are saying that you have two significant figures, since 1.0 has two significant figures.

In other words, in pure mathematics, 10^3, 1×10^3, 1.0×10^3, and so on, all stand for 1,000. But when we are talking about accuracy, the number of digits in the multiplier of the power of 10 tells us the number of significant figures in 1,000.

Example 7

Use scientific notation to show that a length L has been measured as 1000 feet, correct to the nearest hundredth of a foot.

You write 1.00000×10^3.

Using decimal fractions, you indicate accuracy to the nearest hundredth of a foot by writing 1,000.00. Therefore, there are 6 significant figures (the 1 and the five 0's). To indicate 6 significant figures, you write 1.00000, not 1.

Make sure you notice the difference between the nearest *hundred* and the nearest *hundredth*.

Example 8

How many significant figures are there in the numeral 6.4×10^{-3}?

Since there are two digits in the multiplier of the power of 10, the numeral has 2 significant digits.

Perhaps this will make it easier to understand. Suppose you measure a length in millimeters and find that to the nearest tenth of a mm, the length is 6.4 mm. Now, for some reason, you want to write this same result in meters. It becomes 0.0064 meters. It may look more accurate in meters, but it's just another way of writing the same length.

Example 9

Write 6.4×10^{-3} as a decimal fraction.

6.4×10^{-3} is 0.0064. All you do is move the decimal point in 6.4 three places to the left.

We already learned how to do Example 9 earlier in the lesson. Why do it again?

We rework the problem so that we can compare Examples 8 and 9. Based on Example 9, it looks as if the measurement .0064 is very accurate: to the nearest ten-thousandth. But Example 8 tells us to be careful. The answer may still contain only two significant figures.

If you want to indicate that 0.0064 has four significant figures, then use scientific notation and write

$$6.400 \times 10^{-3}$$

4 digits = 4 significant figures

PRACTICE DRILL

1. How many significant figures are there in each of the following?
   (a) 3×10^4 (b) 3.12×10^5 (c) 3.2×10^{-3}
2. Write each of the numbers in (1) in decimal form and tell how accurate each is.

Answers
1. (a) one (b) three (c) two
2. (a) 30,000; accurate to the nearest ten thousand
   (b) 312,000; accurate to the nearest thousand (c) 0.0032; accurate to the nearest ten-thousandth (0.0001)

CHECK THE MAIN IDEAS

1. You can talk about exactly 2 people in a room, but you can never be sure that a piece of string is _____ 2 inches long.
2. The reason is that the 2 inch marking on the ruler has _____.
3. If you are careful, you may be sure that the measurement is accurate to the nearest hundredth of an inch. In this case, you would not call the length 2 but rather _____ inches.
4. When you write 2.00, you say that there are _____ significant figures.
5. You also say that 2.00 guarantees accuracy to the nearest _____.
6. If a length is 2.00 inches, you can be sure that it must be more than 1.995 but less than _____ inches.

Answers
1. exactly   2. thickness
3. 2.00   4. three (3)
5. hundredth (0.01)   6. 2.005

EXERCISE SET 10 (Form A)

Section 10.1

1. What number is named by $2^3 \times 3^2$?
2. Evaluate t^3 when
   (a) $t = 2$ (b) $t = -2$ (c) $t = -\frac{1}{2}$
3. Evaluate m^4 when
   (a) $m = 3$ (b) $m = -3$ (c) $m = -\frac{1}{3}$
4. Evaluate $t^3 - t$ when
   (a) $t = 2$ (b) $t = -2$ (c) $t = -\frac{1}{2}$
5. Evaluate $x^4 + x^3$ when
   (a) $x = 3$ (b) $x = -3$
6. Evaluate x^7 when
   (a) $x = 4$ (b) $x = 10$

Section 10.2

7. What is the value of $\left(\dfrac{x^3 + y^2}{2xy}\right)^0$ when $x = 8$ and $y = -2.5$?
8. Evaluate u^{-3} when
   (a) $u = 2$ (b) $u = -2$ (c) $u = \frac{1}{2}$
9. Evaluate $x^{-3}(y + v)$ when $y = 15$, $x = 2$, and $v = 17$.
10. Evaluate $(x^{-3})^{-2}$ when $x = 2$.
11. Evaluate each of the following when $u = 2$ and $v = 3$.
    (a) $(uv)^{-2}$ (b) uv^{-2} (c) $(u+v)^{-2}$ (d) $u + v^{-2}$ (e) $u + 18v^{-2}$

Answers: Exercise Set 10 (Form A)
1. 72   2. (a) 8 (b) -8 (c) $-\frac{1}{8}$
3. (a) 81 (b) 81 (c) $\frac{1}{81}$
4. (a) 6 (b) -6 (c) $\frac{3}{8}$
5. (a) 108 (b) 54
6. (a) 16,384 (b) 10,000,000
7. 1   8. (a) $\frac{1}{8}$ (b) $-\frac{1}{8}$ (c) 8
9. 4   10. 64   11. (a) $\frac{1}{36}$ (b) $\frac{2}{9}$ (c) $\frac{1}{25}$ (d) $2\frac{1}{9}$ (e) 4

Exercise Set 10 (Form A) 271

Section 10.3

12. (a) Simplify $x^9 x^{-8}$.
    (b) Evaluate $x^9 x^{-8}$ when $x = 2.377$.
13. Simplify $x^9 x^{-8} x^{-3} x^2 x^{-1}$, writing the answer without a negative exponent.
14. Simplify
    (a) $2x^3(3x^4)$ (b) $2x^3(3y^4)$ (c) $2x^3(3x^{-2}y^4)$
15. Find the value of $2x^3 + 3x^4$ when $x = 2$.
16. Find the value of $2x^3(-4x^{-4})$ when $x = \frac{1}{2}$.
17. Write each of the following in place-value notation.
    (a) $(2 \times 10)^3$ (b) 2×10^3 (c) $2 \times 10^{15} \times 3 \times 10^8$
    (d) $2 \times 10^8 \times 3 \times 10^{-15}$
18. Write each of the following in place-value notation.
    (a) $2 \times 10^{15} \times 3 \times 10^{-16} \times 4 \times 10^{-4}$
    (b) $2 \times 10^{15} \times 3 \times 10^{16} \times 4 \times 10^{-27}$
    (c) $2 \times 10^{15} \times 3 \times 10^{-16} + 4 \times 10^{-4}$

Section 10.4

19. For what value of n is each of the following true?
    (a) $8^6 \div 8^{-7} = 8^n$ (b) $(10^{-3})^{-1} = 10^n$ (c) $(10^{-1})^{-3} = 10^n$
20. Evaluate $\frac{x^{-5}}{x^{-3}}$ when $x = -3$.
21. Simplify each of the following using only nonnegative exponents.
    (a) $\frac{x^{-5}}{x^{-3}}$ (b) $\frac{y^3 x^{-5}}{x^{-3}}$ (c) $\frac{y^3 x^{-5}}{x^{-3} y^{-2}}$
22. Simplify $\left(\frac{y^3 x^{-5}}{x^{-3} y^{-2}}\right)^{-2}$, writing the answer without negative exponents.
23. Evaluate $\left(\frac{x^6 y^7}{x^7 y^6}\right)^{-3}$ when $x = 2$ and $y = 3$.
24. (a) Simplify the expression $\frac{2x^7(6x^6)}{3x^6}$.
    (b) Evaluate the above expression when $x = -2$.
25. Evaluate $\frac{2x^7 + 5x^6}{x^6}$ when $x = 7\frac{1}{2}$.
26. Write $\frac{(3 \times 10^4)^3 \times 6 \times 10^{-11}}{81 \times 10^5}$ as a decimal fraction.

Section 10.5

Write each of the following in scientific notation.

27. 234.6×10^5
28. 0.0345×10^7
29. $(32 \times 10^8 \times 8 \times 10^7) \div (4 \times 10^{11})$
30. $3 \times 10^4 + 2 \times 10^2$
31. $32 \times 10^{30} - 18 \times 10^{29}$
32. Express each of the following both in scientific notation and as a decimal fraction.
    (a) $\frac{(80,000)(0.000002)}{0.004}$ (b) $\frac{140(0.000065)}{130,000(0.00035)}$
33. If the mass of the earth is 6×10^{27} grams and each gram weighs 1.1×10^{-6} tons, what is the weight of the earth in tons?

---

12. (a) x (b) 2.377   13. $\frac{1}{x}$
14. (a) $6x^7$ (b) $6x^3 y^4$ (c) $6xy^4$
15. 64   16. -16
17. (a) 8,000 (b) 2,000
(c) 600,000,000,000,000,000,-
000,000 (d) 0.0000006
18. (a) 0.00024 (b) 240,000
(c) 0.6004   19. (a) 13 (b) 3
(c) 3   20. $\frac{1}{9}$ 21. (a) $\frac{1}{x^2}$
(b) $\frac{y^3}{x^2}$ (c) $\frac{y^5}{x^2}$ 22. $\frac{x^4}{y^{10}}$
23. $\frac{8}{27}$ 24. (a) $4x^7$
(b) -512 25. 20
26. 0.0002   27. 2.346×10^7
28. 3.45×10^5 29. 6.4×10^5
30. 3.02×10^4
31. 3.02×10^{31}
32. (a) 40, or 4×10^1
(b) 0.0002, or 2×10^{-4}
33. 6.6×10^{21}, or 6,600,000,-
000,000,000,000,000 tons
34. (a) 4.00×10^2 (b) 4.00×10^7
(c) 4.00×10^{-4}
35. (a) 0.000001 (b) 1.0×10^{-6}
36. (a) 551.3725 cm²
(b) 546.4525 cm²

Section 10.6

34. Write each of the following in scientific notation to indicate that each number has three significant digits.

    (a) 400 (b) 40,000,000 (c) 0.0004

35. (a) Write 1 millimeter in terms of kilometers
    (b) How would you use scientific notation to say that the number of kilometers in (a) is correct to two significant figures?

36. To three significant figures, the length of a rectangle is 32.1 centimeters and the width is 17.1 centimeters.

    (a) What is the greatest area the rectangle can have?
    (b) What is the least area the rectangle can have?

EXERCISE SET 10 (Form B)

Section 10.1

1. What number is named by $5^2 \times 2^5$?
2. Evaluate t^5 when

    (a) $t = 2$ (b) $t = -2$ (c) $t = -\dfrac{1}{2}$

3. Evaluate m^2 when

    (a) $m = 5$ (b) $m = -5$ (c) $m = -\dfrac{1}{5}$

4. Evaluate $t^5 - t$ when

    (a) $t = 3$ (b) $t = -3$ (c) $t = -\dfrac{1}{3}$

5. Evaluate $x^2 + x$ when

    (a) $x = 1$ (b) $x = -1$

6. Evaluate x^6 when

    (a) $x = 3$ (b) $x = 10$

Section 10.2

7. What is the value of $\left(\dfrac{x^3 y^2}{2 + x}\right)^0$ when $x = 3.1$ and $y = -2.7$?
8. Evaluate u^{-2} when

    (a) $u = 3$ (b) $u = -3$ (c) $u = \dfrac{1}{3}$

9. Evaluate $x^{-1}(y + v)$ when $y = 12$, $v = -16$, and $x = 2$.
10. Evaluate $(x^{-2})^{-3}$ when $x = 2$.
11. Evaluate each of the following when $u = 4$ and $v = 2$.

    (a) $(uv)^{-2}$ (b) uv^{-2} (c) $(u+v)^{-2}$ (d) $u+v^{-2}$ (e) $u+3v^{-2}$

Section 10.3

12. (a) Simplify $x^{11} x^{-10}$.
    (b) Evaluate $x^{11} x^{-10}$ when $x = 3.158$.
13. Simplify $x^{15} x^{-3} x^{-11} x^2 x^{-5}$, writing the answer in terms of a single positive power of x.
14. Simplify

    (a) $3x^2(2x^5)$ (b) $3x^2(2y^5)$ (c) $3x^2(2x^{-1}y^2)$

Exercise Set 10 (Form B)

15. Find the value of $3x^2 + 4x^3$ when $x = 2$.
16. Evaluate $3x^2(-3x^{-3})$ when $x = \frac{1}{3}$.
17. Write each of the following in place-value notation.
    (a) $(3 \times 10)^3$ (b) 3×10^3 (c) $3 \times 10^{15} \times 4 \times 10^6$
    (d) $3 \times 10^{-15} \times 4 \times 10^6$
18. Write each of the following in place-value notation.
    (a) $3 \times 10^{15} \times 4 \times 10^{-17} \times 5 \times 10^{-3}$
    (b) $3 \times 10^{15} \times 4 \times 10^{17} \times 5 \times 10^{-30}$
    (c) $3 \times 10^{15} \times 4 \times 10^{-17} + 5 \times 10^{-3}$

Section 10.4

19. For what value of n is each of the following true?
    (a) $7^5 \div 7^{-6} = 7^n$ (b) $(10^{-3})^{-4} = 10^n$ (c) $(10^{-4})^{-3} = 10^n$
20. Evaluate $\frac{x^{-6}}{x^{-4}}$ when $x = -2$.
21. Simplify each of the following, using only nonnegative exponents:
    (a) $\frac{x^{-7}}{x^{-5}}$ (b) $\frac{y^4 x^{-7}}{x^{-5}}$ (c) $\frac{y^4 x^{-7}}{x^{-5} y^{-3}}$
22. Simplify $\left(\frac{y^4 x^{-6}}{x^{-4} y^5}\right)^{-2}$, writing the answer without negative exponents.
23. Evaluate $\left(\frac{x^5 y^7}{x^6 y^8}\right)^{-1}$ when $x = 2$ and $y = 3$.
24. (a) Simplify $\frac{2x^7(6x^8)}{4x^{11}}$.

    (b) Evaluate $\frac{2x^7(6x^8)}{4x^{11}}$ when $x = 2$.
25. Evaluate $\frac{2x^8 + 7x^7}{x^7}$ when $x = 9.5$.
26. Write $\frac{(4 \times 10^3)^3 \times (7 \times 10^{-11})}{64 \times 10^4}$ as a decimal fraction.

Section 10.5

Write each of the following in scientific notation.
27. 123.7×10^5
28. 0.0847×10^7
29. $(27 \times 10^6 \times 9 \times 10^5) \div (81 \times 10^4)$
30. $4 \times 10^4 + 8 \times 10^2$
31. $33 \times 10^{32} - 19 \times 10^{31}$
32. Express each of the following both as decimals and in scientific notation.
    (a) $\frac{60,000(0.00000003)}{0.006}$ (b) $\frac{190(0.000075)}{150,000(0.00095)}$

33. If a gram weighs 1.1×10^{-6} tons, what is the weight of 8×10^{24} grams in tons?

Section 10.6

34. Write each of the following in scientific notation to indicate that each number has four significant figures.
    (a) 8000   (b) 800,000,000   (c) 0.00008

35. (a) Write 3 millimeters in terms of kilometers.
    (b) How would you use scientific notation to say that the number of kilometers in (a) is measured accurately to three significant digits?

36. To three significant figures, the length of a rectangle is 27.6 cm and the width is 18.2 cm.

    (a) What is the greatest area this rectangle can have?
    (b) What is the least area this rectangle can have?

Lesson 11 — Introduction to Polynomials

Overview

There is a cartoon that shows the owner of a television-manufacturing company looking at a new model. There are no wires, tubes, or transistors in the set. So far it seems that the television is only a beautiful cabinet. The owner is gleefully exclaiming, "So far, so good — from here on, it's engineering's baby!"

In a way, up till now in this course, we have built a beautiful cabinet, a gateway to higher mathematics. But the transistors and the wires that we still need are the problems that correspond to the real world. Many of the relationships that we study are very complicated, and to handle them in an exact mathematical way requires further computational skills and knowledge about relationships. The remainder of our course is devoted to developing an understanding of these more complex relationships.

One common relationship that has application and yet is not too complicated for us to study at this time involves the concept of a polynomial. As we shall see in this lesson, the study of polynomials is a natural outgrowth of the development of exponents. We shall define polynomials; and we shall learn how to add, subtract, and multiply them. We shall also see some places in which polynomial relationships occur.

Although the skills may look as if they are getting more advanced, the new ideas are natural outgrowths of the topics we've already studied. Most of the material in the rest of the course is a logical continuation of the study of the structure of arithmetic.

Section 11.1
Monomial Expressions

In Unit 2, we studied linear expressions. But in many important real-life situations, the relationship between the variables is not linear.

OBJECTIVE
To define a monomial and to learn the terminology usually associated with monomials.

Example 1

If the length of a side of a square is s feet and its area is A square feet, then A and s are related by the formula

$$A = s^2 \qquad [1]$$

Use [1] to find the area of a square whose side is 3 feet long.

The area of this square is 9 square feet. According to [1], all you have to do is multiply the length (s) by itself. This gives you 3×3.

Example 2

Use [1] to find the area of a square if the length of a side of the square is 6 feet.

The area of the square is 36 square feet.
You still multiply the length of a side by itself, according to [1]. Therefore, you get 6×6, or 36.

We knew how to solve these problems earlier in the course, but we are making a point. Let's compare Examples 1 and 2. In Example 2, the length of a side of the square is twice the length of the side of the square in Example 1 (that is, 6 feet instead of 3 feet). Yet the area of the square in Example 2 is *four* times that of the square in Example 1 (36 square feet instead of 9 square feet). Doubling s did not double A; it quadrupled A. This means that the relationship between A and s is not linear.

In fact, if we made a chart of A versus s, we would see the following.

s	A $(=s^2)$	Increase in A
1	1	
		3
2	4	
		5
3	9	
		7
4	16	
		9
5	25	
		11
6	36	

Our discussion does not depend on the numbers 3 and 6. It depends on what happens when s is doubled.
When we double s, the new length is $2s$, and when we square $2s$, we get $(2s) \times (2s)$ or $2(2)ss$ or $4s^2$. Since $A = s^2$, $4s^2$ is $4A$ — not $2A$. That is,

$$(2s)^2 = 4s^2 = 4(s^2) = 4A$$

Linear means that A would increase by the same amount every time s increased by 1 (foot). In the chart, notice that as s increases by 1, A increases first by 3, then by 5, then by 7, and so on. The point is that this relationship is not linear.

Example 3

If the length of the side of a cube is s feet and its volume is V cubic feet, then s and V are related by the formula:

$$V = s^3 \qquad [2]$$

What is the volume of a cube if its side is 3 feet long?

Section 11.1 Monomial Expressions

The volume of the cube is 27 cubic feet (ft³).
To use [2], all you have to remember is that s^3 means $s \times s \times s$. In this case, $s = 3$, so s^3 means $3 \times 3 \times 3$ (or 3^3), which is 27.

Example 4

Use [2] to find the volume of a cube if the length of a side of the cube is 6 feet.

The volume of the cube is 216 ft³.
$V = s^3$ and $s = 6$. Hence, $V = 6^3$, or $6 \times 6 \times 6$. Looking at a chart:

s	V (= s^3)	Increase in V		
1	1			
		7		
2	8		12	
		19		6
3	27		18	
		37		6
4	64		24	
		61		6
5	125		30	
		91		
6	216			

Here we see that if we double s to get $2s$, the new volume is $(2s)^3$, which is the same as $2(2)2sss$, or $8s^3$. That is, if we double s, V increases by a factor of 8.

Notice that while the increase in V is not linear, there is still a pattern to how V increases with s.

These examples use forms of exponents. In the last lesson, we saw that if n is any integer, b^n increases by a factor of b every time n increases by 1. For example, if $b = 10$,

n	10^n	Increase in 10^n
1	10	
		90
2	100	
		900
3	1,000	

If we generalize some of the remarks that we made in our discussion of scientific notation in the last lesson, we arrive at the material of this lesson.
To begin with, we saw that scientific notation meant writing a number in the form

$$c \times 10^n \qquad [3]$$

where c is between 1 and 10 and n is any integer.
Let's generalize [3] a little bit. First, let's remove the restriction that c has to be between 1 and 10. We mentioned previously that there are times when it isn't important for c to be less than 10. Second, let's remove the restriction that the base must be 10, because we can study the powers of any number. So we can replace 10 in [3] by a variable such as x.
In other words, we can replace [3] by the more general form

$$cx^n \qquad [4]$$

If we make the additional assumption that n is a nonnegative integer, the expression cx^n in [4] has a special name.

The fact that n can't be negative is more for convenience than necessity. In more advanced algebra courses, the exponent is allowed to be negative.

> **DEFINITION**
> By a monomial in x, we mean any expression of the form
> $$cx^n$$
> where c is any number and n is any **nonnegative integer**.

In other words, a monomial in x is a number times a nonnegative integral power of x.

Let's look at a few examples.

Example 5

Is $7x^5$ a monomial in x?

Yes, it is.
    7 is a number and 5 is a positive integer, so $7x^5$ has the form cx^n with $c = 7$ and $n = 5$.

Example 6

Is $-3x^5$ a monomial in x?

Yes, it is.
    This is of the form cx^n with $c = -3$ and $n = 5$. The definition requires that n be a nonnegative integer, but all c has to be is a number. Nothing in the definition prevents this number from being negative. It doesn't even have to be an integer.

Example 7

Is $7x^{-3}$ a monomial in x?

No, it isn't.
    To be a monomial in x, the expression must have the form cx^n, where n is a *positive* integer. But in this example, $n = -3$, and -3 is a *negative* integer.

Example 8

Is $2x$ a monomial in x?

Yes, it is.
    Remember that x is the same as x^1. Hence $2x = 2x^1$, and this has the form
        $$cx^n$$
with $c = 2$ and $n = 1$.

We want to emphasize that x and x^1 mean the same thing.

Here's an interesting monomial with an unusual value for n.

Example 9

Is 6 a monomial in x?

Yes, it is.
    Remember, there is a definition that $x^0 = 1$. So $6 = 6(1) = 6(x^0) = 6x^0$.
    $6x^0$ has the form cx^n with $c = 6$ and $n = 0$. The point is that 0 is a nonnegative integer, and that is all that's required of the exponent in a monomial.

Actually the definition is $x^0 = 1$, provided that x is not 0.

Section 11.1 Monomial Expressions

Example 10

Is $7t^5$ a monomial in x?

No. It's a monomial in t.
   This problem isn't meant to trick you but rather to emphasize the meaning of a monomial *in* x. A "monomial in x" means that the variable is x. If the variable is t, you say ". . . in t" rather than ". . . in x."

Usually we say "monomial" rather than "monomial in x" or "monomial in t." If we look at $5x^4$, we see that it's a monomial in x. Similarly, it is clear that $5t^7$ is a monomial in t.

   Now that we know what a monomial is, let's study a little bit about the special terminology associated with monomials.

> **TERMINOLOGY**
> In the monomial cx^n, we call c the **coefficient** of x^n. If c is not 0, we call n the **degree** of the monomial. (If c is 0, then cx^n is 0 no matter what n is, and 0 is said to have no degree.)

OPTIONAL
Technically speaking, when we multiply two numbers, each may be called the coefficient of the other. In some books, what we call *the* coefficient is called the **numerical** coefficient.

Let's try a few examples using the terminology.

Example 11

What is the degree of $6x^5$?

The degree of $6x^5$ is 5. The degree is the value of the exponent. 6 is not the degree; it is the coefficient of x^5.

Example 12

What is the coefficient of x^4 in $3x^4$?

The coefficient is 3. The coefficient of x^4 is the number that multiplies it. In this case, the number is 3.

Example 13

What is the coefficient of x^7 in $-3x^7$.

The coefficient is -3 (not 3). Since the exponent modifies only the symbol immediately to its left, you read $-3x^7$ as if it were $(-3)x^7$. In this form it is easy to see that -3 is multiplying x^7.

Example 14

What is the coefficient of the monomial x^4?

The coefficient is 1. That is, $x^4 = 1x^4$, so that 1 multiplies x^4.

REVIEW
Given any expression, we can always multiply it by 1 without changing the value of the expression. Hence, x^n and $1x^n$ have the same meaning.

Example 15

What is the coefficient of x^4 in $-x^4$?

The coefficient is -1. Remember that $-x^4$ means $(-1)x^4$. In this form you can see that -1 is multiplying x^4.

Example 16
What is the degree of $0x^7$?

There is no degree. The definition requires that the coefficient not be 0. $0x^7$ is 0 for all values of x, and by definition, 0 has no degree.

This is a rather tricky point. We accept it just as we accept any other definition. We prefer to look at 0 as a special case, without a degree.
Notice that we have the problem because the coefficient is 0, not because the exponent is 0.

The reason is that if we didn't do this, 0 could have every degree. For example, since $0 = 0x^7$, it appears that 0 has degree 7. But 0 is also $0x^4$, and this looks as though 0 also has degree 4.

Example 17
What is the degree of 6?

The degree of 6 is 0.
As you saw in example 9, $6 = 6(1) = 6(x^0) = 6x^0$. Remember that the degree is defined as the exponent, provided only that the coefficient is not 0. Certainly 6 is different from 0. So by definition, the degree of 6 ($= 6x^0$) is 0.

Now that we know what a monomial is, let's keep sight of the fact that it is just another type of algebraic expression.

We call 6 a constant. Example 17 shows that the degree of any constant, except 0, is 0. Zero is also a constant, but we say (by definition) that it has no degree. Furthermore, $0x^7$, being 0, can be left out without changing anything. On the other hand, 6 can't be left out without changing the value.

Example 18
Evaluate x^3 when $x = 4$.

When $x = 4$, $x^3 = 64$.
To evaluate an expression means to replace the variable by a specified number and see what number the expression becomes. If you replace x by 4, x^3 becomes 4^3, or $4 \times 4 \times 4 = 64$.

Example 19
Evaluate $2x^3$ when $x = 4$.

When $x = 4$, $2x^3 = 128$. Remember that $2x^3$ means $2(x^3)$. That is, first you raise x to the third power and then you multiply by 2. That is, when $x = 4$, $2x^3$ is $2(4^3) = 2(64) = 128$.

GROUPING
There is a difference between $(2x)^3$ and $2(x^3)$, yet without the grouping symbols, both would look like $2x^3$. To help eliminate the use of grouping symbols, we agree that $2(x^3)$ is abbreviated by $2x^3$.
If we want 2 to be included, we have to write $(2x)^3$, because the grouping symbol now tells us that $2x$ is being raised to the third power.

Example 20
Evaluate $(2x)^3$ when $x = 4$.

When $x = 4$, $(2x)^3 = 512$.
The grouping symbols tell you that $2x$ is being raised to the third power. So first you multiply x by 2 to get $2x$, and then you compute $(2x)(2x)(2x)$. Since $x = 4$, $2x = 8$. Therefore, $(2x)^3 = 8^3$, or $8 \times 8 \times 8$, which is 512.

Notice that $(2x)^3$ is a monomial of degree 3, and the coefficient of x^3 is 8. That is,
$$(2x)^3 = (2x)(2x)(2x)$$
$$= 2(2)2xxx$$
$$= 8x^3$$

Remember that the terminology we use for monomials is based on the fact that the monomial is in the form cx^n not in another form such as $(cx)^n$.

Section 11.2 Polynomial Expressions

PRACTICE DRILL

1. Which of the following is not a monomial in x?
   (a) $3x^2$ (b) $\frac{1}{2}x^3$ (c) $2x^{-1}$ (d) $0.37x$ (e) 7
2. Which of the following expressions is a monomial in x of degree 5?
   (a) $3x^2$ (b) $2x^5$ (c) $5x^3$ (d) $2x^{-5}$ (e) 5
3. Evaluate each of the following when $x = 2$.
   (a) $3x^2$ (b) $(3x)^2$ (c) $(3+x)^2$

Answers
1. (c)   2. (b)   3. (a) 12   (b) 36   (c) 25

CHECK THE MAIN IDEAS

1. In general, relationships need not be linear. One nonlinear relationship between x and y is $y = 3x^2$. You call $3x^2$ a _____ in x.
2. A monomial in x is any number multiplied by a power of _____.
3. To be a monomial in x, the power must be a nonnegative _____.
4. $-.73x^4$ is a monomial in x because _____ is a number and the exponent 4 is a nonnegative integer.
5. $0.73x^{-5}$ is not a monomial in x because _____ is not a nonnegative integer.
6. The various parts of a monomial are given special names. For example, in the monomial $40x^5$, 40 is called the _____ of x^5.
7. Since the exponent is 5, you say that the _____ of $40x^5$ is five.
8. To evaluate a monomial such as $3x^2$ for a given value of x, first you square x and then you _____ this result by 3.
9. If you want first to multiply x by 3 and then to square the result, you would write _____.

Answers
1. monomial   2. x
3. integer   4. $-.73$
5. -5 6. coefficient
7. degree   8. multiply
9. $(3x)^2$

Section 11.2
Polynomial Expressions

There are relationships in which the formula requires that we use the sum of two or more monomials. One such relationship occurs when we use the idea of a freely falling body in the absence of air resistance.

OBJECTIVE
To develop the concept of a polynomial as the sum of two or more monomials.

Example 1

If a ball is thrown upward at a speed of 64 feet per second, in the absence of air resistance, it reaches a height of h feet after t seconds according to the formula

$$h = 64t - 16t^2 \qquad [1]$$

What is the height of the ball after 2 seconds?

Remember that you do not have to know why this formula is correct in order to use it. You only have to be able to read it correctly from a mathematical point of view.

After 2 seconds the ball is 64 feet high.

The formula tells you that to find the height when t is 2, you replace t by 2 in formula [1] and see what the resulting value of h is. If you replace t by 2 in [1], you get

$$h = 64(2) - 16(2)^2$$
$$= 128 - 16(4)$$
$$= 128 - 64$$
$$= 64$$

which means that after 2 seconds, the ball is 64 feet high.

What's important about Example 1 is that the formula for h involves the difference of two monomials in t. That is, in this example we must subtract two monomials. Specifically, we subtract $16t^2$ from $64t$. We know that any subtraction problem can be viewed as a form of adddition. So in general terms, in this section our goal is to look at the sum of two or more monomials.

Notice that "mono" means "one." The word "poly" stands for "many." When we add (or subtract) monomials, we call the sum a **polynomial.** For example,

$$3x^4 + 2x^5 + 7 \qquad [2]$$

is a polynomial. It is the sum of the three monomials $3x^4$, $2x^5$, and 7.

Each monomial that makes up a polynomial is called a **term** of the polynomial. For example, $3x^4$, $2x^5$, and 7 are the three terms in the polynomial $3x^4 + 2x^5 + 7$.

To review, we already know that the degree of $3x^4$ is 4, the degree of $2x^5$ is 5, and the degree of $7 \, (= 7x^0)$ is 0.

We define the **degree of a polynomial** as the greatest degree of the monomials (terms) that make up the polynomial. The degree of $3x^4 + 2x^5 + 7$ is 5, because the greatest exponent is 5. That is, of the terms that make up $3x^4 + 2x^5 + 7$, $2x^5$ has the greatest degree (five).

Let's try a few examples to make sure that we are clear about the various definitions.

REVIEW OF GROUPING SYMBOLS

In $64t - 16t^2$, we start grouping at the beginning and continue until we come to a plus or minus sign.

$$(64t) -$$

We then start after the plus or minus sign and repeat the same procedure until we get through the expression.

$$(64t) - (16t^2)$$

Each of the expressions in parentheses is called a *term*.

This agrees with our earlier idea that terms are expressions that are separated by plus and minus signs.

By the commutative property $3x^4 + 2x^5 + 7$ is equal to $2x^5 + 3x^4 + 7$. It is customary to write a polynomial so that we start with the term of greatest degree and then list the terms in order of decreasing exponents.

For example, we usually write

$$2x^5 + 3x^4 + 7$$

rather than

$$3x^4 + 2x^5 + 7$$

This is called the **usual order**.

Example 2

What are the terms in the polynomial

$$8x^2 + 7x + 5x^3 + 9$$

The terms are $8x^2$, $7x$, $5x^3$, and 9 (you can list them in any order). Remember, by definition the terms are the individual monomials whose sum makes up the polynomial.

Example 3

What is the degree of the polynomial

$$8x^2 + 7x + 5x^3 + 9$$

The degree of this polynomial is 3.

By definition, this is the greatest of the degrees of each of the terms (monomials) that make up the polynomial. The degree of $8x^2$ is 2, the degree of $7x$ is 1, the degree of $5x^3$ is 3, and the degree of 9 is 0. $5x^3$ has the greatest degree of all the terms, and this degree is 3. Therefore, you say that the polynomial has degree 3.

That is, $9 = 9x^0$ and $7x = 7x^1$.

Section 11.2 Polynomial Expressions

Example 4

What is the usual order for writing the terms in the polynomial

$$8x^2 + 7x + 5x^3 + 9$$

The usual order is $5x^3 + 8x^2 + 7x + 9$.

In this course, the usual order means that you start with the term that has the greatest degree and continue in decreasing order. That is,

$$5x^3 + 8x^2 + 7x^1 + 9x^0$$

Example 5

What is the degree of $7x^9 + 8x^3 + 6x + 4$?

The degree is 9. The greatest exponent is 9 and occurs in $7x^9$, which is the first term that appears when we use the usual order.

Example 6

What is the degree of $6x^5 + 7x^5 + 3x^4 + 2$?

The degree is 5. Remember that the degree is defined as the value of the greatest exponent. Although the greatest exponent appears twice (once in $6x^5$ and once in $7x^5$), the greatest exponent is still 5.

By the distributive property, $6x^5 + 7x^5 = 13x^5$. That is, $13x^5 = (6+7)x^5 = 6x^5 + 7x^5$. In other words, we may rewrite the polynomial as

$$6x^5 + 7x^5 + 3x^4 + 2 = (6x^5 + 7x^5) + 3x^4 + 2$$
$$= 13x^5 + 3x^4 + 2$$

If you prefer not to use phrases like "distributive property," think of it as six x^5's and seven x^5's is thirteen x^5's.

It is customary to add terms having the same degree and to write the sum as a single term with the common degree.

Example 7

Rewrite $6x^5 + 9x^5 + 3x^5 - 4x^5$ as a single monomial in x.

It is $14x^5$.

By the distributive property,

$$6x^5 + 9x^5 + 3x^5 - 4x^5 = (6 + 9 + 3 - 4)x^5$$
$$= 14x^5$$

In other words, we may think of monomials of the same degree as being *like terms*. We just combine these like terms.

As subtraction is a form of addition, $6x^5 + 9x^5 + 3x^5 - 4x^5$ can be written as the sum $6x^5 + 9x^5 + 3x^5 + (^-4)x^5$. For this reason, we may always think of a polynomial as the **sum** of monomials.

With these ideas in mind, we can now give a thorough review of the various definitions.

> **DEFINITION OF A POLYNOMIAL**
> A polynomial is any sum of monomials.
> The degree of a polynomial is equal to the greatest degree of the monomials that make up the polynomial.
> Each of the monomials is called a term of the polynomial.
> If there is more than one term with the same degree, we combine them into a single term of the common degree.

By stressing *sum*, we have committed ourselves to viewing
$$3x^4 - 2x^2$$
as
$$3x^4 + {}^-2x^2$$
and this means that the coefficient of the x^2 term in $3x^4 - 2x^2$ is $^-2$, not 2.

Just as we can evaluate monomials, we can evaluate polynomials.

Example 8

What is the value of $x^3 + x^2$ when $x = 2$?

When $x = 2$, $x^3 + x^2 = 12$.
You simply replace x by 2 in the expression $x^3 + x^2$ to get

$$2^3 + 2^2 \quad \text{or} \quad 8 + 4 \text{ or } 12$$

In other words, you are adding the two monomials x^3 and x^2, and when $x = 2$, the values of these monomials are 8 and 4.

Don't forget that $2^3 + 2^2$ means $(2^3) + (2^2)$.

Example 9

What is the value of x^5 when $x = 2$?

When $x = 2$, $x^5 = 32$. That is, when you replace x by 2 in x^5, you get 2^5, or $2 \times 2 \times 2 \times 2 \times 2$, or 32.

Compare Examples 8 and 9. When $x = 2$, $x^3 + x^2$ and x^5 have different values. In Example 8, we see that $x^3 + x^2$ is 12 when x is 2; but x^5 is 32 when x is 2. This is another reminder not to combine unlike degrees when we add monomials. That is, $x^3 + x^2$ is *not* the same as writing x^5.

This reminds us that we can always multiply unlike degrees (we add the exponents). But we can add monomials and get single terms only if the monomials have the same degree.

Example 10

By evaluating each expression when $x = 2$, show that the three expressions

$$3x^2 + 2x \quad 5x^2, \quad \text{and } 5x^3$$

have different numerical values.

When $x = 2$, you have

$$3x^2 + 2x = 3(2^2) + 2(2) = 3(4) + 4 = 16$$
$$5x^2 \quad = 5(2^2) = 5(4) = 20$$
$$5x^3 \quad = 5(2^3) = 5(8) = 40$$

If the three expressions were equal, they would have to give the same value when $x = 2$. But no two of these expressions have the same value when $x = 2$. Therefore, the three expressions are unequal.

This means that the expressions $3x^2 + 2x$, $5x^2$, and $5x^3$ are not equivalent.

This emphasizes that you must be careful how you simplify $3x^2 + 2x$. You cannot combine these two monomials by adding the coefficients or the exponents. When in doubt, replace the variable by the same number in each of the expressions that you think are equal. If you get different results (and haven't made a mistake in the arithmetic), the expressions are not equal.

Example 11

What is the value of $3x^4 + 7x^3 - 2x^5$ when $x = 2$?

The value is 40.

As a sum, the polynomial is
$$3x^4 + 7x^3 + {}^-2x^5$$
If you replace x by 2, you get:
$$3(2^4) + 7(2^3) + {}^-2(2^5)$$
$$= 3(16) + 7(8) + {}^-2(32)$$
$$= 48 + 56 + {}^-64$$
$$= 104 + {}^-64 = 40$$

Example 12

What is the value of $x^4 - 2x$ when $x = -3$?

The value is 87 when $x = -3$. $x^4 - 2x$ means $x^4 + {}^-2x$. If you replace x by -3, you get:
$$(-3)^4 + {}^-2(-3) = 81 + 6 = 87$$

It may be easier to read correctly $(-3)^4 + {}^-2(-3)$ than to read $(-3)^4 - 2(-3)$.

PRACTICE DRILL

1. What is the degree of each of the following polynomials in x?

   (a) $3x^2 + 4x$ (b) $3x^2 + 4x^5$ (c) $3x^2 + 4x^5 + 6x^4$
   (d) $9 + 3x$ (e) 7   (f) $4x^5 + 2x^5 + 7x^5$

2. Evaluate each of the following expressions when $x = 2$.

   (a) $2x^3 + 4x$ (b) $6x^4$ (c) $6x^3$ (d) $2x + 4x^3$

Answers

1. (a) 2 (b) 5 (c) 5 (d) 1
   (e) 0 (f) 5   2. (a) 24 (b) 96
   (c) 48 (d) 36

CHECK THE MAIN IDEAS

1. A polynomial in x is a sum of _____ in x.
2. The degree of the polynomial is the _____ degree of the monomials whose sum is the polynomial.
3. In writing a polynomial, it is customary to combine all terms having the same _____.
4. You should not combine terms with different _____ even if the coefficients are the same.

Answers

1. monomials   2. greatest
3. degree   4. degrees

Section 11.3
Adding Polynomials

In the rest of this lesson, we shall study more about the arithmetic of polynomials. In this section, our goal is to see how we may express the sum of two or more polynomials as a single polynomial.

We have already done almost all the work, especially the hard work, in the previous section. The basic idea is that because of the associative and commutative properties of addition, we can rearrange the terms in the sum and combine the like terms.

OBJECTIVE
To be able to express the sum of two or more polynomials as a single polynomial.

Example 1

Find the sum of the two polynomials $5x^2 + 7x + 3$ and $3x^2 + 2x + 5$.

The sum is $8x^2 + 9x + 8$.

We can write the terms in other orders, but we agreed to write the terms in the order of decreasing exponents.

You really have
$$(5x^2 + 7x + 3) + (3x^2 + 2x + 5) \qquad [1]$$
By the associative property of addition, you may omit the grouping symbols. This gives
$$5x^2 + 7x + 3 + 3x^2 + 2x + 5 \qquad [2]$$
By the commutative property (that is, the sum of two or more numbers does not depend on the order in which the numbers are added), you may regroup the terms in [2] as
$$5x^2 + 3x^2 + 7x + 2x + 3 + 5$$
or $\quad (5x^2 + 3x^2) + (7x + 2x) + (3 + 5)$

Then you combine the grouped terms of like degree to get
$$8x^2 + 9x + 8$$

It is common to arrange the polynomials under one another so that the terms with the same degree match. We then combine the coefficients of the terms that have the same degree. For example, in the previous problem we would write

$$\begin{array}{r} 5x^2 + 7x + 3 \\ \underline{3x^2 + 2x + 5} \\ 8x^2 + 9x + 8 \end{array}$$

Keep remembering that when you add like degrees, you add the coefficients and keep the common degree.

In effect, we add the polynomials by grouping the monomials of the same degree together and then adding these monomials.

Example 2

Find the sum of the polynomials $4x^3 + 8x$ and $5x^3 + 2x$.

The sum is $9x^3 + 10x$. You have

$$\begin{array}{r} 4x^3 + 8x \\ \underline{5x^3 + 2x} \\ 9x^3 + 10x \end{array}$$

Because the two monomials that make up the polynomial have different degrees, we cannot combine them into one monomial.

Example 3

Find the sum of $4x^3 + 8x$ and $5x^3 - 2x$.

The sum is $9x^3 + 6x$.
To do the addition, think of $5x^3 - 2x$ as $5x^3 + {}^-2x$. You then have

$$\begin{array}{r} 4x^3 + 8x \\ \underline{5x^3 + {}^-2x} \\ 9x^3 + 6x \end{array}$$

Sometimes a term appears in one polynomial but there is no term of the same degree in the second polynomial. Whenever this happens, keep in mind that if a term is missing, we can always put it in with a coefficient of 0.

For example, 0 is the same as $0x^3$ or $0x^6$ or $0x^7$, etc.

Section 11.3   Adding Polynomials

Example 4

Find the sum of $8x^3 + 7x + 9$ and $4x^2 + 3$.

The sum is $8x^3 + 4x^2 + 7x + 12$.   You write

$$\begin{array}{r} 8x^3 + 0x^2 + 7x + 9 \\ 0x^3 + 4x^2 + 0x + 3 \\ \hline 8x^3 + 4x^2 + 7x + 12 \end{array}$$

In horizontal form we have

$$(8x^3 + 7x + 9) + (4x^2 + 3)$$
$$= 8x^3 + 7x + 9 + 4x^2 + 3$$
$$= 8x^3 + 4x^2 + 7x + 9 + 3$$
$$= 8x^3 + 4x^2 + 7x + 12$$

Once you get the idea, you usually combine several steps into one.

Example 5

Find the sum of the following three polynomials $4x^2 + 8x + 7$, $3x^2 + 9x + 2$, and $5x^2 + 4x + 7$.

The sum is $12x^2 + 21x + 16$.   You have

$$\begin{array}{r} 4x^2 + 8x + 7 \\ 3x^2 + 9x + 2 \\ 5x^2 + 4x + 7 \\ \hline 12x^2 + 21x + 16 \end{array}$$

Example 6

Find the sum of the following three polynomials $4x^2 - 8x - 7$, $3x^2 + 9x - 2$, and $5x^2 - 4x + 7$.

The sum is $12x^2 - 3x - 2$.
   First you rewrite the polynomials in which minus signs appear.

$$4x^2 - 8x - 7 = 4x^2 + {}^-8x + {}^-7$$
$$3x^2 + 9x - 2 = 3x^2 + 9x + {}^-2$$
$$5x^2 - 4x + 7 = 5x^2 + {}^-4x + 7$$

You write the sum as

$$\begin{array}{r} 4x^2 + {}^-8x + {}^-7 \\ 3x^2 + 9x + {}^-2 \\ 5x^2 + {}^-4x + 7 \\ \hline 12x^2 + {}^-3x + {}^-2 \end{array}$$

This can be rewritten as

$$12x^2 - 3x - 2$$

In fact, despite the definition that a polynomial is a sum of monomials, it is customary to write $12x^2 - 3x - 2$ rather than the more cumbersome $12x^2 + {}^-3x + {}^-2$.

Example 7

Find the sum of the following three polynomials $3x^3 - 5x + 6$, $4x^3 + 3x^2 + 7$, and $2x^3 + 5x - 4$.

The sum is $9x^3 + 3x^2 + 17$.
   You have

$$\begin{array}{r} 3x^3 + 0x^2 + {}^-5x + 6 \\ 4x^3 + 3x^2 + 0x + 7 \\ 2x^3 + 0x^2 + 5x + {}^-4 \\ \hline 9x^3 + 3x^2 + 0x + 9 \\ = 9x^3 + 3x^2 + 0 + 9 \\ = 9x^3 + 3x^2 + 9 \end{array}$$

Example 8

Find the sum of $3x^3 - 5x^2 + 7$, $5x^4 + 2x^3 + 2x - 9$, and $6x^2$.

You have

$$\begin{array}{r} 0x^4 + 3x^3 + {}^-5x^2 + 0x + 7 \\ 5x^4 + 2x^3 + 0x^2 + 2x + {}^-9 \\ 0x^4 + 0x^3 + 6x^2 + 0x + 0 \\ \hline 5x^4 + 5x^3 + 1x^2 + 2x + {}^-2 \end{array}$$
$= 5x^4 + 5x^3 + x^2 + 2x - 2$

REMINDER

We don't want to combine, for example, $3x^3$ and $5x^4$ into a single monomial, because these two terms have different degrees.

We use the terms with 0 coefficients only to help see how we add like terms. You may prefer to use the hoirzontal form, which is often easier once you get used to adding polynomials.

PRACTICE DRILL

Write each of the following sums as a single polynomial in x.

1. $(6x^2 + 5x + 7) + (2x^2 + 3x + 2)$
2. $(6x^2 + 5x + 7) + (2x^2 + 3x^4 + 2)$
3. $(6x^2 + 5x + 7) + (2x + 3)$
4. $(8x^3 + 5x + 7) + (6x^2 + 3x + 2)$
5. $(8x^3 - 5x^2 + x + 7) + (x^3 + 6x^2 - 3x - 8)$

Answers

1. $8x^2 + 8x + 9$
2. $3x^4 + 8x^2 + 5x + 9$
3. $6x^2 + 7x + 10$
4. $8x^3 + 6x^2 + 8x + 9$
5. $9x^3 + x^2 - 2x - 1$

CHECK THE MAIN IDEAS

1. In adding polynomials, you add terms that have the same _____.
2. To add two terms that have the same degree, you keep the common degree and add the two _____; for example, $4x^2 + 5x^2 = 9x^2$.
3. If a term of any particular degree is missing from a polynomial, you may always assume that it's present with a _____ of 0.

Answers

1. degree 2. coefficients
3. coefficient

Section 11.4
Subtracting Polynomials

Every time we have defined subtraction in this course, we have described it as the inverse of addition (or the change-making method).

For example, to find the difference

$$6 - (-4) \qquad [1]$$

we asked the question, "What number must be added to -4 to get 6?" One way to get the answer is to observe that $-4 + 4$ is 0 and that $0 + 6$ is 6. That is,

$$\begin{array}{c} -4 \rightarrow 0 \rightarrow 6 \\ 4 + 6 = 10 \end{array}$$

We can also get this answer by changing the subtraction to addition and changing the sign of the number being subtracted. That is,

$$6 - (-4) = 6 + (+4)$$

In this section, we shall use the same principle to subtract one polynomial from another. Let's begin by looking at the difference of two monomials.

OBJECTIVE

To be able to express the difference of two polynomials as a polynomial.

Section 11.4 Subtracting Polynomials

Example 1

What polynomial do you get if you subtract $2x^3$ from $5x^2$?

You get $5x^2 - 2x^3$.
  You are taking $2x^3$ from $5x^2$. But you can't write the difference as a single monomial because the two monomials have different degrees. Therefore, the answer is written as $5x^2 - 2x^3$.

Remember that we can write the difference in the form of a sum if we prefer.
$$5x^2 - 2x^3 = 5x^2 + (-2)x^3$$
$$= 5x^2 + {}^-2x^3$$

  Example 1 reminds us that we can subtract or add monomials with like degrees only to get a sum that is a single monomial.

Example 2

Express $2x^3 - 5x^3$ as a single monomial.

$2x^3 - 5x^3 = -3x^3$ or $(-3)x^3$.
  By the distributive property, you have
$$2x^3 - 5x^3 = (2-5)x^3 = (-3)x^3$$

As a check, notice $-3x^2 + 5x^3 = 2x^3$.

NOTE
Since the exponent applies only to the number immediately to its left, we may write $(-3)x^3$ as ${}^-3x^3$. That is, ${}^-3x^3$ means we cube x and then multiply by negative 3.

Example 3

Express $2x^3 - (-5)x^3$ as a single monomial.

$2x^3 - (-5)x^3 = 7x^3$.
  You want to find out what number you must add to $(-5)x^3$ (or $-5x^3$) to get $2x^3$. By change-making, you have
$$\begin{aligned}{}^-5x^3 &\to 0(0x^3) \to 2x^3\\ 5x^3\ (+)\ &2x^3\quad = 7x^3\end{aligned}$$

Just as $5 + {}^-5 = 0$, so does $5x^3 + {}^-5x^3 = 0$.

  We can also find the answer by using the rule for subtracting signed numbers. To subtract ${}^-5x^3$ from $2x^3$:

STEP 1: Write the problem as $2x^3 - {}^-5x^3$.

Order is important. Don't write ${}^-5x^3 - 2x^3$ because that would be subtracting $2x^3$ from ${}^-5x^3$.

STEP 2: Leave the first number as it is.

$2x^3$

STEP 3: Change the subtraction to addition.

$2x^3 +$

STEP 4: Change the sign of the coefficient of the number being subtracted.

$2x^3 + {}^+5x^3 = 2x^3 + 5x^3$

STEP 5: Do the resulting addition problem.

$2x^3 + 5x^3 = 7x^3$

Example 4

Write $4x^6 - {}^-7x^6$ as a single monomial.

$4x^6 - {}^-7x^6 = 11x^6$.
  If you use the "add the opposite" rule, you have

STEP 1: Leave the first number alone: $4x^6$.
STEP 2: Change the minus to a plus: $4x^6 +$.
STEP 3: Change the sign of the coefficient of the term that's being subtracted:
    $4x^6 + {}^+7x^6$, or $4x^6 + 7x^6 = 11x^6$.

You can also find the monomial that must be added to $-7x^6$ to give $4x^6$. Using the change-making method, you have

$$-7x^6 \to 0 \to 4x^6$$
$$7x^6(+)4x^6 = 11x^6$$

The same principle can be applied to the difference of any two polynomials.

Example 5

Use change-making to express

$$(6x^5 + 4x^3) - (2x^5 - 5x^3)$$

as a single polynomial.

It is $4x^5 + 9x^3$.
You want to find out what you must add to $2x^5 - 5x^3$ to get $6x^5 + 4x^3$. That is,

$$\begin{array}{r} 2x^5 - 5x^3 \\ +\ ?\quad\ ? \\ \hline 6x^5 + 4x^3 \end{array}$$

To replace the question marks, remember that you know from addition

$$2x^5 + 4x^5 = 6x^5 \quad \text{and} \quad {}^-5x^3 + 9x^3 = 4x^3$$

That is,

$$\begin{array}{r} 2x^5 + {}^-5x^3 \\ + 4x^5 + {}^+9x^3 \\ \hline 6x^6 + 4x^3 \end{array}$$

Actually, we solved this problem by doing two separate problems involving subtraction of monomials. We found that

$$6x^5 - 2x^5 = 4x^5$$
$$\text{and } 4x^3 - {}^-5x^3 = 9x^3$$

In summary, we solved

$$(6x^5 + 4x^3) - (2x^5 - 5x^3)$$

by subtracting $2x^5$ from $6x^5$ and $^-5x^3$ from $4x^3$.
This gives us a start in applying the "add the opposite" idea to this problem.

STEP 1: Rewrite the subtraction problem so that each polynomial is the sum of monomials. $\quad(6x^5 + 4x^3) - (2x^5 + {}^-5x^3)$

STEP 2: Leave the first polynomial alone. $\quad(6x^5 + 4x^3)$

STEP 3: Replace the minus sign by a plus sign. $\quad(6x^5 + 4x^3) +$

STEP 4: Change the sign of the coefficient in each term of the second polynomial. $\quad(6x^5 + 4x^3) + (^-2x^5 + {}^+5x^3)$

STEP 5: Do the resulting addition problem.
$$\begin{array}{r} 6x^5 + 4x^3 \\ + {}^-2x^5 + 5x^3 \\ \hline 4x^5 + 9x^3 \end{array}$$

For any subtraction problem, we can use the change-making process or we can use the "add the opposite" rule. The important thing is not the method but our understanding of how to find the polynomial that must be added to the second one to get the first. Let's try some problems by the "add the opposite" rule and check our results by addition.

Section 11.4   **Subtracting Polynomials**   291

Example 6

Express $(8x^4 - 2x^3 + 7) - (2x^4 - 5x^3 - 6)$ as a single polynomial.

The answer is $6x^4 + 3x^3 + 13$.

STEP 1: $(8x^4 + {}^-2x^3 + 7) - (2x^4 + {}^-5x^3 + {}^-6)$
STEP 2: $(8x^4 + {}^-2x^3 + 7)$
STEP 3: $(8x^4 + {}^-2x^3 + 7) +$
STEP 4: $(8x^4 + {}^-2x^3 + 7) + ({}^-2x^4 + 5x^3 + 6)$
STEP 5: $\quad 8x^4 + {}^-2x^3 + 7$
$\quad\quad + {}^-2x^4 + 5x^3 + 6$
$\quad\quad\overline{6x^4 + 3x^3 + 13}$

CHECK
$6x^4 + 3x^3 + 13$
$+ 2x^4 + {}^-5x^3 + {}^-6$
$\overline{8x^4 + {}^-2x^3 + 7}$
$= 8x^4 - 2x^3 + 7$

Example 7

Express $(8x^4 + 4x^3 + 5) - (6x^4 - 2x^3 - 3x^2 - 7)$ as a single polynomial.

The answer is $2x^4 + 6x^3 + 3x^2 + 12$.

STEP 1: $(8x^4 + 4x^3 + 5) - (6x^4 + {}^-2x^3 + {}^-3x^2 + {}^-7)$
STEP 2: $(8x^4 + 4x^3 + 5)$
STEP 3: $(8x^4 + 4x^3 + 5) +$
STEP 4: $(8x^4 + 4x^3 + 5) + ({}^-6x^4 + 2x^3 + 3x^2 + 7)$
STEP 5: $\quad 8x^4 + 4x^3 + 0x^2 + 0x + 5$
$\quad\quad + {}^-6x^4 + 2x^3 + 3x^2 + 0x + 7$
$\quad\quad\overline{2x^4 + 6x^3 + 3x^2 + 12}$

CHECK
$2x^4 + 6x^3 + 3x^2 + 12$
$+ 6x^4 + {}^-2x^3 + {}^-3x^2 + {}^-7$
$\overline{8x^4 + 4x^3 + 0x^2 + 5}$
$= 8x^4 + 4x^3 + 5$

PRACTICE DRILL

Simplify each of the following.
1. $6x^3 - 2x^3$ 2. $6x^3 - {}^-2x^3$ 3. $6x^3 - 2x^2$
4. $(6x^3 - 2x^2) - (2x^3 + 4x^2)$
5. $(6x^3 + 4x^2 + 7) - (2x^3 - 3x^2 + 9)$
6. $(6x^3 + 4x^2 + 7) - (2x^3 - 3x^2 - 5x + 9)$

Answers
1. $4x^3$ 2. $8x^3$
3. $6x^3 - 2x^2$ or $6x^3 + {}^-2x^2$
4. $4x^3 - 6x^2$ 5. $4x^3 + 7x^2 - 2$
6. $4x^3 + 7x^2 + 5x - 2$

CHECK THE MAIN IDEAS

1. In terms of the addition of polynomials, $5x^3 - 2x^3$ means the polynomial you want to add to (a) _____ to get (b) _____.
2. Therefore, $5x^3 - 2x^2 = 3x^3$, since $2x^3 + $ _____ $= 5x^3$.
3. Similarly, $(6x^3 - 2x^2) - (5x^3 + 3x^2) = x^3 - 5x^2$ because $(5x^3 + 3x^2) + $ _____ $= 6x^3 - 2x^2$.
4. A short cut for computing $(6x^3 - 2x^2) - (5x^3 + 3x^2)$ is to replace the minus sign by a _____ sign and then to change the sign of each term in the polynomial $5x^3 + 3x^2$.
5. $(6x^3 - 2x^2) - (5x^3 + 3x^2)$ is the same as $(6x^3 - 2x^2) + $ _____ . In this way, any subtraction problem involving polynomials may be interpreted as a form of addition of polynomials.

Answers
1. (a) $2x^3$ (b) $5x^3$ 2. $3x^3$
3. $x^3 - 5x^2$ 4. plus
5. ${}^-5x^3 + {}^-3x^2$

Section 11.5
Multiplication of Polynomials

OBJECTIVE
To be able to express the product of two or more polynomials as a single polynomial.

Many formulas used in mathematics require that two or more variables be multiplied. For example, to find the area of a rectangle, we must multiply the length of the rectangle by the width. In turn, these formulas often require that we know how to multiply one polynomial by another.

Example 1

Express the area A of a rectangle in terms of the width W of the rectangle if the length of the rectangle is 3 feet longer than the width.

The area of the rectangle (in square feet) is given by

$$A = W(W + 3) \qquad [1]$$

You know that the area is obtained by multiplying the length and the width. Since W denotes the width and $L = W + 3$, you have

$$A = LW$$
$$= (W + 3)W$$

If you wish, you may use the fact that multiplication is commutative to rewrite $(W + 3)W$ as $W(W + 3)$.

W and $W + 3$ are first degree (linear) polynomials in W. Therefore, Example 1 shows that to find A we must multiply two polynomials in W.

Example 1 is not a very complicated illustration. Sometimes we may be called upon to multiply polynomials of higher degree. In such cases, the computation becomes more involved, but the basic idea remains the same. We may use the distributive property in trying to express the product of two or more polynomials as a single polynomial.

Example 2

Express the product $W(W + 3)$ as a single polynomial in W.

$W(W + 3) = W^2 + 3W$.

All you do is use the distributive property. That is, you multiply the number outside the parentheses (W) by the first number inside the parentheses (W) and add to that the product of the outside number and the second inside number (3).

$$W(W + 3) = WW + W3$$
$$= W^2 + 3W$$

By a single polynomial, we mean a sum of monomials in which no two different terms have the same degree. $W(W + 3)$ is not a sum of monomials. It is the product of a monomial and a polynomial.

Example 2 shows us how we may use the distributive property to write the product of two first-degree polynomials as a single second-degree polynomial.

Perhaps you feel that W isn't really a polynomial — that it's just a monomial. You may wonder how the distributive property would help if you wanted to express $(W + 2)(W + 3)$ as a single polynomial. Remember how you handled this in Lesson 4? You treated $W + 3$ as a single number and then used the distributive property to obtain

$$(W + 2)(W + 3) = W(W + 3) + 2(W + 3) \qquad [2]$$

If we prefer to be more intuitive, we can use the area of a rectangle to obtain the same result. That is,

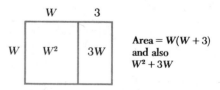

Area = $W(W + 3)$ and also $W^2 + 3W$

To review, if we let V denote $W + 3$, $(W + 2)(W + 3)$ becomes $(W + 2)V$. By the distributive property, this is $WV + 2V$ or $W(W + 3) + 2(W + 3)$.

Section 11.5   Multiplication of Polynomials

Now you can apply the distributive property on the right side of [2] to obtain

$$(W+2)(W+3) = W(W+3) + 2(W+3)$$
$$= W^2 + 3W + 2W + 2(3)$$
$$= W^2 + 5W + 6$$

We do not leave the answer as

$$W^2 + 3W + 2W + 6$$

because in this form, we still have two different terms of the same degree, $3W$ and $2W$.

Example 3

Express $(x+5)(x+2)$ as a single polynomial.

$(x+5)(x+2) = x^2 + 7x + 10$.   Pictorially:

	x	5
x	x^2	$5x$
2	$2x$	10

Without using a picture, you have by the distributive property

$$(x+5)(x+2) = x(x+2) + 5(x+2)$$
$$= x^2 + 2x + 5x + 10$$

Adding like terms you get
$$x^2 + 7x + 10$$

The product consists of terms each of which is the product of a term from the first set of parentheses and a term from the second set of parentheses. That is,

$$(x+5)(x+2)$$

Example 4

Express $(x+5)(x-2)$ as a single polynomial.

$(x+5)(x-2) = x^2 + 3x - 10$.
One way to do this is

$$(x+5)(x-2) = x(x-2) + 5(x-2)$$
$$= x^2 - 2x + 5x - 10$$
$$= x^2 + 3x - 10$$

If you prefer to work with sums, write $x - 2$ as $x + {}^-2$. Then

$$(x+5)(x-2)$$
$$= (x+5)(x+{}^-2)$$
$$= x(x+{}^-2) + 5(x+{}^-2)$$
$$= x^2 + {}^-2x + 5x + 5({}^-2)$$
$$= x^2 + ({}^-2+5)x + {}^-10$$
$$= x^2 + 3x - 10$$

Example 5

Express $(x-5)(x-2)$ as a single polynomial.

$$(x-5)(x-2) = x^2 - 7x + 10$$
$$(x-5)(x-2) = x(x-2) - 5(x-2)$$
$$= x^2 - 2x - 5x + 10$$
$$= x^2 - 7x + 10$$

In other words,

$$(x-5)(x-2) = (x+{}^-5)(x+{}^-2)$$
$$= x(x+{}^-2) + {}^-5(x+{}^-2)$$
$$= x^2 + {}^-2x + {}^-5x + {}^-5({}^-2)$$
$$= x^2 + ({}^-2+{}^-5)x + {}^+10$$
$$= x^2 + {}^-7x + {}^+10$$
$$= x^2 - 7x + 10$$

What happens if the polynomials we're multiplying are not linear? Suppose we want to express $(2x^3 + 5)(4x^6 + 7)$ as a single polynomial. We still use the approach that we used in Examples 3, 4, and 5. We take all possible terms that consist of the product of a term from the first set of parentheses and a term from the second set of parentheses. For example,

$$(2x^3 + 5)(4x^6 + 7) = 2x^3(4x^6 + 7) + 5(4x^6 + 7)$$
$$= 2x^3(4x^6) + 2x^3 7 + 5(4x^6) + 5(7)$$

The rest of the problem involves expressing the product of two monomials such as $2x^3$ and $4x^6$ as a single monomial. So let's study this aspect next.

Example 6

Express $2x^3(4x^6)$ as a single monomial in x.

$2x^3(4x^6) = 8x^9$.

By the commutative property of multiplication, you can rearrange the factors so that numbers are next to numbers and variables are next to variables. That is,

$$2x^3(4x^6) = 2(4)x^3 x^6$$

Now, $2(4) = 8$, and by the rule for multiplying like bases, $x^3 x^6 = x^9$. Therefore, $2x^3(4x^6) = 2(4)x^3 x^6 = 8x^9$.

Remember that $x^3 x^6 = x^{3+6}$

Example 7

Express the product of $3x^2$ and $^-7x^5$ as a single monomial in x.

$3x^2(^-7x^5) = ^-21x^7$. Namely

$$3x^2(^-7x^5) = 3(^-7)x^2 x^5$$
$$= 3(^-7)x^{2+5}$$
$$= ^-21x^7$$

To multiply two monomials in x, we multiply the coefficients and add the degrees.

Now we can finish the problem we introduced just before Example 6.

Example 8

Write $(2x^3 + 5)(4x^6 + 7)$ as a single polynomial.

$(2x^3 + 5)(4x^6 + 7) = 8x^9 + 20x^6 + 14x^3 + 35$. Starting out as you did previously,

$$(2x^3 + 5)(4x^6 + 7) = 2x^3(4x^6 + 7) + 5(4x^6 + 7)$$
$$= 2x^3(4x^6) + 2x^3(7) + 5(4x^6) + 5(7)$$
$$= 2(4)x^3 x^6 + 2(7)x^3 + 5(4)x^6 + 5(7)$$
$$= 8x^9 \quad\quad + 14x^3 + 20x^6 + 35$$
$$= 8x^9 \quad\quad + 20x^6 + 14x^3 + 35$$

We may use this procedure to multiply any two (or more) polynomials.

Section 11.5 Multiplication of Polynomials

Example 9

Express $(5x^3 + 4x^2)(6x^4 + 7x^5)$ as a single polynomial.

$(5x^3 + 4x^2)(6x^4 + 7x^5) = 35x^8 + 58x^7 + 24x^6$. Using the same principle,

$$(5x^3 + 4x^2)(6x^4 + 7x^5)$$
$$= 5x^3(6x^4 + 7x^5) + 4x^2(6x^4 + 7x^5)$$
$$= 5x^3(6x^4) + 5x^3(7x^5) + 4x^2(6x^4) + 4x^2(7x^5)$$
$$= 5(6)x^3x^4 + 5(7)x^3x^5 + 4(6)x^2x^4 + 4(7)x^2x^5$$
$$= 30x^{3+4} + 35x^{3+5} + 24x^{2+4} + 28x^{2+5}$$
$$= 30x^7 + 35x^8 + 24x^6 + 28x^7$$
$$= 35x^8 + 30x^7 + 28x^7 + 24x^6$$
$$= 35x^8 + 58x^7 + 24x^6$$

The order in which you do things is not important. What is important is that you should have all possible products of a term from the first set of parentheses and a term from the second set of parentheses.

First we rewrite the terms in order of decreasing exponents. Then we group the two terms of degree 7. Since the text uses the usual order for writing polynomials, it will be easier for you to check your answer with the text if you use the usual order.

The same ideas work even when a polynomial consists of more than two terms. The principle is still to take all possible terms that are the product of a term from the first set of parentheses and a term from the second set of parentheses.

Example 10

Write $(x + 2)(x^2 + 3x + 5)$ as a single polynomial.

$(x + 2)(x^2 + 3x + 5) = x^3 + 5x^2 + 11x + 10$. That is,

$$(x + 2)(x^2 + 3x + 5)$$
$$= x(x^2 + 3x + 5) + 2(x^2 + 3x + 5)$$
$$= x(x^2) + x(3x) + x5 + 2x^2 + 2(3x) + 2(5)$$
$$= x^3 + 3x^2 + 5x + 2x^2 + 6x + 10$$
$$= x^3 + 3x^2 + 2x^2 + 5x + 6x + 10$$
$$= x^3 + 5x^2 + 11x + 10$$

Remember the generalized distributive property from Lesson 4? It states

$$a(b + c + d) = ab + ac + ad$$

Some of the lines in the solution of Example 10 are hard to read because they are a bit long. For this reason, we introduce a new form similar to the one used in ordinary multiplication.
We write

$$\begin{array}{r} x^2 + 3x + 5 \\ \underline{x + 2} \end{array}$$

We then multiply each term in the top line by 2.

$$\begin{array}{r} x^2 + 3x + 5 \\ \underline{3x + 2} \\ 2x^2 + 6x + 10 \end{array}$$

Unless you're using a calculator the chances are that you will solve a problem like 345 × 123 by writing the problem in the form

$$\begin{array}{r} 345 \\ \times\,123 \\ \hline 1035 \\ 690 \\ 345 \\ \hline 42435 \end{array}$$

Next we multiply each term in the top line by x, and we write the terms so that like terms are combined.

$$\begin{array}{r} x^2 + 3x + 5 \\ \underline{x + 2} \\ 2x^2 + 6x + 10 \\ \underline{x^3 + 3x^2 + 5x} \\ x^3 + 5x^2 + 11x + 10 \end{array}$$

Let's try this new format on another problem.

We've still formed all possible products where the first factor is from the first set of parentheses and the second factor is from the second set of parentheses. By using the vertical format and collecting like terms, we use much less length to a line. We still use as many lines, but it is easier to see everything that happens.

Example 11

Express $(2x^3 + 5x^2 + 6x)(3x^2 + 7x + 6)$ as a single polynomial.

It is $6x^5 + 29x^4 + 65x^3 + 72x^2 + 36x$.

STEP 1: Write one factor under the other. Since multiplication is commutative, you may write whichever factor you wish on top.

$$\begin{array}{r} 2x^3 + 5x^2 + 6x \\ 3x^2 + 7x\phantom{{}+5x^2} + 6 \\ \hline \end{array}$$

STEP 2: Multiply each term in the top line by 6.

$$\begin{array}{r} 2x^3 + 5x^2 + 6x \\ 3x^2 + 7x + 6 \\ \hline 12x^3 + 30x^2 + 36x \end{array}$$

STEP 3: Now multiply each term in the top line by $7x$, and write the result so that like terms match up.

$$\begin{array}{r} 2x^3 + 5x^2 + 6x \\ 3x^2 + 7x + 6 \\ \hline 12x^3 + 30x^2 + 36x \\ 14x^4 + 35x^3 + 42x^2 \end{array}$$

STEP 4: Repeat the process, but now multiply each term in the top line by $3x^2$.

$$\begin{array}{r} 2x^3 + 5x^2 + 6x \\ 3x^2 + 7x + 6 \\ \hline 12x^3 + 30x^2 + 36x \\ 14x^4 + 35x^3 + 42x^2 \\ 6x^5 + 15x^4 + 18x^3 \\ \hline \end{array}$$

STEP 5: Now, add like terms.

$6x^5 + 29x^4 + 65x^3 + 72x^2 + 36x$

You could first multiply by $3x^2$, then by $7x$, and then by 6 to obtain

$$\begin{array}{r} 2x^3 + 5x^2 + 6x \\ 3x^2 + 7x + 6 \\ \hline 6x^5 + 15x^4 + 18x^3 \\ + 14x^4 + 35x^3 + 42x^2 \\ + 12x^3 + 30x^2 + 36x \\ \hline 6x^5 + 29x^4 + 65x^3 + 72x^2 + 36x \end{array}$$

This section may seem rather long. It takes a lot of space to illustrate the multiplication of polynomials. Yet the entire section is just one application of the distributive property, together with the rules for adding polynomials.

Before we end this Section, let's see what happens when we have the product of more than two polynomials. In this case, we do the same things that we do in ordinary arithmetic.

In ordinary arithmetic, we multiply two numbers at a time. For example, to find $2 \times 3 \times 4$, first we multiply 2 and 3 to get 6, and then we mutliply 6 and 4 to get 24.

Example 12

Express the product of $x + 5$, $x + 2$, and x^2 as a single polynomial.

$(x + 5)(x + 2)x^2 = x^4 + 7x^3 + 10x^2$.

By the associative property, you may assume that the factors are grouped, for example, as

$$[(x + 5)(x + 2)]x^2$$

But from Example 3, you already know that

$$(x + 5)(x + 2) = x^2 + 7x + 10$$

Section 11.5 Multiplication of Polynomials

Therefore
$$(x+5)(x+2)x^2 = [(x+5)(x+2)]x^2$$
$$= [x^2 + 7x + 10]x^2$$
$$= x^2 x^2 + 7x x^2 + 10 x^2$$
$$= x^4 + 7x^3 + 10x^2$$

In vertical form:

$$\begin{array}{r} x + 5 \\ x + 2 \\ \hline x^2 + 5x \\ +2x + 10 \\ \hline x^2 + 7x + 10 \\ x^2 \\ \hline x^4 + 7x^3 + 10x^2 \end{array}$$

If you feel the need for more practice in multiplying polynomials, the exercise set contains many such problems.

Because division is the inverse of multiplication, we must understand how to multiply two polynomials before we can find the quotient of two polynomials. Dividing polynomials is the subject of Lesson 12.

PRACTICE DRILL

Write each of the following as a single polynomial in simplest form.
1. $x(x+3)$
2. $(x+2)(x+3)$
3. $2x^2(3x^3)$
4. $(x^2 + x)(2x^3 + 3)$

Answers
1. $x^2 + 3x$ 2. $x^2 + 5x + 6$
3. $6x^5$
4. $2x^5 + 2x^4 + 3x^2 + 3x$

CHECK THE MAIN IDEAS

1. The key to multiplying polynomials is that while you cannot add terms with different degrees, you can _____ such terms.
2. This comes from the rule of exponents, which says that to multiply like bases, you keep the common base and add the _____. For example, $x^2(x^3) = x^5$.
3. This idea may then be combined with the _____ property to show, for example, that $x^2(x^3 + x^5) = x^5 + x^7$.
4. You may think of multiplication of polynomials in terms of geometry. For example, you may think of $(x+2)(x+3)$ as being the _____ of a rectangle whose dimensions are $x + 2$ by $x + 3$.

Answers
1. multiply 2. exponents
3. distributive 4. area

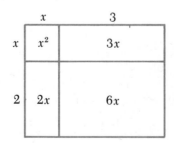

EXERCISE SET 11 (Form A)

Section 11.1

1. What is the coefficient of x^2 in the monomial $-x^2$?
2. When $x = 3$, what is the value of
   (a) $-x^2$ (b) $(-x)^2$
3. Use the distributive property (or any other method) to express $3t^2 + 4t^2$ as a single monomial in t.
4. When $t = 3$, what is the value of
   (a) $3t^2 + 4t^2$ (b) $7t^2$ (c) $14t^2$ (d) $7t^4$
5. Write $8t^3 - 14t^3 + 16t^3 - 4t^3$ as a single monomial in t.

Section 11.2

6. Given the polynomial $7x^2 + 1 - 2x^4$,
   (a) What is the degree of this polynomial?
   (b) What is the coefficient of x^4 in this polynomial?
7. Evaluate each of the following expressions when $x = 1$.
   (a) $7x^2 + 1 - 2x^4$ (b) $7(x^2 + 1) - 2x^4$
8. Evaluate each of the following expressions when $x = -1$.
   (a) $2x^2 + 3x^3$ (b) $5x^5$ (c) $2(x^2 + 3)x^3$
9. Evaluate $x^3 - 2x^2 - x + 1$ when
   (a) $x = 0$ (b) $x = 1$ (c) $x = -1$ (d) $x = \frac{1}{2}$
10. Use the distributive property to write $2x^4(3x^2 + 2x + 1)$ as a polynomial in x.
11. Write $x^3 - 2x(x^3 - 1)$ as a polynomial without the grouping symbols.

Section 11.3

Find the sum of the following.
12. $6x + 5$ and $3x - 8$
13. $5x^2 - 7$ and $2x^3 - 6x^2 + 7x$
14. $6x^4 - 7$, $x^5 - 6x^4 + 3x^3 + 4$, and $2x^2 - 3x^3 + 3$
15. What is the value of $(6x^4 - 7) + (x^5 - 6x^4 + 3x^3 + 4) + (2x^2 - 3x^3 + 3)$ when $x = 2$?
16. Write $x^2[2(x + 1) - 3(x - 4)]$ as a single polynomial in x.

Section 11.4

17. Subtract $6x + 5$ from $3x - 8$, and write the answer as a polynomial in x.
18. Subtract $3x - 8$ from $6x + 5$, and write the answer as a polynomial in x.
19. Write $(6x + 5) - (3x - 8)$ as a single polynomial in x.
20. What polynomial in x must be added to $(3x - 8)$ to get $6x + 5$ as the sum?
21. Subtract $5x^3 - 3x + 1$ from $2x^3 + x^2 - 1$.
22. Write $(8x^3 + 3x + 1) - (3x^3 - 9x^2 + 4x + 7)$ as a single polynomial in x.
23. Write as a single polynomial in x.
    $$6x^3 + 7x + 2 + 8x^2 - 5x + 4 - [(8x^3 + 3x + 1) - (3x^3 - 9x^2 + 4x + 7)].$$
24. Simplify $\{(x + w) - [(w - x) - v]\} - 2x$.
25. Evaluate $\{(x + w) - [(w - x) - v]\} - 2x$ when $x = 4.18$, $w = -2.78$, and $v = 3.21$.

Answers: Exercise Set 1
Form A
1. -1
2. (a) -9 (b) 9
3. $7t^2$
4. (a) 63 (b) 63 (c) 126 (d) 567
5. $6t^3$
6. (a) 4 (b) -2
7. (a) 6 (b) 12
8. (a) -1 (b) -5 (c) -8
9. (a) 1 (b) -1 (c) -1 (d) $\frac{1}{8}$
10. $6x^6 + 4x^5 + 2x^4$
11. $x^3 - 2x^4 + 2x$, or $-2x^4 + x^3 + 2x$
12. $9x - 3$
13. $2x^3 - x^2 + 7x - 7$
14. $x^5 + 2x^2$ 15. 40
16. $14x^2 - x^3$ or $-1x^3 + 14x^2$ (or $-x^3 + 14x^2$)
17. $-3x - 13$ or $-3x + -13$
18. $3x + 13$ 19. $3x + 13$
20. $3x + 13$
21. $-3x^3 + x^2 + 3x - 2$
22. $5x^3 + 9x^2 - x - 6$
23. $x^3 - x^2 + 3x + 12$
24. v 25. 3.21
26. (a) $x^2 - 3x + 2$ (b) $x^2 - (a + b)x + ab$
27. (a) $9x^4 - 16$ (b) $n^2 - m^2$ (c) $16x^6 - 9y^4$
28. (a) $x^6 + 2x^3 + 1$ (b) $p^2 + 2pq + q^2$ (c) $4x^8 + 12x^4y^2 + 9y^4$
29. (a) $3x^2 + 5x - 2$ (b) $9x^2 - 12x + 4$ (c) $-6x^2 + 17x - 6$
30. (a) $x^3 - 1$ (b) $m^3 - n^3$
31. (a) $x^3 - 5x^2 + 7x - 3$ (b) $-x^3 + 2x^2 + 3x$ (c) $2x^3 - 7x^2 + 4x - 3$
32. $13x^2 - 26x + 10$
33. 10 34. (a) $x^2 - 4x + 4$ (b) $x^3 - 6x^2 + 12x - 8$ (c) $2x^3 - 13x^2 + 25x - 18$
35. (a) $12x^3 - 8x^2 - 3x + 2$ (b) $16x^4 - 1$ (c) $12x^4 - 20x^3 + 5x^2 + 5x - 2$

Exercise Set 11 (Form B)

Section 11.5

Perform the indicated operations.

26. (a) $(x-1)(x-2)$ (b) $(x-a)(x-b)$
27. (a) $(3x^2+4)(3x^2-4)$ (b) $(n+m)(n-m)$ (c) $(4x^3+3y^2)(4x^3-3y^2)$
28. (a) $(x^3+1)^2$ (b) $(p+q)^2$ (c) $(2x^4+3y^2)^2$
29. (a) $(3x-1)(x+2)$ (b) $(3x-2)^2$ (c) $(3x-1)(x+2)-(3x-2)^2$
30. (a) $(x-1)(x^2+x+1)$ (b) $(m-n)(m^2+mn+n^2)$
31. (a) $(x-3)(x^2-2x+1)$ (b) $x[3-x(x-2)]$
    (c) $(x-3)(x^2-2x+1)-x[3-x(x-2)]$
32. Simplify $3x^2-2\{x-x[x+4(x-3)]-5\}$.
33. Evaluate $3x^2-2\{x-x[x+4(x-3)]-5\}$ when $x=2$.
34. Write as a polynomial in x
    (a) $(x-2)^2$ (b) $(x-2)^3$ (c) $2(x-2)^3-(x-2)^2-3(x-2)-4$

    (Note that each of the above expressions is a polynomial in $x-2$ since each term is a power of $x-2$ multiplied by a constant.)
35. Write each of the following products as a polynomial in x.

    (a) $(2x+1)(2x-1)(3x-2)$
    (b) $(2x+1)(2x-1)(4x^2+1)$
    (c) $(2x+1)(2x-1)(3x^2-5x+2)$

EXERCISE SET 11 (Form B)

Section 11.1

1. What is the coefficient of x^3 in the monomial $-x^3$?
2. When $x=2$, what is the value of

    (a) $-x^4$ (b) $(-x)^4$

3. Write $5t^2+6t^2$ as a single monomial in t.
4. When $t=2$, what is the value of

    (a) $5t^2+6t^2$ (b) $11t^2$ (c) $22t^2$ (d) $11t^4$

5. Write $15t^4-8t^4-7t^4-2t^4+6t^4$ as a single monomial in t.

Section 11.2

6. In the polynomial $8x^4-7x^5+9x^3$,

    (a) What is the degree of the polynomial?
    (b) What is the coefficient of x^5?

7. When $x=1$, what is the value of

    (a) $9x^2+2-3x^4$ (b) $9(x^2+2)-3x^4$

8. Evaluate each of the following expressions when $x=-1$.

    (a) $3x^4+4x^5$ (b) $7x^9$ (c) $3(x^4+4)x^5$

9. Evaluate $x^4+2x^3-x^2-2x+7$ when

    (a) $x=0$ (b) $x=1$ (c) $x=-2$ (d) $x=\frac{1}{2}$

10. Write $4x^3(8x^2+3x+2)$ as a polynomial in x.
11. Rewrite $4x^4-3x(x^3-2)$ as a polynomial in x without grouping symbols.

Section 11.3
12. Find the sum of $8x + 7$ and $5x - 9$.
13. Find the sum of $6x^2 - 9$ and $3x^4 - 5x^2 + 7x$.
14. Find the sum of $8x^4 - 9$, $x^5 - 8x^4 + 2x^2 + 4$, and $2x^3 - 2x^2 + 5$.
15. Evaluate $(8x^4 - 9) + (x^5 - 8x^4 + 2x^4 + 4) + (2x^3 - 2x^2 + 5)$ when $x = 2$.
16. Write $x^2[3(x + 2) - 2(x - 5)]$ as a polynomial in x.

Section 11.4
17. What polynomial in x do you get if you subtract $8x + 7$ from $5x - 9$?
18. What polynomial in x do you get if you subtract $5x - 9$ from $8x + 7$?
19. Write $(8x + 7) - (5x - 9)$ as a polynomial in x.
20. What polynomial in x must be added to $5x - 9$ to give $8x + 7$ as the sum?
21. What polynomial in x will you get if you subtract $6x^4 - 4x^2 + 3$ from $4x^4 - 2x^3 + 5x^2 - 3$?
22. Write $7x^3 + 4x - 2 - (6x^3 - 5x^2 + 3x + 3)$ as a polynomial in x.
23. Write as a single polynomial in x.

$$2x^3 + 2x + x^2 + 6x^2 + x + 5 - [7x^3 + 4x - 2 - (6x^3 - 5x^2 + 3x + 3)]$$

24. Simplify $\{(x + w) - [(v + x) - w]\} - 2w$.
25. Evaluate $(x + w) - [(v + x) - w] - 2w$ when $x = 2.91$, $w = 2.34$, and $v = 7.3$.

Section 11.5
Perform the indicated operations.
26. (a) $(x + 2)(x + 3)$ (b) $(x + m)(x + n)$
27. (a) $(2x^4 + 3)(2x^4 - 3)$ (b) $(a + b)(a - b)$ (c) $(2x^4 + 3y)(2x^4 - 3y)$
28. (a) $(x^4 + 2)^2$ (b) $(m + n)^2$ (c) $(3x^4 + 2y)^2$
29. (a) $(4x - 1)(x + 2)$ (b) $(2x - 1)^2$ (c) $(4x - 1)(x + 2) - (2x - 1)^2$
30. (a) $(x - 2)(x^2 + 2x + 4)$ (b) $(a - b)(a^2 + ab + b^2)$
31. (a) $(x - 5)(x^2 - 2x + 2)$ (b) $x[2x^2 - x(x - 2)]$
    (c) $(x - 5)(x^2 - 2x + 2) - x[2x^2 - x(x - 2)]$
32. Simplify $2x^2 - 3\{x - x[x + 2(x - 3)] - 1\}$.
33. Evaluate $2x^2 - 3\{x - x[x + 2(x - 3)] - 1\}$ when $x = 2$.
34. Write each of the following as a polynomial in x.
    (a) $(x - 3)^2$ (b) $(x - 3)^3$ (c) $(x - 3)^3 - 3(x - 3)^2 + 3(x - 3) + 3$
35. Write each of the following products as a polynomial in x.
    (a) $(3x + 2)(3x - 2)(x + 1)$
    (b) $(3x + 2)(3x - 2)(9x^2 + 4)$
    (c) $(3x + 2)(3x - 2)(3x^2 - 2x + 1)$

Lesson 12: Division of Polynomials

Overview Long division has caused people problems for a long time. There is an old story about three college football players who are complaining about their math courses. The first one says, "Algebra gives me a lot of trouble." The second one says "Algebra was easy for me, but geometry gives me a lot of trouble." The third one asked, "Did either of you guys ever hear of long division?"

Just as long division is a stumbling block in ordinary arithmetic, it can be a stumbling block in the arithmetic of polynomials. For this reason, we have elected to treat division of polynomials separately from the other arithmetic operations with polynomials.

Division is a special case of a more general topic known as factoring. As we shall see in the next unit, the idea of factoring is very important in algebra. In this lesson, we introduce factoring. As the name suggests, factoring is the opposite of multiplication. Remember, a factor is a number that is being multiplied by another. Factoring means that we want to know what numbers we have to multiply to get a given number. This is not always easy. If we want the product of 71 and 73, it is not hard to multiply the two numbers and get 5,183. On the other hand, it is not so easy to be given 5,183 and determine that it is the product of 71 and 73. This process is what we call factoring. In this lesson, we introduce a few basic points about factoring.

Section 12.1
Dividing a Polynomial by a Monomial

When we add, subtract, or multiply integers, the result is always an integer. But the quotient of two integers need not be an integer. For example, 3 and 5 are integers, but neither $3 \div 5$ nor $5 \div 3$ is an integer.

This is also true for polynomials. In the last lesson, we found that the sum, difference, or product of two polynomials is always a polynomial. But finding the quotient of two polynomials can present a much more complicated problem.

Let's begin at a reasonably logical starting point: with the quotient of two monomials. We already know how to multiply two monomials.

OBJECTIVE
To be able to express the quotient of a polynomial divided by a monomial as an algebraic expression.

Example 1

Find the product of $3x^2$ and $4x^5$.

The product is $12x^7$. All you have to do is multiply the coefficients and add the exponents. That is,

$$3x^2(4x^5) = 3(4)x^2x^5 = 12x^{2+5} = 12x^7$$

Because division is the inverse of multiplication, we can use our knowledge of multiplication of polynomials to find the quotient of polynomials. Let's look at Example 1 from a different point of view.

first $\div$ second = third

means

first = second $\times$ third.

Example 2

Express $12x^7 \div 3x^2$ as a monomial in x.

$12x^7 \div 3x^2 = 4x^5$.

To find what you have to multiply $3x^2$ by to get $12x^7$, you have to know what to multiply 3 by to get 12, and what to multiply x^2 by to get x^7. That is,

That is, what must we multiply $3x^2$ by to get $12x^7$?

To get 12 you multiply 3 by ⟶ 4
To get x^7 you multiply x^2 by ⟶ x^5
Therefore, to get $12x^7$ you multiply $3x^2$ by ⟶ $4x^5$

NOTE

$3x^2$ is an algebraic expression that has a numerical value for each numerical value of x. Remember, we have agreed that we must not divide by 0. Therefore, in Example 2, we can divide only so long as $3x^2$ is not 0. In Example 2, we are assuming that x is not equal to 0.

Example 3

For what value of m is $16m^4 \div 2m$ not defined?

$16m^2 \div 2m$ is undefined when $m = 0$. That is, when $m = 0$, $2m = 0$, and you are not allowed to divide by 0.

Example 4

Assuming that m is not 0, express $16m^4 \div 2m$ as a single monomial.

When $m \neq 0$, $16m^4 \div 2m = 8m^3$. (If $m = 0$, then $16m^4 \div 2m$ is $0 \div 0$, which is "forbidden.")

Rather than writing "m is not equal to 0," it is customary to write "$m \neq 0$." That is, a slash through the equal sign means "unequal."

Section 12.1 Dividing a Polynomial by a Monomial

To get 16, multiply 2 by ⟶ 8
To get m^4, multiply m by ⟶ m^3
So to get $16m^4$, multiply $2m$ by ⟶ $8m^3$

Notice that the dividend can be 0, it's the divisor that must be different from 0.

It is all right if the dividend (numerator) is 0. The divisor (denominator) musn't be 0. For example, $15 \div 0$ is undefined, but $0 \div 15$ is 0.

Example 5

For what value of n is the expression $16n^2 \div (n - 1)$ undefined?

This expression is undefined only when $n = 1$.
 You cannot divide by 0. Since the divisor is $n - 1$, you cannot divide when $n - 1$ is equal to 0. But the only time $n - 1 = 0$ is when $n = 1$.
 The dividend is $16n^2$, which is 0 when $n = 0$. But when $n = 0$, the expression $16n^2 \div (n - 1)$ becomes

$$16(0^2) \div (0 - 1) \quad \text{or} \quad 0 \div {}^-1$$

This is 0 (since $^-1 \times 0 = 0$).

 If we want to, we can write the quotient of two polynomials in the form of a common fraction. For example, we can write $12x^7 \div 3x^2$ as $\dfrac{12x^7}{3x^2}$, and then we can do the division by reducing the fraction to lowest terms.

numerator ⟶ n ⟵ dividend
over ⟶ *⟵ divided by*
denominator ⟶ d ⟵ divisor

$$\frac{12x^7}{3x^2} = \frac{3x^2(4x^5)}{3x^2(1)} = \frac{3x^2(4x^5)}{3x^2(1)} = \frac{4x^5}{1} = 4x^5$$

Example 6

Write $x^2 \div x^7$ as a single power of x.

$x^2 \div x^7$ is equal to x^{-5}.
 By the rule for dividing like bases, you have

$$x^2 \div x^7 = x^{2-7} = x^{-5}$$

By writing division in the form of a fraction, you have

$$x^2 \div x^7 = \frac{x^2}{x^7} = \frac{x^2(1)}{x^2(x^5)} = \frac{x^2(1)}{x^2(x^5)} = \frac{1}{x^5} = x^{-5}$$

Because we cannot divide by 0, we assume that $x \neq 0$ in Example 6. In this text we will assume that x is limited to values that don't require division by 0.

 Remember, to be a monomial in x, the expression must have an exponent that is a nonnegative integer. In other words, x^{-5} is not a monomial *in* x. Example 6 shows us that the quotient of two monomials in x need not be a monomial in x. For the quotient of two monomials in x to be a monomial in x, the degree of the numerator must be no less than the degree of the denominator.

NOTE
x^{-5} is a monomial in x^{-1}. That is, $x^{-5} = (x^{-1})^5$. This is another reason for specifying that x^{-5} is not a monomial *in* x.

Example 7

Write $\dfrac{12x^8}{24x^5}$ as a fraction in lowest terms.

In lowest terms, it is $\dfrac{x^3}{2}$, which may be written as $\dfrac{1}{2}x^3$.
 You may cancel just as you do with a numerical common fraction. Or, you can use the fact that when you multiply common fractions, you multiply the

numerators and the denominators, to obtain

$$\frac{12x^8}{24x^5} = \left(\frac{12}{24}\right)\left(\frac{x^8}{x^5}\right) = \frac{1}{2}x^{8-5} = \frac{1}{2}x^3$$

In this case, the quotient is a monomial.

Remember that for cx^n to be a monomial in x, n must be a nonnegative integer, but c can be any real number, not necessarily an integer.

Example 8

Write $\frac{24x^5}{12x^8}$ as a fraction in lowest terms.

$\frac{24x^5}{12x^8} = \frac{2}{x^3}$. That is,

$$\frac{24x^5}{12x^8} = \left(\frac{24}{12}\right)\left(\frac{x^5}{x^8}\right) = 2x^{5-8} = 2x^{-3} = \frac{2}{x^3}$$

In this case, the quotient of the two monomials in x is not a monomial in x.

Remember that the exponent affects only the number immediately to the left of it. If you had $(2x)^{-3}$, then the value would be

$$\frac{1}{(2x)^3} \quad \text{or} \quad \frac{1}{8x^3}$$

To summarize, when we divide two monomials, first we make sure that we use values of the variable that do not make us divide by 0. Then we can find the quotient, but it will be a monomial in x only if the degree of the numerator (that is, the exponent in the numerator) is at least as great as the degree of the denominator.

In this lesson, our goal is to be able to find the quotient of any two polynomials. So far, we have learned how to find the quotient of two monomials. Now we'd like to extend our results to cover dividing any polynomial by a monomial.

Our approach is to show that whenever we have a polynomial divided by a monomial, we may look at it as several problems of the type in which we divide a monomial by a monomial. Let's start with some examples, the first two of which are a review.

Example 9

Write $6x^5 \div 3x^2$ as a monomial in x.

$6x^5 \div 3x^2 = 2x^3$. You may write the problem as $\frac{6x^5}{3x^2}$ and then cancel to obtain $2x^3$. Or, you can write the fractions as

$$\left(\frac{6}{3}\right)\left(\frac{x^5}{x^2}\right) = 2x^{5-2} = 2x^3$$

CHECK
$2x^3(3x^2) = 2(3)x^3x^2 = 6x^{3+2}$
$= 6x^5$

Example 10

Write $12x^8 \div 3x^2$ as a monomial in x.

$12x^8 \div 3x^2 = 4x^6$.

        To get 12, multiply 3 by ⟶ 4
        To get x^8, multiply x^2 by ⟶ x^6
        To get $12x^8$, multiply $3x^2$ by ⟶ $4x^6$

Now let's combine the results of Examples 9 and 10 to do Example 11.

Section 12.1 Dividing a Polynomial by a Monomial

Example 11

Express $(6x^5 + 12x^8) \div 3x^2$ as a polynomial in x.

The quotient is $2x^3 + 4x^6$.

    Notice that you get the answer by adding the answers for Examples 9 and 10. But do you see the reason for adding these two answers?

    The problem may be easier to understand if you rewrite it in terms of a common fraction. The problem looks like

$$\frac{6x^5 + 12x^8}{3x^2} \quad [1]$$

But this is the same as

$$\frac{6x^5}{3x^2} + \frac{12x^8}{3x^2} \quad [2]$$

    The first term in [2] is the same problem that you did in Example 9. The second term in [2] is the same problem that you did in Example 10.

    Let's try another problem.

Example 12

Express $(12x^4 + 18x^6) \div 6x^3$ as a polynomial in x.

The quotient is $2x + 3x^3$. First rewrite the problem as

$$\frac{12x^4 + 18x^6}{6x^3}$$

By the rule for adding fractions with the same denominator, you know that:

$$\frac{12x^4 + 18x^6}{6x^3} = \frac{12x^4}{6x^3} + \frac{18x^6}{6x^3}$$
$$= 2x + 3x^3$$

    The method of adding fractions with a common denominator extends to any number of terms.

Example 13

Express $\dfrac{6x^4 + 12x^3 + 4x^2 + 10}{2}$ as a single polynomial.

It is $3x^4 + 6x^3 + 2x^2 + 5$. You divide each term in the numerator by 2.

$$\frac{6x^4}{2} + \frac{12x^3}{2} + \frac{4x^2}{2} + \frac{10}{2} = \frac{6x^4 + 12x^3 + 4x^2 + 10}{2}$$

Furthermore,

$$\frac{6x^4}{2} = 3x^4$$
$$\frac{12x^3}{2} = 6x^3$$
$$\frac{4x^2}{2} = 2x^2$$
$$\frac{10}{2} = 5$$

As the problem stands, we have the quotient of a polynomial in x divided by a monomial. The question asks us to rewrite this as a single polynomial in x.

This is why it's important to know how to add numerical fractions. The same principles apply to algebraic fractions. Whether we use numbers or algebraic expressions, the rule for adding fractions with common denominators is that we keep the common denominator and add the numerators. Therefore

$$\frac{6x^5}{3x^2} + \frac{12x^8}{3x^2} = \frac{6x^5 + 12x^8}{3x^2}$$

If we read this equation from right to left, we get equation [1]. In general,

$$\frac{b}{d} + \frac{c}{d} = \frac{b+c}{d}$$

is the same as saying

$$\frac{b+c}{d} = \frac{b}{d} + \frac{c}{d}$$

As we read from right to left, we are adding common denominators, which means that we add the numerators and keep the common denominator.

Notice that to divide $6x^4$ by 2, we simply divide the 6 by 2. That is true because

$$\frac{6x^4}{2} = \left(\frac{6}{2}\right)x^4 = 3x^4$$

Example 14

Express $(3x^4 + 9x^3 + 15x^5) \div 3x$ as a polynomial in x.

The quotient is $x^3 + 3x^2 + 5x^4$. You write the quotient as

$$\frac{3x^4 + 9x^3 + 15x^5}{3x} = \frac{3x^4}{3x} + \frac{9x^3}{3x} + \frac{15x^5}{3x}$$
$$= x^3 + 3x^2 + 5x^4$$

If we prefer what we call the usual order, we write the terms in the order of decreasing degree.

$$5x^4 + x^3 + 3x^2$$

So far, we have been getting a polynomial in x as our quotient. However, if the degree of any term in the dividend (numerator) is less than the degree of the monomial in the divisor (denominator), then we won't get a polynomial.

Example 15

Express $(3x^4 + 6x^2) \div 3x^3$ as a sum.

It is $x + \frac{2}{x}$, or $x + 2x^{-1}$. Again, you may write

$$\frac{3x^4 + 6x^2}{3x^3} = \frac{3x^4}{3x^3} + \frac{6x^2}{3x^3}$$
$$= \left(\frac{3}{3}\right)\left(\frac{x^4}{x^3}\right) + \left(\frac{6}{3}\right)\left(\frac{x^2}{x^3}\right)$$
$$= x^{4-3} + 2x^{2-3}$$
$$= x + 2x^{-1}$$
$$= x + \frac{2}{x}$$

$x + 2x^{-1}$ is not a polynomial in x.

Remember, $2x^{-1}$ means $2(x^{-1})$, not $(2x)^{-1}$. Therefore, $2x^{-1}$ means $2\left(\frac{1}{x}\right)$ or $\frac{2}{x}$; not $\frac{1}{2x}$.

PRACTICE DRILL

Write each of the following as a single polynomial.
1. $x^5 \div x^3$
2. $24x^5 \div x^3$
3. $16x^4 \div x^3$
4. $(24x^5 + 16x^4) \div x^3$
5. $(24x^5 + 16x^4) \div 8x^3$

Answers
1. x^2 2. $24x^2$ 3. $16x$
4. $24x^2 + 16x$ 5. $3x^2 + 2x$

CHECK THE MAIN IDEAS

1. Though the sum, difference, and product of two polynomials are always a polynomial, the _____ of two polynomials need not be a polynomial.
2. For example, $6x^5 \div 6x^6 = \frac{1}{x}$ or x^{-1}, which is not a polynomial because _____ is not a nonnegative integer.
3. There are times, however, when the quotient of two polynomials is a polynomial. For example, $6x^5 \div 2x^3$ means the expression you must _____ $2x^3$ to get $6x^5$.
4. Since $2x^3(3x^2) = 6x^5$, then $6x^5 \div 2x^3 =$ _____, which is a polynomial.
5. To divide a polynomial by a monomial, you divide each _____ of the polynomial by the monomial.
6. There is one precaution you must take in evaluating the quotient of two polynomials. When you replace the variable by a number, the polynomial represents a number, so you must make sure that you are not dividing by _____.

Answers
1. quotient 2. -1
3. multiply by 4. $3x^2$
5. term 6. 0 (zero)

Section 12.2
An Introduction to Factoring

OBJECTIVE
To know the definition of factoring and to be able to use the concept to simplify certain algebraic expressions.

If we want to find the area of a rectangle, we multiply its length and width. If we want to find the distance an object travels, we multiply its speed by the time it travels at that speed. Evidently, we need to know the product of two numbers quite often.

There are also many times when we want to know how to write a number as the product of two or more numbers. This happens when we try to reduce a fraction to lowest terms.

Example 1

Write $\frac{24}{30}$ in lowest terms.

$\frac{24}{30} = \frac{4}{5}$. Since $24 = 6 \times 4$ and $30 = 6 \times 5$, you may write

$$\frac{24}{30} = \frac{6 \times 4}{6 \times 5}$$

You may cancel 6 as a common factor from the numerator and denominator to obtain

$$\frac{24}{30} = \frac{4}{5}$$

In Example 1, we did not start with the fact that $6 \times 4 = 24$. Instead, we started with 24 and tried to write it as the product of two numbers in a way that made it easier to see how to reduce $\frac{24}{30}$ to lowest terms. Occasionally, however, we may have to recognize 24 as a product of two other numbers.

Example 2

Write $\frac{24}{32}$ in lowest terms.

$\frac{24}{32} = \frac{3}{4}$.

In this example, you must make use of the facts that $8 \times 3 = 24$ and $8 \times 4 = 32$. You write

$$\frac{24}{32} = \frac{8 \times 3}{8 \times 4}$$

Now you cancel the common factor, 8, from both the numerator and denominator to obtain

$$\frac{24}{32} = \frac{3}{4}$$

Rather than decide in advance whether to write 24 as 6×4 or 8×3, remember that 24 is always $2 \times 2 \times 2 \times 3$. So if we write $\frac{24}{30}$ as $\frac{2 \times 2 \times 2 \times 3}{2 \times 3 \times 5}$, we see that 2×3, or 6, can be cancelled from both the numerator and denominator of $\frac{24}{30}$. And since $32 = 2 \times 2 \times 2 \times 2 \times 2$, $\frac{24}{32}$ means $\frac{2 \times 2 \times 2 \times 3}{2 \times 2 \times 2 \times 2}$, which means that we can cancel $2 \times 2 \times 2$, or 8, from both the numerator and denominator of $\frac{24}{32}$.

DEFINITION
To **factor** a number means to write it as a product of two or more numbers.

The concept of factoring is used in algebra as well as in arithmetic.

Example 3

Write $(x^2 + 1)(2x + 3)$ as a single polynomial.

$(x^2 + 1)(2x + 3) = 2x^3 + 3x^2 + 2x + 3$.
   That is,
$$(x^2 + 1)(2x + 3) = x^2(2x + 3) + 1(2x + 3)$$
$$= 2x^3 + 3x^2 + 2x + 3 \qquad [1]$$

If we read [1] from left to right, we are expressing the product of two polynomials as a single polynomial. What happens if we read [1] from right to left? Then we start with a polynomial and express it as the product of two other polynomials.

Example 4

Use Example 3 to factor $2x^3 + 3x^2 + 2x + 3$.

$2x^3 + 3x^2 + 2x + 3 = (x^2 + 1)(2x + 3)$. [2]

To get [2], all you have to do is reverse both sides of the equality expressed in [1]. You have expressed the given polynomial as the product of other polynomials.

There may be other ways to factor this polynomial, just as there are several ways to factor 24. Equation [2] shows one possible factorization.

It is not so easy to see how to do this, unless we've already seen Example 3. We'll show how later. Now, we want simply to illustrate what factoring means.

One of the easiest forms of factoring comes from the distributive property, which says that if b, c, and d are any numbers, then
$$b(c + d) = bc + bd \qquad [3]$$
If we read [3] from right to left, we get
$$bc + bd = b(c + d) \qquad [4]$$
Equation [4] is one way of factoring a sum.

Example 5

Use [4] to factor $3m + 3n$.

$3m + 3n = 3(m + n)$.

If you don't want to think of this as being the distributive property from right to left, then think of it this way. You noticed that 3 was a factor of each term, so you removed it, and multiplied it by what was left. This gave you $3(m + n)$.

Even though 3 appears in both $3m$ and $3n$, you remove it only once as a common factor. That is, when you write $3(m + n)$, the distributive property tells you that the 3 multiplies both the m and the n.

If you're not sure if $3m + 3n$ is $3(m + n)$ or $6(m + n)$, you can always replace m and n by numbers and see what happens.

This is the distributive property in a different order. The distributive property says that
$$3(m + n) = 3m + 3n$$
If we reverse the two sides of this equality, we get
$$3m + 3n = 3(m + n)$$

In general, we try to avoid 0 and 1 as check numbers. One reason is that 1^n is always 1 and 0^n is always 0 (except when $n = 0$). Further, $1n$ is always n and $0n$ is always 0. Because of these very special properties, strange things can happen when we use 0 or 1 as check numbers.

Section 12.2 An Introduction to Factoring

Example 6

Find the value of each of the three expressions $3m + 3n$, $3(m + n)$, and $6(m + n)$ when $m = 1$ and $n = 2$.

If you replace m by 1 and n by 2, you have

$$3m + 3n = 3(1) + 3(2) = 3 + 6 = 9$$
$$3(m + n) = 3(1 + 2) = 3(3) = 9$$
$$6(m + n) = 6(1 + 2) = 6(3) = 18$$

From this problem, you know that the expressions $3m + 3n$ and $6(m + n)$ have different values when $m = 1$ and $n = 2$. This tells you that $3m + 3n$ *cannot* be a synonym for $6(m + n)$.

Example 6 is a device that we shall use often in the remainder of this course. Whenever there is any doubt about whether or not two expressions are equal, we shall compute the value of each for a given value of the variable to see if the expressions are equal for this value of the variable.

> The fact that $3m + 3n$ and $3(m + n)$ have the same value when $m = 1$ and $n = 2$ is not ironclad proof that these two expressions are equal. It could be a coincidence. But the fact that $3m + 3n$ and $6(m + n)$ are different when $m = 1$ and $n = 2$ is sufficient proof that $3m + 3n$ and $6(m + n)$ must be **unequal** expressions.

Example 7

Use the fact that $bc + bd = b(c + d)$ to factor $9p + 9q$.

The answer is $9p + 9q = 9(p + q)$.

STEP 1: Since 9 is a common factor in each term, write it outside the parentheses. $9(\ \ \ )$

STEP 2: You must multiply 9 by p to get $9p$, so put p inside the parentheses. $9(p +\ \ )$

STEP 3: You must multiply 9 by q to get $9q$, so put q inside the parentheses. $9(p + q)$

We must remember that we can't apply the distributive property unless a factor is common to each term.

Example 8

Factor $6m + 9n$.

$6m + 9n = 3(2m + 3n)$.

Clearly, $6m$ is divisible by 6, but $9n$ is not. If you break $6m$ and $9n$ into prime factors, you get

$$6m = 3(2)m \qquad [5]$$
$$9n = 3(3)n \qquad [6]$$

From [5] and [6] you see that 3 is a factor of both terms. So you may remove 3, place it outside the parentheses, and get $3(2m + 3n)$.

> To check that you've factored correctly, multiply the factors you get for your answer and see if this product equals the original expression given in the problem. For example,
>
> $$3(2m + 3n) = 3(2m) + 3(3n)$$
> $$= 6m + 9n$$

STEP 1: $\qquad 6m + 9n = 3(\quad\quad)$ [7]

To get $6m$, you must multiply 3 by $2m$.

STEP 2: Place $2m$ inside the parentheses.

$$6m + 9n = 3(2m +\ \ ) \qquad [8]$$

STEP 3: To get $9n$, you must multiply 3 by $3n$. So write $3n$ inside the parentheses, too.

$$6m + 9n = 3(2m + 3n)$$

You must be careful to remember that factoring involves multiplication.

Example 9

Factor $3 + 3m$.

$3 + 3m = 3(1 + m)$.   That is, 3 means $3(1)$, so 3 is a factor of both terms.   In fact, $3 = 3 \times 1$ and $3m = 3 \times m$.   You factor out 3 and write

$$3(1 + m)$$

As a check:

$$3(1 + m) = 3(1) + 3m = 3 + 3m$$

Do not say that when you remove 3 as a factor from 3, nothing is left! In factored form, $3 = 3 \times 1$, and when you remove 3 as a factor, the factor 1 is left.

We shall use the idea of removing a common factor in the next section. Let's conclude this section by seeing how our discussion gives us another way to reduce the quotient of a polynomial and a monomial to lowest terms. Let's do some preliminary examples first.

Example 10

Factor $6m^2 + 9m^3$.

$6m^2 + 9m^3 = 3m^2(2 + 3m)$.
  Using prime factors, you have

$$6m^2 = 2(3)mm \quad [9]$$
$$9m^3 = 3(3)mmm \quad [10]$$

CHECK
$$3m^2(2 + 3m)$$
$$= 3m^2(2) + 3m^2(3m)$$
$$= 3(2)m^2 + 3(3)m^2m$$
$$= 6m^2 + 9m^3$$

Comparing [9] and [10], you see that 3 is a common factor of both terms, and m appears twice as a common factor.
  If you get confused, remove the factors one at a time. For example,

$$\begin{aligned}6m^2 + 9m^3 &= 2(3)mm + 3(3)mmm \\ &= 3(2mm + 3mmm) \\ &= 3m(2m + 3mm) \\ &= 3mm(2 + 3m) \\ &= 3m^2(2 + 3m)\end{aligned}$$

With a little practice, you see that

$$6m^2 = 3m^2(2) \quad \text{and} \quad 9m^3 = 3m^2(3m)$$

so that

$$\begin{aligned}6m^2 + 9m^3 &= [3m^2(2) + 3m^2(3m)] \\ &= 3m^2[2 + 3m]\end{aligned}$$

The terms now left in the parentheses have no factors in common except for 1, which is a factor of every number.

Example 11

Factor $30m^2n^3p + 24m^4nr$.

$30m^2n^3p + 24m^4nr = 6m^2n(5n^2p + 4m^2r)$.
  Begin by breaking each term into prime factors.

$$30m^2n^3p = 2(3)5mmnnnp$$
$$24m^4nr = 2(2)2(3)mmmmnr$$

Section 12.2 An Introduction to Factoring 311

Therefore

$$30m^2n^3p + 24m^4nr$$
$$= 2(3)5mmnnnp + 2(2)2(3)mmmmnr$$
$$= 2[(3)5mmnnnp + 2(2)3mmmmnr]$$
$$= 2(3)[5mmnnnp + 2(2)mmmmnr]$$
$$= 2(3)m[5mnnnp + 2(2)mmmnr]$$
$$= 2(3)mm[5nnnp + 2(2)mmnr]$$
$$= 2(3)mmn[5nnp + 2(2)mmr]$$
$$= 6m^2n[5n^2p + 4m^2r]$$

As a check:

$$6m^2n(5n^2p + 4m^2r)$$
$$= 6m^2n(5n^2p) + 6m^2n(4m^2r)$$
$$= 30m^2n^3p + 24m^4nr$$

In each step, we take one prime factor that appears in each term and place that factor outside the grouping symbols. We then delete that factor from each of the terms in which it appears. We do this each time that factor occurs in *both* terms.

As soon as that factor no longer appears in both terms, we stop removing it, even though it may still appear in one term.

Example 12

Express $\dfrac{6m^2 + 9m^3}{3m}$ as a polynomial in m.

It is $2m + 3m^2$ (or $3m^2 + 2m$).

From Example 10, you know that the numerator may be written as $3m^2(2 + 3m)$. The fraction may be written as

$$\frac{3m^2(2 + 3m)}{3m}$$

Since $3m$ is a factor of both the numerator and the denominator, you cancel it from both to obtain $m(2 + 3m)$.

Technically, although $m(2 + 3m)$ is correct, it is not the answer to Example 12 because the answer is supposed to be written as a polynomial in m (not as the product of two polynomials); so you write $m(2 + 3m)$ as $2m + 3m^2$.

$$\frac{3m^2(2 + 3m)}{3m}$$
$$= \frac{3mm(2 + 3m)}{3m(1)}$$
$$= \frac{3m[m(2 + 3m)]}{3m(1)}$$
$$= \frac{[m(2 + 3m)]}{1} = m(2 + 3m)$$

That is how we reduce polynomial fractions. We write them in factored form and then cancel the common factors. It is important that we cancel *common* factors. A mistake that beginners often make in reducing fractions is to reduce, for example, $\dfrac{3 + 6}{3}$ by canceling the 3 from numerator and denominator. That is, they may write

$$\frac{3 + 6}{3} = \frac{\cancel{3} + 6}{\cancel{3}}$$

We can see that this is wrong by evaluating both sides of the equality. In the first case, we have $\dfrac{3 + 6}{3} = \dfrac{9}{3} = 3$; and in the second case, we have $\dfrac{\cancel{3} + 6}{\cancel{3}} = \dfrac{1 + 6}{1} = \dfrac{7}{1} = 7$. But 3 and 7 are unequal.

We should factor $3 + 6$ as $3(1 + 2)$, in which case the fraction would become

$$\frac{3(1 + 2)}{3} = \frac{3(1 + 2)}{3(1)}$$

Now we can cancel 3 as a common factor of the numerator and denominator to get $\dfrac{(1 + 2)}{1}$, or 3.

Let's conclude this section with one more example.

Just as in arithmetic, we can cancel the common factor 3 in $\dfrac{4 \times 3}{5 \times 3}$ to get $\dfrac{4}{5}$; but we can't cancel 3 in $\dfrac{4 + 3}{5 + 3}$ since here 3 is a *term* of both the numerator and the denominator but not a *factor* of either the numerator or the denominator.

With specific numbers, it's easier just to add — that is, replace $3 + 6$ by 9. But when we work with variables, we don't have such simple replacements.

Example 13

Write $\dfrac{30m^2n^3p + 24m^4nr}{6mn}$ in lowest terms.

In lowest terms, the fraction is $m(5n^2p + 4m^2r)$, or $5n^2mp + 4m^3r$.

All you have to do is factor the numerator, which you already did in Example 11 (if a different problem had been given, you would begin by factoring the numerator).

You get $\dfrac{6m^2n(5n^2p + 4m^2r)}{6mn}$. You may cancel $6mn$ as a common fractor from both numerator and denominator to get the correct answer. That is,

$$\dfrac{6m^2n(5n^2p + 4m^2r)}{6mn} = \dfrac{\cancel{6m}mn(5n^2p + 4m^2r)}{\cancel{6mn}}$$

This example offers us something new to think about — a fact that plays a role in more advanced courses. In this expression, there are *four* variables: m, n, p, and r. In many real-life problems, we have to deal with several variables.

There is more to the skill of factoring than what has been presented in this section. We shall analyze some more ideas in the next lesson.

With the information we now have, we are ready to return to the main point of this lesson, to continue with our study of finding the quotient of two polynomials.

PRACTICE DRILL

Factor each of the following.
1. $8a + 8b$ 2. $8a + 16b$ 3. $8a + 12b$ 4. $8a + 12a^2$

Write in lowest terms.

5. $\dfrac{8a + 8b}{8}$ 6. $\dfrac{8a + 8a^2}{8a}$ 7. $\dfrac{8a^3 + 12a^4}{4a^2}$

Answers
1. $8(a + b)$ 2. $8(a + 2b)$
3. $4(2a + 3b)$ 4. $4a(2 + 3a)$
5. $a + b$ 6. $1 + a$
7. $2a + 3a^2$

CHECK THE MAIN IDEAS

1. Multiplication starts with one or more numbers and goes on to find their product. If you start with the product and want to find what numbers had to be multiplied to get this product, you call the process _____.
2. You may want to factor in arithmetic when you reduce a _____ to lowest terms.
3. In algebra, a form of factoring occurs when you read $a(b + c) = ab + ac$ from _____ to left.
4. When you write $a(b + c) = ab + ac$, you are using the _____ property.
5. When you write $ab + ac = a(b + c)$, you have _____ $ab + ac$.

Answers
1. factoring   2. (common) fraction   3. right
4. distributive   5. factored

Section 12.3
Dividing Polynomials by Factoring

In the last lesson, we learned that $(x + 2)(x + 3)$ is equal to $x^2 + 5x + 6$. Another way of saying this is

$$(x^2 + 5x + 6) \div (x + 2) = (x + 3) \qquad [1]$$

In this section, we shall learn how to perform the division in [1] without knowing in advance that $(x + 2)(x + 3)$ is $x^2 + 5x + 6$. We shall see how we are able to rewrite $x^2 + 5x + 6$ so that $x + 2$ will be a factor.

The procedure is very much the same as the method we use when we divide whole numbers. For example, to show that $2,821 \div 13 = 217$, we use "rapid" subtraction to show that

OBJECTIVE
To be able to use factoring to write the quotient of two polynomials as a polynomial.

Section 12.3 Dividing Polynomials by Factoring

$$2{,}821 = 200(13) + 10(13) + 7(13) \qquad [2]$$

$$\begin{array}{r} 217 \\ 13\overline{)2{,}821} \\ -\,2{,}600 \\ \hline 221 \\ -\,130 \\ \hline 91 \\ -\,91 \\ \hline 0 \end{array}$$

Then we factor 13 from the right side of [2] to obtain

$$2{,}821 = (200 + 10 + 7)(13) = 217(13)$$

This is exactly what we do with polynomials, but it may look more complicated because we use letters in place of numbers.

Let's work the problem in the form of a puzzle. What polynomial in x must go inside the parentheses in order that

$$(\quad)(x+2) = x^2 + 5x + 6$$

be true?

We know the answer is $x + 3$, but let's ignore the answer so that we can develop the method.

STEP 1: To get x^2 (in the expression $x^2 + 5x + 6$) you must multiply x by x. So as your first guess, put x in the parentheses. This gives

$$(x)(x+2) \qquad [3]$$

STEP 2: Find the polynomial named by [3] and subtract this from $x^2 + 5x + 6$. That is, $x(x+2) = x^2 + 2x$, and $x^2 + 5x + 6 - (x^2 + 2x) = 3x + 6$. Since subtraction is the inverse of addition, you have shown that

$$x^2 + 5x + 6 = x(x+2) + 3x + 6 \qquad [4]$$

In other words, $x(x+2) = x^2 + 2x$, so at least we have the first term of $x^2 + 5x + 6$ correct. In fact, we must add $3x + 6$ to $x^2 + 2x$ to get $x^2 + 5x + 6$, which we're showing in Step 2.

STEP 3: Look at the right side of [4]; it is the sum of three terms, the first of which, $x(x+2)$, already has $x + 2$ as a factor. So you go to work on the other two terms, $3x$ and 6, proceeding as in Step 1. That is, you want to get $3x$ and you know that you must multiply x by 3 to do this. You write $3(x+2)$.

STEP 4: In this case, $3(x+2) = 3x + 6$, so there's no more to do. You just go back to [4] in Step 2 and rewrite $3x + 6$ as $3(x+2)$ to obtain

$$x^2 + 5x + 6 = x(x+2) + 3(x+2) \qquad [5]$$

STEP 5: Now you use your knowledge of factoring by observing that $x + 2$ is a common factor on the right side of [5]. You factor it to obtain

$$x^2 + 5x + 6 = (x+2)(x+3)$$

We've filled in the parentheses with $x + 3$.

Example 1

Express $(x^2 + 7x + 10) \div (x + 5)$ as a polynomial in x.

The quotient is $(x + 2)$.

1. You multiply $x + 5$ by x to get the x^2 term.
2. But $x(x+5) = x^2 + 5x$, and you want the product to be $x^2 + 7x + 10$. So you subtract $x^2 + 5x$ from $x^2 + 7x + 10$ to get $2x + 10$.
3. You want to get $x + 5$ as a factor of $2x + 10$, so you write $x + 5$ and then multiply it by 2 to get the required $2x$. In fact, $2(x+5) = 2x + 10$, which is exactly what you want.
4. In summary, you've shown that

$$\begin{aligned} x^2 + 7x + 10 &= x^2 + 5x + 2x + 10 \\ &= x(x+5) + 2(x+5) \end{aligned} \qquad [6]$$

That is,

$$x^2 + 7x + 10 = x(x+5) + 2x + 10$$

That is,

$$\begin{aligned} x^2 + 7x + 10 &= x(x+5) + 2x + 10 \\ &= x(x+5) + 2(x+5) \end{aligned}$$

5. $(x+5)$ is a common factor of the two terms on the right side of [6], so you factor $x + 5$ from the right side of [6] to obtain $x^2 + 7x + 10 = (x+5)(x+2)$.

6. Hence

$$\frac{x^2 + 7x + 10}{x + 5} = \frac{(x+5)(x+2)}{(x+5)}$$

and you cancel the common factor $x + 5$ to obtain $x + 2$ as the answer.

Example 2

Express $(6x^2 + 11x + 4) \div (2x + 1)$ as a polynomial in x.

The quotient is $3x + 4$.
   Remember, you want to start with $2x + 1$ and find what you have to multiply it by to get $6x^2 + 11x + 4$.

1. Write $2x + 1$, and ask, "What must I multiply $2x + 1$ by to make the term $6x^2$ appear?"

   Answer: $3x$

2. However, $3x(2x + 1) = 6x^2 + 3x$, and you want the answer to be $6x^2 + 11x + 4$. So now you ask the question, "What must I add to $6x^2 + 3x$ to get $6x^2 + 11x + 4$?" So far, you've shown that

   $$6x^2 + 11x + 4 = 3x(2x + 1) + 8x + 4 \qquad [7]$$

   Answer: $8x + 4$. We subtract $6x^2 + 3x$ from $6x^2 + 11x + 4$ to get $8x + 4$.

   To finish the problem, you want to write $8x + 4$ in [7] so that $2x + 1$ is a factor.

3. Write $2x + 1$ and ask, "What must I multiply $2x + 1$ by to make the term $8x$ appear?"

   Answer: 4

4. Now $4(2x + 1) = 8x + 4$. What must you add to $8x + 4$ to get $8x + 4$? The answer is nothing, so you have finished. You have shown

   $$6x^2 + 11x + 4 = 3x(2x + 1) + 4(2x + 1) \qquad [8]$$

5. You remove $2x + 1$ as a common factor on the right side of [8] to obtain

   $$6x^2 + 11x + 4 = (2x + 1)(3x + 4) \qquad [9]$$

   Remember that we can always multiply $3x + 4$ by $2x + 1$ to check that we do get $6x^2 + 11x + 4$ as the product.

   If you now divide $6x^2 + 11x + 4$ by $2x + 1$, in effect you cancel $2x + 1$ from the right side of [9]. You get $3x + 4$ as the answer.

   In the next problems, you will see that just as with ordinary numerical division, you may get a remainder.

   Let's divide $12x^2 + 10x + 5$ by $3x + 1$ and see what happens. We begin by observing that we have to multiply $3x + 1$ by $4x$ to get the $12x^2$ term. Then we have to find what we need to add to $4x(3x + 1)$ to get $12x^2 + 10x + 5$. The answer is $6x + 5$.
   So far, we have

   $$12x^2 + 10x + 5 = 4x(3x + 1) + 6x + 5 \qquad [10]$$

   $4x(3x + 1) = 12x^2 + 4x$

   That is,
   $$12x^2 + 10x + 5 - (12x^2 + 4x)$$
   $$= 12x^2 + 10x + 5 + {}^-12x^2 + {}^-4x$$
   $$= 12x^2 + {}^-12x^2 + \underbrace{10x + {}^-4x} + 5$$
   $$= 6x + 5$$

   Next, we want to rewrite the right side of [10] so that $3x + 1$ is a factor of $6x + 5$. We multiply $3x + 1$ by 2 to get the $6x$ term, and we see that $2(3x + 1) = 6x + 2$, not $6x + 5$. However, we subtract $6x + 2$ from $6x + 5$, and we get 3. This means that

   $$6x + 5 = 2(3x + 1) + 3$$

   CHECK
   $$2(3x + 1) + 3$$
   $$= 6x + 2 + 3 = 6x + 5$$

We may rewrite [10] as

$$12x^2 + 10x + 5 = 4x(3x + 1) + 2(3x + 1) + 3 \qquad [11]$$

   The right side of [11] has $3x + 1$ as a factor in the first two terms. The third term is the number 3, which has a lower degree than $3x + 1$. We cannot divide any further and still get a polynomial as a quotient. But we can factor $3x + 1$ from the first two terms on the right side of [11] and leave the answer as

   Remember that the degree of a non-zero constant is 0 and the degree of $3x + 1$ is 1.

Section 12.3   Dividing Polynomials by Factoring   315

$$12x^2 + 10x + 5 = 4x(3x + 1) + 2(3x + 1) + 3$$

$$= (3x + 1)(4x + 2) + 3$$

We say that $3x + 1$ "goes into" $12x^2 + 10x + 5$, $(4x + 2)$ times *with a remainder of 3*.

The basic method for dividing polynomials remains the same when the polynomials become more complicated — that is, even if the polynomials have higher degrees.

CHECK

$$(3x + 1)(4x + 2) + 3$$
$$= 12x^2 + 4x + 6x + 2 + 3$$
$$= 12x^2 + 10x + 5$$

Example 3

Express $(2x^3 + 5x^2 + 7x + 4) \div (x + 1)$ as a polynomial in x.

The quotient is $2x^2 + 3x + 4$.

First you write $x + 1$ and multiply this by $2x^2$ to get the $2x^3$ term. But $2x^2(x + 1) = 2x^3 + 2x^2$. You subtract this from $2x^3 + 5x^2 + 7x + 4$ to see that you have $3x^2 + 7x + 4$ left. That is,

$$2x^3 + 5x^2 + 7x + 4 = 2x^2(x + 1) + 3x^2 + 7x + 4$$

Now you do the same thing to $3x^2 + 7x + 4$. You multiply $x + 1$ by $3x$ to get the $3x^2$ term. But $3x(x + 1) = 3x^2 + 3x$. You subtract this from $3x^2 + 7x + 4$ to see that you have $4x + 4$ left. That is,

$$2x^3 + 5x^2 + 7x + 4 = 2x^2(x + 1) + 3x(x + 1) + 4x + 4$$
$$= 2x^2(x + 1) + 3x(x + 1) + 4(x + 1) \quad [12]$$

All this work is just to rewrite $2x^3 + 5x^2 + 7x + 4$ in a way that has $x + 1$ as a factor.

The right side of [12] now contains $x + 1$ as a factor common to each term. You may factor to obtain

$$2x^3 + 5x^2 + 7x + 4 = (x + 1)(2x^2 + 3x + 4) \quad [13]$$

To divide by $x + 1$, you cancel this factor from the right side of [13] to get the correct answer.

If we look at [13], we can see that $(2x^3 + 5x^2 + 7x + 4) \div (2x^2 + 3x + 4)$ is $x + 1$. There is no reason why the method of this section can't be used for divisors whose degree is more than 1.

As a check, you can show that

$$(x + 1)(2x^2 + 3x + 4)$$
$$= 2x^3 + 5x^2 + 7x + 4$$

Example 4

Show that $2x^3 + 5x^2 + 7x + 4 \div 2x^2 + 3x + 4$ is $x + 1$.

Begin by writing $2x^2 + 3x + 4$ and multiplying this by x to get the $2x^3$ term. But $x(2x^2 + 3x + 4) = 2x^3 + 3x^2 + 4x$. You can use this information to rewrite $2x^3 + 5x^2 + 7x + 4$ as $2x^3 + 3x^2 + 4x + 2x^2 + 3x + 4$. That is,

$$2x^3 + 5x^2 + 7x + 4 = x(2x^2 + 3x + 4) + 2x^2 + 3x + 4$$

That is,

$$5x^2 = 3x^2 + 2x^2$$
$$7x = 4x + 3x$$

In turn, this may be rewritten as

$$x(2x^2 + 3x + 4) + 1(2x^2 + 3x + 4)$$

any you may now remove $2x^2 + 3x + 4$ as a common factor to obtain

$$(2x^2 + 3x + 4)(x + 1)$$

This is the same result that we obtained in [13], but we've done the problem by dividing by $2x^2 + 3x + 4$ rather than by $x + 1$. This method becomes awkward because there are so many terms to keep track of. In the next section, we study a more compact approach.

PRACTICE DRILL

Use the method of this section to write each of the following as a single polynomial.
1. $x^2 + 2x + 1 \div x + 1$
2. $x^2 + 16x + 48 \div x + 4$
3. $x^3 - 3x^2 + 3x - 1 \div x - 1$
4. $x^3 + 3x^2 + 3x + 1 \div x^2 + 2x + 1$

Answers
1. $x + 1$ 2. $x + 12$
3. $x^2 - 2x + 1$ 4. $x + 1$

CHECK THE MAIN IDEAS

1. In finding the quotient $2{,}821 \div 13$, you try to find the factor that when multiplied by _____ gives 2,821.
2. Similarly, you may look at $x^2 + 4x + 3 \div x + 1$ as trying to find the factor you must multiply _____ by to get $x^2 + 4x + 3$ as the product.
3. One method is to write the divisor $x + 1$ and multiply it by _____ to get the x^2 term in the product.
4. Since $x(x + 1) = x^2 + x$, you may rewrite the product as $x^2 + 4x + 3 = x^2 + x + 3x + 3$, or $x(x + 1) + $ _____.
5. You may factor $3x + 3$ and rewrite it as $3($ _____$)$.
6. Hence, $x^2 + 4x + 3 = x(x + 1) + 3(x + 1)$, and you may now factor $x + 1$ on the right side of the equation to get $x^2 + 4x + 3 = (x + 1)($ _____$)$.
7. Therefore, $x^2 + 4x + 3 \div x + 1$ is the same as $(x + 1)(x + 3) \div (x + 1)$, which is the same as _____.

Answers
1. 13   2. $x + 1$ 3. x
4. $3x + 3$ 5. $x + 1$
6. $x + 3$ 7. $x + 3$

Section 12.4
The Division Algorithm

We first learned to divide whole numbers as a form of rapid subtraction. Once we understood how to do this, we learned a convenient "recipe" for doing this. We shall now do the same thing for polynomials.

In the last section, we learned the idea behind how we can divide any polynomial by any other non-zero polynomial. In this section, we shall use the last section to derive a procedure that is more convenient.

Let's return to the problem $(x^2 + 5x + 6) \div (x + 2)$, which we discussed in the last section. We shall copy the form that we use for ordinary numerical division. We make the division form with $x + 2$ on the outside and $x^2 + 5x + 6$ on the inside.

$$x + 2 \overline{\smash{)}x^2 + 5x + 6}$$

OBJECTIVE
To be able to find the quotient of two polynomials in a more compact way.

The precise mathematical term for "recipe" is **algorithm**.

STEP 1: Divide the first term in the dividend by the first term in the divisor. Colloquially, you say, "x goes into x^2, x times." Write the quotient above the first term in the dividend.

$$\begin{array}{r} x \phantom{{}+5x+6} \\ x + 2 \overline{\smash{)}x^2 + 5x + 6} \end{array}$$

In this method, it is crucial that all polynomials be in decreasing order of degree (just as in ordinary place value, the denominations decrease by a factor of 10 as we move from left to right).

STEP 2: Multiply the quotient you have so far (which is just x) by the divisor $(x + 2)$ and subtract this result from the dividend.

$$\begin{array}{r} x \phantom{{}+5x+6} \\ x + 2 \overline{\smash{)}x^2 + 5x + 6} \\ \underline{x^2 + 2x} \\ 3x + 6 \end{array}$$

This is exactly what we were doing when we asked what we must multiply $x + 2$ by to get x^2.

This is just a compact way of doing what we did before. That is, we looked at $x(x + 2)$ and subtracted this from $x^2 + 5x + 6$ to see what was left.

Remember that when you subtract, you change the sign of each coefficient and add.

Section 12.4 The Division Algorithm

STEP 3: You begin again with Step 1, but now start with $3x + 6$ in the dividend. That is, take the first term of the divisor (which is still x) and divide it into the first term of the new dividend, which is $3x$. The quotient is 3, and you write it in the quotient above the term $5x$.

Now we're finding out how to rewrite $3x + 6$ so that $x + 2$ is a factor.

$$\begin{array}{r} x + 3 \\ x + 2 \overline{\smash{)}x^2 + 5x + 6} \\ \underline{x^2 + 2x } \\ 3x + 6 \end{array}$$

STEP 4: Now repeat Step 2, but multiply $x + 2$ by the 3. This gives you $3x + 6$, and you subtract this from the new dividend to obtain

$$\begin{array}{r} x + 3 \\ x + 2 \overline{\smash{)}x^2 + 5x + 6} \\ \underline{x^2 + 2x } \\ 3x + 6 \\ \underline{3x + 6} \\ 0 \end{array}$$

This is exactly the same way we solved the problem in the last section, but we have used a lot less space. All that must be written is the work in Step 4. The first three steps are shown so that you can see how to get the final form.

To check,

$$\begin{array}{r} (x + 3)(x + 2) = x^2 + 3x \\ + 2x + 6 \\ \hline x^2 + 5x + 6 \end{array}$$

Since this way seems so much more convenient than the method of the last section, why did we use the more complicated method? The last section explains in a step by step way why *we do what we do. Once we know* why, *this section gives us a quicker way of seeing the "how" of the method.*

To practice the new form, let's redo each problem that we did in the examples of the last section. By repeating the same problems, we shall gain a better feeling that this method is just a more compact form of the method we used in the last section.

Example 1

Perform the division $x + 5\overline{\smash{)}x^2 + 7x + 10}$.

$$\begin{array}{r} x + 2 \\ x + 5 \overline{\smash{)}x^2 + 7x + 10} \\ \underline{x^2 + 5x } \\ 2x + 10 \\ \underline{2x + 10} \\ 0 \end{array}$$

STEP 1: We divide x^2 by x to get x.
STEP 2: We multiply x by $x + 5$ to get $x^2 + 5x$.
STEP 3: We subtract $x(x + 5)$ from the dividend to form a new dividend.

Then we repeat the same steps with the new dividend.

That is,

STEP 1: $$x + 5 \overline{\smash{)}x^2 + 7x + 10}^{x}$$

STEP 2: $$\begin{array}{r} x \\ x + 5 \overline{\smash{)}x^2 + 7x + 10} \\ x^2 + 5x \end{array}$$

STEP 3: $$\begin{array}{r} x \\ x + 5 \overline{\smash{)}x^2 + 7x + 10} \\ \underline{x^2 + 5x } \\ 2x + 10 \end{array}$$

These steps match up with the ones we used when we solved this problem as Example 1 of the previous section.

STEP 4: $$\begin{array}{r} x + 2 \\ x + 5 \overline{\smash{)}x^2 + 7x + 10} \\ \underline{x^2 + 5x } \\ 2x + 10 \end{array}$$

317

STEP 5:
$$\begin{array}{r} x+2 \\ x+5{\overline{\smash{\big)}\,x^2+7x+10}} \\ \underline{x^2+5x} \\ 2x+10 \\ \underline{2x+10} \\ 0 \end{array}$$

Example 2

Perform the division $2x+1\overline{)6x^2+11x+4}$.

$$\begin{array}{r} 3x+4 \\ 2x+1{\overline{\smash{\big)}\,6x^2+11x+4}} \\ \underline{6x^2+3x} \\ 8x+4 \\ \underline{8x+4} \\ 0 \end{array}$$

STEP 1: Divide the first term of the dividend by the first term of the divisor, and write this quotient above the first term in the divisor.

$$\begin{array}{r} 3x \\ 2x+1{\overline{\smash{\big)}\,6x^2+11x+4}} \end{array}$$

STEP 2: Multiply this first term of the quotient by the divisor and write the product under the corresponding (like) terms of the dividend.

$$\begin{array}{r} 3x \\ 2x+1{\overline{\smash{\big)}\,6x^2+11x+4}} \\ 6x^2+3x \end{array}$$

STEP 3: Perform the subtraction to obtain the new dividend.

$$\begin{array}{r} 3x \\ 2x+1{\overline{\smash{\big)}\,6x^2+11x+4}} \\ \underline{6x^2+3x} \\ 8x+4 \end{array}$$

STEP 4: Repeat the previous steps, using $8x+4$ as the new dividend.

$$\begin{array}{r} 3x+4 \\ 2x+1{\overline{\smash{\big)}\,6x^2+11x+4}} \\ \underline{6x^2+3x} \\ 8x+4 \end{array}$$

$$\begin{array}{r} 3x+4 \\ 2x+1{\overline{\smash{\big)}\,6x^2+11x+4}} \\ \underline{6x^2+3x} \\ 8x+4 \\ \underline{8x+4} \\ 0 \end{array}$$

Example 3

Perform the division $3x+1\overline{)12x^2+10x+5}$.

STEP 1:
$$\begin{array}{r} 4x \\ 3x+1{\overline{\smash{\big)}\,12x^2+10x+5}} \\ 12x^2+4x \end{array}$$

Section 12.4   The Division Algorithm

STEP 2:
$$\begin{array}{r} 4x \\ 3x+1\overline{)12x^2+10x+5} \\ \underline{12x^2+4x} \\ 6x+5 \end{array}$$

STEP 3:
$$\begin{array}{r} 4x\ +\ 2 \\ 3x+1\overline{)12x^2+10x+5} \\ \underline{12x^2+4x} \\ 6x+5 \end{array}$$

STEP 4:
$$\begin{array}{r} 4x\ +\ 2 \\ 3x+1\overline{)12x^2+10x+5} \\ \underline{12x^2+4x} \\ 6x+5 \\ \underline{6x+2} \\ 3 \end{array}$$

This ends the problem, because if we divide the new dividend (3) by $3x$, we no longer get a monomial in x (the exponent will be negative). That is, just as the quotient of two integers need not be an integer, the quotient of two polynomials in x need not be a polynomial in x.

Step 4 represents the solution, which you may write as:

$$\begin{array}{r} 4x\ +\ 2\quad R=3 \\ 3x+1\overline{)12x^2+10x+5} \\ \underline{12x^2+4x} \\ 6x+5 \\ \underline{6x+2} \\ 3 \end{array}$$

Example 4

Perform the division $x+1\overline{)2x^3+5x^2+7x+4}$.

$$\begin{array}{r} 2x^2+3x\ +4 \\ x+1\overline{)2x^3+5x^2+7x+4} \\ \underline{2x^3+2x^2} \\ 3x^2+7x+4 \\ \underline{3x^2+3x} \\ 4x+4 \\ \underline{4x+4} \\ 0 \end{array}$$

Sometimes we just write $3x^2+7x$ rather than $3x^2+7x+4$. That is, we bring down the terms one at a time, as in regular numerical division.

This is exactly what we did in the other problems, except that the process had to be repeated one more time. Let's look at this problem in more detail.

STEP 1: Divide the first term in the dividend by the first term in the divisor and write this quotient above the first term in the dividend.

$$\begin{array}{r} 2x^2 \\ x+1\overline{)2x^3+5x^2+7x+4} \end{array}$$

STEP 2: Multiply this term of the quotient by the divisor and subtract the result from the dividend.

$$\begin{array}{r} 2x^2 \\ x+1\overline{)2x^3+5x^2+7x+4} \\ \underline{2x^3+2x^2} \\ 3x^2+7x+4 \end{array}$$

STEP 3: Repeat Steps 1 and 2 with $3x^2 + 7x + 4$ as the dividend.

$$\begin{array}{r} 2x^2 + 3x \\ x+1\overline{\smash{)}2x^3 + 5x^2 + 7x + 4} \\ \underline{2x^3 + 2x^2} \\ 3x^2 + 7x + 4 \\ \underline{3x^2 + 3x} \\ 4x + 4 \end{array}$$

STEP 4: Repeat Steps 1 and 2 with $4x + 4$ as the dividend.

$$\begin{array}{r} 2x^2 + 3x \;\; + 4 \\ x+1\overline{\smash{)}2x^3 + 5x^2 + 7x + 4} \\ \underline{2x^3 + 2x^2} \\ 3x^2 + 7x + 4 \\ \underline{3x^2 + 3x} \\ 4x + 4 \\ \underline{4x + 4} \\ 0 \end{array}$$

Notice that Step 4 is what we see if we are doing the problem without explaining the steps.

Up till now, our dividends have not missed any powers of x. Should a power be missing, we can always put it in with a 0 coefficient. For example, $3x^2 + 7$ is equivalent to $3x^2 + 0x + 7$. The idea of using 0 as a coefficient is sometimes helpful.

Example 5

Perform the division $x + 1\overline{\smash{)}x^3 + 1}$.

Before starting this problem, rewrite $x^3 + 1$ as $x^3 + 0x^2 + 0x + 1$. This will help to keep track of the necessary terms.

$$\begin{array}{r} x^2 \\ x+1\overline{\smash{)}x^3 + 0x^2 + 0x + 1} \\ \underline{x^3 + 1x^2} \\ -\;x^2 + 0x + 1 \end{array}$$

Then divide $-x^2$ by x to get $-1x$ or ^-x, and continue as before.

$$\begin{array}{r} x^2 + {}^-x \\ x+1\overline{\smash{)}x^3 + 0x^2 + 0x + 1} \\ \underline{x^3 + 1x^2} \\ -1x^2 + 0x + 1 \\ \underline{-1x^2 - 1x} \\ x + 1 \end{array}$$

By putting $0x^2$ in the dividend, we save a place for the x^2 in $x^3 + x^2$. This makes it easier for us not to put the term x^2 in the wrong place.

Remember that $-x = -1x$, $x = 1x$, and that when we subtract, we add the opposite.

Now complete the problem as follows:

$$\begin{array}{r} x^2 + {}^-x \;\; + 1 \\ x+1\overline{\smash{)}x^3 + 0x^2 + 0x + 1} \\ \underline{x^3 + 1x^2} \\ -1x^2 + 0x + 1 \\ \underline{-1x^2 - 1x} \\ x + 1 \\ \underline{x + 1} \\ 0 \end{array}$$

Recall that $x^2 + {}^-x + 1$ means the same as $x^2 - x + 1$.

CHECK

$(x + 1)(x^2 - x + 1)$
$= x(x^2 - x + 1) + 1(x^2 - x + 1)$
$= x^3 - x^2 + x + x^2 - x + 1$
$= x^3 \underbrace{- x^2 + x^2}_{0} + \underbrace{x - x}_{0} + 1$
$= x^3 + 1$

In other words, $x^3 + 1$ may be written as $(x + 1)(x^2 - x + 1)$. We can check the result by multiplying the polynomials $x + 1$ and $x^2 - x + 1$.

When we divide two polynomials, there is a good chance that there will be a

Section 12.4   The Division Algorithm   321

non-zero remainder. Sometimes the remainder can be more helpful than the other part of the quotient. Let's examine one such situation.

Example 6

Evaluate $x^4 - 6x^3 + 2x^2 - 9x - 1$ when $x = 5$.

The value is -121.

The obvious method is to replace x by 5 in the expression to obtain

$$5^4 - 6(5^3) + 2(5^2) - 9(5) - 1$$

or $\quad 625 - 750 + 50 - 45 - 1 = -121$

However, notice that when $x = 5$, $x - 5 = 0$, and 0 times any number is still 0. If you do long division, you can simply divide $x^4 - 6x^3 + 2x^2 - 9x - 1$ by $x - 5$.

As shown in the margin, the remainder is -121, which is the correct answer. You do not have to compute 5^4, and $6(5^3)$, and so on. You have shown by long division that

$$x^4 - 6x^3 + 2x^2 - 9x - 1 = (x - 5)(x^3 - x^2 - 3x - 24) - 121$$

When you replace x by 5, the right side of the equation is simply -121, since $x - 5 = 0$ and $0(x^3 - x^2 - 3x - 24) = 0$.

The greater the degree of the polynomial, the more work you save by the division.

This completes our study of the arithmetic of polynomials. In the next unit, we'll see how this arithmetic can be used to solve certain algebraic equations.

> This also happens in whole-number arithmetic. For example, if we want to see whether a number is even or odd, we divide it by two. The *remainder* tells us the answer. If the remainder is 1, the number is odd; and if the remainder is 0, the number is even.

$$\begin{array}{r} x^3 - x^2 - 3x - 24 \\ x - 5 \overline{)x^4 - 6x^3 + 2x^2 - 9x - 1} \\ \underline{x^4 - 5x^3} \\ -x^3 + 2x^2 \\ \underline{-x^3 + 5x^2} \\ -3x^2 - 9x \\ \underline{-3x^2 + 15x} \\ -24x - 1 \\ \underline{-24x + 120} \\ -121 \end{array}$$

PRACTICE DRILL

Perform the indicated long division and show the remainder, if any.

1. $x + 1 \overline{)x^2 + 2x + 1}$ 2. $x^2 + 2x + 1 \overline{)6x^3 + 12x^2 + 6x}$
3. $x^2 + 2x + 1 \overline{)6x^3 + 13x^2 + 8x + 1}$ 4. $x + 1 \overline{)x^2 + 5x + 6}$
5. $x^2 + 2x + 1 \overline{)6x^3 + 13x^2 + 9x + 3}$

Answers
1. $x + 1$ 2. $6x$ 3. $6x + 1$
4. $x + 4$ with a remainder of 2
5. $6x + 1$ with a remainder of $x + 2$

CHECK THE MAIN IDEAS

1. In the last section, you found a method for determining the expression that had to be multiplied by one given polynomial to find another given _____. This method is called division of polynomials.
2. This is a cumbersome method, because it takes lots of space. In this section, you have found a more convenient method for solving the _____ problem.
3. This is very similar to long division of numbers. When you write

$$\begin{array}{r} 217 \\ 13 \overline{)2{,}821} \\ \underline{2{,}600} \\ 221 \\ \underline{130} \\ 91 \\ \underline{91} \end{array}$$

this is a good way of showing that there are _____ 13's in 2,821.

Answers
1. polynomial   2. same
3. 217

EXERCISE SET 12 (Form A)

Section 12.1

1. Write $\dfrac{8x^4 + 4x^3}{x^2}$ as a polynomial in x.
2. Write $x^{-2}(8x^4 + 4x^3)$ as a polynomial in x.
3. Write $\dfrac{8x^4 + 12x^3 + 6x^2}{2x}$ as a polynomial in x.
4. For what value(s) of x are $\dfrac{8x^4 + 12x^3 + 6x^2}{2x}$ and $4x^3 + 6x^2 + 3x$ *not* equal?
5. Evaluate $\dfrac{8x^5 + 16x^6}{8x^5}$ when $x = 2.37$.

Section 12.2

Factor out all common factors.

6. $5x + 5y$
7. $5x + 5$
8. $5(c + d) + 7(c + d)$
9. $m(c + d) + n(c + d)$
10. $6m(2m + n) + 7n(2m + n)$
11. $6m(2m + n) + 8n(2m + n)$
12. (a) $6c + 3cd$ (b) $2cd + cd^2$ (c) $(6c + 3cd) + (2cd + cd^2) = 6c + 5cd + cd^2$
13. Factor $40m^2n^3p + 24m^3n^2p$.
14. Evaluate $\dfrac{40m^2n^3p + 24m^3n^2p}{8m^2n^2p}$ when $m = 5.7$, $n = 3.2$, and $p = 4.1$.

Sections 12.3 and 12.4

Perform the indicated divisions by the method of Section 12.3 or the method of Section 12.4.

15. $(x^2 + 9x + 14) \div (x + 7)$
16. $(x^2 - 11x - 26) \div (x + 2)$
17. $(x^2 + 2x - 8) \div (x - 2)$
18. $(x^2 - 9) \div (x - 2)$
19. $(x^3 - 8) \div (x - 2)$
20. $(2x^3 + 3x^2 + 2x + 3) \div (x^2 + 1)$
21. $(3x^3 - 4x^2 - 7x - 2) \div (x^2 - 2x - 1)$
22. (a) $(6x^2 + x - 10) \div (2x + 3)$ (b) Simplify $(6x^2 + x - 10) - (2x + 3)(3x - 4)$
23. $(x^2 + 9x + 7) \div (x + 14)$
24. Use the previous problem to rewrite $\dfrac{x^2 + 9x + 7}{x + 14}$ in a form such that x appears only to the first power.
25. Check the answer in the previous problem by letting $x = 63$ in both expressions.
26. Use division to find the remainder when $2x^5 - 6x^4 + 5x^3 - 9x^2 - 17x + 4$ is divided by $x - 3$.
27. Use the result of the previous problem to find the value of c if
$$2x^5 - 6x^4 + 5x^3 - 9x^2 - 17x + 4 = (x - 3)(2x^4 + 5x^2 + 6x + 1) + c$$
28. Use the result of the previous problem to evaluate
$$2x^5 - 6x^4 + 5x^3 - 9x^2 - 17x + 4$$
when $x = 3$.
29. What is the remainder when $x^2 + 4$ is divided by $x + 2$?

Answers: Exercise Set 12
Form A

1. $8x^2 + 4x$ 2. $8x^2 + 4x$
3. $4x^3 + 6x^2 + 3x$ 4. $x = 0$
5. 5.74 6. $5(x + y)$
7. $5(x + 1)$ 8. $12(c + d)$
9. $(m + n)(c + d)$
10. $(6m + 7n)(2m + n)$
11. $2(3m + 4n)(2m + n)$
12. (a) $3c(2 + d)$ (b) $cd(2 + d)$
(c) $c(3 + d)(2 + d)$
13. $8m^2n^2p(5n + 3m)$
14. 33.1 15. $x + 2$
16. $x - 13$ 17. $x + 4$
18. $x + 3$ 19. $x^2 + 2x + 4$
20. $2x + 3$ 21. $3x + 2$
22. (a) $3x - 4$, $R = 2$ (b) 2
23. $x - 5$, $R = 77$
24. $x - 5 + \dfrac{77}{x + 14}$
25. both expressions equal 59
26. 7 27. 7 28. 7
29. 8

EXERCISE SET 12 (Form B)

Section 12.1

1. Write $\dfrac{5x^4 + 3x^5}{x^3}$ as a polynomial in x.

2. Write $x^{-3}(5x^4 + 3x^5)$ as a polynomial in x.

3. Write $\dfrac{6x^5 + 9x^4 + 12x^2}{3x}$ as a polynomial in x.

4. For what value(s) of x are $\dfrac{6x^5 + 9x^4 + 12x^2}{3x}$ and $2x^4 + 3x^3 + 4x$ *not* equal?

5. Evaluate $\dfrac{7x^6 + 14x^7}{7x^6}$ when $x = 4.5$.

Section 12.2

Factor out all common factors.

6. $8x + 8y$ 7. $8x + 8$ 8. $9(m + n) + 8(m + n)$
9. $c(m + n) + d(m + n)$ 10. $8r(3x + 2y) + 7s(2x + 3y)$
11. $8r(3x + 2y) + 8s(2x + 3y)$
12. (a) $9c + 3cd$ (b) $6cd + 2cd^2$ (c) $9c + 3cd + 6cd + 2cd^2$
13. $48m^3n^5p + 60m^4n^4p$
14. Evaluate $\dfrac{48m^3n^5p + 60m^4n^4p}{12m^3n^4p}$ when $n = 2.5$ and $m = 5.2$.

Sections 12.3 and 12.4

Perform the indicated division by the method of Section 12.3 or the method of Section 12.4.

15. $(x^2 + 11x + 24) \div (x + 8)$ 16. $(x^2 - 7x - 18) \div (x + 2)$
17. $(x^2 + 5x - 14) \div (x - 2)$ 18. $(x^2 - 16) \div (x - 4)$
19. $(x^3 - 1) \div (x - 1)$ 20. $(2x^3 + 5x^2 + 4x + 10) \div (x^2 + 2)$
21. $(3x^3 - 11x^2 + 12x - 4) \div (x^2 - 3x + 2)$
22. (a) $(12x^2 + 34x + 20) \div (3x + 7)$
    (b) Simplify $(12x^2 + 34x + 20) - (3x + 7)(4x + 2)$.
23. $(x^2 + 9x + 9) \div (x + 12)$
24. Use the previous problem to rewrite $\dfrac{x^2 + 9x + 9}{x + 12}$ in a form such that x appears only to the first power.
25. Check the answer in the previous problem by letting $x = 33$ in both expressions.
26. Use division to find the remainder when $2x^5 - 8x^4 + 3x^3 - 12x^2 + 7x + 2$ is divided by $x - 4$.
27. Use the result of the previous problem to find the value of c if
$$2x^5 - 8x^4 - 3x^3 - 12x^2 + 7x + 2 = (x - 4)(2x^4 + 3x^2 + 7) + c$$
28. Use the result of the previous problem to evaluate
$$2x^5 - 8x^4 + 3x^3 - 12x^2 + 7x + 2$$
when $x = 4$.
29. What is the remainder when $x^2 + 9$ is divided by $x + 3$?

More on Nonlinear Equations

unit 5

Lesson 13

An Introduction to Quadratic Equations

Overview There is an old story of a father who asked his daughter to explain to him Einstein's Theory of Relativity in simple terms. The daughter thought for a while and said, "If you're in the company of a handsome man for an hour, it seems like only a minute. But if you sit on a hot stove for a minute, it seems like an hour!" The father looked up in amazement and exclaimed, "And for this Einstein made a living!"

When we try to simplify, we run the risk of oversimplifying. The next statement is possibly an oversimplification. But we may say that our course in elementary algebra has finished. The major objectives have been accomplished. We know what an equation is. We know how to solve a linear equation. If the equation is not linear, we know how we may use trial and error to approximate the solutions.

However, there are many other, nonlinear relationships that can be studied effectively in the same detail that we studied the linear equation. Most of these equations belong properly to a second course in algebra, yet it seems appropriate to study one more equation in detail before this course ends.

This lesson is devoted to the quadratic equation. These equations involve second-degree polynomials, and they occur very often. Important in their own right, they also afford us an excellent review of many topics we've already studied.

Section 13.1
Using Arithmetic to Solve Equations

Early in this course, we discussed undoing equations.

Example 1

For what value of x does

$$3x + 4 = 19 \qquad [1]$$

OBJECTIVE
Given an equation like
$$x^3 + x = 7$$
to use arithmetic to help find values of x that are a solution of the equation.

$3x + 4 = 19$ when $x = 5$.
You undo the equation in order to isolate x.

$$\begin{array}{rl} 3x + 4 = & 19 \\ -4 & -4 \\ \hline 3x = & 15 \end{array} \qquad [2]$$

You solve [1] by dividing both sides of [2] by 3 to get $x = 5$.

If we did not know how to undo equations we could have solved Example 1 by the use of ordinary arithmetic. We would evaluate the expression $3x + 4$ for different values of x until we found one for which $3x + 4 = 19$. For example, we could make up a table as follows:

x	$3x$	$3x + 4$
1	3	7
2	6	10
3	9	13
4	12	16
→ 5	15	19 ←

We don't normally use trial and error for linear equations because the undoing method is quite simple. More importantly, the undoing method tells us that $x = 5$ is the only solution. With trial and error, we may miss some solutions.

The arithmetic method for solving equations may be used even if the equation is not linear.

The trial-and-error method is less attractive if the solution involves values of x that are not whole numbers.

Example 2

Find the value of x for which

$$3x + 4 = 15 \qquad [3]$$

$3x + 4 = 15$ when $x = \frac{11}{3} \left(= 3\frac{2}{3} \right)$. This is easy enough to get by the undoing method. You proceed just as in Example 1.

$$\begin{array}{rl} 3x + 4 = & 15 \\ -4 & -4 \\ \hline 3x = & 11 \end{array}$$

So $x = \frac{11}{3}$.

CHECK
$$3\left(\frac{11}{3}\right) + 4 = 11 + 4 = 15$$

If you prefer to do the work horizontally, you can write
$$(3x + 4) + {}^-4 = 15 + {}^-4$$
$$3x + (4 + {}^-4) = 11$$
$$3x + 0 = 11$$
$$3x = 11$$

In the table following Example 1, we see that when $x = 3$, $3x + 4 = 13$, which is less than 15, and when $x = 4$, $3x + 4 = 16$, which is greater than 15. So we can assume that the answer lies between 3 and 4, and by trial and error we could

Section 13.1 **Using Arithmetic to Solve Equations**

eventually find that $x = 3\frac{2}{3}$. But we have already seen that there are many times when an important relationship is not linear, and in such cases we can still use the arithmetical trial-and-error method. No matter how complicated the equation becomes, we can always try to find solutions by careful guesswork.

In the linear case, we may even use the logic that since 15 is two-thirds of the way between 13 and 16, the value of x must be two-thirds of the way between 3 and 4. But we shall see that these ideas don't work well with nonlinear equations.

Example 3

By replacing x by 1, 2, 3, and so on, find a solution of the equation

$$x^2 + 3x + 4 = 44 \qquad [4]$$

One solution of [4] is $x = 5$. With a chart, you find

x	x^2	$3x$	$x^2 + 3x$	$x^2 + 3x + 4$
1	1	3	4	8
2	4	6	10	14
3	9	9	18	22
4	16	12	28	32
→ 5	25	15	40	44 ←
6	36	18	54	58

To try to solve [4] by the undoing method is harder than it may seem. We can begin by subtracting 4 from both sides of [4] to get

$$\begin{array}{r} x^2 + 3x + 4 = 44 \\ -4 \quad -44 \\ \hline x^2 + 3x \quad = 40 \end{array} \qquad [5]$$

But since x^2 and $3x$ have different degrees, we can't write the left side of [5] as a monomial.

Equation [4] is fairly easy, because as we see in the chart, we are able to find a solution using only whole-number values for x. This isn't always the case.

Example 4

Use the previous chart to show that the equation

$$x^2 + 3x + 4 = 50 \qquad [5]$$

has a solution between $x = 5$ and $x = 6$.

Using the last two lines of the chart, you have

x	x^2	$3x$	$x^2 + 3x$	$x^2 + 3x + 4$
5	25	15	40	44
→				50 ←
6	36	18	54	58

Because 50 is between 44 and 58, we assume that x is between 5 and 6.

If we want a better estimate, we can continue the same idea with values of x such as 5.1, 5.2, 5.3, and so on.

x	x^2	$3x$	$x^2 + 3x$	$x^2 + 3x + 4$
5	25	15	40	44
5.1	26.01	15.3	41.31	45.31
5.2	27.04	15.6	42.64	46.64
5.3	28.09	15.9	43.99	47.99
5.4	29.16	16.2	45.36	49.36 ←
→ 5.5	30.25	16.5	46.75	50.75

This new chart gives us enough information to conclude that a value of x for which $x^2 + 3x + 4 = 50$ lies between 5.4 and 5.5. Using the chart, we see that when $x = 5.4$, the expression is 49.36, which is less than 50; but when $x = 5.5$, the expression is 50.75, which is greater than 50.

If we want an answer correct to the nearest tenth, we can compute the value of $x^2 + 3x + 4$ when $x = 5.45$. This gives us:

x	x^2	$3x$	$x^2 + 3x$	$x^2 + 3x + 4$
5.4	29.16	16.2	45.36	49.36
→ 5.45	29.7025	16.35	46.0525	50.0525 ←
5.5	30.25	15.5	46.75	50.75

This chart shows that 5.4 is too small to be the right answer (we knew this from the previous chart) and that 5.45 is too big. The answer we want is between 5.4 and 5.45, and it is closer to 5.4 than it is to 5.5. To the nearest tenth, one solution of $x^2 + 3x + 4 = 50$ is 5.4.

The last few charts are basically the same but in this chart, we have to square numbers with two digits after the decimal point.

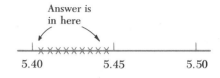

Example 5

By evaluating $x^3 + x$ when $x = 1$ and when $x = 2$, show that the equation $x^3 + x = 7$ has a solution between $x = 1$ and $x = 2$.

When $x = 1$, $x^3 + x = 1^3 + 1 = 1 + 1 = 2$, which is less than 7. When $x = 2$, $x^3 + x = 2^3 + 2 = 8 + 2 = 10$, which is greater than 7. If $x = 1$ is too small and $x = 2$ is too big, you know that a solution lies between $x = 1$ and $x = 2$.

Example 6

Use a chart to find a better estimate for the solution of $x^3 + x = 7$.

A value of x for which $x^3 + x = 7$ lies between 1.7 and 1.8.
The chart may be made as follows:

x	x^3	$x^3 + x$
1.0	1.000	2.000
1.1	1.331	2.431
1.2	1.728	2.928
1.3	2.197	3.497
1.4	2.744	4.144
1.5	3.375	4.875
1.6	4.096	5.696
1.7	4.913	6.613
→ 1.8	5.832	7.632 ← 7

From the chart, you see that $x = 1.7$ is too small to be correct (since 6.613 is less than 7) but 1.8 is too great to be the correct answer (since 7.632 is greater than 7). The number you want must lie between 1.7 and 1.8.

Do you see how helpful the hand-calculator is? It is very cumbersome to raise a number like 1.3 to the third power. We must compute

$$1.3 \times 1.3 \times 1.3$$

Example 7

Use the chart method to find a solution of the equation $x^3 + x = 7$ correct to the nearest tenth.

To the nearest tenth, a solution of $x^3 + x = 7$ is $x = 1.7$.

Section 13.1 Using Arithmetic to Solve Equations

The chart may look like

x	x^3	$x^3 + x$
1.70	4.913000	6.613000
1.75	5.359375	7.109375
1.80	5.832000	7.632000

Since 7 is between 6.613 and 7.109375, the value of x you want is between 1.7 and 1.75. It is closer to 1.7 than to 1.8.

We can continue to use the chart method to get better approximations. For example,

x	x^3	$x^3 + x$
1.70	4.913000	6.613000
1.71	5.000211	6.710211
1.72	5.088448	6.808448
1.73	5.177717	6.907717
1.74	5.268024	7.008024

This tells us that our answer is between 1.73 and 1.74. We could continue to use this same idea to find the answer to any degree of accuracy we desire, if we have enough patience. Modern calculators have made it much easier to find numerical approximations for solutions of many kinds of algebraic equations.

Our studies have become rather theoretical, so let's end this section with some problems that apply to the real world.

Example 8

A ball is propelled vertically upward with a speed of 160 feet per second. It reaches a height of h feet at the end of t seconds according to the formula

$$h = 160t - 16t^2 \qquad [7]$$

How high is the ball after 2 seconds?

We say "propelled" rather than "thrown" because 160 feet per second is almost 110 miles per hour — hardly a realistic velocity for throwing a ball vertically upward.

After two seconds, the ball is at a height of 256 feet.
All you have to do is replace t by 2 in [7]. You get

$$\begin{aligned} h &= 160(2) - 16(2)^2 \\ &= 320 - 16(4) \\ &= 320 - 64 \\ &= 256 \text{ (feet)} \end{aligned}$$

Example 9

Under the same conditions, how high is the ball after 3 seconds?

After 3 seconds, the ball reaches a height of 336 feet.
This time you replace t by 3 in [7] to obtain

$$\begin{aligned} h &= 160(3) - 16(3)^2 \\ &= 480 - 16(9) \\ &= 480 - 144 \\ &= 336 \text{ (feet)} \end{aligned}$$

Example 10

Use the results of the two previous examples to estimate a time at which the ball will be 300 feet high.

The ball will be at a height of 300 feet at some time during the third second. That is, after the second second, the ball is fewer than 300 feet high (256 feet); and after the third second, it's more than 300 feet high (336). It must reach a height of 300 feet some time during this period.

However, as powerful as the arithmetic process is, there are problems that arise. Some of these problems are presented in the next section.

> This may sound like a simple point, but it has a far-reaching application in higher mathematics. It is called continuity. If something is moving upward and goes from a height of 256 feet to 336 feet, it must pass through every height in between these two. Since 300 is between 256 and 336, at some time the ball must be at a height of 300 feet.

PRACTICE DRILL

1. Find the value of $x^2 + 3x$ when
   (a) $x = 3$ (b) $x = 4$
2. Use the answers in Problem 1 to estimate between what two consecutive whole numbers there is a solution of the equation $x^2 + 3x = 25$.
3. The height h of an object at the end of t seconds is given by $h = 100t - 16t^2$, where h is in feet. Find the height of the object after (a) 2 seconds and (b) 3 seconds.
4. Use the answers to Problem 3 to find an approximate time that the object is 150 feet high.

Answers

1. (a) 18 (b) 28
2. x is greater than 3 but less than 4
3. (a) 136 feet (b) 156 feet
4. between 2 and 3 seconds (closer to 3 seconds than to 2)

CHECK THE MAIN IDEAS

1. You can always solve a ─────── equation by the undoing method.
2. Given the equation $4x + 3 = 31$, you can ─────── 3 from both sides to obtain $4x = 28$.
3. Then you ─────── both sides by 4 to get $x = 7$. You may check your answer by observing that $4(7) + 3 = 31$.
4. If the equation is not linear, you may have to use (a) ─────── and (b) ───────.
5. Suppose you want to find a solution of $x^4 + x = 50$. If you replace x by 2, $x^4 + x =$ ───────, which is less than 50.
6. If you replace x by 3, $x^4 + x =$ ───────, which is greater than 50.
7. Since $x^4 + x$ is less than 50 when $x = 2$ but greater than 50 when $x = 3$, you know that one solution of the equation must lie between the consecutive whole numbers (a) ─────── and (b) ───────.
8. You can continue to use trial and error to get a better estimate of the solution. If you let $x = 2.5$, you find that $x^4 + x = 41.5625$, which is less than 50. This tells you that the answer is between 2.5 and 3. To the nearest whole number, the answer is ───────.
9. To get an even better estimate, you can find that when $x = 2.6$, $x^4 + x = 48.2976$, and when $x = 2.7$, $x^4 + x = 55.8441$. You may say a solution lies between the consecutive tenths (a) ─────── and (b) ───────.
10. As you look for more and more accuracy, the arithmetic becomes more and more complicated. The problem is not so bad as it used to be before the availability of ───────.

Answers

1. linear   2. subtract
3. divide   4. (a) trial
(b) error   5. 18   6. 84
7. (a) 2 (b) 3   8. 3
9. (a) 2.6 (b) 2.7
10. (hand-held) calculators

Section 13.2
Disadvantages of Arithmetical Solutions

OBJECTIVE
To show some flaws in using arithmetic methods to solve algebra problems.

In the last section we studied a problem in which a ball was propelled vertically upward at a speed of 160 feet per second. We saw that it reached a height of h feet after t seconds according to the formula

$$h = 160t - 16t^2 \qquad [1]$$

Example 1

According to [1], at what height is the ball after 2 seconds?

The ball is at a height of 256 feet. All you do is replace t by 2 and solve for h.

Can we be sure from Example 1 that if the ball is at a height of 256 feet, it has been in the air for only 2 seconds? Let's try the following example.

Is it possible that the ball is at the same height more than once?

Example 2

Use equation [1] to determine the height of the ball at the end of 8 seconds.

After 8 seconds, the ball is 256 feet high.
You replace t by 8 in [1] and solve the resulting equation for h. You get

$$\begin{aligned} h &= 160(8) - 16(8)^2 \\ &= 1{,}280 - 16(64) \\ &= 1{,}280 - 1{,}024 \\ &= 256 \text{ (feet)} \end{aligned}$$

Finding one answer to a problem does not guarantee that there are no other answers.

Examples 1 and 2 tell us that there are at least two times ($t = 2$ and $t = 8$) at which the ball reaches a height of 256 feet. If we replace h by 256 in [1], we can say that the equation

$$256 = 160t - 16t^2 \qquad [2]$$

has at least two solutions ($t = 2$ and $t = 8$).

It shouldn't be too hard to see why there are two different times at which the ball is at a height of 256 feet. It reaches that height once on the way up and once on the way down.

NOTE

Although it isn't important for our purposes, it is interesting to know that in this problem the ball stays in the air for 10 seconds and reaches its greatest height after 5 seconds. The following chart illustrates the point.

t	t^2	$160t$	$16t^2$	$160t - 16t^2 \; (= h)$
0	0	0	0	0
1	1	160	16	144
2	4	320	64	256
3	9	480	144	336
4	16	640	256	384
5	25	800	400	400
6	36	960	576	384
7	49	1,120	784	336
8	64	1,280	1,024	256
9	81	1,440	1,296	144
10	100	1,600	1,600	0

The ball reaches its greatest height (400 feet) after 5 seconds. It is at the same height — less than 400 feet — at two different times: once as it rises towards its greatest height, and once as it comes back down.

In this section, we have shown how an equation involving a second-degree polynomial can have more than one solution. In practice, we have shown that equation [2] has at least two solutions.

Equation [2] isn't an exception. In the last section, we showed by a chart that one solution of the equation

$$x^2 + 3x + 4 = 44 \qquad [3]$$

is $x = 5$.

If we evaluate $x^2 + 3x + 4$ when $x = -8$, we get $(-8)^2 + 3(-8) + 4 = 64 - 24 + 4 = 44$. Now we know that equation [3] also has at least two solutions. More importantly, we see that by trial and error, we may neglect to try negative values of x. And even if we do use negative values, we may lose patience before we try $x = -8$.

Furthermore, if equation [3] has two solutions, can we be sure that there aren't others? For example, one equation may have three or more solutions. Other equations may have no solutions. Let's look at some sample problems.

In fact, there are only two solutions, but we haven't proven this yet (we shall in section 13.5). All we know definitely is that at least $t = 2$ and $t = 8$ are solutions of [2].

We'll learn how to "choose" $x = -8$ before the end of this lesson.

Our real-life experience tells us that the ball is at the same height twice: once on the way up and once on the way down. So there shouldn't be more than two answers to Examples 1 and 2.

Example 3

Evaluate $x^3 - x$ when (a) $x = 0$, (b) $x = 1$, and (c) $x = -1$.

In all three cases, the value of the expression is 0.
   (a) If you replace x by 0, you have

$$0^3 - 0 = 0 - 0 = 0$$

   (b) If you replace x by 1, you have

$$1^3 - 1 = 1 - 1 = 0$$

   (c) If you replace x by -1, you have

$$(-1)^3 - (-1) = -1 + 1 = 0$$

Example 4

From Example 3, what can you say definitely about the number of solutions to the equation $x^3 - x = 0$?

There must be at least three solutions, for in Example 3, you have already shown that when $x = 0$, 1, or -1, $x^3 - x = 0$.

There may or may not be more but we have found three definite solutions in doing Example 3.

Let's try the same idea with an example involving more computation.

Example 5

Evaluate the expression $x^3 - 6x^2 + 11x + 1$ when (a) $x = 1$, (b) $x = 2$, and (c) $x = 3$.

For all three cases, the value is 7.
   (a) If you replace x by 1, you have

$$1^3 - 6(1)^2 + 11(1) + 1 = 1 - 6 + 11 + 1 = 7$$

   (b) When $x = 2$, you have

$$2^3 - 6(2)^2 + 11(2) + 1$$
$$= 8 - 6(4) + 22 + 1 = 8 - 24 + 22 + 1 = 7$$

   (c) When $x = 3$, you have

$$3^3 - 6(3)^2 + 11(3) + 1$$
$$= 27 - 6(9) + 33 + 1 = 27 - 54 + 33 + 1 = 7$$

Don't worry about how to pick these numbers. You'll learn most of the techniques by the end of this unit.

Section 13.2   Disadvantages of Arithmetical Solutions

Example 6

What may you conclude from Example 5 concerning the solutions of the equation $x^3 - 6x^2 + 11x + 1 = 7$?

You may conclude that there are at least three solutions. In Example 5, you showed that $x^3 - 6x^2 + 11x + 1$ is equal to 7 when x is replaced by either 1, 2, or 3.

We begin to see that when we have equations that are not linear, it is possible to have more than one solution, and it is not always too easy to determine how many solutions an equation may have.

The only way a linear equation has more than one solution is if it is an identity. For example,
$$x + 1 = 1 + x$$
is true for all values of x.

Example 7

How many real numbers, x, are there for which $x^2 + 1 = 0$?

There are none.
The square of any real number is nonnegative, so x^2 is at least 0. Therefore, $x^2 + 1$ is always at least as great as $0 + 1$, or 1. In other words, there is no real value of x for which $x^2 + 1 = 0$.

Example 7 tells us that no matter what real value we choose for x, $x^2 + 1$ will always be at least as great as 1. This means that every guess will give us an answer that is greater than 0. We shall never find a value for x which makes $x^2 + 1$ less than 0. So this time, the trial-and-error method doesn't place x between a value that's too big and one that's too small.

REVIEW
By definition, a real number is one whose square is nonnegative. Remember, when we multiply a negative number by itself, the result is positive.

Example 8

Show that the equation $x^4 + 7 = 3$ has no solutions.

Since 4 is even, you know that x^4 cannot be negative. Therefore, $x^4 + 7$ must be at least as great as $0 + 7$, or 7. If that is true, it is impossible that there be a real number x for which $x^4 + 7 = 3$. In other words, for any value of x, $x^4 + 7$ must be at least as great as 7.

Let's not lose sight of the main point. This section is not meant to teach algebra. It is meant to show some problems we may have if we rely only on trial and error. In summary:

1. There may be more than one number that's a solution to an algebraic equation.

2. There may be no numbers that are solutions to the equation.

There is another important reason for using algebra. In earlier lessons, we saw that if we had a formula like
$$C = 3p + 5$$
we could solve for p in terms of C.
That is,
$$C - 5 = 3p \qquad \text{so } p = \frac{C - 5}{3}$$

We may want to invert formulas even when the relationships are not linear.

Any real number raised to an even power is nonnegative. This is a generalization of the fact that the square of a real number is nonnegative. For example,
$$x^6 = (x^3)^2$$
and the square of (x^3) is nonnegative.

In this case, how shall we know by trial and error if we've found them all?

In this case, how shall we know by trial and error that there are no solutions?

For example, given a formula such as
$$h = 160t - 16t^2$$
we may want to know how to express t in terms of h.

This is often a very difficult problem. Even in advanced algebra courses, this type of inversion can cause a great deal of trouble. But, at an elementary level, we still can handle certain nonlinear equations. For the rest of this lesson and the next, we shall be interested in solving **quadratic equations.**

A quadratic equation is one that involves polynomials of degree two. Remember, we may use arithmetical trial and error regardless of the type of equation, but after the pitfalls we have seen in this section, we know that the algebraic approach may be helpful.

PRACTICE DRILL

1. Find the value of $x^2 - 3x$ when
   (a) $x = 4$ (b) $x = 5$

2. Use the answers you obtained for Problem 1 to estimate a solution of the equation $x^2 - 3x = 7$.

3. Find the value of $x^2 - 3x$ when
   (a) $x = -1$ (b) $x = -2$

4. Use the answers to Problem 3 to estimate another solution of $x^2 - 3x = 7$.

5. For what value(s) of x does $x^2 + 7 = 4$?

Answers
1. (a) 4 (b) 10   2. $4 < x < 5$
3. (a) 4 (b) 10
4. $-2 < x < -1$
5. no real values

CHECK THE MAIN IDEAS

1. When a linear equation is written in the form $bx = c$ and $b \neq 0$, the equation has exactly _____ solution. In fact, the solution is $x = \frac{c}{b}$.

2. If the equation is not _____, it may have more than one solution or it may have no solutions.

3. For example, when $x = 5$, $x^2 - 12x + 35 = $ _____.

4. When $x = 7$, $x^2 - 12x + 35 = $ _____.

5. From Problems 3 and 4, you see that the equation $x^2 - 12x + 35 = 0$ has at least (a) _____ solutions, $x = 5$ and $x = $ (b) _____. Can you be sure that there aren't even more solutions?

6. As for an equation that has no real solutions, notice that for any real number x, x^2 must be at least as great as _____.

7. Hence, $x^2 + 7$ must be at least as great as _____.

8. Since 4 is less than 7, the equation $x^2 + 7 = 4$ has _____ real solutions. Obviously, you could look forever using trial and error but never find a solution, because there is none.

9. In some cases, it is possible that even though you haven't found an answer, the equation does have at least one solution. In other words, in the real world, when you rely on trial and error, you cannot be sure whether a problem has no _____ or whether it has some but you don't know enough to be able to find them.

Answers
1. one   2. linear   3. 0
4. 0   5. (a) two (b) 7
6. 0   7. 7   8. no
9. answers

Section 13.3
Divisors of Zero

A fact that may not seem very important has a very deep effect on our ability to find solutions of certain equations.

> If the product of two or more numbers is zero, at least one of the numbers must be zero.

We already know that the product of zero and any number is zero. This new fact means that the *only* way a product can be zero is for at least one of the numbers to be zero.

Example 1

In the expression $(x - 1)(x - 2)$, what are the two factors?

The two factors are $x - 1$ and $x - 2$.
You are multiplying two expressions. One of the expressions is $x - 1$, and the other is $x - 2$.

Example 2

For what value(s) of x does $(x - 1)(x - 2) = 0$?

$(x - 1)(x - 2) = 0$ if and only if either $x = 1$ or $x = 2$.
You already know that the only way the product of two numbers can be zero is if one of the numbers is zero. The only two numbers being multiplied are $x - 1$ and $x - 2$. So either $x - 1$ is 0, or $x - 2$ is 0. If $x - 1 = 0$, then $x = 1$. If $x - 2 = 0$, then $x = 2$. Therefore, the solutions are 1 and 2.
As a check, you may let $x = 1$ and see that in this case $(x - 1)(x - 2) = (1 - 1)(1 - 2)$, which equals $0(-1)$. This equals 0, because 0 times any number is 0. Similarly, if you let $x = 2$, then $(x - 1)(x - 2)$ becomes $(2 - 1)(2 - 2) = 1(0) = 0$.
Because the only way that a product can be 0 is if one of the factors is 0; and since $x - 1$ and $x - 2$ are the only two factors; you have shown that $x = 1$ and $x = 2$ are the *only* two solutions of $(x - 1)(x - 2) = 0$.

Let's try a few more examples to make sure that we understand this simple but very important idea.

Example 3

For what value of n does $3n = 0$?

$3n = 0$ if and only if $n = 0$.
The only way that a product can be 0 is if one of the factors is 0. In this case, the only two factors are 3 and n, and 3 is certainly not 0. Therefore, n must be 0.

Example 4

Find all values of q for which $(q + 1)(q + 2) = 0$.

In this case, q must be either -1 or -2.
You know that the product of $q + 1$ and $q + 2$ is 0, hence at least one of these two expressions must be 0.

OBJECTIVE

To solve certain equations by using the fact that a product of numbers can be zero if and only if at least one of the factors is zero.

PROOF

Suppose $bc = 0$ and b is not 0. Multiply both sides of the equation by $\frac{1}{b}$:

$$\frac{1}{b}(bc) = \frac{1}{b}(0)$$

Therefore $\left(\frac{1}{b}b\right)c = 0$

$$1c = 0 \quad \text{or} \quad c = 0$$

So if b isn't 0, c is. This means at least one of the numbers is 0.

Remember, anything enclosed within parentheses is considered to be one number. Hence, both $x - 1$ and $x - 2$ are considered as numbers, and they are multiplying one another.

That is, $x - 1 = 0$ and $x - 2 = 0$ are both linear equations that we already know how to solve.

In light of some earlier remarks, this point is very important. It tells us that the equation $(x - 1)(x - 2) = 0$ has only the two solutions $x = 1$ and $x = 2$. There are no other solutions that we may have missed. In other words, if x is neither 1 nor 2, $(x - 1)(x - 2)$ will *not* equal 0.

We knew before that if $n = 0$ then, $3n = 0$. Now we're saying that $3n = 0$ only if $n = 0$. That is, $3n$ won't be 0 for any other value of n. In summary, if n is 0, then $3n = 0$; and if n is not 0, neither is $3n$.

If $q + 1 = 0$, then $q = -1$; while if $q + 2 = 0$, then $q = -2$. That is,

$$q + 1 = 0 \qquad q + 2 = 0$$
$$\underline{-1 \quad -1} \qquad \underline{-2 \quad -2}$$
$$q \quad = -1 \qquad q \quad = -2$$

You may write this more compactly as

$$\underbrace{(q + 1)}_{0}\underbrace{(q + 2)}_{0} = 0$$

If neither $q + 1$ nor $q + 2$ is 0, the product can't be zero. So $q = -1$ and $q = -2$ give all the answers to the equation.

Example 5

Find all values of x for which $(x + 1)(x - 2)(2x - 3) = 0$.

In this case, either $x = -1$ or $x = 2$ or $x = \frac{3}{2}$.

Again, you use the same idea. In this case, you have a product of three factors that is equal to zero. The three factors are $(x + 1)$, $(x - 2)$, and $(2x - 3)$. Either $x + 1 = 0$, or $x - 2 = 0$, or $2x - 3 = 0$. Solving these linear equations gives you the answers -1, 2, and $\frac{3}{2}$. No other numbers can be solutions, since no other values of x make either $x + 1$, $x - 2$, or $2x - 3$ equal to 0.

Note that x cannot be -1, 2, and $\frac{3}{2}$ all at the same time. Our answer to Example 5 indicates that we have a solution of the equation if x is any one of these three numbers. For example, if $x = 2$, then $x + 1 = 3$, $x - 2 = 0$, and $2x - 3 = 1$. In this case, $(x + 1)(x - 2)(2x - 3) = 3(0)(1)$, and this product is 0 because one of the three factors is 0.

If $x = -1$, then $x - 1$ is 0, and that makes the entire product 0. That is, with $x = -1$, $x + 1 = 0$, $x - 2 = -3$, and $2x - 3 = 2(-1) - 3 = -5$. Hence

$$(x + 1)(x - 2)(2x - 3) = 0(-3)(-5) = 0$$

That is,

$$\underbrace{(x + 1)}_{0}\underbrace{(x - 2)}_{0}\underbrace{(2x - 3)}_{0} = 0$$

Remember, all it takes is one factor equal to 0.

Example 6

Find all the (real) solutions of $(x^2 + 1)(x^2 + 3) = 0$.

There are no solutions.
The only way that a product of numbers can be zero is for one of the numbers to be zero. But in this problem, the only factors are $x^2 + 1$ and $x^2 + 3$. You know that $x^2 + 1$ is at least as great as 1, and $x^2 + 3$ is at least as great as 3. Therefore, both factors are always greater than 0. Since neither factor can equal 0, their product cannot be zero.

Remember that for any real number x, its square is nonnegative.

We are now ready to try to use this information to solve polynomial equations. We try to write the equation so that one side is 0 and the other side is factored.

Example 7

Write $(x - 1)(x - 2)$ as a single polynomial in x.

$(x - 1)(x - 2) = x^2 - 3x + 2$.
You already learned this in the previous unit. You simply multiply two first-degree polynomials.

REVIEW

$(x - 1)(x - 2)$
$= x(x - 2) + {}^-1(x - 2)$
$= x^2 - 2x - x + 2$
$= x^2 - 3x + 2$

Section 13.3   **Divisors of Zero**

Example 8

Use the results of Examples 2 and 7 to find all solutions of the equation

$$x^2 - 3x + 2 = 0 \quad [1]$$

The only two solutions of this equation are $x = 1$ and $x = 2$.

The technique is to notice in the last example that $x^2 - 3x + 2$ means the same thing as $(x-1)(x-2)$. Hence you can replace $x^2 - 3x + 2$ by $(x-1)(x-2)$ to obtain

$$(x-1)(x-2) = 0 \quad [2]$$

In example 2, you saw that the only solutions to this equation are $x = 1$ and $x = 2$.

To check, notice that if $x = 1$, $x^2 - 3x + 2 = 1^2 - 3(1) + 2 = 1 - 3 + 2 = 0$. If $x = 2$, $2^2 - 3(2) + 2 = 4 - 6 + 2 = 0$.

Since equations [1] and [2] are equivalent, there can be no more solutions to equation [1] because there are no more solutions to equation [2].

The two key points are that we want one side of the equation to be a product and the other side to be 0.

When we say that $x^2 - 3x + 2$ and $(x - 1)(x - 2)$ are the same, we mean that both expressions give the same value for a given value of x.

That is, $x^2 - 3x + 2$ and $(x - 1)(x - 2)$ are synonyms as shown in Example 7. Hence $x^2 - 3x + 2 = 0$ has the same solution as $(x - 1)(x - 2) = 0$.

If the product is not 0, there isn't too much we know about the individual factors. For example, if $bc = 24$, it may be that $b = 6$ and $c = 4$ or that $b = 8$ and $c = 3$. More generally, if b is any non-zero number, c is just $\dfrac{24}{b}$.

Example 9

Write $(x+1)(x-2)(2x-3)$ as a single polynomial.

It is $2x^3 - 5x^2 - x + 6$. From the previous unit we know that

$$(x+1)(x-2)(2x-3) = [(x+1)(x-2)](2x-3)$$
$$= (x^2 - x - 2)(2x - 3)$$

Example 10

Use Example 9 to find all the solutions of the equation

$$2x^3 - 5x^2 - x + 6 = 0 \quad [3]$$

The solutions are $x = -1$, $x = 2$, and $x = \dfrac{3}{2}$. From Example 9, you know that $2x^3 - 5x^2 - x + 6$ and $(x+1)(x-2)(2x-3)$ are equivalent expressions. So you may replace the left side of [3] by $(x+1)(x-2)(2x-3)$ to obtain

$$(x+1)(x-2)(2x-3) = 0 \quad [4]$$

You solved [4] in Example 5. That is the solutions for [4] are given by $x = -1$, $x = 2$, and $x = \dfrac{3}{2}$. Because [3] and [4] are equivalent, they have the same solutions.

By now, it is easy to understand that once we know that $x^2 - 3x + 2$ may be rewritten as $(x-1)(x-2)$, it is not hard to solve the equation

$$x^2 - 3x + 2 = 0$$

All we do is rewrite the equation as

$$(x-1)(x-2) = 0$$

and use the fact that if a product is 0, then at least one of the factors must be 0.

REVIEW

$$\begin{array}{r} x^2 - x - 2 \\ 2x - 3 \\ \hline 2x^3 - 2x^2 - 4x \\ -3x^2 + 3x + 6 \\ \hline 2x^3 - 5x^2 - x + 6 \end{array}$$

Or:

$$[(x+1)(x-2)](2x-3)$$
$$= [x^2 + x - 2x - 2](2x-3)$$
$$= (x^2 - x - 2)(2x - 3)$$
$$= (x^2 - x - 2)2x + (x^2 - x - 2)(-3)$$
$$= 2x^3 - 2x^2 - 4x - 3x^2 + 3x + 6$$
$$= 2x^3 + (^-2x^2 - 3x^2) + (^-4x + 3x) + 6$$
$$= 2x^3 - 5x^2 - x + 6$$

What may not be so easy to determine is how we know that $x^2 - 3x + 2$ can be rewritten as $(x-1)(x-2)$. The next section deals with this topic.

PRACTICE DRILL

1. Name the factors in each of the following expressions.
   (a) $x(x-4)$ (b) $(x-1)(x-3)$ (c) $(x^2+1)x$ (d) $(2x+3)(x+1)$
2. Find all the solutions of each of the following.
   (a) $x(x-4) = 0$ (b) $(x-1)(x-3) = 0$
   (c) $(x^2+1)x = 0$ (d) $(2x+3)(x+1) = 0$
3. Express $(x-5)(x-7)$ as a single polynomial.
4. Use Problem 3 to find all solutions of the equation $x^2 - 12x + 35 = 0$.

Answers
1. (a) x and $x-4$ (b) $x-1$ and $x-3$ (c) x^2+1 and x (d) $2x+3$ and $x+1$
2. (a) $x=0$ and $x=4$ (b) $x=1$ and $x=3$ (c) $x=0$ (d) $x=-\frac{3}{2}$ and $x=-1$
3. $x^2 - 12x + 35$
4. $x=5$ and $x=7$

CHECK THE MAIN IDEAS

1. In this section, the main idea is that if the _____ of two or more numbers is 0, then at least one of the numbers must be 0.
2. The only way that $(x-1)(x-3)$ can be 0 is if either $x-3$ or $x-1$ is equal to _____.
3. As a check, if you let $x = 1$ in $(x-1)(x-3)$, you get $(1-1)(1-3)$ or $0(-2)$, which equals _____. Notice that the fact that one factor is 0 is enough to make the entire product 0.
4. This result does not apply to addition. If the sum of two or more numbers is 0, neither number has to be 0. In fact, the sum of any signed number and its opposite is _____.
5. Given an equation like $(x-1) + (x-3) = 0$, you cannot conclude that $x-1$ or $x-3$ is equal to _____.
6. You must add $x-1$ and $x-3$ to get $2x-4$, and for this to equal 0, x must equal _____.

Answers
1. product 2. 0 3. 0
4. 0 5. 0 6. 2

Section 13.4
Factoring Expressions of the Form $x^2 + bx + c$

Let's look at what happens if we multiply, for example, $(x+3)(x+4)$. We get
$$(x+3)(x+4) = x(x+4) + 3(x+4)$$
$$= x^2 + 4x + 3x + 12$$
$$= x^2 + 7x + 12$$

Interestingly, the coefficient of the term containing x is the sum of 3 and 4, or 7; and the constant term (that is, the coefficient of x^0) is 12, which is 3×4.

Let's generalize this. We let r and s stand for any numbers, and we look at the expression $(x+r)(x+s)$. We have

$$(x+r)(x+s) = x(x+s) + r(x+s)$$
$$= x^2 + sx + rx + rs$$
$$= x^2 + (r+s)x + rs \qquad [1]$$

The right side of [1] shows us that we add r and s to get the coefficient of x and that we multiply r and s to get the constant term. We can use this information to multiply things like $(x+3)(x+4)$ very quickly. All we need do is write down x^2 and then add 4 and 3 to get the term $7x$; then we multiply 4 and 3 to get the term 12.

OBJECTIVE
Under certain conditions, to be able to write
$$x^2 + bx + c$$
as the product of two linear polynomials in x.

Section 13.4 **Factoring Expressions of the Form $x^2 + bx + c$**

Example 1
Write down the product $(x + 5)(x + 6)$ by the above method.

$$(x + 5)(x + 6) = x^2 + 11x + 30$$

That is, you have $x^2 + (6 + 5)x + 6(5)$. You add the 5 and 6 to get the coefficient of x; and you multiply 5 and 6 to get the constant term.

Let's try another problem to check the idea.

Of course, we get the same answer the long way.

$$\begin{aligned}(x + 5)(x + 6) &= x(x + 6) + 5(x + 6)\\ &= x^2 + 6x + 5x + 30\\ &= x^2 + 11x + 30\end{aligned}$$

Example 2
What second-degree polynomial is named by $(x + 7)(x + 8)$?

$$(x + 7)(x + 8) = x^2 + 15x + 56$$

The coefficient of x is $7 + 8$, and the constant term is the product of 7 and 8. Therefore

$$\begin{aligned}(x + 7)(x + 8) &= x^2 + (7 + 8)x + 7(8)\\ &= x^2 + 15x + 56\end{aligned}$$

The idea still works where there are minus signs instead of plus signs; that is, when some of the terms are negative.

Example 3
What second-degree polynomial is named by $(x - 3)(x + 5)$?

$$(x - 3)(x + 5) = x^2 + 2x - 15$$

You may think of $x - 3$ as being $x + {}^-3$. In terms of [1], this gives you $r = -3$ and $s = 5$. The sum of r and s is ${}^-3 + 5$, which is 2; and the product of ${}^-3$ and 5 is -15. In other words,

$$\begin{aligned}(x - 3)(x + 5) &= (x + {}^-3)(x + 5)\\ &= x^2 + ({}^-3 + 5)x + {}^-3(5)\\ &= x^2 + (2)x + (-15)\\ &= x^2 + 2x - 15\end{aligned}$$

Once you understand, you may be able to look at $(x - 3)(x + 5)$ and think of the numbers as -3 and 5. In this case, the sum is 2 and the product is -15. But you can always do the problem the long way.

Now all we have to do is use the process in reverse to be able to factor certain second-degree polynomials. To get the idea, let's rework the first three examples from a different point of view.

Example 4
Write $x^2 + 11x + 30$ as the product of two linear (first-degree) polynomials.

$x^2 + 11x + 30 = (x + 5)(x + 6)$ [Or, since multiplication is commutative, you may also write the product as $(x + 6)(x + 5)$.]

You know from the previous discussion that you are looking for the form

$$(x + r)(x + s)$$

where $r + s$ is 11 (that is, the coefficient of x) and rs is 30 (that is, the constant term). This is something like a riddle. You want two numbers whose sum is 11 and whose product is 30. Since 6 and 5 work, you may replace r by 5 and s by 6 (or r by 6 and s by 5) to obtain $(x + 5)(x + 6)$.

We can use trial and error very nicely here if we assume that the two numbers are whole numbers. In that case, the only possible pairs are 1 and 30, 2 and 15, 3 and 10, and 5 and 6. No other pair of whole numbers has 30 as its product.

Of these possibilities, only the pair 5 and 6 has 11 as the sum. However, there are many, many pairs of numbers whose sum is 11, therefore, it's best to solve the riddle by writing the possible products first.

Example 5

Write $x^2 + 15x + 56$ as the product of two linear polynomials.

$$x^2 + 15x + 56 = (x + 7)(x + 8)$$

In this case, you want two numbers whose product is 56 and whose sum is 15. The only pairs of whole numbers whose product is 56 are 56 and 1, 28 and 2, 14 and 4, and 7 and 8. Of these pairs, only the pair 7 and 8 has 15 as the sum. Hence, $r = 7$ and $s = 8$, and you have

$$(x + 7)(x + 8) = x^2 + 15x + 56$$

Example 6

Write $x^2 + 2x - 15$ as the product of two linear polynomials.

$$x^2 + 2x - 15 = (x + 5)(x - 3)$$

Now you want two numbers whose product is -15 and whose sum is 2. The only pairs of integers whose product is -15 are 15 and -1, -15 and 1, 5 and -3, and -5 and 3. Of these, only the pair 5 and -3 has 2 as the sum. Therefore, $r = 5$ and $s = -3$, so that $(x + r)(x + s) = (x + 5)(x + {}^-3)$, or $(x + 5)(x - 3)$.

Remember that for the product of two numbers to be negative, the numbers must have opposite signs.

Let's try a few we haven't seen before.

Example 7

Write $x^2 + 13x + 36$ as the product of two first-degree polynomials.

$$x^2 + 13x + 36 = (x + 4)(x + 9)$$

The answer is going to have the form $(x + r)(x + s)$, where $r + s = 13$ (the coefficient of x) and $rs = 36$ (the constant term).

You want two numbers whose product is 36 and whose sum is 13. Restricting your attention to whole numbers, the only pairs whose product is 36 are 36 and 1, 18 and 2, 12 and 3, 9 and 4, and 6 and 6. Of these, only the pair 4 and 9 has 13 as its sum. Hence, the solution is given by either $(x + 4)(x + 9)$ or $(x + 9)(x + 4)$.

Example 8

Write $x^2 + 37x + 36$ as the product of two first-degree polynomials.

$$x^2 + 37x + 36 = (x + 36)(x + 1)$$

Now you want two numbers whose product is still 36 but whose sum is 37. Looking at the pairs in the previous example, you see that only the pair 36 and 1 adds up to 37. So

$$x^2 + 37x + 36 = (x + 36)(x + 1)$$

Example 9

Write $x^2 - 15x + 36$ as the product of two first-degree polynomials.

$$x^2 - 15x + 36 = (x - 3)(x - 12)$$

You still want two numbers whose product is 36, but now you want the sum to be $^-15$.

If you restrict your attention to integers, the only possible pairs are $^-36$ and $^-1$,

We are looking for two negative numbers. The fact that the product is positive means that the two numbers must have the same sign; and the fact that the sum is negative means that the common sign must be negative.

Section 13.4   **Factoring Expressions of the Form** $x^2 + bx + c$

⁻18 and ⁻2, ⁻12 and ⁻3, ⁻9 and ⁻4, and ⁻6 and ⁻6. Of these, only the pair ⁻12 and ⁻3 has ⁻15 as its sum. Hence $x^2 - 15x + 36$ (that is, $x^2 + {}^-15x + 36$) is equal to $(x + {}^-3)(x + {}^-12)$, or $(x - 3)(x - 12)$.

Example 10

Write $x^2 - 5x - 24$ as the product of two first-degree polynomials.

$$x^2 - 5x - 24 = (x - 8)(x + 3)$$

Remember that $x^2 - 5x - 24$ is the same as $x^2 + {}^-5x + {}^-24$. You want two numbers whose product is ⁻24 but whose sum is ⁻5. The fact that the product is negative means that the two numbers have the opposite signs. If you look at the integers, the possibilities are ⁻24 and 1, 24 and ⁻1, 12 and ⁻2, ⁻12 and 2, 8 and ⁻3, ⁻8 and 3, 6 and ⁻4, and ⁻6 and 4. Of these possibilities, only the pair ⁻8 and 3 has ⁻5 as its sum. Hence, the answer is given by $(x + {}^-8)(x + {}^+3)$ or $(x - 8)(x + 3)$.

> Be careful. Don't pick 8 and ⁻3 — its sum is 5, not ⁻5.

Sometimes the problems may look a little more complicated, but the same idea always applies.

Example 11

Write $x^2 - 9$ as the product of two linear polynomials.

$$x^2 - 9 = (x + 3)(x - 3)$$

Perhaps you think the x term is missing. But whenever a term is missing, you may assume it's there with a coefficient of 0. In other words, you may look at $x^2 - 9$ as meaning $x^2 + 0x - 9$, or $x^2 + 0x + {}^-9$.

Now you want two numbers whose product is ⁻9 but whose sum is 0 (since the coefficient of x is 0). If the sum is 0, the two numbers must be opposites; and since the product of these opposites is ⁻9, the numbers must be 3 and ⁻3. So the answer is given by $(x + 3)(x + {}^-3)$, or $(x + 3)(x - 3)$.

> CHECK
> $(x + 3)(x - 3)$
> $= x(x - 3) + 3(x - 3)$
> $= x^2 - 3x + 3x - 9$
> $= x^2 - 9$

Example 12

Write $x^2 - 81$ as the product of two linear polynomials.

$$x^2 - 81 = (x + 9)(x - 9)$$

If you rewrite $x^2 - 81$ as $x^2 + 0x + {}^-81$, you see that you want two numbers whose product is ⁻81 and whose sum is 0. That is, you are looking for a pair of opposites whose product is ⁻81. This pair is 9 and ⁻9. Therefore, the required factorization is $(x + 9)(x + {}^-9)$, or $(x + 9)(x - 9)$.

> Do you see the pattern for this type of factoring? If b is any real number,
> $$(x + b)(x - b) = x^2 - b^2$$
> or, to emphasize factoring,
> $$x^2 - b^2 = (x + b)(x - b)$$

This approach can be used to see that it is not always possible to factor a second-degree polynomial into the product of two first-degree polynomials.

Example 13

Show that there cannot be real numbers r and s such that $(x + r)(x + s) = x^2 + 81$.

If you rewrite $x^2 + 81$ as $x^2 + 0x + 81$, you see that the product of r and s must be 81, while their sum must be 0. If the product of two numbers is positive, the numbers must have the same sign. But if the numbers have the same sign, their sum can't be 0.

> That is, the sum of two negatives is negative and the sum of two positives is positive.

In higher algebra, we work with new numbers, called imaginary numbers, whose squares can be negative. But in this course, we use only real numbers. Therefore, Example 13 shows us a case in which a second-degree polynomial behaves like a prime number: it can't be broken down into a product of smaller polynomials.

In the next section, we shall use the fact that certain polynomials can be written as a product of lower-degree polynomials to help solve equations involving second-degree polynomials.

PRACTICE DRILL

Write each of the following as the product of two linear polynomials in x.
1. $x^2 + 5x + 4$
2. $x^2 - 5x + 4$
3. $x^2 + 3x - 4$
4. $x^2 + 4x - 5$
5. $x^2 + 4x + 4$
6. $x^2 - 100$

Answers
1. $(x+4)(x+1)$, or $(x+1)(x+4)$
2. $(x-1)(x-4)$
3. $(x+4)(x-1)$
4. $(x+5)(x-1)$
5. $(x+2)(x+2)$, or $(x+2)^2$
6. $(x+10)(x-10)$

CHECK THE MAIN IDEAS

1. Suppose you want to write $x^2 + 5x + 4$ as the product of two linear polynomials in x. You want to find two numbers whose sum is (a) _____ and whose product is (b) _____.
2. Two such numbers are 1 and 4. This means that you may write $x^2 + 5x + 4$ as _____.
3. In general, there are fewer ways that a number can be written as the product of two integers than as the sum of two integers, so you usually work with the product first when you deal with larger numbers. To factor $x^2 - 5x - 150$, you want two numbers whose _____ is $^-150$.
4. The fact that the product is negative means that the two factors have _____ signs.
5. If the two numbers you need are to be _____, the only possible pairs are 1 and $^-150$, $^-1$ and 150, 2 and $^-75$, $^-2$ and 75, 3 and $^-50$, $^-3$ and 50, 6 and $^-25$, $^-6$ and 25, 10 and $^-15$, $^-10$ and 15.
6. Since you want the pair whose sum is _____, the only possibility is that the pair must be 10 and $^-15$.
7. That is, $x^2 - 5x - 150 = (x+10)($_____$)$.

Answers
1. (a) 5 (b) 4
2. $(x+1)(x+4)$ 3. product
4. opposite (different)
5. integers   6. $^-5$
7. $x - 15$

Section 13.5
Solving Quadratic Equations by Factoring

In Section 13.1, we dealt with equations of the form

$$x^2 + 3x + 4 = 44 \qquad [1]$$

and

$$256 = 160t - 16t^2 \qquad [2]$$

Equations [1] and [2] are special cases of quadratic equations.

OBJECTIVE

To solve certain equations of the form

$$x^2 + bx + c = 0$$

by factoring the second degree polynomial.

DEFINITION

A quadratic equation is an equation in which both sides are polynomials, the greatest degree of which is 2.

Alternatively, it is an equation in which every term is a monomial, of which the greatest degree is 2.

Section 13.5 Solving Quadratic Equations by Factoring

Example 1

Is $3x^2 + 5x + 4 = {}^-6x^2 + 7x - 8$ a quadratic equation?

Yes, it is. Both sides of the equation are polynomials and the greatest degree of any term is 2.

Example 2

Is $3x^2 + 5x + 4 = 0$ a polynomial?

Yes, it is. Both sides are polynomials (remember that 0 is a special case of a polynomial), and the greatest degree of any term is 2 ($3x^2$).

Example 3

Is $3x^2 + 4x = x^3$ a quadratic equation?

No, it isn't. It is an equation in which both sides are polynomials, but x^3 has degree 3. To form a quadratic equation, the greatest degree of any term must be 2.

Example 4

Is $3x^2 + x^{-1} = 2x + 7$ a quadratic equation?

No, it isn't.
 No exponent is greater than 2, but since x^{-1} appears, the left side is not a polynomial in x. Remember, to form a quadratic equation in x, both sides of the equation must be polynomials in x (with the greatest degree of any term being 2); and $3x^2 + x^{-1}$ is not a polynomial in x.

In this section, we use the results of the previous two sections to solve certain quadratic equations.

Example 5

Find all solutions of the quadratic equation

$$x^2 + 5x + 6 = 0 \qquad [3]$$

The solutions are $x = {}^-2$ and $x = {}^-3$.
 In the last section, you learned how to factor $x^2 + 5x + 6$. You have to find two numbers whose sum is 5 and whose product is 6. The two numbers are 2 and 3. Hence $x^2 + 5x + 6 = (x + 2)(x + 3)$, and you can rewrite the given equation as

$$(x + 2)(x + 3) = 0 \qquad [4]$$

Equation [4] is easy to solve by the method of Section 13.3. One of the two factors must be 0. If $x + 2 = 0$, then $x = {}^-2$. And if $x + 3 = 0$, then $x = {}^-3$.

We have put together the results of Sections 13.3 and 13.4 to solve the equation in Example 5. Let's try a few other examples.

NOTE
Linear and quadratic equations are special cases of **polynomial equations.** Any equation in which both sides are polynomials is called a polynomial equation.
 In a polynomial equation, the greatest degree of the terms is called the degree of the polynomial equation. A linear equation is a polynomial equation of degree 1. A quadratic equation is a polynomial equation of degree 2.

In other words, $3x^2 + 4x = x^3$ is a polynomial equation of degree 3.

REVIEW
A polynomial in x is any sum of monomials in x. A monomial in x is any term of the form cx^n, where c is a number and n is a whole number ($^-1$ is an integer, but not a whole number).

CHECK
$$(-2)^2 + 5(-2) + 6$$
$$= 4 - 10 + 6 = 0$$
$$(-3)^2 + 5(-3) + 6$$
$$= 9 - 15 + 6 = 0$$

Example 6

Find all solutions of the quadratic equation

$$x^2 - 5x + 6 = 0 \qquad [5]$$

The solutions are $x = 2$ and $x = 3$.

In this case, you want two numbers whose product is 6 but whose sum is $^-5$. The two numbers are $^-2$ and $^-3$. You may replace [5] by

$$(x - 2)(x - 3) = 0 \qquad [6]$$

From [6], you see that either $x - 2 = 0$ or $x - 3 = 0$, therefore, either $x = 2$ or $x = 3$.

Example 7

Find all solutions of the quadratic equation

$$x^2 + 5x - 6 = 0 \qquad [7]$$

The solutions are $x = 1$ and $x = ^-6$.

Now you want two numbers whose product is $^-6$ and whose sum is 5. The two numbers are 6 and $^-1$. You may rewrite [7] as

$$(x - 1)(x + 6) = 0 \qquad [8]$$

This tells you that either $x - 1$ or $x + 6$ must be 0. If $x - 1 = 0$, then $x = 1$. If $x + 6 = 0$, then $x = ^-6$.

Example 8

Find all solutions of $x^2 + 6x + 9 = 0$.

The only solution is $x = ^-3$.

In this case, you want two numbers whose product is 9 and whose sum is 6. This is possible only if each of the two numbers is 3. That is, $3 + 3 = 6$ and $3 \times 3 = 9$.

Therefore, you may rewrite the equation as

$$(x + 3)(x + 3) = 0 \quad \text{or} \quad (x + 3)^2 = 0$$

Now the only way either factor can be 0 (since the factors are the same) is if $x + 3 = 0$; and $x + 3 = 0$ if and only if $x = ^-3$.

Example 8 shows us that it is possible for a quadratic equation to have only one solution.

Another special case of a quadratic equation that will be useful in the next lesson is the following.

Example 9

Find all solutions of

$$x^2 - 16 = 0 \qquad [9]$$

The solutions are given by either $x = 4$ or $x = ^-4$. You may think of [9] as being

$$x^2 + 0x - 16 = 0$$

In the next two examples, we've tried to make the equations look very much like [3]. We want to show that just a little change in the problem can make a big change in the answers.

CHECK

$2^2 - 5(2) + 6 = 4 - 10 + 6 = 0$
$3^2 - 5(3) + 6 = 9 - 15 + 6 = 0$

No other values of x can solve the equation, since the only way [6] can be solved is if either $(x - 2)$ or $(x - 3)$ is 0.

CHECK

$1^2 + 5(1) - 6$
$= 1 + 5 - 6 = 0$
$(-6)^2 + 5(-6) + 6$
$= 36 - 30 + 6 = 0$

A quadratic whose two linear factors are equal (such as in this example) is called a perfect square. Perfect squares play a very important role in the next lesson.

When a solution appears twice, some people call the solution a root of multiplicity two, or a double root.

In many cases [9] is given in the alternative form

$$x^2 = 16 \qquad [10]$$

In this form, it is often easy to guess that $x = 4$ is a solution, but we may miss the fact that $x = ^-4$. If we subtract 16 from both sides of [10], we get $x^2 - 16 = 0$, which is the equation in Example 9.

Section 13.5   Solving Quadratic Equations by Factoring

You want two numbers whose product is ⁻16 and whose sum is 0. The two numbers are 4 and ⁻4. Therefore, [9] may be replaced by

$$(x+4)(x-4)=0$$

from which you can see that either $x = {}^-4$ or $x = 4$.

The method we have used so far requires that one side of the equation be 0 (by tradition, it is the right side, but this is optional); and that the coefficient of x^2 be 1.
In other words, so far we have assumed that the quadratic equation had the special form

$$x^2 + bx + c = 0 \qquad [11]$$

where b and c are constants. If the quadratic equation does not have this form, we can always rewrite it so that it does. But for our present method to work, we must have the equation written in the form of [11].
Let's illustrate the last few remarks with some specific examples.

For this reason, we call [11] the standard form for a quadratic equation.

Example 10

Find all solutions of the quadratic equation

$$x^2 + 5x + 6 = 20 \qquad [12]$$

The solutions are given by $x = 2$ and $x = {}^-7$.
To make the right side of [12] equal to 0, subtract 20 (add ⁻20) from both sides of [12] to obtain

$$\begin{array}{r} x^2 + 5x + 6 = 20 \\ + {}^-20 \quad {}^-20 \\ \hline x^2 + 5x - 14 = 0 \end{array} \qquad [13]$$

CHECK
$$2^2 + 5(2) + 6$$
$$= 4 + 10 + 6 = 20$$
$$(-7)^2 + 5(-7) + 6$$
$$= 49 - 35 + 6 = 20$$

To factor the left side of [13], you must find two numbers whose product is ⁻14 and whose sum is 5. The two numbers are 7 and ⁻2.
Therefore, you may rewrite [13] as

$$(x+7)(x-2)=0 \qquad [14]$$

Since the product of $x + 7$ and $x - 2$ is 0, either $x + 7 = 0$ or $x - 2 = 0$. If $x + 7 = 0$, then $x = {}^-7$; and if $x - 2 = 0$, then $x = 2$.

It is crucial that in [14], the right side be 0. Otherwise, we have no hold on the value of either factor.
For example, in [12], we had the right to replace

$$x^2 + 5x + 6 \text{ by } (x+2)(x+3)$$

But it's not that easy to solve

$$(x+2)(x+3) = 20$$

because the product doesn't equal 0.

Let's try another.

Example 11

Find all solutions of the equation

$$x^2 + 7x + 12 = 30 \qquad [15]$$

The solutions are given by $x = 2$ and $x = {}^-9$.
You subtract 30 from both sides of [15] to obtain

$$x^2 + 7x - 18 = 0 \qquad [16]$$

Now you want two numbers whose sum is 7 and whose product is ⁻18. This gives

$$(x+9)(x-2) = 0$$

Now we're ready to see how we could solve $x^2 + 3x + 4 = 44$ in the last section.

That is,

$$\begin{array}{r} x^2 + 7x + 12 = 30 \\ -30 \quad -30 \\ \hline x^2 + 7x - 18 = 0 \end{array}$$

Again, we could have rewritten [15] as

$$(x+3)(x+4) = 30$$

but we want one side of the equation to equal 0.

Example 12

Find all solutions of

$$x^2 + 3x + 4 = 44 \qquad [1]$$

The solutions are $x = {}^-8$ and $x = 5$.

You subtract 44 from both sides of [1] to get

$$x^2 + 3x - 40 = 0 \qquad [17]$$

Then, to factor the left side of [17], you must find two numbers whose product is $^-40$ and whose sum is 3. The two numbers are 8 and $^-5$.

Therefore, you may rewrite [17] as

$$(x + 8)(x - 5) = 0 \qquad [18]$$

Equation [18] is true only if $x + 8 = 0$ (in which case $x = {}^-8$) or if $x - 5 = 0$ (in which case $x = 5$).

Notice that we have now answered a question we raised in Section 13.2. We have used algebra to show that [1] *can't* have solutions other than the two we've already found.

We have handled the case in which the right side of the quadratic equation is not 0. What happens if the coefficient of x^2 is not 1?

CHECK
$$(-8)^2 + 3(-8) + 4$$
$$= 64 - 24 + 4 = 44$$
$$5^2 + 3(5) + 4$$
$$= 25 + 15 + 4 = 44$$

It's not always very easy to find the two numbers we want, especially when they aren't integers. In the next lesson, we'll see a method that avoids this problem.

We already have checked that $x = -8$ and $x = 5$ are solutions of [1]. But we also know that [1] is equivalent to [18]; and we've already shown that [18] can have only these two solutions, since no other value of x makes either $x + 8$ or $x - 5$ equal to 0.

Example 13

Find all solutions of the equation

$$2x^2 - 10x + 12 = 0 \qquad [19]$$

The solutions are $x = 2$ and $x = 3$.

By the method of the previous lesson, you may rewrite [19] as

$$2(x^2 - 5x + 6) = 0$$
or $\quad 2(x - 2)(x - 3) = 0 \qquad [20]$

[20] tells you that $x = 2$ or $x = 3$.

That is, 2 is common to all three terms on the left side of [19].

Once we factored the 2 in this problem, the remaining factor became a quadratic whose coefficient of x^2 was 1. We can use this idea almost all the time. Let's try one more.

Example 14

Find all solutions of the equation

$$3x^2 + 12x + 9 = 0 \qquad [21]$$

The solutions are $x = {}^-1$ and $x = {}^-3$.

The method is to factor 3 from the left side of [21] to obtain

$$3(x^2 + 4x + 3) = 0 \qquad [22]$$

Then in [22], you replace $x^2 + 4x + 3$ by $(x + 1)(x + 3)$ to obtain

$$3(x + 1)(x + 3) = 0 \qquad [23]$$

Hence either $3 = 0$ (which is impossible), or $x + 1 = 0$, or $x + 3 = 0$. This leads to $x = {}^-1$ and $x = {}^-3$ as the only solutions.

The point is that $x^2 + 4x + 3$ is written so that the coefficient of x^2 is 1.

Sometimes we must remove a common factor and rewrite the equation so that one side is 0, in the same problem. This is exactly what happens if we want to solve

$$256 = 160t - 16t^2 \qquad [2]$$

To make the right side 0, we must add $16t^2 - 160t$ to both sides of [2] to get

$$16t^2 - 160t + 256 = 0 \qquad [24]$$

Then we factor 16 from the left side of [24].

$$16(t^2 - 10t + 16) = 0 \qquad [25]$$

See? $t^2 - 10t + 16$ has 1 as the coefficient of t^2.

To factor $t^2 - 10t + 16$, we need two numbers whose product is 16 and whose sum is -10. The two numbers are $^-8$ and $^-2$. Hence [25] becomes

$$16(t - 2)(t - 8) = 0 \qquad [26]$$

From [26] it is easy to see that t is either 2 or 8.

This agrees with what we saw previously, but now we understand the technique that allows us to find the correct answers without having to use too much trial and error.

In the next section, we shall study a little bit about practical applications: that is, word problems. Once we translate the word problem into a quadratic equation, we proceed just as we did in this section.

PRACTICE DRILL

Use factoring to solve each of the following quadratic equations.
1. $x^2 - 12x + 35 = 0$ 2. $x^2 - 12x + 36 = 0$ 3. $x^2 - 12x + 32 = 0$
4. $x^2 - 49 = 0$ 5. $x^2 - 12x + 35 = 8$

Answers
1. $x = 5$ and $x = 7$ 2. $x = 6$
3. $x = 4$ and $x = 8$ 4. $x = 7$ and $x = ^-7$ 5. $x = 3$ and $x = 9$

CHECK THE MAIN IDEAS

1. In this section, you learn to use _____ to solve certain quadratic equations.
2. To solve the quadratic equation

$$x^2 - 12x + 35 = 0 \qquad [1]$$

you factor the left side by finding two numbers whose sum is _____ and whose product is 35. The two numbers are $^-5$ and $^-7$.

3. This means that you may rewrite [1] as

$$(x - 5)(\text{_____}) = 0 \qquad [2]$$

4. From [2] you see that the solutions of [1] are _____.
5. The method for solving [1] depended on the fact that one side of the equation is (a) _____. If the equation were

$$x^2 - 12x + 35 = 8 \qquad [3]$$

you could still write $x^2 - 12x + 35$ as $(x - 5)(x - 7)$. But this is not helpful because the right side of [3] is not (b) _____.

6. The only way you can be sure of the numerical value of at least one of the factors occurs when the product is _____.
7. To solve equation [3], you must rewrite it so that one side is 0. This can be done by subtracting _____ from both sides to get

$$x^2 - 12x + 27 = 0 \qquad [4]$$

Answers
1. factoring   2. $^-12$
3. $x - 7$ 4. $x = 5$ or $x = 7$
5. (a) 0   (b) 0   6. 0
7. 8   8. 0   9. 0
10. 9

8. The right side of [4] is 0, so it is correct to factor the left side. Since $^-3 \times {}^-9 = 27$ and $^-3 + {}^-9 = {}^-12$, [4] may be rewritten as

$$(x - 3)(x - 9) = \underline{\qquad} \qquad [5]$$

9. From [5] you know that either $x - 3$ or $x - 9$ must equal _____.
10. Since [5] and [3] have the same solutions, you now know that the solutions of the equation $x^2 - 12x + 35 = 8$ are $x = 3$ and $x = $ _____.

Section 13.6
Some Applications

In our study of linear equations, we saw that there are real-life situations in which we had to solve linear equations in order to solve a particular problem or to use a certain formula. In this section, we want to look at a few situations that lead us to quadratic equations.

OBJECTIVE
To be able to solve by factoring certain quadratic equations that occur in mathematical relationships.

Example 1

The product of two consecutive integers is 12. What are the two integers?

The two integers are either 3 and 4 or $^-3$ and $^-4$.
If you call one integer n, the next consecutive integer is $n + 1$. The product of these two integers is 12, so the equation is

$$n(n + 1) = 12 \qquad [1]$$
or $\qquad n^2 + n = 12 \qquad [2]$

Subtract 12 from both sides of [2] to obtain

$$n^2 + n - 12 = 0$$
or $\qquad (n + 4)(n - 3) = 0 \qquad [3]$

From [3] you see that either $n = -4$ (in which case $n + 1 = -3$) or $n = 3$ (in which case $n + 1 = 4$).

If we restricted the solutions to whole numbers rather than integers, the solution $n = {}^-4$ would be incorrect because it's negative. But nothing in Example 1 states that we can't have a negative integer. In certain real situations, the variable must be nonnegative.

Remember, consecutive integers are a pair of integers that differ by 1, such as 5 and 6 or $^-2$ and $^-1$.

Notice that there are two sets of answers in this problem, unlike the linear case, in which there is usually one answer.

Example 2

The area of a rectangle is 12 square feet. The length is one foot longer than the width. What are the dimensions of the rectangle?

The rectangle is 4 feet long and 3 feet wide.
You let w stand for the width, so $w + 1$ stands for the length. Because the area of a rectangle is the product of the length and the width, you have

$$w(w + 1) = 12 \qquad [4]$$

Just as in Example 1, this leads to

$$w^2 + w = 12$$
$$w^2 + w - 12 = 0$$
$$(w + 4)(w - 3) = 0$$

But now you accept only the fact that $w = 3$. That is, if $w + 4 = 0$, then $w = {}^-4$, but a rectangle can't have a negative width. If $w = 3$, then $w + 1 = 4$; the dimensions of the rectangle are 3 feet by 4 feet.

Length and width cannot be negative. No length is shorter than nothing.

Using w instead of n, this equation is the same as [1], except that while n can be negative, w can't be.
To emphasize that w is nonnegative, we often rewrite equation [4] as $w(w + 1) = 12$, (and) $w \geq 0$.

Example 3

Find two consecutive even numbers whose product is 24.

The numbers are either 4 and 6 or $^-6$ and $^-4$.

Consecutive even numbers differ by 2 (because there is an odd number between every pair of even numbers). So, if you let e denote the lesser of the two even numbers, $e + 2$ denotes the greater. The product is 24, so you have

$$e(e + 2) = 24 \quad [5]$$
$$e^2 + 2e = 24$$
$$e^2 + 2e - 24 = 0 \quad [6]$$

You may factor the left side of [6] to obtain

$$(e + 6)(e - 4) = 0 \quad [7]$$

which means that either $e + 6 = 0$ or $e - 4 = 0$. If $e + 6 = 0$, then $e = {^-6}$ and $e + 2 = {^-4}$. If $e - 4 = 0$, then $e = 4$ and $e + 2 = 6$.

Example 4

The length of a rectangle is 2 feet longer than the width. What are the dimensions of the rectangle if its area is 24 square feet?

The dimensions are 4 feet by 6 feet.
You let w stand for the width, and $w + 2$ for the length. You get

$$w(w + 2) = 24$$

and w must be positive. Then just as you worked from [5] to [7], you get that $w = 4$, so that $w + 2 = 6$. You ignore the value $w = {^-6}$, because as the width of a rectangle, w can't be negative.

Right triangles also give us a chance to work with quadratic equations. **A right triangle** is a triangle in which one of the angles is 90° (that is, the triangle has a square corner). There is an interesting relationship among the lengths of the three sides in a right triangle. The relationship is known as the **Pythagorean Theorem**. In terms of the first illustration, the relationship is given by

$$c^2 = a^2 + b^2 \quad [8]$$

The expressions a and b are called the legs (arms or sides) of the right triangle, and c is called the **hypotenuse** (the longest side, which is opposite the right angle, is called the hypotenuse).

Example 5

If the legs of a right triangle are 3 feet long and 4 feet long, what is the length of the hypotenuse?

The hypotenuse is 5 feet long.
All you do is replace a by 3 and b by 4 (or a by 4 and b by 3) in [8] to obtain

$$c^2 = 3^2 + 4^2$$
$$= 9 + 16 = 25 \quad [9]$$

Equation [9] tells you that the square of c is 25, so that either $c = 5$ or $c = {^-5}$. But c can't be -5, since a length can never be negative.

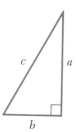

There is a nice way to visualize the Pythagorean Theorem. Imagine a square the length of whose side is $a + b$. The second figure shows us one way to partition this square (where T denotes the area of the triangle in the previous figure). The last figure shows us another way to partition the same square.

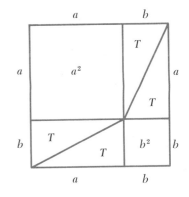

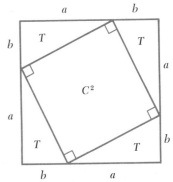

From the second figure, the area is $a^2 + b^2 + 4T$. From the last figure, the area is also $c^2 + 4T$. Since a given square has only one area, then

$$a^2 + b^2 + 4T = c^2 + 4T$$

If we now subtract $4T$ from both sides, we get $a^2 + b^2 = c^2$.

Example 6

The hypotenuse of a right triangle is 13 inches long. One of its legs is 5 inches long. How long is the other leg?

The other leg is 12 inches long.
  In the formula $c^2 = a^2 + b^2$, c stands for the hypotenuse. You may use either a or b to stand for the five-inch leg. Suppose you let $b = 5$. Then the equation becomes

$$13^2 = a^2 + 5^2$$
so
$$169 = a^2 + 25 \quad [10]$$

$13^2 = 13 \times 13 = 169$
$5^2 = 5 \times 5 = 25$

You subtract 25 from both sides of equation [10] to isolate a^2. You get

$$\begin{aligned} 169 &= a^2 + 25 \\ -25 & \quad -25 \\ \hline 144 &= a^2 \end{aligned}$$

That is, $a \times a = 144$.

  The square of both 12 and $^{-}12$ is 144, but again you disregard $^{-}12$, because length must be nonnegative.

REVIEW
To solve $a^2 = 144$, we may rewrite the equation as
$$a^2 - 144 = 0$$
so that
$$(a + 12)(a - 12) = 0$$
We then get $a = 12$ or $a = {}^-12$.

  Now that we know a bit more about the Pythagorean Theorem, let's see how it leads to quadratic equations.

Example 7

The hypotenuse of a right triangle is 5 feet long. One leg is 1 foot longer than the other. What is the length of the shorter length?

The shorter length is 3 feet long.
  Let b stand for the shorter length. Then the greater length must be $b + 1$. That is, $a = b + 1$. Using the Pythagorean Theorem, you have

$$\begin{aligned} c^2 &= a^2 + b^2 \\ 5^2 &= (b+1)^2 + b^2 \\ 25 &= (b^2 + 2b + 1) + b^2 \\ 25 &= 2b^2 + 2b + 1 \quad [11] \end{aligned}$$

REVIEW
$\begin{aligned}(b+1)^2 &= (b+1)(b+1) \\ &= (b+1)b + (b+1)1 \\ &= b^2 + b + b + 1 \\ &= b^2 + 2b + 1\end{aligned}$

To write [11] in a more desirable form, subtract 25 from both sides to obtain

$$\begin{aligned} 0 &= 2b^2 + 2b - 24 \quad [12] \\ &= 2(b^2 + b - 12) \\ &= 2(b + 4)(b - 3) \quad [13] \end{aligned}$$

Notice that [11] and [13] are equivalent equations, but [13] is easier to solve.

  From [13] you see that either $b + 4 = 0$ or $b - 3 = 0$, so either $b = {}^-4$ or $b = 3$. Since length is nonnegative, you conclude that b must be 3, and $b + 1 = 4$. This tells us that the legs are 3 feet and 4 feet, while the hypotenuse is 5 feet.

This is the same triangle that we discussed in Example 5, but now we had to use algebra.

  As far as nongeometric examples are concerned, we've already mentioned that quadratic equations occur when we deal with freely falling objects. To review, let's try another example of this type.

Example 8

An object projected vertically upward with a speed of 480 feet per second reaches a height of h feet after t seconds, given by the formula

$$h = 480t - 16t^2 \quad [14]$$

How long is the object in flight?

Section 13.6 Some Applications

The object is in flight for 30 seconds.

The object stops moving when it's back on the ground. Since $h = 0$ when the object is on the ground, you replace h by 0 in [14] to obtain

$$0 = 480t - 16t^2$$
$$0 = 16t(30 - t) \qquad [15]$$

For [15] to be true, either $t = 0$ or $30 - t = 0$. If $t = 0$, the object hasn't left the ground yet. If $30 - t = 0$, then $t = 30$.

Example 9

In Example 8, at what time will the object reach a height of 3,200 feet?

The object will reach this height twice (once on the way up and once on the way down). The two times are 10 seconds and 20 seconds.
This time you replace h by 3,200 in [14] to obtain

$$3,200 = 480t - 16t^2 \qquad [16]$$

Now if you add $16t^2 - 480t$ to both sides of [16], you get

$$16t^2 - 480t + 3,200 = 0$$
$$16(t^2 - 30t + 200) = 0$$
$$16(t - 10)(t - 20) = 0 \qquad [17]$$

Notice that once we have equation [16], we get to equation [17] by the method of the previous section.

There are additional examples in the exercise set to illustrate the fact that just as linear equations occur when we try to solve certain real-life problems, quadratic equations occur when we try to solve other real-life problems.

PRACTICE DRILL

1. The product of two consecutive integers is 56. What are the two integers?
2. The product of two consecutive whole numbers is 56. What are the numbers?
3. The product of two consecutive odd numbers is 99. What are the numbers?
4. The area of a rectangle is 60 square feet. The length of the rectangle is 4 feet longer than the width. What is the width of the rectangle?

Answers

1. 7 and 8 or $^-8$ and $^-7$
2. 7 and 8   3. 9 and 11 or $^-11$ and $^-9$ 4. 6 feet

CHECK THE MAIN IDEAS

1. In this section, you learn how certain practical problems can be translated into _____ equations.
2. Earlier in the course

$$A = WL \qquad [1]$$

   represented the formula for the area of a _____.
3. If the length (L) is 4 feet longer than the width (W), you may rewrite [1] as

$$A = W(\underline{}) \qquad [2]$$

4. By the distributive property, [2] may be rewritten as

$$A = \underline{} + 4W \qquad [3]$$

5. If you know that $A = 60$, equation [3] becomes

$$\underline{} = W^2 + 4W \qquad [4]$$

6. You can subtract _____ from both sides of [4] to get

$$0 = W^2 + 4W - 60 \qquad [5]$$

Answers

1. quadratic   2. rectangle
3. $W + 4$ 4. W^2 5. 60
6. 60   7. $W + 10$ 8. 6
9. quadratic   10. section

7. Since $10 + {}^-6 = 4$ and $10 \times {}^-6 = {}^-60$, you may rewrite [5] as
$$(W - 6)(\underline{\hspace{1in}}) = 0 \qquad [6]$$
8. Because the width can't be negative, you conclude from [6] that $W = \underline{\hspace{1in}}$.
9. You started with formula [1] for the area of a rectangle; and when you replaced the variables by the given information, you derived a \underline{\hspace{1in}} equation.
10. Once you have the quadratic equation, it does not matter where it comes from. You solve it by the method of the previous \underline{\hspace{1in}}.

EXERCISE SET 13 (Form A)

Section 13.1

1. Evaluate $x^3 - 3x + 1$ when
   (a) $x = {}^-2$ (b) $x = {}^-1$

2. Since $x^3 - 3x + 1$ is negative when $x = {}^-2$ but positive when $x = {}^-1$, the equation $x^3 - 3x + 1 = 0$ must have at least one solution. Locate one solution between two consecutive integers.

3. Find the value of $\dfrac{x^3 + 2x + 1}{x^2 + 1}$ when
   (a) $x = 2$ (b) $x = 3$

4. Since the value of $\dfrac{x^3 + 2x + 1}{x^2 + 1}$ is less than 3 for $x = 2$ and more than 3 for $x = 3$, locate between two consecutive integers a value of x for which $\dfrac{x^3 + 2x + 1}{x^2 + 1} = 3$.

Section 13.2

5. Evaluate $x^4 + 2x^3 - 2x^2 + 2x + 4$ when
   (a) $x = 0$ (b) $x = 1$ (c) $x = 3$ (d) $x = {}^-3$

6. Use the previous problem to name at least two values of x for which
$$x^4 + 2x^3 - 2x^2 + 2x + 4 = 7$$

7. Evaluate $x^3 - 2x^2 - 9x + 20$ when
   (a) $x = 1$ (b) $x = -1$ (c) $x = 2$ (d) $x = 3$ (e) $x = {}^-3$

8. Use the previous problem to name at least three values of x for which
$$x^3 - 2x^2 - 9x + 20 = 2$$

9. (a) For what (real) value of x does $x^3 + 8 = 0$?
   (b) Are there any (real) values of x for which $x^2 + 8 = 0$?
   (c) Are there any (real) values of x for which
   $$3x^2 + 4(x - 2) = 2x^2 + 3(x - 2) + x - 10$$

Section 13.3

Find all (real) values of x, if any, that satisfy the following equations.

10. $(x - 5)(x + 6) = 0$ 11. $x(x - 5)(x + 6) = 0$
12. $(2x - 1)(3x + 5)(x - 2)(x - 5) = 0$

Answers: Exercise Set 13
Form A
1. (a) ${}^-1$ (b) 3   2. ${}^-2$ and ${}^-1$
3. (a) 2.6 (b) 3.4   4. 2 and 3
5. (a) 4 (b) 7 (c) 127 (d) 7
6. 1 and ${}^-3$ 7. (a) 10 (b) 26 (c) 2 (d) 2 (e) 2
8. ${}^-3$, 2, and 3   9. (a) $x = {}^-2$ (b) No (c) No   10. $x = 5$ or ${}^-6$ 11. $x = 0, 5,$ or ${}^-6$
12. $x = \dfrac{1}{2}, -\dfrac{5}{3}, 2,$ or 5
13. $x = \dfrac{1}{2}, 2,$ or 5
14. no real values of x
15. $x = {}^-1$ or ${}^-8$ 16. $x = 4$
17. $(x + 7)(x + 8)$
18. $(x - 7)(x - 8)$
19. $(x + 18)(x - 2)$
20. $(x + 8)(x + 8)$, or $(x + 8)^2$
21. $(x + 8)(x - 8)$
22. $x(x + 7)(x + 8)$
23. (a) $(u - 4)(u - 1)$ (b) $u^2 - 5u + 4$ (c) $(x + 2)(x - 2)(x + 1)(x - 1)$
24. $x = 15$ or ${}^-2$
25. $x = {}^-15$ or 2
26. $x = 5$ or 8   27. $x = 7$
28. $x = 4$ or 10
29. $x = 0, 4,$ or 10
30. $x = {}^-2$ or 16
31. $x = {}^-15$ or 1
32. $x = {}^-2$ or 2
33. 5 and 8   34. ${}^-2$
35. 10 and 11
36. (a) $A = 2W^2 + 6W$ (b) 1,980 square feet (c) 10 feet (d) 26 feet   37. 8 meters and 15 meters   38. (a) 2,256 feet (b) after 10 seconds and also after 40 seconds (c) 50 seconds
39. (a) 360 watts (b) 110 volts (c) either 4 amps or 18 amps

Exercise Set 13 (Form A)

13. $(2x-1)(3x^2+5)(x-2)(x-5) = 0$
14. $(x^2+1)(x^2+8) = 0$
15. $(x+1)^2(x+8) = 0$
16. $(x-5)+(x-3) = 0$

Section 13.4
Write each of the following as the product of two linear polynomials in x.

17. $x^2 + 15x + 56$
18. $x^2 - 15x + 56$
19. $x^2 + 16x - 36$
20. $x^2 + 16x + 64$
21. $x^2 - 64$
22. Write $x^3 + 15x^2 + 56x$ as a product of three linear polynomials in x.
23. (a) Write $u^2 - 5u + 4$ as the product of two linear polynomials in u.
    (b) Recall that $x^4 = (x^2)^2$. What expression do you get if you replace x^2 by u in the expression $x^4 - 5x^2 + 4$?
    (c) Use the results of (a) and (b) to express $x^4 - 5x^2 + 4$ as the product of four linear polynomials in x.

Section 13.5
Use factoring to find all real values of x that satisfy each of the following equations.

24. $x^2 - 13x - 30 = 0$
25. $x^2 + 13x - 30 = 0$
26. $x^2 - 13x + 40 = 0$
27. $x^2 - 14x + 49 = 0$
28. $x^2 - 14x + 40 = 0$
29. $x^3 - 14x^2 + 40x = 0$
30. $x^2 - 14x + 40 = 72$
31. $x^2 + 14x = 15$
32. $x^4 - 2x^2 - 8 = 0$

Section 13.6
33. Find all numbers such that the square of the number is 40 less than 13 times the number.
34. The square of a negative number is 30 more than 13 times the number. What is the number?
35. The product of two consecutive whole numbers exceeds 9 times the smaller by 20. What are the two numbers?
36. The length L of a rectangle is 6 feet more than twice the width W of the rectangle.

    (a) Express the area of the rectangle as a second degree polynomial in W.
    (b) What is the area of the rectangle if its width is 30 feet?
    (c) What is the width of the rectangle if its area is 260 square feet?
    (d) What is the length of the rectangle if its area is 260 square feet?

37. The hypotenuse of a right triangle is 17 meters. One of the legs is 7 meters longer than the other. What are the lengths of the two legs?
38. A projectile launched vertically upward at a speed of 800 feet per second reaches a height of h feet after t seconds according to the rule

$$h = 800t - 16t^2$$

   (a) What is its height at the end of 3 seconds?
   (b) At what time does it reach a height of 6,400 feet?
   (c) After how many seconds does the projectile return to the ground?

39. A formula for electric power is
$$P = VI - RI^2$$
where
P = power (watts)
V = voltage (volts)
I = current (amperes)
R = resistance (ohms)

(a) What is the power if $V = 110$ volts, $R = 5$ ohms, and $I = 4$ amperes (amps)?
(b) What is the voltage if $P = 360$ watts, $R = 5$ ohms, and $I = 4$ amps?
(c) What is the current if $P = 360$ watts, $V = 110$ volts, and $R = 5$ ohms?

EXERCISE SET 13 (Form B)

Section 13.1

1. Evaluate $x^3 - 4x + 9$ when
   (a) $x = {}^-2$ (b) $x = {}^-3$

2. Since $x^3 - 4x + 9$ is positive when $x = {}^-2$ and negative when $x = {}^-3$, the equation $x^3 - 4x + 9$ must have at least one solution. Locate one solution between two consecutive integers.

3. Find the value of $\dfrac{x^3 - 2x + 1}{x + 1}$ when
   (a) $x = 3$ (b) $x = 4$

4. Since the value of $\dfrac{x^3 - 2x + 1}{x + 1}$ is more than 5 when $x = 3$ and less than 12 when $x = 4$, locate between two consecutive integers a value of x for which $\dfrac{x^3 - 2x + 1}{x + 1} = 8$.

Section 13.2

5. Evaluate $x^4 - x^3 - x^2 - x + 8$ when
   (a) $x = 0$ (b) $x = 1$ (c) $x = {}^-1$ (d) $x = 2$

6. Use the previous problem to find at least two values of x for which
$$x^4 - x^3 - x^2 - x + 8 = 10$$

7. Evaluate $x^3 - 5x^2 + 6x + 4$ when x equals
   (a) 0   (b) ${}^-1$ (c) 1   (d) 2   (e) 3

8. Use the previous problem to find at least three values of x for which
$$x^3 - 5x^2 + 6x + 4 = 4$$

9. (a) For what (real) value of x does $x^3 + 1 = 0$?
   (b) Are there any (real) values of x for which $x^2 + 1 = 0$?
   (c) Are there any (real) values of x for which
$$4x^2 + 3(x - 5) = 3x^2 + 2(x - 7) + x - 2$$

Exercise Set 13 (Form B)

Section 13.3

Find all real values of x that are solutions of the following equations.

10. $(x-4)(x+5)=0$
11. $x(x-4)(x+5)=0$
12. $(4x-3)(3x+2)(x-4)(x+1)=0$
13. $(4x-3)(3x^2+2)(x-4)(x+1)=0$
14. $(x^2+3)(x^2+4)=0$
15. $(x+3)^2(x+4)=0$
16. $(x-5)+(x-7)=0$

Section 13.4

Write each of the following as the product of two linear polynomials in x.

17. $x^2+17x+72$
18. $x^2-17x+72$
19. $x^2+12x-45$
20. $x^2+12x+36$
21. x^2-36
22. Write x^3+17x^2+72x as the product of three linear polynomials in x.
23. (a) Write $u^2-10u+9$ as the product of two linear polynomials in u.
    (b) By replacing x^2 by u, rewrite x^4-10x^2+9 as a polynomial in u.
    (c) Use the results of (a) and (b) to express x^4-10x^2+9 as the product of four linear polynomials in x.

Section 13.5

Use factoring to find all real values of x that are solutions of each of the following equations.

24. $x^2-15x-34=0$
25. $x^2+15x-34=0$
26. $x^2-15x+44=0$
27. $x^2-16x+64=0$
28. $x^2-16x+63=0$
29. $x^3-16x^2+63x=0$
30. $x^2-16x+63=15$
31. $x^2+16x=17$
32. $x^4-8x^2-9=0$

Section 13.6

33. Find all numbers for which the square of the number is 44 less than 15 times the number.
34. The square of a negative number is 34 more than 15 times the number. What is the number?
35. The product of two consecutive whole numbers exceeds 8 times the smaller by 60. What are the two whole numbers?
36. The length L of a rectangle is 9 cm more than three times the width W of the rectangle.

    (a) Express the area A of the rectangle as a polynomial in W.
    (b) What is the area of the rectangle if its width is 20 cm?
    (c) What is the width of the rectangle if its area is 210 cm²?
    (d) What is the length of the rectangle if its area is 210 cm²?

37. The hypotenuse of a right triangle is 13 feet. One of the legs is 7 feet longer than the other. What are the lengths of the two legs?
38. A projectile launched vertically upward at a speed of 320 feet per second reaches a height of h feet after t seconds according to the rule

$$h = 320t - 16t^2$$

   (a) What is its height after 2 seconds?
   (b) At what time(s) does it reach a height of 1,200 feet?
   (c) After how many seconds does the projectile return to the ground?

39. A formula for electric power is
$$P = VI - RI^2$$
where
P = power (watts)
V = voltage (volts)
I = current (amps)
R = resistance (ohms)

(a) What is the power if $V = 120$ volts, $I = 5$ amps, and $R = 6$ ohms?
(b) What is the voltage if $P = 350$ watts, $I = 5$ amps, and $R = 10$ ohms?
(c) What is the current if $P = 350$ watts, $V = 120$ volts, and $R = 10$ ohms?

Lesson 14

The General Solution of a Quadratic Equation

Overview There is an old story about a long-distance swimmer who tried to swim the English Channel. He got so tired one mile from his destination that he had to turn around and swim back!

In a way, we have reached that same point of decision. We don't have to swim back, but we shouldn't stop just yet. We need not have started to work with quadratic equations. However, since we have begun the work, we should try to finish what we've started. This lesson completes the study of quadratic equations. By the end of this lesson, we shall be able to find, exactly, all the solutions of any quadratic equation. A byproduct of our study will be a better understanding of square roots and how to work with them.

When this lesson is finished, we shall be able to solve any linear or quadratic equation. Once we can do this, we shall be prepared to handle the mathematical aspects of many quantitative situations. In many ways, the material of these last three lessons teaches as much about mathematical maturity as it does about quadratic equations.

Section 14.1
More on Square Roots

OBJECTIVE
To review the concept of a square root and to be able to estimate the square root of a given number.

A prerequisite skill for solving any given quadratic equation is the ability to work with square roots. The reason for this will become clear as we go through the lesson.

First, let's see what makes it important to learn anything more about equations.

Example 1 (Review)

Use the method of the previous lesson to solve the quadratic equation

$$x^2 + 3x + 4 = 44 \qquad [1]$$

You see that the solutions are $x = 5$ and $x = -8$.

You solve [1] by first subtracting 44 from both sides to obtain

$$x^2 + 3x - 40 = 0 \qquad [2]$$

Then you find two numbers whose product is -40 and whose sum is 3 (8 and -5). You rewrite [2] as

$$(x + 8)(x - 5) = 0 \qquad [3]$$

Next, you solve [3] by using the fact that for a product to be 0, at least one of the factors must be 0. This tells you that either $x + 8 = 0$ or $x - 5 = 0$. From this you get $x = -8$ and $x = 5$ as the two solutions.

Suppose we try this same method on another problem we presented in Section 13.1. Let's see what happens.

Example 2

Rewrite

$$x^2 + 3x + 4 = 50 \qquad [4]$$

so that the right side is 0.

All you do is subtract 50 from both sides of equation [4] to get

$$\begin{aligned} x^2 + 3x + 4 &= 50 \\ -50 &-50 \\ \hline x^2 + 3x - 46 &= 0 \end{aligned} \qquad [5]$$

We proceed from [4] to [5] in exactly the same way that we go from [1] to [2].

If we now want to use [5] to help us solve [4], we use the same theory from Lesson 13. We try to find two numbers whose product is -46 and whose sum is 3. But this may not be as easy as it seems.

We're still looking for two (real) numbers with a given sum and a given product, but now the numbers aren't so easy to find. In fact, we cannot be sure that they even exist. This new difficulty makes this lesson different from the last.

Example 3

Show that there is no pair of *integers* whose product is -46 and whose sum is 3.

There are only two ways in which 46 can be written as a product of whole numbers. They are 1×46 and 2×23.

Because you want the product to be -46 (and not 46), you know that the two factors must have opposite signs. If you restrict yourselves to integers, the only possible pairs are 1 and -46, -1 and 46, 2 and -23, and -2 and 23. But none of these pairs has 3 as the sum.

$$\begin{aligned} 1 + {-}46 &= {-}45 \\ {-}1 + 46 &= 45 \\ 2 + {-}23 &= {-}21 \\ {-}2 + 23 &= 21 \end{aligned}$$

Section 14.1   More on Square Roots   361

Example 3 tells us only that there is no pair of *integers* whose product is −46 and whose sum is 3. It doesn't imply that there is no pair of *real* numbers with these two properties. By the end of this lesson, we shall find the two numbers, and the *exact* solution of

$$x^2 + 3x + 4 = 50 \qquad [4]$$

There are many steps involved in finding the solution of [4]. One of the steps requires that we be familiar with square roots.

> We studied square roots in our discussion of area in Lesson 1.

Example 4

What *positive* number is named by $\sqrt{9}$?

$\sqrt{9}$ is another name for 3.
You want that positive number that when multiplied by itself equals 9. You know that $3 \times 3 = 9$.

We also know that $-3 \times -3 = 9$. That's why we stressed the word "positive" in Example 4. If we want to be technical, it is incorrect to ask "What *number* when multiplied by itself equals 9?" There are two numbers with this property.

> REVIEW
> By $\sqrt{b}$ we mean the number that when multiplied by itself equals b. That is,
> $$\sqrt{b} \times \sqrt{b} = b$$
> At the time we made the above definition, we were using positive numbers only. (Signed numbers were discussed in Lesson 3.)
> REVIEW
> Multiplying a number by itself is called squaring the number.

Example 5

Use the method of the previous lesson to help solve the equation $x^2 = 9$.

First you subtract 9 from both sides to get

$$x^2 - 9 = 0$$

Then you rewrite this as

$$(x + 3)(x - 3) = 0$$

to conclude that $x = -3$ and $x = 3$ are the only two solutions.

> Notice that the equation $x^2 = 9$ is the equation we'd write if we wanted to find "what number when multiplied by itself equals 9?"

To avoid the problem of dealing with both answers, we agree to let $\sqrt{9}$ mean 3. If we want the answer to be -3, we shall write $-\sqrt{9}$.

> If we want both answers, we'll write
> $$\pm\sqrt{9}$$
> which is an abbreviation for
> $$x = \sqrt{9} \text{ or } x = -\sqrt{9}$$
> In other words, to solve $x^2 = 9$, we take the square root of both sides but write them as $x = \pm\sqrt{9}$ to allow for both correct answers (3 and −3).

Example 6

Find all solutions of the equation

$$x^2 = 81 \qquad [6]$$

The solutions are given by $x = \pm\sqrt{81}$; that is, either $x = 9$ or $x = -9$.
You subtract 81 from both sides of [6] to obtain

$$x^2 - 81 = 0 \qquad [7]$$

Then you find two numbers whose product is −81 and whose sum is 0 (9 and −9). You can then rewrite [7] as

$$(x + 9)(x - 9) = 0 \qquad [8]$$

From [8] it follows that either $x = 9$ or $x = -9$.

The idea works even when we aren't dealing with perfect squares.

Example 7

Find all the solutions of the equation

$$x^2 = 41 \qquad [9]$$

The solutions are given by $x = \pm\sqrt{41}$. That is, either $x = \sqrt{41}$ or $x = -\sqrt{41}$.

To get the answer, you take the square root of both sides. That is, you rewrite [9] as

$$x^2 - 41 = 0$$

This is the same as

$$(x + \sqrt{41})(x - \sqrt{41}) = 0$$

In theory, we solve [9] just as we solved [6], but now the two numbers we need (that is, the pair whose product is -41 and whose sum is 0) are no longer integers. The fact that $\sqrt{41}$ is not so simple a number as, for example, $\sqrt{81}$ does not make it any less important or less exact.

Since $6 \times 6 = 36$ (which is less than 41) and $7 \times 7 = 49$ (which is greater than 41), we know that $\sqrt{41}$ is a real number between 6 and 7.

We can use the numerical method of Section 13.1 to approximate $\sqrt{41}$ to as great a number of decimal-place accuracy as we desire. Or, we can use a calculator to find that

$$\sqrt{41} = 6.403124 \ldots$$

(The dots mean that the decimal does not end, but the calculator runs out of spaces sooner or later.)

Example 8

Find the length of the hypotenuse of a right triangle if its legs are 4 feet long and 5 feet long.

The length of the hypotenuse is $\sqrt{41}$ feet.

Remember that the answer is obtained by using the Pythagorean Theorem (Lesson 13), which states that

$$c^2 = a^2 + b^2 \qquad [10]$$

In this problem, you replace a by 4 and b by 5 (or a by 5 and b by 4) in [10] to get

$$c^2 = 4^2 + 5^2 = 16 + 25$$
$$c^2 = 41 \qquad [11]$$

If the answer seems unnatural, notice that we can draw the triangle to scale. We can draw a 4-foot length and a 5-foot length at right angles to each other and actually measure the hypotenuse.

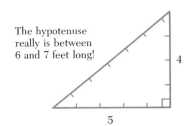

The hypotenuse really is between 6 and 7 feet long!

Equation [11] is exactly the same equation as equation [9]. We simply changed the name of the variable.

In this example, we do not write the answer as $c = \pm\sqrt{41}$ because length cannot be negative. In other words, equation [11] should really be written as

$$c^2 = 41 \quad \text{and} \quad c > 0 \qquad [12]$$

and in this case the correct answer is $c = \sqrt{41}$. To write $c = \pm\sqrt{41}$ as a solution for [12] is wrong.

Now let's make a brief review so that we don't lose the basic message of this unit. An example like $x^2 = 41$ shows us that we may have to use the concept of square roots to find two numbers with a given sum and a given product.

Many quadratic equations require a more subtle use of this fact than we needed in the equation $x^2 = 41$, but the need to know how to handle square roots is the same. For this reason, we use this section to be sure that we feel comfortable with the idea of square roots.

Perhaps we should study some crucial ideas more closely. Let's treat them in the form of notes.

There are several reasons why we let $\sqrt{41}$ mean only the positive square root. But one of the reasons is in Example 8. We often deal with situations (such as length or weight) where only positive numbers are possible.

In more advanced algebra courses, we study the idea of a *complex number*. The square of a complex number can be negative. But the notion of a complex number goes beyond anything we need in this course.

NOTE

We can't take the square root of a negative number. That is, there is no real number whose square is negative.

Section 14.1 More on Square Roots

Example 9

Find all real numbers r for which $r^2 = -4$.

There are none.
   If r is a real number, it is either positive or negative or 0. If it is 0, its square is also 0 (and therefore not -4). If it is positive, its square is also positive (and therefore not -4). If it is negative, its square is still positive (and therefore still not -4). So there is no real number whose square is a negative real number.

It's always important to remember that the square of a negative number is positive.

NOTE

A positive number always has two square roots, and they are opposites. For example, there are two numbers, 3 and -3, whose square is 9.

REVIEW

$x = \pm 3$ is shorthand for "either $x = 3$ or $x = {}^-3$." More generally, $x = \pm c$ means $x = c$ or $x = {}^-c$.

NOTE

Zero is the only number that has only one square root: 0 itself.

Example 10

Find all the solutions of $r^2 = 0$.

In this case, r must be 0.
   Since r^2 means $r \times r$, you see that $r \times r = 0$. Because at least one of the factors must be 0, and both factors are r, then r must equal 0.

PRACTICE DRILL

For what value of x is each of the following true?
1. $x^2 - 81 = 0$
2. $x^2 = 121$
3. $x^2 - 53 = 0$
4. $x^2 = 57$
5. $x^2 + 9 = 0$
6. $x^2 = -11$
7. $x^2 = 0$

Answers
1. $x = \pm 9$ (either $x = 9$ or $x = -9$)
2. $x = \pm 11$ (either $x = 11$ or $x = -11$)
3. $x = \pm\sqrt{53}$
4. $x = \pm\sqrt{57}$
5. no solutions
6. no solutions
7. $x = 0$

CHECK THE MAIN IDEAS

1. In the last lesson you saw that to factor $x^2 + bx + c$, you have to find two numbers whose sum is _____ and whose product is c.
2. In some cases, no such real numbers exist. In other other cases, they exist but are hard to find. For example, given $x^2 + 4x + 2$, you want two numbers whose sum is 4 and whose product is 2. The only pairs of integers whose product is 2 are 2 and 1 and _____.
3. Neither of these pairs adds up to 4, so if the required numbers exist, they are not _____.
4. To learn how to handle this situation, you must study square roots. The square root of a number b is the number that when multiplied by itself equals _____.
5. Any positive number has _____ square roots.
6. They have the same magnitude but _____ signs.
7. The square of both 3 and _____ is 9.
8. When you write $\sqrt{9}$, you mean only the number _____.
9. If you mean -3, you write _____.
10. Since the square of a real number can't be negative, then $\sqrt{-9}$ is not a _____ number. That is, $-\sqrt{9}$ and $\sqrt{-9}$ don't mean the same thing.

Answers
1. b
2. -2 and -1
3. integers
4. b
5. two
6. opposite (different, or unequal)
7. -3
8. 3
9. $-\sqrt{9}$
10. real

Section 14.2
Arithmetic Involving Square Roots

Now that we know what square roots are and how to approximate their numerical values, we need to apply the usual principles of arithmetic to square roots. This arithmetic, as we shall see later in the lesson, is very important when we try to find the exact answers to any quadratic equation.

OBJECTIVE
To apply the rules of arithmetic to expressions involving square roots.

Example 1

Find the sum of $3\sqrt{2}$ and $5\sqrt{2}$.

The sum is $8\sqrt{2}$.
You know by the distributive property that $3\sqrt{2} + 5\sqrt{2} = (3+5)\sqrt{2} = 8\sqrt{2}$. This is just a special case of

$$3n + 5n = (3+5)n = 8n$$

with $n = \sqrt{2}$.

Do not confuse $\sqrt{2} + \sqrt{2}$ with $\sqrt{2} \times \sqrt{2}$. $\sqrt{2} \times \sqrt{2} = 2$, but $\sqrt{2} + \sqrt{2} = 2\sqrt{2}$. When we add like denominations, we add the coefficients and keep the common denomination.

Example 2

Find the sum of $16\sqrt{141}$ and $19\sqrt{141}$.

The sum is $35\sqrt{141}$.
This is another special case of the distributive property

$$16n + 19n = (16+19)n = 35n$$

with $n = \sqrt{141}$. In other words,

$$16\sqrt{141} + 19\sqrt{141} = (16+19)\sqrt{141} = 35\sqrt{141}$$

Though an expression such as $8\sqrt{2}$ is an exact quantity, in real-life situations we often prefer a numerical approximation. How accurate the approximation must be depends on the nature of the problem we're solving.

Example 3

What is the value of $8\sqrt{2}$ if you accept the fact that $\sqrt{2} = 1.41$?

The value is 11.28. All you do is replace $\sqrt{2}$ by 1.41. Hence $8\sqrt{2} = 8(1.41) = 11.28$.

The fact that $\sqrt{2}$ is approximately 1.41 may be found from either trial and error, tables, or a calculator.

Example 4

What is the value of $8\sqrt{2}$ if you accept the fact that $\sqrt{2} = 1.4142$?

The value is 11.3136. You replace $\sqrt{2}$ by 1.4142 and follow the rest of Example 3.

We get different answers in Examples 3 and 4 because we use different approximations. However, if we are interested in getting the answer to the nearest tenth, both numbers give us 11.3.

Example 5

What is the value of $8\sqrt{2}$ if you accept the fact that $\sqrt{2} = 1.4142135$?

The value is 11.313708. You act just as you did in the two previous examples. The difference is that the given value of $\sqrt{2}$ comes from a calculator.

If we can be sure only of three significant figures, we can't get an answer any more accurate than the others. We still round off to 11.3.

Section 14.2 Arithmetic Involving Square Roots

Example 6

Find the sum of $3\sqrt{2}$ and $5\sqrt{2}$, and express the answer correct to the nearest tenth.

To the nearest tenth, the sum is 11.3.
You already learned in Example 1 that the sum is $8\sqrt{2}$. In the last three examples, you saw that to the nearest tenth, $8\sqrt{2}$ equals 11.3.

When we add square roots, the coefficients do not have to be whole numbers.

Example 7

Find the sum of $\frac{1}{2}\sqrt{2}$ and $\frac{1}{3}\sqrt{2}$.

The sum is $\frac{5}{6}\sqrt{2}$.

Using the distributive property, you may write

$$\frac{1}{2}\sqrt{2} + \frac{1}{3}\sqrt{2} = \left(\frac{1}{2} + \frac{1}{3}\right)\sqrt{2} = \frac{5}{6}\sqrt{2}$$

However, we must be careful when we add square roots of different numbers. Let's check this idea with Examples 8 and 9.

Example 8

What is the value of $\sqrt{5}$ to the nearest tenth?

To the nearest tenth, $\sqrt{5} = 2.2$.
You get this result by trial and error, tables, use of the hand calculator, and so on.

answer is in this range
- $(2.2)^2 = 4.84$
- $\leftarrow 5$
- $(2.25)^2 = 5.0625$
- $(2.3)^2 = 5.29$

Example 9

Find the sum of $\sqrt{2}$ and $\sqrt{3}$, and express the sum correct to the nearest tenth.

To the nearest tenth, $\sqrt{2} + \sqrt{3} = 3.1$.
You know that $\sqrt{2} = 1.41 \ldots$ and $\sqrt{3} = 1.73 \ldots$. So

$$\sqrt{2} + \sqrt{3} = 1.41\ldots + 1.73\ldots = 3.14\ldots$$

and to the nearest tenth, $\sqrt{2} + \sqrt{3} = 3.1$.

Together, Examples 8 and 9 tell us that $\sqrt{2} + \sqrt{3}$ is not the same as $\sqrt{5}$. We must avoid the desire to abbreviate an expression such as $\sqrt{2} + \sqrt{3}$ by writing $\sqrt{5}$. $\sqrt{2}$ and $\sqrt{3}$ are unlike terms, so we can't write $\sqrt{2} + \sqrt{3}$ as a single square root.

To save space, it is conventional to write $\frac{1}{2}\sqrt{2}$ as $\frac{\sqrt{2}}{2}$. As a result, we often see $\frac{1}{2}\sqrt{2} + \frac{1}{3}\sqrt{2}$ written as

$$\frac{\sqrt{2}}{2} + \frac{\sqrt{2}}{3}.$$

We use common denominators, just as when we deal with whole numbers to write this as

$$\frac{\sqrt{2}}{2} = \frac{3\sqrt{2}}{6}$$

$$\frac{\sqrt{2}}{3} = \frac{2\sqrt{2}}{6}$$

so that

$$\frac{1}{2}\sqrt{2} + \frac{1}{3}\sqrt{2} = \frac{3\sqrt{2}}{6} + \frac{2\sqrt{2}}{6}$$

$$= \frac{3\sqrt{2} + 2\sqrt{2}}{6}$$

$$= \frac{5\sqrt{2}}{6}$$

It is customary to carry more decimal-place accuracy than we need in the final answer so that we keep as much accuracy as possible before we round off to get the final answer.

Until you feel certain of how to handle square roots, do the problem and check both expressions to see if they give the same result. If the two numbers are unequal, you made a mistake if you called them equal.

Example 10

Find the sum of $\sqrt{7}$ and $\sqrt{3}$, correct to the nearest hundredth.

To the nearest hundredth, $\sqrt{7} + \sqrt{3}$ is 4.38. You find that $\sqrt{7} = 2.6458\ldots$ and $\sqrt{3} = 1.7320\ldots$. Hence, $\sqrt{7} + \sqrt{3} = 4.3778\ldots$, which is 4.38 to the nearest hundredth.

Again, notice that the sum is not $\sqrt{10}$. To the nearest hundredth, $\sqrt{10} = 3.16$.

What about the product of $\sqrt{7}$ and $\sqrt{3}$? After doing Example 10, we cannot simply assume that we multiply 7 and 3 to obtain $\sqrt{21}$. That approach didn't work when we found the sum of $\sqrt{7}$ and $\sqrt{3}$.

But it turns out that it is true that

$$\sqrt{7} \times \sqrt{3} = \sqrt{21} \qquad [1]$$

Why is [1] true? Perhaps the easiest way to find out is to show that when we square $\sqrt{7} \times \sqrt{3}$ we get 21: that's the definition of $\sqrt{21}$. This is an occasion when the structure of arithmetic that we studied in Lesson 4 is extremely useful. We have

$$(\sqrt{7} \times \sqrt{3}) \times (\sqrt{7} \times \sqrt{3}) = (\sqrt{7} \times \sqrt{7}) \times (\sqrt{3} \times \sqrt{3})$$
$$= 7 \times 3$$
$$= 21$$

*We often refer to the "square root sign" as a **radical**.*

It is often easier to multiply than to add. For example, when we want to express the sum of two common fractions as a common fraction, we have to pick a common denominator. But when we want to multiply common fractions, we don't need common denominators.

Whether or not we say it, we often use the commutative and associative properties of multiplication (or addition).

We can restate [1] in much more general terms. If b and c are any two nonnegative (real) numbers, then

$$\sqrt{b} \times \sqrt{c} = \sqrt{b \times c}$$
$$\text{or} \quad (\sqrt{b})(\sqrt{c}) = \sqrt{bc} \qquad [2]$$

*Because we call the square-root sign a **radical**, we call what's under the radical the **radicand**. For example, in $\sqrt{7}$, 7 is called the radicand. So the product of two square roots is the square root of the product of the radicands.*

Example 11

Express $\sqrt{5} \times \sqrt{2}$ as a single square root.

$\sqrt{5} \times \sqrt{2} = \sqrt{10}$.

You multiply 5 and 2 to get 10, and you take the square root of the product.

Notice that by symmetry, [2] can be written as

$$\sqrt{b \times c} = \sqrt{b} \times \sqrt{c} \qquad [3]$$

Symmetry means that either side of an equality can be written first.

Equation [3] suggests a way to simplify square roots. For example, suppose we have

$$\sqrt{4 \times 5} \qquad [4]$$

By [3], this may be rewritten as $\sqrt{4} \times \sqrt{5}$, and since $\sqrt{4} = 2$, we have

$$\sqrt{4 \times 5} = \sqrt{4} \times \sqrt{5} = 2 \times \sqrt{5} = 2\sqrt{5}$$

Section 14.2 Arithmetic Involving Square Roots

Prime factorization can also be applied toward simplifying radicals. For example, if we want to find $\sqrt{20}$, we can factor 20 as $2 \times 2 \times 5$ and write

$$\begin{aligned}\sqrt{20} &= \sqrt{2 \times 2 \times 5} \\ &= \sqrt{2 \times 2} \times \sqrt{5} \\ &= 2 \times \sqrt{5} \\ &= 2\sqrt{5}\end{aligned}$$

A quick way to get from $\sqrt{2 \times 2 \times 5}$ to $2\sqrt{5}$ is to remember that every time a factor appears *twice* in the radicand, it can be taken outside once.
For example, $\sqrt{3 \times 3 \times 7} = 3\sqrt{7}$. The reason is that by [3], we can write

$$\begin{aligned}\sqrt{(3 \times 3) \times 7} &= \sqrt{3 \times 3} \times \sqrt{7} \\ &= 3 \times \sqrt{7}\end{aligned}$$

Example 12

Write $\sqrt{72}$ as a multiple of $\sqrt{2}$.

$\sqrt{72} = 6\sqrt{2}$.

If you recognize that 36 is a perfect square and that $72 = 36 \times 2$, you have

$$\sqrt{72} = \sqrt{36 \times 2} = \sqrt{36} \times \sqrt{2} = 6 \times \sqrt{2} = 6\sqrt{2}$$

You can also use prime factorization to see that $72 = 8 \times 9 = 2 \times 2 \times 2 \times 3 \times 3$. Therefore

$$\begin{aligned}\sqrt{72} &= \sqrt{2 \times 2 \times 2 \times 3 \times 3} \\ &= 2 \times \sqrt{2 \times 3 \times 3} \\ &= 2 \times 3 \times \sqrt{2} = 6\sqrt{2}\end{aligned}$$

In exponent notation,

$$72 = 2^3 \times 3^2$$

CHECK

$$\begin{aligned}(6\sqrt{2})^2 &= (6\sqrt{2})(6\sqrt{2}) \\ &= 6(6)\sqrt{2}(\sqrt{2}) \\ &= 36 \quad (2) \quad = 72\end{aligned}$$

HAND CALCULATOR

$$\begin{aligned}\sqrt{72} &= 8.485281 \\ 6\sqrt{2} &= 6(1.4142135) \\ &= 8.485281\end{aligned}$$

Example 13

Write $2\sqrt{3} + 5\sqrt{12} + 4\sqrt{27}$ as a multiple of $\sqrt{3}$.

The sum is $24\sqrt{3}$. You have

$$\sqrt{12} = \sqrt{2 \times 2 \times 3} = 2\sqrt{3}$$

and

$$\sqrt{27} = \sqrt{3 \times 3 \times 3} = 3\sqrt{3}$$

Hence

$$\begin{aligned}&2\sqrt{3} + 5\sqrt{12} + 4\sqrt{27} \\ &= 2\sqrt{3} + 5(2\sqrt{3}) + 4(3\sqrt{3}) \\ &= 2\sqrt{3} + 10\sqrt{3} + 12\sqrt{3} \\ &= (2 + 10 + 12)\sqrt{3} \\ &= 24\sqrt{3}\end{aligned}$$

We should emphasize that when we multiply we do not need like square roots. For example, suppose we take $\sqrt{2} \times \sqrt{3}$ and multiply it by itself:

$$\begin{aligned}&(\sqrt{2} \times \sqrt{3}) \times (\sqrt{2} \times \sqrt{3}) \\ &= (\sqrt{2} \times \sqrt{2}) \times (\sqrt{3} \times \sqrt{3}) \\ &= 2 \times 3 = 6\end{aligned}$$

If the square of $\sqrt{2} \times \sqrt{3}$ is 6, then $\sqrt{2} \times \sqrt{3} = \sqrt{6}$. In other words, when we multiply square roots, we multiply the radicands and keep the square root.

As a preview of some of the work we'll be doing in the rest of the lesson, let's conclude this section with a few examples that show us how the material is connected to the idea of solving quadratic equations.

Example 14

Find the sum of $2+\sqrt{3}$ and $2-\sqrt{3}$.

The sum is 4. That is,

$$(2+\sqrt{3})+(2-\sqrt{3})$$
$$=(2+\sqrt{3})+(2+{}^-\sqrt{3})$$
$$=(2+2)+\underbrace{(\sqrt{3}+{}^-\sqrt{3})}_{0}=4$$

Remember that for *any* number n, $n+{}^-n=0$.

Example 15

Find the product of $2+\sqrt{3}$ and $2-\sqrt{3}$.

The product is 1. That is,

$$(2+\sqrt{3})(2-\sqrt{3})=2^2-(\sqrt{3})^2$$
$$=4-3=1$$

These two examples show us a pair of numbers ($2+\sqrt{3}$ and $2-\sqrt{3}$) whose sum is 4 and whose product is 1. In the last lesson, to solve the equation

$$x^2+4x+1=0$$

by factoring, we had to find two numbers whose product is 1 and whose sum is 4. There are no integers with these properties, but the pair of numbers $2+\sqrt{3}$ and $2-\sqrt{3}$ does fit the properties.

Again, remember that for *any* numbers n and m,

$$(n+m)(n-m)=n^2-m^2$$

To review:

$$(n+m)(n-m)$$
$$=n(n-m)+m(n-m)$$
$$=n^2\underbrace{-nm+mn}_{0}-m^2$$

That is, the only pair of positive integers whose product is 1 is 1 and 1. In this case the sum is 2, not 4.

Example 16

Show that $3+\sqrt{11}$ and $3-\sqrt{11}$ are a pair of numbers whose sum is 6 and whose product is $^-2$.

The sum is

$$\begin{array}{r}3+\sqrt{11}\\+\,3-\sqrt{11}\\\hline 6\end{array}$$

The product is

$$(3+\sqrt{11})(3-\sqrt{11})=3^2-(\sqrt{11})^2$$
$$=9-11=-2$$

These are exactly the two numbers we'd have to find if we wanted to use factoring to solve the equation

$$x^2+6x-2=0$$

because $(x^2+6x-2)=(x+r)(x+s)$ where $r+s=6$ and $rs={}^-2$.

Practice the material in this section. It will be useful later in the lesson, when you attempt to find a way to solve all quadratic equations.

PRACTICE DRILL

What number is named by each of the following?
1. $(\sqrt{5}+2)(\sqrt{5}-2)$
2. $(5+\sqrt{2})(5-\sqrt{2})$
3. $(\sqrt{5}+\sqrt{2})(\sqrt{5}-\sqrt{2})$
4. $(\sqrt{5}+2)+(\sqrt{5}-2)$
5. $(5+\sqrt{2})+(5-\sqrt{2})$
6. $(\sqrt{5}+\sqrt{2})+(\sqrt{5}-\sqrt{2})$

Answers
1. 1 2. 23 3. 3
4. $2\sqrt{5}$ 5. 10 6. $2\sqrt{5}$
7. 3.14 8. 0.32
9. 2.4393 (or rounded off, 2.44)

If you assume that $\sqrt{2} = 1.41$ and $\sqrt{3} = 1.73$, find the value of:
7. $\sqrt{3} + \sqrt{2}$ 8. $\sqrt{3} - \sqrt{2}$ 9. $\sqrt{3} \times \sqrt{2}$

CHECK THE MAIN IDEAS

1. When you square $\sqrt{5}$, you get _____.
2. This should not be confused with doubling $\sqrt{5}$, which is _____.
3. In general, $m\sqrt{5} + n\sqrt{5} = (\underline{})\sqrt{5}$. That is, you add the coefficients and keep the common radicand.
4. You must be careful about adding different radicands. For example, $\sqrt{9} = 3$ and $\sqrt{16} = 4$. Therefore, $\sqrt{9} + \sqrt{16} = $ _____.
5. But $\sqrt{9 + 16} = $ _____. Therefore, $\sqrt{9} + \sqrt{16}$ is not equal to $\sqrt{9 + 16}$.
6. Once you know how to add and multiply square roots, you may apply the regular rules of arithmetic. For example, for any real numbers b and c, $(b+c)(b-c) = b^2 - c^2$. Therefore, $(5 + \sqrt{2})(5 - \sqrt{2}) = 5^2 - (\underline{})^2$, or $25 - 2$, which is 23.
7. If you add $5 + \sqrt{2}$ and $5 - \sqrt{2}$, you get _____.
8. Combining the results of problems 6 and 7, you have two numbers whose sum is 10 and whose product is 23. The two numbers are _____.
9. Once you know this, you can factor $x^2 + 10x + $ _____. The factors are $(x + 5 + \sqrt{2})(x + 5 - \sqrt{2})$.

Answers
1. 5   2. $2\sqrt{5}$ 3. $m + n$
4. 7   5. 5   6. $\sqrt{2}$
7. 10   8. $5 \pm \sqrt{2}$ 9. 23

Section 14.3
Completing the Square

Until now, we've used a single strategy to solve quadratic equations. We've written one side of the equation as zero and factored the other side. But in any game, there are times when we must use more than a single strategy — times when one strategy works better than another.

Example 1
Find all solutions of the quadratic equation
$$(x - 3)^2 = 16 \qquad [1]$$

The solutions are $x = -1$ and $x = 4$.
You can solve this problem just as you would do in the previous lesson. Namely, rewrite $(x - 3)^2$ as $(x - 3)(x - 3)$ or $x^2 - 6x + 9$. Then subtract 16 from both sides of [1] to obtain

$$\begin{array}{r} x^2 - 6x + 9 = 16 \\ -16 -16 \\ \hline x^2 - 6x - 7 = 0 \end{array}$$

or $\qquad (x - 7)(x + 1) = 0$

But why go through all this work? Equation [1] is set up in a way that almost begs you to take the square root of both sides. If you do this, you get
$$x - 3 = \pm\sqrt{16} \qquad [2]$$

OBJECTIVE
To learn what must be added to $x^2 + bx$ to make it the square of a linear polynomial.

Perhaps you're not used to thinking of algebra as a game. But is this such a farfetched idea? You are given an equation, and the strategy is to solve the equation, using as the "rules of the game," the rules of arithmetic.

Equation [1] is also a special form of a quadratic equation. In this case, the special form equates two perfect squares. The left side is the square of a linear polynomial, and the right side is the square of an integer.

Remember that $\pm\sqrt{16}$ is an abbreviation for "$+\sqrt{16}$ or $-\sqrt{16}$."

That is, $x - 3 = 4$, or $x - 3 = -4$. If $x - 3 = 4$, then $x = 7$; while if $x - 3 = -4$, then $x = -1$. See how quickly you are able to get the answer with square roots?

Let's try a few more examples.

Example 2

Find all solutions of the equation

$$(x + 5)^2 = 36 \qquad [3]$$

The solutions are $x = -11$ and $x = 1$.
 You take the square root of both sides to obtain

$$x + 5 = \pm\sqrt{36} \qquad [4]$$

Another way of writing [4] is

$$x + 5 = 6 \quad \text{or} \quad x + 5 = -6 \qquad [5]$$

Now if $x + 5 = 6$, then $x = 1$; and if $x + 5 = -6$, then $x = -11$.

We cannot take the square root of both sides in [3] by writing $x + 5 = 6$. If we do, we still get the answer $x = 1$, but we lose the answer $x = -11$. We must continue to keep in mind that every positive number has *two* square roots.

Example 3

Find all solutions of the equation:

$$(x - 4)^2 = 7 \qquad [6]$$

The solutions are given by $x = 4 \pm \sqrt{7}$. That is, either $x = 4 + \sqrt{7}$ or $x = 4 - \sqrt{7}$.
 Just as in the other examples, you take the square root of both sides to obtain

$$x - 4 = \pm\sqrt{7} \qquad [7]$$

You now add 4 to both sides of [7] to get the desired result.

$$x = 4 \pm \sqrt{7} \qquad [8]$$

Equation [8] is an abbreviation for the two separate equations

$$x = 4 + \sqrt{7} \qquad [9]$$
or $\qquad x = 4 - \sqrt{7} \qquad [10]$

Example 4

Find the positive solution of

$$(x - 3)^2 = 11 \qquad [11]$$

and express the answer correct to the nearest hundredth.

To the nearest hundredth, the positive solution is $x = 6.32$.
 You take the square root of both sides of [11] to get

$$x - 3 = \pm\sqrt{11} \qquad [12]$$

Then you add 3 to both sides of [12] to get

$$x = 3 \pm \sqrt{11} \qquad [13]$$

REVIEW
If we let u and v denote any algebraic expression (including numbers), then $u^2 = v^2$ if either $u = v$ or $u = \bar{v}$.

CHECK
$(-11 + 5)^2 = (-6)^2 = 36$
$(1 + 5)^2 = 6^2 = 36$

Notice that both the square of 6 and the square of -6 equal 36.

You might want to use the $\pm$ on both sides of [4] when you take the square root. That is, you may want to write

$$\pm(x + 5) = \pm 6$$

But writing $u = \pm v$ means the same thing as $\pm u = \pm v$. In either case $u = v$ or $u = -v$.

There is no difference in theory between writing $\pm\sqrt{36}$ and $\pm\sqrt{7}$. The only difference is that 36 is a perfect square, so that we may replace $\sqrt{36}$ by 6.

To the beginner, there is something unreal about $\sqrt{7}$. But it's a real number: it's the number whose square is 7.
 Using a hand calculator, we find that $\sqrt{7} = 2.6457513$. If we round this value off to 2.646, we see from [9] and [10] that

$$x = 4 + 2.646 = 6.646$$
or $\qquad x = 4 - 2.646 = 1.354$

As a rough check (since the calculator rounds off), we have

$$(6.646 - 4)^2 = (2.646)^2$$
$$= 7.001316$$
$$(1.354 - 4)^2 = (-2.646)^2$$
$$= 7.001316$$

Section 14.3 Completing the Square

If you use a calculator, tables, or trial and error, you find that to the nearest hundredth, $\sqrt{11} = 3.32$; hence, you may rewrite [13] as $x = 3 + 3.32$; or $x = 3 - 3.32$. That is, $x = 6.32$ or $x = -0.32$; but only 6.32 is positive.

NOTATION
To the nearest hundredth, $\sqrt{11} = 3.32$; but to be correct, we write $\sqrt{11} \approx 3.32$.

NOTE

Since 3.32 is only an approximation for $\sqrt{11}$, we cannot expect that $x = 6.32$ will be the exact solution of [11]. However, we can expect that no number (to the nearest hundredth) will give us a better approximation.

To check this, we notice that 6.32 is between the two hundreths 6.31 and 6.33. If we replace x by 6.32, 6.31, and 6.33 in the left side of [11], we get

$$(6.31 - 3)^2 = (3.31)^2 = 10.9561$$
$$\searrow 0.0439$$
$$11.0000$$
$$\nearrow 0.0224$$
$$(6.32 - 3)^2 = (3.32)^2 = 11.0224$$
$$(6.33 - 3)^2 = (3.33)^2 = 11.0889$$

From this chart, it is easy to see that $x = 6.32$ is closer to the right answer than either $x = 6.31$ or $x = 6.32$.

Now let's return to the major theme of this unit. So far in this lesson, we have shown that some quadratic equations are written in the form

$$(x + b)^2 = c \qquad [14]$$

If they are written in the form [14], they are rather easy to solve.

Supposing the equation is not written in this form? For example, one problem we want to solve before the end of this unit is

$$x^2 + 3x + 4 = 50 \qquad [15]$$

That is, in this section we have already studied

$(x - 3)^2 = 16 \; (= 4^2)$
$(x + 5)^2 = 36 \; (= 6^2)$
$(x - 4)^2 = 7 \;\; (= \sqrt{7}^2)$
$(x - 3)^2 = 11$

Equation [15] leads to the main point of this section, which is completing the square. But just what is meant by completing the square?

To understand completing the square, let's return to the product

$$(x + r)(x + s) \qquad [16]$$

We already know that [16] is the same as

$$x^2 + (r + s)x + rs \qquad [17]$$

An interesting case occurs when r and s are equal. In that case, $r + s = r + r = 2r$ and $rs = r \cdot r = r^2$. That is,

$$(x + r)^2 = x^2 + 2rx + r^2 \qquad [18]$$

If we want to emphasize factoring, we can reverse the sides of [18] to get

$$x^2 + 2rx + r^2 = (x + r)^2 \qquad [19]$$

Now let's look at the left side of [19] to find a way of remembering to add r^2 to make $x^2 + 2rx$ a perfect square.

We get r^2 by squaring r, and we get r by taking half the coefficient of $x (2r)$. This gives us a system. To convert $x^2 + 2rx$ into a perfect square:

For this reason, $x^2 + 2rx + r^2$ is called a **perfect square**. It is the square of $(x + r)$. The sum of three terms is called a **trinomial**. So $x^2 + 2rx + r^2$ is a perfect square trinomial.

STEP 1: Take half the coefficient of x.

Since the coefficient is $2r$, half is r.

STEP 2: Square this result.

In this case, we get r^2.

STEP 3: Add this result to $x^2 + 2rx$.

That is, form $x^2 + 2rx + r^2$.

STEP 4: The resulting expression is equal to $(x + r)^2$.

Steps 1, 2, and 3 are called completing the square. The process adds on to $x^2 + 2rx$ the amount necessary to make it $(x + r)^2$.

Let's try to complete some squares.

Example 5

What must you add to $x^2 + 10x$ to complete the square?

You must add 25.
 The coefficient of x is 10. Half of it is 5. So you square 5, which gives you 25. $x^2 + 10x + 25$ is a perfect square. It is $(x + 5)^2$. That is, $(x + 5)(x + 5) = x^2 + 10x + 25$.

CHECK
$(x + 5)^2 = x^2 + (5 + 5)x + 5^2$
$= x^2 + 10x + 25$

Example 6

What must you add to $x^2 + 16x$ to complete the square?

You must add 64.
 You take the coefficient of x (which is 16) and divide it by 2 (that is, you take half of it). This gives you 8. Then you square 8 to get 64. So $x^2 + 16x + 64$ is a perfect square. It is $(x + 8)^2$.

 If the coefficient of x is negative, we still proceed in the same way.

Again, the number we add on to x is half the coefficient of x in the expression $x^2 + 16x$.

Example 7

What must you add on to $x^2 - 6x$ to complete the square?

You add 9.
 The coefficient of x is now -6, and half of that is -3. The square of -3 is 9. So $x^2 - 6x + 9 = (x - 3)^2$.

 Sometimes we have to add on fractions to complete the square.

That is,

$x^2 - 6x + 9 = x^2 + {}^-6x + 9$
$= (x + {}^-3)^2$
$= (x - 3)^2$

Example 8

What must you add onto $x^2 + 9x$ to complete the square?

To complete the square, you must add $\frac{81}{4}$.

 In this case, the coefficient of x is 9. Half of this is $4\frac{1}{2}$ or $\frac{9}{2}$; and if you square this, you get $\frac{81}{4}$. That is $x^2 + 9x + \frac{81}{4} = \left(x + \frac{9}{2}\right)^2$.

That is, $\frac{9}{2} \times \frac{9}{2} = \frac{9 \times 9}{2 \times 2} = \frac{81}{4}$

 Additional examples are left for the exercises. In the next section, we shall see how this idea allows us to solve any quadratic equation.

PRACTICE DRILL

Find all the solutions of
1. $(x - 5)^2 = 16$ 2. $(x + 3)^2 = 25$ 3. $(x - 2)^2 = 7$
4. Use the approximation that the square root of 3 is 1.732 to find the solutions of $(x - 5)^2 = 3$.

Complete the square in each of the following.
5. $x^2 + 12x$ 6. $x^2 - 12x$ 7. $x^2 + 13x$

Answers
1. $x = 1$ or $x = 9$ 2. $x = 2$ or $x = -8$ 3. $x = 2 \pm \sqrt{7}$
4. 3.268 and 6.732   5. $x^2 + 12x + 36$ 6. $x^2 - 12x + 36$
7. $x^2 + 13x + \left(\frac{13}{2}\right)^2$, or $x^2 + 13x + \frac{169}{4}$

Section 14.4 How to Solve Any Quadratic Equation

CHECK THE MAIN IDEAS

1. So far, the method for solving quadratic equations requires finding two numbers that have a given sum and a given product. But there is another method that is helpful if the equation can be written in the form:
$$(x - b)^2 = c^2 \qquad [1]$$
If [1] is true, then you can take the _____ of both sides.

2. This gives you $x - b =$ _____.

3. Suppose you have the equation $(x - 5)^2 = 16$. If you take the square root of both sides you find that $x - 5 =$ _____.

4. Once you know that $x - 5 = \pm 4$, you know that x is $5 +$ (a) _____ or $5 -$ (b) _____. In other words, either $x = 9$ or $x = 1$.

5. Similarly, if you take the square root of both sides of $(x - 2)^2 = 7$, you get that $x - 2 =$ _____.

6. The equation isn't always given in the form $(x + b)^2 = c^2$. To get this form, you have to complete the _____.

7. Starting with $x^2 + bx$, you complete the square first by taking half of _____.

8. Then you square this and add it to $x^2 + bx$. Since the square of $\frac{b}{2}$ is $\frac{b^2}{4}$, you get $x^2 + bx +$ _____.

9. This is called completing the square because $x^2 + bx + \frac{b^2}{4}$ is the square of the linear polynomial _____.

Answers
1. square root 2. $\pm c$
3. ± 4 4. (a) 4 (b) 4
5. $\pm \sqrt{7}$ 6. square
7. b (or the coefficient of x)
8. $\frac{b^2}{4}$ 9. $x + \frac{b}{2}$

Section 14.4
How to Solve Any Quadratic Equation

In this section, we learn to solve any quadratic equation by the method of completing the square.

Example 1

Solve the equation
$$x^2 + 8x = 9 \qquad [1]$$
by completing the square.

The solutions are $x = 1$ and $x = -9$.

Look at the left side of [1]. The coefficient of x^2 is 1. So you can complete the square by taking half the coefficient of x, squaring it, and adding it to $x^2 + 8x$. The coefficient of x is 8; half of it is 4; and the square of 4 is 16. You add 16 to $x^2 + 8x$ to get $x^2 + 8x + 16$.

But whatever you add to one side of an equation, you must also add to the other side in order to maintain the equality. So you add 16 to both sides of [1] to get

$$x^2 + 8x + 16 = 9 + 16$$
or
$$(x + 4)^2 = 25 \qquad [2]$$

You now solve [2] as in the previous section, by taking the square root of both sides.

$$x + 4 = \pm 5 \qquad [3]$$

OBJECTIVE
To show how to use the method of completing the square to find *all* (real) solutions of *any* quadratic equation.

For our first example we've deliberately chosen an equation that can be solved by the method of a previous lesson — by factoring. This gives us an easy way to check our answer and to see that there are times when the same problem can be solved in different ways.

We usually leave out the explanation and simply add 16, so that we write

$$x^2 + 8x = 9$$
$$x^2 + 8x + 16 = 25$$
$$(x + 4)^2 = 25$$
$$x + 4 = \pm 5$$
$$x = -4 \pm 5$$

That is, either $x + 4 = 5$ or $x + 4 = -5$. If $x + 4 = 5$, then $x = 1$; if $x + 4 = -5$, then $x = -9$.

Once we add 16 to both sides of [1] to get [2], we proceed just as we did in the previous section. In other words, all we've done is put the equation into a form that we were able to solve in the previous section.

We could have solved equation [1] by subtracting 9 from both sides to get

$$x^2 + 8x - 9 = 0$$
or $(x + 9)(x - 1) = 0$

This is how factoring works. In this case, we want two numbers whose product is -9 and whose sum is 8.

Hence either $x + 9 = 0$ or $x - 1 = 0$.

It is possible that based on Example 1, factoring seems easier to use than completing the square. The true value of completing the square is clear when the factoring becomes more difficult. For example, if we make just a slight variation in Example 1, the factoring method becomes very difficult, but the completing-the-square method remains the same.

Example 2

Solve the equation

$$x^2 + 8x = 10 \qquad [4]$$

To get [4], we replaced 9 by 10 on the right side of [1].

The solutions are given by $x = -4 \pm \sqrt{26}$.

You still complete the square in the same way: by adding 16 to both sides of [4]. You get

$$x^2 + 8x + 16 = 26 \quad \text{or} \quad (x + 4)^2 = 26$$

You then take the square root of both sides to get

$$x + 4 = \pm\sqrt{26} \qquad [5]$$

As far as completing the square is concerned, the only difference in this example is that 26 is not a perfect square.

Finally, you subtract 4 from both sides of [5] to get $x = -4 \pm \sqrt{26}$.

NOTE

The answers $x = -4 \pm \sqrt{26}$ are exact. We can get numerical approximations by computing the square root of 26 to any decimal-place accuracy we desire.

We could try to use factoring to solve equation [4]. We subtract 10 from both sides to get

$$x^2 + 8x - 10 = 0 \qquad [6]$$

But to solve [6] by factoring, we need two numbers whose product is -10 and whose sum is 8. There is no such pair of *integers*. In other words, any problem that can be solved by factoring can also be solved by completing the square. But sometimes problems that can be solved by completing the square are very difficult to solve by factoring.

On a calculator, the square root of 26 is 5.0990195. So if we were satisfied with an approximation to the nearest 0.01, we would use 5.10. That is, the approximate answers would be

$x = -4 + 5.10 = 1.10$
and $x = -4 - 5.10 = -9.10$

The final 0 is kept to remind us that we are accurate to the nearest 0.01.

Example 3

Solve the equation

$$x^2 + 3x = 46 \qquad [7]$$

The answers are given by

$$x = \frac{-3 \pm \sqrt{193}}{2} \qquad [8]$$

Section 14.4 How to Solve Any Quadratic Equation

To complete the square, you take half of 3 and square it. This gives you $\left(\frac{3}{2}\right)^2$ or $\frac{9}{4}$, which you add to both sides of [8].

This leads to

$$x^2 + 3x + \frac{9}{4} = 46 + \frac{9}{4} = \frac{184}{4} + \frac{9}{4}$$

or

$$\left(x + \frac{3}{2}\right)^2 = \frac{193}{4} \qquad [9]$$

Now you take the square root of both sides of [9] to obtain

$$x + \frac{3}{2} = \pm\sqrt{\frac{193}{4}} = \pm\frac{\sqrt{193}}{2} \qquad [10]$$

[10] Notice that in theory nothing new is present in this example. It is true, however, that the amount of arithmetic is greater and a bit more complicated.

You then subtract $\frac{3}{2}$ from both sides of [10] to obtain

$$x = -\frac{3}{2} \pm \frac{\sqrt{193}}{2}$$

or

$$x = \frac{-3 \pm \sqrt{193}}{2} \qquad [11]$$

Again, we remember that $\sqrt{193}$ is an exact number (it's exactly the number that when multiplied by itself equals 193). Usually, however, we prefer a decimal approximation. On a calculator, we find that $\sqrt{193} = 13.892443$.

If we wish, we can rewrite [11] as

$$x \approx \frac{-3 \pm 13.892443}{2} \qquad [12]$$

From [12], we see that within the accuracy of a calculator, either

$$x = 5.4462215$$
$$\text{or} \quad x = -8.4462215 \qquad [13]$$

That is,

$$x = \frac{-3 + 13.892443}{2}$$
or
$$x = \frac{-3 - 13.892443}{2}$$

Equations [11] and [13] give us much more insight about what happened when we used trial and error to find an approximate solution of $x^2 + 3x + 4 = 50$. In fact, [11] gives us the two exact answers, which we can approximate to any decimal-place accuracy that we want.

Example 4

Solve the equation

$$x^2 + 3x = 75 \qquad [14]$$

The solutions are given by

$$x = \frac{-3 \pm \sqrt{309}}{2} \qquad [15]$$

Again, you complete the square by adding $\frac{9}{4}$ to both sides of [14]. You get

$$x^2 + 3x + \frac{9}{4} = 75 + \frac{9}{4} = \frac{300}{4} + \frac{9}{4}$$

or

$$\left(x + \frac{3}{2}\right)^2 = \frac{309}{4} \qquad [16]$$

Then you take the square root of both sides of [16] to get

$$x + \frac{3}{2} = \pm \frac{\sqrt{309}}{2} \qquad [17]$$

Subtracting $\frac{3}{2}$ from both sides of [17], you get

$$x = -\frac{3}{2} \pm \frac{\sqrt{309}}{2}.$$

Example 5

Find the value of the positive solution of [14] correct to the nearest 0.001.

The solution is 7.289. To the nearest thousandth, the square root of 309 is 17.578, and you use this in [15].

Up to now, the coefficient of x^2 has always been 1. If the coefficient of x^2 is not 1, we do not have to make great changes. All we have to do is divide both sides of the equation by the coefficient of x^2 to get an equivalent equation in which the coefficient of x^2 is 1.

That is,

$$x = \frac{-3 + 17.578}{2}$$
$$= \frac{14.578}{2}$$
$$= 7.289$$

Example 6

Find all solutions of the equation

$$4x^2 + 8x = 5 \qquad [18]$$

The solutions are $x = \frac{1}{2}$ and $x = -\frac{5}{2}$.

You divide both sides of [18] by 4 to get

$$x^2 + 2x = \frac{5}{4} \qquad [19]$$

Remember to divide $8x$ by 4. In terms of the distributive property, think of [18] as being

$$4(x^2 + 2x) = 5$$

Now you can complete the square on the left side of [19] by adding 1. To keep the equation balanced, you must add 1 to both sides. You get

$$x^2 + 2x + 1 = \frac{5}{4} + 1 = \frac{9}{4}$$

and then divide both sides of the equation by 4.

or $\quad (x + 1)^2 = \frac{9}{4} \qquad [20]$

Next you take the square root of both sides of [20] to obtain

$$x + 1 = \pm \frac{3}{2}$$

or $\quad x = -1 \pm \frac{3}{2} \qquad [21]$

That is,

$$x = -1 + \frac{3}{2} = \frac{1}{2}$$

or $\quad x = -1 - \frac{3}{2} = -\frac{5}{2}$

CHECK

$$4\left(\frac{1}{2}\right)^2 + 8\left(\frac{1}{2}\right) = 4\left(\frac{1}{4}\right) + 4$$
$$= 1 + 4 = 5$$

$$4\left(-\frac{5}{2}\right)^2 + 8\left(-\frac{5}{2}\right)$$
$$= 4\left(\frac{25}{4}\right) - 20 = 25 - 20 = 5$$

Section 14.4 How to Solve Any Quadratic Equation   377

Example 7

Find all solutions of the equation

$$2x^2 - 5x = 18 \qquad [22]$$

The solutions are $x = 4\frac{1}{2}$ and $x = -2$.

You begin by dividing both sides of [22] by 2 to get

$$x^2 - \frac{5}{2}x = 9 \qquad [23]$$

Now you complete the square by adding $\frac{25}{16}$ to both sides of [23]. You get

$$x^2 - \frac{5}{2}x + \frac{25}{16} = 9 + \frac{25}{16} = \frac{144}{16} + \frac{25}{16}$$

or

$$\left(x - \frac{5}{4}\right)^2 = \frac{169}{16} \qquad [24]$$

The theory remains the same. The only difference in [23] is that when we take half the coefficient of x and square it, we have to deal with fractions instead of whole numbers.

If you take the square root of both sides of [24], you get

$$x - \frac{5}{4} = \pm \frac{13}{4}$$

$$x = \frac{5}{4} \pm \frac{13}{4}$$

The point is that once we divide both sides of the equation by the coefficient of x^2, we get the same kind of equation that we were solving earlier in the section.

That is, either $x = \frac{5}{4} + \frac{13}{4} \left(= \frac{18}{4} \right)$ or $x = \frac{5}{4} - \frac{13}{4} \left(= -\frac{8}{4} \right)$.

Of course, sometimes things work out less conveniently and we get more complicated radicals. Suppose we replace the 18 on the right side of [22] by 6.

Example 8

Find all solutions of the equation

$$2x^2 - 5x = 6 \qquad [25]$$

The solutions are given by $x = \frac{5 \pm \sqrt{73}}{4}$.

Just as in Example 7, you begin by dividing both sides of the equation by 2. You get

$$x^2 - \frac{5}{2}x = 3$$

To complete the square, you add $\frac{25}{16}$ to both sides of the equation

$$x^2 - \frac{5}{2}x + \frac{25}{16} = 3 + \frac{25}{16} = \frac{48}{16} + \frac{25}{16}$$

so that

$$\left(x - \frac{5}{4}\right)^2 = \frac{73}{16}$$

Now you take the square root of both sides to obtain

$$x - \frac{5}{4} = \pm \frac{\sqrt{73}}{4}$$

or

$$x = \frac{5}{4} \pm \frac{\sqrt{73}}{4} = \frac{5 \pm \sqrt{73}}{4} \qquad [26]$$

NOTE

If we prefer a numerical approximation, we find that $\sqrt{73} = 8.544\ldots$. We may rewrite [26] as $x = \dfrac{5 \pm 8.544\ldots}{4}$. That is, the solutions of [25] are given (approximately) by

$$x = \frac{5 + 8.544}{4} = 3.386$$

or

$$x = \frac{5 - 8.544}{4} = -0.886$$

CHECK

$2(3.386)^2 - 5(3.386)$
$= 2(11.4649960) - 16.93$
$= 22.92992 - 16.93$
$= 5.99992$ (which is about 6)

$2(-0.886)^2 - 5(-0.886)$
$= 2(0.78499) + 4.43$
$= 1.56998 + 4.43$
$= 5.99998$ (which is also about 6)

Examples 6, 7, and 8 show us why we may assume that the coefficient of x^2 is always 1. When it isn't, we just divide both sides of the equation by the coefficient, and the resulting equation (which has the same solutions as the original one) has the coefficient of x equal to 1.

By now we can see that the method of completing the square follows the same steps regardless of the numbers we use. It is typical in mathematics that whenever a method works in general, we try to state it in general terms rather than redo the problem each time we see new numbers.

For the quadratic equation, instead of dealing specifically with an equation like

$$x^2 + 4x - 6 = 0 \qquad [27]$$

we deal with the general equation

$$x^2 + 2bx + c = 0 \qquad [28]$$

Let's see how we can solve [28] to find the value(s) of x in terms of b and c.

To help see what we're doing in [28], we think of [27] as a special case of [28], in which $2b = 4$ and $c = -6$.

STEP 1: Add ^-c to both sides of [28]. You get

$$x^2 + 2bx + c + {}^-c = 0 + {}^-c$$
$$x^2 + 2bx + \underbrace{(c + {}^-c)}_{0} = {}^-c$$
$$x^2 + 2bx = {}^-c \qquad [29]$$

STEP 2: Complete the square by adding b^2 to both sides of [29]. You get

$$x^2 + 2bx + b^2 = {}^-c + b^2 \,(= b^2 + {}^-c)$$
or $\quad (x + b)^2 = b^2 - c \qquad [30]$

STEP 3: Take the square root of both sides of [30].

$$x + b = \pm\sqrt{b^2 - c} \qquad [31]$$

STEP 4: Add ^-b to both sides of [31].
$$x = \pm\sqrt{b^2 - c} \qquad [32]$$

Equation [32] shows us how to write down the answer just by looking at the equation. For example, in

$$x^2 + 4x - 6 = 0 \qquad [27]$$

$2b = 4$ and $c = -6$. If $2b = 4$, then $b = 2$. So we simply replace b by 2 and c by -6 in [32] to obtain

$$x = -2 \pm \sqrt{2^2 - {}^-6}$$
$$x = -2 \pm \sqrt{4 + 6}$$
$$x = -2 \pm \sqrt{10} \qquad [33]$$

This happened in our study of linear equations. Instead of talking about solving a specific equation like $3x + 7 = 2x - 9$, we talked about the general equation $bx + c = dx + e$.

The use of $2b$ rather than b in [28] helps to make the final formula and the arithmetic a bit easier. Notice that any number can be written in the form $2b$. For example, $5 = 2\left(\dfrac{5}{2}\right)$, so $5 = 2b$, where $b = \dfrac{5}{2}$.

NEW VOCABULARY

Instead of talking about the values of x that satisfy an equation, we often talk about the **roots** of an equation. For example, given the equation $(x - 3)(x - 5) = 0$, we say that the roots of this equation are 3 and 5.

Remember that $x^2 + 4x - 6$ is the same as $x^2 + 4x + {}^-6$. We write [27] in this way because [32] assumes that the equation is $x^2 + bx + c = 0$, and not $x^2 + bx - c = 0$.

It is important to understand that getting [33] from [32] saves us the trouble of having to start from scratch with the method of completing the square to solve

Section 14.4 How to Solve Any Quadratic Equation

[27]. For example, the usual way to solve [27] is by using the same steps we used to solve [28]. That is,

STEP 1: $x^2 + 4x = 6$
STEP 2: $x^2 + 4x + 4 = 6 + 4 = 10$ or $(x + 2)^2 = 10$
STEP 3: $x + 2 = \pm\sqrt{10}$
STEP 4: $x = {}^-2 \pm \sqrt{10}$ [34]

Before closing this section, we should realize that many people like to write the general quadratic equation in the form

$$ax^2 + bx + c = 0 \qquad [35]$$

In this form, we find x by the formula

$$x = \frac{-b \pm \sqrt{b^2 - 4ac}}{2a} \qquad [36]$$

The relatively complicated form of [36] leads us to prefer [28] to [35].

Equations [33] and [34] are the same. The reason that we stressed completing the square rather than memorizing [32] is that the proof of [32] requires us to understand how to complete the square.

If we understand how to complete the square, we can always derive formula [32] if we need it. But just memorizing [32] without understanding it will not help our understanding of mathematics.

PRACTICE DRILL

Use completing the square to find all the solutions of the following quadratic equations.

1. $x^2 + 16x = 17$
2. $x^2 + 6x = 27$
3. $x^2 - 6x - 12 = 0$
4. $x^2 + 4x = -4$
5. $x^2 + 4x = -7$
6. $x^2 + 4x = 7$

Answers
1. $x = 1$ or $x = -17$
2. $x = 3$ or $x = -9$
3. $x = 3 \pm \sqrt{21}$
4. $x = -2$
5. no real values of x.
6. $x = -2 \pm \sqrt{11}$

CHECK THE MAIN IDEAS

1. The idea of completing the square can be applied to solving any quadratic equation. Suppose you have the equation

$$x^2 + 4x - 7 = 0 \qquad [1]$$

You can begin by adding (a) _____ to both sides of [1] to obtain

$$x^2 + 4x = \text{(b)} \underline{\qquad} \qquad [2]$$

2. By the method of the previous section, you can complete the square on the left side of [2] by adding _____.

3. Whatever you add to one side of the equation, you must also add to the other side. So if you complete the square, [2] becomes

$$x^2 + 4x + 4 = \underline{\qquad} \qquad [3]$$

4. Since $x^2 + 4x + 4 = (x + 2)^2$, equation [3] becomes $(x + 2)^2 = 11$. Taking the square root of both sides, you get

$$x + 2 = \underline{\qquad} \qquad [4]$$

5. From equation [4] you conclude that x is either $-2 +$ (a) _____ or $-2 -$ (b) _____.

6. The key to how many solutions there are to the equation

$$x^2 + 4x = b \qquad [5]$$

depends on the size of b. To complete the square, you add _____ to both sides of [5].

7. In particular, if b is less than -4, then the right side of [5] is _____.

8. This means that [5] has no real solutions, because the square of a real number cannot be _____.

Answers
1. (a) 7 (b) 7
2. 4
3. 11
4. $\pm\sqrt{11}$
5. (a) $\sqrt{11}$ (b) $\sqrt{11}$
6. 4
7. negative
8. negative
9. $x + 2$
10. negative

9. For example, if the equation were $x^2 + 4x = -7$, when you added 4 to both sides, you would have $x^2 + 4x + 4 = -7 + 4 = -3$. That is,
$$(\underline{\qquad})^2 = -3 \qquad [6]$$
10. But [6] can't have a real solution, because the left side of [6] is never \underline{\qquad}.

Section 14.5
Graphing a Quadratic Relationship

OBJECTIVE
To be able to graph

Many quadratic equations appear to be quite similar, yet they have very different solutions. For example, look at the three quadratic equations

$$x^2 + 4x + 3 = 0 \qquad [1]$$
$$x^2 + 4x + 4 = 0 \qquad [2]$$
$$x^2 + 4x + 5 = 0 \qquad [3]$$

$$y = ax^2 + bx + c$$

and to be able to use the graph to explain why every quadratic equation has either no, one, or two (real) solutions.

These three equations are identical except for the constant term on the left side of each equation, which is slightly different.
Yet the three equations have very different forms of solutions. Let's review.

Example 1
Find all the solutions of
$$x^2 + 4x + 3 = 0 \qquad [1]$$

There are two solutions: $x = -3$ and $x = -1$.
You can readily factor the left side of [1] to obtain
$$(x + 1)(x + 3) = 0 \qquad [4]$$
From [4] you see that either $x + 1 = 0$ or $x + 3 = 0$.

Whether we use factoring or completing the square, we get the same answer. The crucial point is that [1] possesses two (real) solutions.

Example 2
Find all solutions of
$$x^2 + 4x + 4 = 0 \qquad [2]$$
The only solution is $x = -2$.
You can factor [2] to obtain
$$(x + 2)(x + 2) = 0 \qquad [5]$$
From [5] you see that either $x + 2 = 0$ or $x + 2 = 0$. That is, x must equal -2.

Perhaps you have already noticed that [2] represents a perfect square. That is,
$$x^2 + 4x + 4 = (x + 2)^2$$
The important point is that [2] possesses only one solution, and not two, as in [1].

Example 3
Find all the real solutions of
$$x^2 + 4x + 5 = 0 \qquad [3]$$

There are no real solutions.
You may rewrite the left side of [3] as
$$x^2 + 4x + 4 + 1 = 0 \qquad [6]$$
Since you have completed the square, [6] may be written as
$$(x + 2)^2 + 1 = 0 \qquad [7]$$

Section 14.5   Graphing a Quadratic Relationship   381

But the fact that x is real means that $(x + 2)^2$ is nonnegative, so that the left side of [7] is at least as great as 1. Because no number as great as 1 can equal 0, equation [7] has no solutions.

If we compare these three examples, we find that although the equations [1], [2], and [3] are similar, their solutions are very different. This can be visualized readily in terms of graphs.

We have already studied graphs of linear relationships. The same ideas apply to nonlinear relationships. For example, we already know that we can think of the equation $y = x$ as being the set of points (x,y) in the plane for which $y = x$. The graph of the relationship $y = x$ is shown in the illustration.

We can also think about $y = x^2$ as being the set of points (x,y) for which $y = x^2$. This set of points is a bit harder to describe, because the points do not lie on the same line.

If we used the method of the previous section to solve [7], we would add -1 to both sides to obtain $(x + 2)^2 = -1$. Then we'd point out that if x was real, the left side must be nonnegative, and so could never equal -1 for any real value of x.

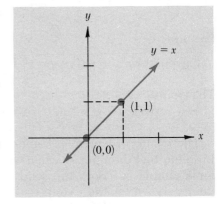

$(0,0)$ and $(1,1)$ satisfy $y = x$.
Since $y = x$ is a line, we can draw the entire graph just by drawing the line which passes through $(0,0)$ and $(1,1)$.

Example 4

Show that $(0,0)$ satisfies $y = x^2$.

If you replace x by 0 and y by 0, you get $0 = 0^2$, or $0 = 0$, which is a true statement.

Example 5

Show that $(1,1)$ satisfies $y = x^2$.

If you replace x and y by 1, you get $1 = 1^2$, or $1 = 1$, which is a true statement.

Example 6

Show that $(-1,1)$ satisfies $y = x^2$

Now you replace x by -1 and y by 1 to get $1 = (-1)^2$, or $1 = 1$. This is a true statement.

Examples 4, 5, and 6 show that the points $(0,0)$, $(1,1)$, and $(-1,1)$ all satisfy $y = x^2$. As a result, the graph of $y = x^2$ cannot be a straight line (see the illustration).

Notice that both $(1,1)$ and $(-1,1)$ belong to the curve $y = x^2$. This is true because x and $-x$ have the same square, x^2. This idea holds true in general.

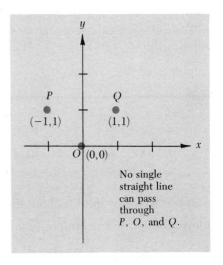

No single straight line can pass through P, O, and Q.

The big difference is that when a graph is a straight line, we need to know only two points. If the graph is "curved," we have to know many points, because then two points aren't enough to determine the entire shape.

Example 7

What point on the curve $y = x^2$ has its x-coordinate equal to 2?

The point $(2,4)$.
All you do is replace x by 2 in the equation $y = x^2$. You get $y = 2^2$, or 4. That is, the point you want is the one whose y-coordinate is 4 when the x-coordinate is 2.

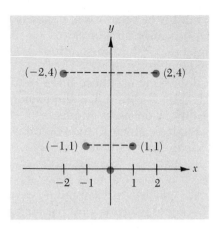

Example 8

What point on the curve $y = x^2$ has its x-coordinate equal to -2?

The point $(-2,4)$.
You replace x by -2 in $y = x^2$ to obtain $y = (-2)^2$, or $y = 4$.

The results of Examples 7 and 8 are shown in the top figure. Now we have a problem that didn't exist when we dealt with straight lines. Look at the figure for Example 8. Both the solid curve and the dotted curve pass through the points shown for the examples. How do we know which of the two curves, if either, is correct? The exact answer requires material that is beyond the scope of this course.

However, we can make a few arguments at our level to help explain why the solid line is the correct graph of the equation $y = x^2$.

The most straightforward argument is to sketch some additional points. That is, we can pick a value of x at random and then let $y = x^2$. We then locate the resulting point (x,y) on the graph.

Let's graph another quadratic to check our understanding.

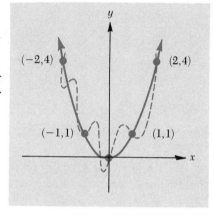

Example 9

Sketch the graph of $y = x^2 + 4x + 3$.

This graph is shown in the illustration.
You can pick values of x at random. When you choose x, you find y from the relationship $y = x^2 + 4x + 3$. For example,

x	x^2	$4x$	$x^2 + 4x$	$x^2 + 4x + 3$	(x,y)
-3	9	-12	-3	0	$(-3,0)$
-2	4	-8	-4	-1	$(-2,-1)$
-1	1	-4	-3	0	$(-1,0)$
0	0	0	0	3	$(0,3)$
1	1	4	5	8	$(1,8)$

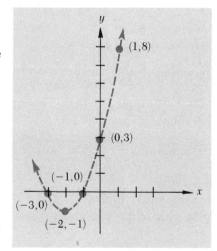

If you have any doubts about the graph drawn in the illustration, you can locate more points. For example, you can pick $x = \frac{1}{2}$. Then $x^2 = \frac{1}{4}$, $4x = 2$, and $x^2 + 4x + 3 = 5\frac{1}{4}$. This tells you that $\left(\frac{1}{2}, 5\frac{1}{4}\right)$ is also a point on the graph.

We can add credibility to the graph shown in the illustration by using the results of completing the square. Since $x^2 + 4x + 4 = (x+2)^2$, we rewrite $x^2 + 4x + 3$ as $x^2 + 4x + 4 + {}^-4 + 3$, or, $(x^2 + 4x + 4) + ({}^-4 + 3) = (x+2)^2 + {}^-1$. Notice how we use $b + {}^-b = 0$ (in this case, $4 + {}^-4 = 0$) to rewrite equations in more desirable forms.

Section 14.5 Graphing a Quadratic Relationship

Since $(x+2)^2$ is nonnegative, $(x+2)^2 - 1$ is at least as great as -1. That is, $(x+2)^2$ is 0 only when $x = -2$. Otherwise, $(x+2)^2$ is positive.

This tells us that the lowest point on the curve $y = x^2 + 4x + 3$ is $(-2, -1)$, which is shown in the illustration.

> **DEFINITION**
> The graph of $y = x^2 + bx + c$ is called a **parabola**. The lowest point of the parabola is called its **vertex**.

Example 10

Use the method of completing the square to find the lowest point (vertex) of $y = x^2 + 4x + 5$.

The lowest point is $(-2, 1)$.

You complete the square by rewriting $x^2 + 4x + 5$ as $x^2 + 4x + (4+1)$, or $(x^2 + 4x + 4) + 1$. That is, $x^2 + 4x + 5 = (x+2)^2 + 1$.

Now, $(x+2)^2$ is at least as great as 0 and equals 0 only if $x = -2$. Hence $(x+2)^2 + 1$ is least when $x = -2$, at which time $(x+2)^2 + 1$ is equal to $(-2+2)^2 + 1$, or 1.

The graph of $y = x^2 + 4x + 5$ is shown in the figure for Examples 10 and 11.

We know that we must add 4 to $x^2 + 4x$ to complete the square. This is what gives us the hint to rewrite 5 as $4 + 1$.

There is a nice connection between graphs and quadratic equations. Let's look at an equation such as $x^2 + 4x + 3 = 0$. This is the same as $x^2 + 4x + 3 = y$, if we replace 0 by y. That is, if we want to find the x-intercepts of the parabola $y = x^2 + 4x + 3$, we replace y by 0 and solve the resulting equation for x. In this case, the resulting equation is $x^2 + 4x + 3 = 0$. The roots, -1 and -3, are indeed the x-intercepts of the parabola, as shown in the illustration for Example 9.

This gives us a simple way to see if a quadratic equation has none, one, or two (real) roots.

Remember that for any curve we find the x-intercepts by replacing y by 0 in the equation of the curve.

Example 11

Use the graph of $y = x^2 + 4x + 5$ to explain why the equation

$$x^2 + 4x + 5 = 0 \quad [3]$$

can have real solutions.

If you complete the square, you may rewrite $y = x^2 + 4x + 5$ as

$$y = x^2 + 4x + 4 + 1$$
or $$y = (x+2)^2 + 1 \quad [8]$$

$x + 2$ is 0 only when $x = -2$. Otherwise, $(x+2)^2$ is greater than 0. So the lowest point of $y = x^2 + 4x + 5$ occurs when $x = -2$, at which time $y = 1$ [that is, $y = (-2+2)^2 + 1$ or $0 + 1$]. Hence $(-2, 1)$ is the lowest point on the parabola $y = x^2 + 4x + 5$ (see the illustration).

Therefore, there is no point on this parabola for which $y = 0$. That is, the equation $0 = x^2 + 4x + 5$ has no solutions.

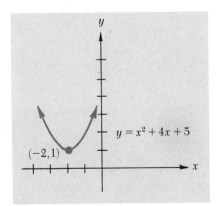

Again, remember that if you have any trouble seeing how to draw the graph, just pick a value for x and then solve equation [3] for the corresponding y value. Once you've found enough points, you can sketch the parabola that passes through these points.

Example 12

Sketch the graph of $y = (x + 2)^2$ and use this result to show why the equation

$$x^2 + 4x + 4 = 0$$

has only one solution.

$(x + 2)^2$ will be 0 only when $x = -2$, so $(-2, 0)$ is the lowest point on this parabola. See the illustration.

In summary, there is a very interesting relationship between the solutions of the equation

$$ax^2 + bx + c = 0$$

and the graph

$$y = ax^2 + bx + c$$

The relationship is that the solutions of $ax^2 + bx + c = 0$ give us the x-coordinates of the points at which the curve (parabola) $y = ax^2 + bx + c$ meets the x-axis.

For example, $x^2 + 4x + 3 = 0$ has two real solutions because $y = x^2 + 4x + 3$ crosses the x-axis twice. $x^2 + 4x + 4 = 0$ has one solution because the parabola $y = x^2 + 4x + 4$ meets the x-axis at only one point. The equation $x^2 + 4x + 5 = 0$ has no real solutions because the parabola $y = x^2 + 4x + 5$ never meets the x-axis.

The illustration shows the three possibilities.

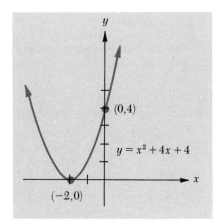

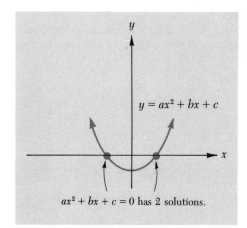

(a) $ax^2 + bx + c = 0$ has 2 solutions.

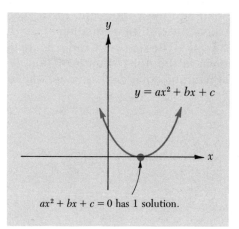

(b) $ax^2 + bx + c = 0$ has 1 solution.

(c) $ax^2 + bx + c = 0$ has no (real) solutions.

Section 14.5 Graphing a Quadratic Relationship

OPTIONAL NOTE ON CURVE SKETCHING

Instead of sketching $y = x$ or $y = x^2$, we may have to sketch $y = x^3$. The procedure is always the same. We pick a value for x and solve for the corresponding value of y. For example, if we pick x to be $-3, -2, -1, 0, 1, 2,$ or 3, we come up with the following table.

x	$x^3 \; (=y)$	Point on $y = x^3$
-3	-27	$(-3, -27)$
-2	-8	$(-2, -8)$
-1	-1	$(-1, -1)$
0	0	$(0, 0)$
1	1	$(1, 1)$
2	8	$(2, 8)$
3	27	$(3, 27)$

The results are shown in the first figure. We have chosen a different scale for the y-axis so that the points fit into our picture.

We need advanced mathematics so that we can guess more easily how the space between points must be filled in.

The difficulty is that once the curve need not be a straight line, we cannot be sure how various points are joined. For example, the second figure shows two different curves that pass through the points in the first figure (one solid, one dotted). The hard part is to find out why the curve shown in the third figure is the correct sketch of $y = x^3$.

But even when we can't be sure, we can continue to locate point after point until we feel comfortable that we have enough points located to trust the graph we want to make.

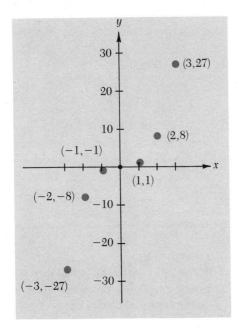
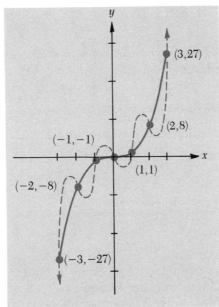
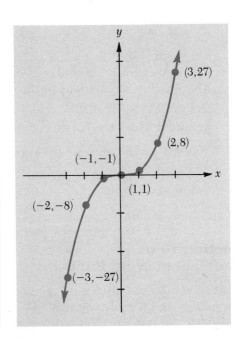

PRACTICE DRILL

What are the coordinates of the point P in Problems 1–3 if
1. P is on the parabola $y = x^2$ and the x-coordinate of P is 7.
2. P is on the parabola $y = x^2 + 1$ and the x-coordinate of P is 3.
3. P is on the parabola $y = (x + 1)^2$ and the x-coordinate of P is -1.
4. What is the lowest point on the parabola $y = (x - 3)^2 + 4$?

Answers
1. (7,49) 2. (3,10)
3. (−1,0) 4. (3,4)

CHECK THE MAIN IDEAS

1. The graph of a linear relationship is called a _____.
2. The name given to the graph of a quadratic relationship is a _____.
3. You locate points on a parabola in the same way that you locate points on a line. If you want the point on the parabola $y = x^2$ whose x-coordinate is 7, you replace x by 7 in $y = x^2$ to find that $y = $ _____. That is, the point (7,49) is on the parabola $y = x^2$.
4. It is more complicated to graph a parabola than a straight line because it only takes _____ points to determine the exact location of the line.
5. However, you can use completing the square to find the _____ point of a parabola whose equation has the form $y = x^2 + bx + c$.
6. The idea rests on the fact that a square of a real number is nonnegative and is 0 only if the number is 0. Suppose you have the parabola $y = x^2 + 4x + 5$. Since $x^2 + 4x + 4$ is $(x + 2)^2$, you may rewrite the equation of the parabola as $y = (x^2 + 4x + 4) + $ _____.
7. You may write the equation of the parabola as $y = (x + 2)^2 + 1$. The smallest number that $(x + 2)^2$ can be is 0, and this happens when $x = $ _____.
8. When $x = -2$, $(x + 2)^2 + 1$ equals $0 + 1$, or 1. Therefore, the lowest point on $y = x^2 + 4x + 5$ is _____.
9. If you sketch the parabola $y = ax^2 + bx + c$, you can get a good idea about the number of solutions of the quadratic equation $ax^2 + bx + c = 0$. The values of x for which $ax^2 + bx + c = 0$ are the x-coordinates of the points at which the parabola $y = ax^2 + bx + c$ intersects the _____-axis.
10. Since a parabola meets the x-axis at either one, two, or (a) _____ points, a quadratic equation has either one, two, or (b) _____ solutions.

Answers
1. (straight) line 2. parabola
3. 7^2, or 49 4. two
5. lowest 6. 1 7. -2
8. (−2,1) 9. x
10. (a) no (b) no

Section 14.6
More about Word Problems

At the end of the last lesson, we saw how certain word problems required that we understand the idea of quadratic equations in order to find exact solutions. This idea still applies. But now we are able to solve a larger variety of quadratic equations.

OBJECTIVE
To revisit word problems whose solutions require the ability to solve quadratic equations by completing the square.

Example 1

The area of a rectangle is 15 square feet. What are its dimensions if the length is 2 feet greater than the width?

The dimensions are 3 feet by 5 feet.

Section 14.6 More about Word Problems 387

Let w denote the width, so $w + 2$ denotes the length.   Therefore

$$w(w + 2) = 15$$
$$w^2 + 2w = 15$$
$$w^2 + 2w - 15 = 0 \qquad [1]$$

You then factor [1] to obtain

$$(w + 5)(w - 3) = 0$$

You conclude that either $w = 3$ or $w = -5$ (you discard $w = -5$, because the width of a rectangle is nonnegative).

Suppose the area of the rectangle had been 17 square feet instead of 15 square feet? The idea remains the same, but we would have to use completing the square to get the exact answer.

Example 2

Express to the nearest hundredth the dimensions of a rectangle if the area of the rectangle is 17 square feet and the length is 2 feet greater than the width.

The dimensions are 3.24 feet by 5.24 feet.
  Letting w denote the width, you have

$$w(w + 2) = 17$$
$$w^2 + 2w = 17 \qquad [2]$$

Now you add 1 to both sides of [2] to complete the square. You get

$$w^2 + 2w + 1 = 17 + 1 = 18$$
$$(w + 1)^2 = 18 \qquad [3]$$

Then you take the square root of both sides of [3] to get

$$w + 1 = \pm\sqrt{18} \qquad [4]$$

From [4], you have

$$w = -1 \pm \sqrt{18} \qquad [5]$$

We can write $\sqrt{18}$ as $3\sqrt{2}$ (that is, $\sqrt{18} = \sqrt{3 \times 3 \times 2} = 3\sqrt{2}$) But if we're going to use the calculator or tables, it's just as easy to use $\sqrt{18}$.

Since width can't be negative, you may drop the minus sign ($-1 - \sqrt{18}$ is negative) in [5] to obtain

$$w = -1 + \sqrt{18} \qquad [6]$$

The length exceeds the width by 2, so you add 2 to the right side of [6] and find that the length is $1 + \sqrt{18}$.

This is the exact answer, but to get two decimal-place accuracy, you rewrite $\sqrt{18}$ as 4.24. Then the width equals $-1 + 4.24 = 3.24$ feet and the length equals $1 + 4.24 = 5.24$ feet.

There are many other types of examples, but let's return to Example 8 in Section 13.1 to see how far we've come since we used trial and error to solve quadratic equations.

Example 3

A ball reaches a height of h feet after t seconds according to the rule

$$h = 160t - 16t^2 \qquad [7]$$

At what time does the ball reach a height of 300 feet?

It reaches a height of 300 feet after 2.5 and 7.5 seconds.
You begin by replacing h by 300 in [7]. This gives you

$$300 = 160t - 16t^2 \qquad [8]$$

To make the coefficient of t^2 positive, multiply both sides of [8] by -1 to get

$$-300 = -160t + 16t^2 \qquad [9]$$

Interchanging both sides of [9] to get $16t^2 - 160t = -300$, you can then divide both sides by 16.

$$t^2 - 10t = -\frac{300}{16} = -\frac{75}{4} \qquad [10]$$

You add 25 to both sides of [10] to complete the square and get

$$t^2 - 10t + 25 = 25 - \frac{75}{4} = \frac{25}{4}$$

or $\quad (t-5)^2 = \left(\frac{5}{2}\right)^2$

Then you take the square root of both sides to find that $t - 5 = \pm 2.5$, so $t = 2.5$ or $t = 7.5$.

We now know the exact time at which the ball reaches a height of 300 feet. There is no longer any need to use trial and error. This is what we mean when we say that we can solve any quadratic equation exactly. Even a problem with no answer will not bother us.

Example 4

According to the information in Example 3, when does the ball reach a height of 480 feet?

It never gets that high.
You replace h by 480 in [7] to get

$$480 = 160t - 16t^2$$

Then you divide both sides by -16 to get

$$-30 = -10t + t^2$$

or $\quad t^2 - 10t = -30 \qquad [11]$

> We could divide by 16, but then the coefficient of t^2 would be -1 rather than 1.

To complete the square, you add 25 to both sides of [11] to get

$$t^2 - 10t + 25 = -30 + 25 = -5$$

or $\quad (t-5)^2 = -5 \qquad [12]$

There is no real value of t that satisfies equation [12] because the left side of [12] is nonnegative while the right side is negative.

> What happens is that the ball is in the air for 10 seconds. It reaches its greatest height at 5 seconds; and when $t = 5$,
>
> $$h = 160(5) - 16(5^2)$$
> $$= 800 - 400$$
> $$= 400$$
>
> This means the ball's greatest height is 400 feet. It never reaches a height of 480 feet.

This completes our study of quadratic equations. Completing the square has helped us solve those equations we couldn't solve by the factoring method of the previous lesson.

1. One number exceeds another by 2. What is the smaller number if their product is 16?
2. The product of two positive numbers is 16. One number is 2 more than the other. What is the smaller number?
3. A rectangle has an area of 60 square feet. If the length of the rectangle exceeds the width by 6 feet, what is the width of the rectangle?

Answers
1. $-1 \pm \sqrt{17}$ 2. $-1 + \sqrt{17}$
3. The width is $-3 + \sqrt{69}$ (or approximately 5.3 feet)

CHECK THE MAIN IDEAS

1. The types of problem in Section 13.6 and this section are similiar. The only difference is that in this section, you solve the quadratic equations by _____ rather than by factoring.
2. For example, both $60 = x^2 + 6x$ and $60 = x^2 + 4x$ are examples of _____ equations.
3. The major difference is that there is no pair of _____ whose sum is 6 and whose product is -60.
4. This is why you must _____. Otherwise, there is no difference between this section and Section 16.6. You get the same kind of equation, but now you can solve any quadratic equation.

Answers
1. completing the square
2. quadratic   3. integers
4. complete the square

EXERCISE SET 14 (Form A)

Unless otherwise specified, leave the square roots of nonperfect squares in radical form. That is, unless otherwise specified, write $\sqrt{10}$ rather than 3.16, and so on.

Section 14.1

1. Can $x^2 + 4x + 1$ be written in the form $(x + r)(x + s)$ where r and s are integers?
2. (a) For what value(s) of t will $3t^2 = 2t^2 + 9$?
   (b) Using the answer to (a), what is the value of $14 + t$?
3. Suppose $d = t^2 - 4$.
   (a) What is the value of d when $t = -5$?
   (b) For what real value of t will $d = -5$?
4. Suppose $h = 3 - \sqrt{t}$.
   (a) For what value(s) of t will $h = -1$?
   (b) What is the value of h when $t = -1$?
5. One leg of a right triangle is 15 cm and the other leg is 18 cm. What is the length of the hypotenuse to the nearest (a) centimeter?   (b) millimeter?
6. A freely falling body falls d feet in t seconds according to the rule $d = 16t^2$. How long does it take for the body to fall
   (a) 1,600 feet   (b) 100 feet   (c) 112 feet

Section 14.2

7. Express $\sqrt{3} + 2\sqrt{27} + 4\sqrt{48} - 4\sqrt{75}$ as a multiple of $\sqrt{3}$.
8. Rewrite the answer to the previous problem as a decimal correct to the nearest tenth.

Answers: Exercise Set 14 (Form A)
1. no   2. (a) $t = \pm 3$ ($t = 3$ or -3) (b) 11 or 17
3. (a) 21   (b) None
4. (a) 16   (b) No real value
5. (a) 23   (b) 23.4
6. (a) 10 seconds   (b) 2.5 seconds
(c) $\sqrt{7}$ seconds (approximately 2.65 seconds)   7. $3\sqrt{3}$ 8. 5.2

9. For any nonnegative number x, write $\sqrt{x} + 2\sqrt{9x} + 4\sqrt{16x} - 4\sqrt{25x}$ as a multiple of $\sqrt{x}$.
10. When $x = 2$, evaluate
    (a) $\sqrt{4x+1} + \sqrt{5x+6}$ (b) $\sqrt{9x+7}$
11. Combine like terms to simplify $3\sqrt{2} + 3\sqrt{3} + 7\sqrt{2} - 4\sqrt{3}$.
12. Rewrite the answer to the previous problem in decimal form to the nearest tenth.
13. Write each of the following as whole numbers.
    (a) $(\sqrt{7})(\sqrt{7})$ (b) $(\sqrt{7})(\sqrt{63})$ (c) $(7+\sqrt{3})(7-\sqrt{3})$
14. Evaluate $(\sqrt{x+y})(\sqrt{x-y})$ when $x = 10$ and $y = 8$.

Section 14.3
15. What must you add to $x^2 - 10x$ in order to complete the square?
16. Find all real values of x for which
    (a) $x^2 - 10x = 1575$ (b) $x^2 - 10x = 29$ (c) $x^2 - 10x = -29$
17. Express the solutions of $x^2 - 10x = 29$ in decimal form, correct to the nearest tenth.

Section 14.4
Find all real values of x for which
18. $x^2 + 8x = 33$
19. $x^2 + 8x = 30$
20. $x^2 + 8x + 30 = 0$
21. $2x^2 + 8x = 30$
22. (a) $x^2 = x + 12$ (b) $x = \sqrt{x+12}$
23. $x = \sqrt{30 - 8x}$
24. (a) $(2x-5)(x+2) = (x-2)(x+1) + 17$
    (b) $(2x-5)(x+2) = x^2 + (x-2)(x+1) - 8$
    (c) $(2x-5)(x+2) = x^2 + (x-2)(x+1) + 8$
25. $x^3 - x(x^2 - x + 2) = 15$

Section 14.5
26. (a) For what (real) value of x is $(x-2)^2 + 8$ the least?
    (b) What is the least value that the expression $(x-2)^2 + 8$ can have?
    (c) What is the lowest point on the parabola $y = x^2 - 4x + 12$?
    (d) Use (a) to explain why $x^2 - 4x + 12 = 0$ doesn't have any real solutions.
27. (a) For what real values of x does $x^2 + 4x = 2x + 3$?
    (b) At what points does the line $y = 2x + 3$ meet the parabola $y = x^2 + 4x$?
28. At what point(s) does the line $y = 2x - 3$ meet the parabola $y = x^2 + 4x$?
29. At what point(s) does the line $y = 2x - 1$ meet the parabola $y = x^2 + 4x$?
30. At what points does the parabola $y = x^2 + 4x$ cross the x-axis?
31. What it the lowest point on the parabola $y = x^2 + 4x$?

---

9. $3\sqrt{x}$
10. (a) 7   (b) 5
11. $10\sqrt{2} - \sqrt{3}$
12. 12.4
13. (a) 7   (b) 21   (c) 46
14. 6
15. 25
16. (a) $x = 5 \pm 40$ ($x = -35$ or 45)
    (b) $x = 5 \pm \sqrt{54}$ ($x = 5 \pm 3\sqrt{6}$)
    (c) there is none
17. $x = -2.3$ or $x = 12.3$
18. $x = 3$ or -11
19. $x = -4 \pm \sqrt{46}$
20. there are none
21. $x = -2 \pm \sqrt{19}$
22. (a) $x = -3$ or $x = 4$
    (b) $x = 4$
23. $x = -4 + \sqrt{46}$
24. (a) $x = -5$ or $x = 5$ ($x = \pm 5$)
    (b) all values of x (c) no values of x
25. $x = -3$ or $x = 5$
26. (a) $x = 2$ (b) 8   (c) (2,8)
    (d) $x^2 - 4x + 12 \geq 2$ for all real x (and 0 is less than 2)
27. (a) $x = 1$ or $x = -3$
    (b) $(-3, -3)$ and $(1, 5)$
28. they don't meet (see graph for Problem 32)
29. $(-1, -3)$ (see graph for Problem 32)
30. $(0, 0)$ and $(-4, 0)$
31. $(-2, -4)$

Exercise Set 14 (Form A)

32. In the space provided, sketch the parabola $y = x^2 + 4x$.

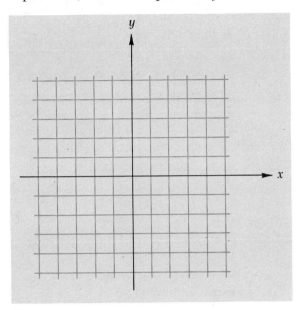

Section 14.6

33. The hypotenuse of a right triangle has a length of 15 cm. The longer leg is 7 centimeters longer than the shorter leg. What is the length of the shorter leg?
34. Write the answer to the previous problem in decimal form correct to the nearest millimeter.
35. A projectile launched with a vertical speed of 320 feet per second reaches a height of h feet after t seconds according to the rule $h = 320t - 16t^2$.

    (a) At what time(s) is the projectile at a height of 1,440 feet?
    (b) Express your answers in (a) in decimal form correct to the nearest tenth of a second.

36. The length of a rectangle is 12 inches longer than 3 times the width.

    (a) What is the width of the rectangle if the area is 210 square inches?
    (b) Express the answer in (a) correct to the nearest tenth of an inch.

37. In the formula $P = VI - RI^2$, P denotes power (watts), V denotes voltage (volts), R denotes resistance (ohms), and I denotes current (amps). Let $v = 80$ volts, $R = 5$ ohms, and $P = 205$ watts.

    (a) What is the value of the current?
    (b) Express your answer to (a) in decimal form to the nearest tenth of an amp.

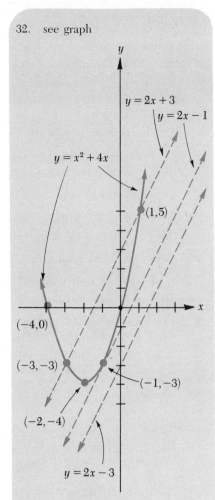

32. see graph

33. $\frac{1}{2}(\sqrt{401} - 7)$
34. 6.5 cm (65 mm)
35. (a) $t = (10 \pm \sqrt{10})$ seconds
    (b) at either 6.8 or 13.2 seconds
36. (a) $\sqrt{74} - 2$ (b) 6.6 inches
37. (a) $I = (8 \pm \sqrt{23})$ amps
    (b) either 3.2 or 12.8 amps

Lesson 14 The General Solution of a Quadratic Equation

EXERCISE SET 14 (Form B)

Unless otherwise specified, leave the square roots of nonperfect squares in radical form. That is, unless otherwise specified, write $\sqrt{10}$ rather than 3.16, and so on.

Section 14.1

1. Are there integers m and n for which $x^2 + 5x + 3 = (x+m)(x+n)$?
2. (a) For what values of t does $4t^2 = 3t^2 + 16$?
   (b) Using the answer to (a), what is the value of $17 + t$?
3. Suppose $d = t^2 - 1$.
   (a) What is the value of d when $t = -3$?
   (b) For what real value of t will $d = -3$?
4. Suppose $h = 4 - \sqrt{t}$.
   (a) For what real value of t will $h = -5$?
   (b) Is h a real number when $t = -5$?
5. The legs of a right triangle are 10 cm and 20 cm. To the nearest millimeter, what is the length of the hypotenuse?
6. If a freely falling body falls d feet in t seconds according to the formula $d = 16t^2$, how long does it take the body to fall
   (a) 400 feet   (b) 225 feet   (c) 176 feet

Section 14.2

7. Express $\sqrt{2} + 3\sqrt{8} + 4\sqrt{18} - 2\sqrt{50}$ as a multiple of $\sqrt{2}$.
8. Write the answer to the previous problem as a decimal correct to the nearest tenth.
9. Assuming that x is a nonnegative number, rewrite $\sqrt{x} + 3\sqrt{4x} + 4\sqrt{9x} - 2\sqrt{25x}$ as a multiple of $\sqrt{x}$.
10. When $x = 7$, evaluate
    (a) $\sqrt{5x+1} + \sqrt{9x+1}$ (b) $\sqrt{14x+2}$
11. Combine like terms to simplify $5\sqrt{2} - 5\sqrt{3} + 3\sqrt{2} + 7\sqrt{3}$.
12. Rewrite the answer to the previous problem in decimal form to the nearest tenth.
13. Write each of the following as whole numbers:
    (a) $(\sqrt{5})(\sqrt{5})$ (b) $(\sqrt{5})(\sqrt{20})$ (c) $(5+\sqrt{2})(5-\sqrt{2})$
14. Evaluate $(\sqrt{x}+y)(\sqrt{x}-y)$ when $x = 13$ and $y = 5$.

Section 14.3

15. What must you add to $x^2 - 12x$ to complete the square?
16. Find all real values of x for which
    (a) $x^2 - 12x = 864$ (b) $x^2 - 12x = 37$ (c) $x^2 - 12x = -37$
17. Express the solutions of $x^2 - 12x = 37$ in decimal form correct to the nearest tenth.

Section 14.4
Find all real values of x for which

18. $x^2 + 16x = 80$
19. $x^2 + 16x = 70$
20. $x^2 + 16x + 70 = 0$
21. $2x^2 + 16x = 70$

Exercise Set 14 (Form B)

22. (a) $x^2 = x + 20$ (b) $x = \sqrt{x + 20}$
23. $x = \sqrt{70 - 16x}$
24. (a) $(2x + 3)(x - 2) = (x - 4)(x + 3) + 42$
    (b) $(2x + 3)(x - 2) = x^2 + (x - 4)(x + 3) + 6$
    (c) $(2x + 3)(x - 2) = x^2 + (x - 4)(x + 3) - 6$
25. $x^3 - x(x^2 - x + 7) = 18$

Section 14.5

26. (a) For what real value of x is $(x - 1)^2 + 4$ the least?
    (b) What is the least value $(x - 1)^2 + 4$ can have if x is real?
    (c) What is the lowest point on the parabola $y = x^2 - 2x + 5$?
    (d) Use part (a) to explain why there are no real values of x for which $x^2 - 2x + 5 = 0$.
27. (a) For what real values of x does $x^2 - 2x = 2x - 3$?
    (b) At what points does the line $y = 2x - 3$ meet the parabola $y = x^2 - 2x$?
28. At what points does the line $y = 2x - 5$ meet the parabola $y = x^2 - 2x$?
29. At what points does the line $y = 2x - 4$ meet the parabola $y = x^2 - 2x$?
30. At what points does the parabola $y = x^2 - 2x$ cross the x-axis?
31. What is the lowest point on the parabola $y = x^2 - 2x$?
32. In the space provided, sketch the parabola $y = x^2 - 2x$.

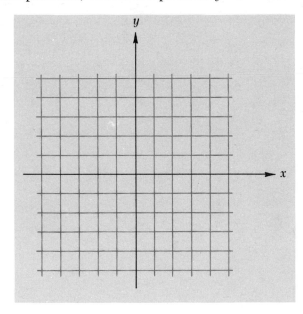

Section 14.6

33. The hypotenuse of a right triangle is 13 cm. One leg is 9 centimeters shorter than the other leg. What is the length of the shorter leg?
34. Write the answer to the previous problem in decimal form correct to the nearest millimeter.
35. A projectile launched with a vertical speed of 800 feet per second, reaches a height of h feet after t seconds according to the rule $h = 800t - 16t^2$.

    (a) At what times is the projectile at a height of 8,000 feet?
    (b) Express your answers in (a) correct to the nearest tenth of a second.

36. The length of a rectangle is 6 cm longer than twice the width of the rectangle.
    (a) If the area of the rectangle is 210 square centimeters, what is the width of the rectangle?
    (b) Express the answer in (a) correct to the nearest millimeter.

37. Power (P), voltage (V), resistance (R), and current (I) are related by the formula $P = VI - RI^2$. Suppose $P = 108$ watts, $V = 60$ volts, and $R = 6$ ohms.
    (a) What is the value of the current?
    (b) Express your answer in (a) as a decimal correct to the nearest tenth of an ampere.

Lesson 15

Algebraic Fractions

Overview The story is told of the wealthy woman who as a token of her appreciation increased her charitable donations from one-tenth of her income to one-twentieth of her income! This story illustrates a relationship that causes many arithmetic students some problems. The bigger the denominator is, the less the number that is named by the common fraction. Such relationships are common in mathematics. Sir Isaac Newton showed that gravitational attraction between two objects decreased as the distance between them increased.

Such relationships are neither linear nor quadratic. Therefore, to apply mathematics to the real world requires that we go beyond what we've already learned in this course. There is really no well-defined line between elementary algebra and more advanced algebra. In one sense, higher mathematics is just a more detailed development of the topics we've already introduced in this course.

At all levels, mathematical analysis deals with expressing relationships and solving equations. Two methods for solving equations have been studied in our course. One is to use trial and error and the other is to use the rules of arithmetic.

Trial and error can be used no matter how complicated the equation is. Unfortunately, trial and error leads to problems that can be answered properly only in more advanced courses. For example, how do we know that an equation has a solution? If it does, how do we know that there is only one? If there is more than one solution, how do we know we've found them all?

When possible, we prefer to use the rules of arithmetic to find a system that gives us the exact answer. So far we have learned to do this with linear and quadratic equations. We should now be able to solve any linear or quadratic equation.

Another frequently occuring equation that we have not dealt with yet is the fractional equation. This type of equation involves quotients of polynomials. By using our knowledge of common fractions, we can often convert a fractional equation into either a linear or a quadratic equation. Once we do this we can solve the equation. This lesson shows how to solve fractional equations.

Section 15.1
Fractions Revisited

In solving equations, we often need to know the rules of arithmetic for fractions.

Example 1

Express $\dfrac{3}{x} + \dfrac{5}{x}$ as a single term.

As a single term, it's $\dfrac{8}{x}$.

The rule for adding fractions with common denominators is to add the numerators (in this case, $3 + 5$) and to keep the common denominator (in this case, x).

Example 1 shows us that the rule for adding fractions is the same whether we deal with numbers or with algebraic expressions.

> **OBJECTIVE**
> To apply the arithmetic of common fractions to algebraic expressions and equations.

> Remember that we do not allow a denominator to be 0. For this reason, it is understood that x may represent any number except 0.

Example 2

Rewrite $\dfrac{3}{x}$ as an equivalent fraction whose denominator is $5x$.

$\dfrac{3}{x}$ has the same value as $\dfrac{15}{5x}$.

The rule is that if you multiply the numerator and denominator of a fraction by the same non-zero number, the value of the fraction remains the same. To replace x by $5x$ in the denominator means that you multiplied x by 5. To keep the value of the fraction the same, you must also multiply the numerator by 5. That is,

$$\dfrac{3}{x} = \dfrac{3(5)}{x(5)} = \dfrac{15}{5x}$$

Sometimes the amount of work increases, but the basic idea stays the same.

> As a rough check, we may look at both expressions for a given value of x. For example, when $x = 4$,
>
> $$\dfrac{3}{x} = \dfrac{3}{4}$$
>
> and $\dfrac{15}{5x} = \dfrac{15}{20} = \dfrac{3}{4}$
>
> The rules for fractions remain the same whether the fractions are numerical or algebraic.

Example 3

Rewrite $\dfrac{x+2}{x+3}$ as a fraction whose denominator is $x^2 + 7x + 12$.

$$\dfrac{x+2}{x+3} = \dfrac{x^2 + 6x + 8}{x^2 + 7x + 12}$$

Remember that you can multiply numerator and denominator by the same non-zero number without changing the value of the fraction. The question is, what must you multiply $x + 3$ by to get $x^2 + 7x + 12$? You should remember enough about factoring to know that $x^2 + 7x + 12 = (x + 3)(x + 4)$.

This means that you must multiply the denominator by $x + 4$ to get $x^2 + 7x + 12$. Therefore, you must also multiply the numerator by $x + 4$. This gives you

$$\dfrac{x+2}{x+3} = \dfrac{(x+2)(x+4)}{(x+3)(x+4)} = \dfrac{x^2 + 6x + 8}{x^2 + 7x + 12}$$

> **DEFINITION**
> The quotient of two polynomials (in x) is called a **rational expression** (in x), or an **algebraic fraction** (in x). This parallels the definition that the quotient of two integers is called a rational number and the numeral that denotes this number is called a fraction.

Section 15.1 Fractions Revisited

NOTE

Technically, the correct answer should state that x is not equal to -4. That is, we are allowed to multiply numerator and denominator only by the same *non-zero* number; but $x+4$ is 0 when $x=-4$.

Example 4

Write $\dfrac{x+1}{x+4}$ in a form whose denominator is $x^2+7x+12$.

$$\frac{x+1}{x+4}=\frac{x^2+4x+3}{x^2+7x+12} \quad \text{where } x \neq -3$$

In Example 3, you reviewed the fact that $x^2+7x+12=(x+3)(x+4)$. So to get from $x+4$ to $x^2+7x+12$, you must multiply by $x+3$. To keep the value of the fraction the same, you must also multiply the numerator by $x+3$. That is,

$$\frac{x+1}{x+4}=\frac{(x+1)(x+3)}{(x+4)(x+3)}=\frac{x^2+4x+3}{x^2+7x+12}$$

Why would we ever want to rewrite fractions this way? The same problem occurs when we work with numerical fractions. For example, why would we want to write $\dfrac{1}{2}$ as $\dfrac{2}{4}$ or $\dfrac{15}{30}$? One answer is that often we need to use common denominators.

This may happen when we use algebraic fractions. As shown in the next example, we may want to add two algebraic fractions that have different denominators. To make things a bit simpler, we use the same fractions we used in Examples 3 and 4.

That is, we should write

$$\frac{x+2}{x+3}=\frac{x^2+6x+8}{x^2+7x+12}, x \neq -4$$

If we replace x by -4 on the left side, we get

$$\frac{-4+2}{-4+3}=\frac{-2}{-1}=2$$

On the right side, we get $\dfrac{0}{0}$, since $x+4$ is then 0 and it is a factor of both the numerator and denominator.

Remember that since $x+3$ must not equal 0, x must not equal -3.

However, $x=-1$ is acceptable. In that case, the numerator is 0 but the denominator isn't. When the numerator is 0 and the denominator isn't, the value of the fraction is 0. For example,

$$\frac{0}{3}=0$$

Example 5

Express the sum of $\dfrac{x+2}{x+3}$ and $\dfrac{x+1}{x+4}$ as a single fraction.

The sum is $\dfrac{2x^2+10x+11}{x^2+7x+12}$.

Since the two fractions have different denominators, first you want to rewrite them so that they have the same denominator. A sure way of getting a common denominator is to multiply the given denominators. That is, $(x+3)(x+4)$ is certainly divisible by both $x+3$ and $x+4$.

From the two previous examples, you already know that

$$\frac{x+2}{x+3}=\frac{x^2+6x+8}{x^2+7x+12}$$

and

$$\frac{x+1}{x+4}=\frac{x^2+4x+3}{x^2+7x+12}$$

So you may rewrite the sum as

$$\frac{x^2+6x+8}{x^2+7x+12}+\frac{x^2+4x+3}{x^2+7x+12}$$

Now that you have a common denominator, you may add the numerators and keep the common denominator. This gives you

$$\frac{(x^2+6x+8)+(x^2+4x+3)}{x^2+7x+12}$$

In adding fractions, we must have common denominators. It is not crucial that we have the least common denominator.

The final step is to use the rules for adding polynomials to rewrite the numerator in a simpler form: that is, you combine like terms to obtain

$$\frac{2x^2 + 10x + 11}{x^2 + 7x + 12}$$

Example 5 was a bit easier because we had already done much of the work in Examples 3 and 4. There is a relatively quick way of adding two fractions with different denominators that involves the idea of *cross multiplication*. Let's look at another example to see what's happening.

Example 6

For any expressions b, c, d, and e, express

$$\frac{b}{c} + \frac{d}{e}$$

as a single fraction.

The sum is $\frac{be + cd}{ce}$. Since the two denominators are c and e, ce is a common denominator. Therefore

$$\frac{b}{c} = \frac{be}{ce} \quad \text{and} \quad \frac{d}{e} = \frac{cd}{ce}$$

Using common denominators, you have

$$\frac{b}{c} + \frac{d}{e} = \frac{be}{ce} + \frac{cd}{ce} = \frac{be + cd}{ce}$$

It's understood that neither c nor e can be 0. Example 6 is actually the rule for adding common fractions.

Let's see how the shortcut works.

Example 7

Write $\frac{2}{x} + \frac{x}{3}$ as a single fraction.

The sum is $\frac{6 + x^2}{3x}$. The 6 comes from multiplying the numerator of the first (2) by the denominator of the second (3). x^2 is the product of the numerator of the second (x) and the numerator of the first (x). $3x$ is the product of the two denominators (x and 3).

Without the shortcut, we have

$$\frac{2}{x} = \frac{3(2)}{3(x)} = \frac{6}{3x}$$
$$\frac{x}{3} = \frac{x(x)}{3(x)} = \frac{x^2}{3x}$$

Then

$$\frac{2}{x} + \frac{x}{3} = \frac{6}{3x} + \frac{x^2}{3x}$$
$$= \frac{6 + x^2}{3x}$$

Further drill involving addition of fractions is left for the exercise set. The other three operations are relatively easy and are also left for the exercise set. For example, we already know that subtraction is the inverse of addition. Therefore, there is no need to discuss subtraction once we know how to add.

Multiplication of fractions does not involve common denominators. All we do to multiply two or more fractions is multiply numerators and multiply denominators. The invert-and-multiply rule takes care of division, once we know how to multiply fractions.

But there is one useful procedure in fractional equations that helps us to get rid of denominators. When two fractions are equal, the numerator of the first times the denominator of the second equals the numerator of the second times

Section 15.1   Fractions Revisited

the denominator of the first. Using symbols, if $\frac{b}{c}=\frac{d}{e}$, then $cd=be$. In other words,

$$\frac{b}{c}=\frac{be}{ce} \quad \text{and} \quad \frac{d}{e}=\frac{cd}{ce}$$

When two fractions are equal and they have the same denominators, they also have the same numerator; hence, $be = cd$.

For example, $\frac{4}{5}=\frac{12}{15}$ and $4(15) = 5(12)$.

Example 8

For what value of x does

$$\frac{3}{x}=\frac{x}{12} \qquad [1]$$

When x is either 6 or -6.
  The numerator of the first (3) times the denominator of the second (12) is 36. The numerator of the second (x) times the denominator of the first (x) is x^2. Therefore, the equation is the same as

$$x^2 = 36 \qquad [2]$$

  To solve [2], you take the square root of both sides and find that $x = \pm 6$. As a check, notice that when $x = 6$, [1] becomes $\frac{3}{6}=\frac{6}{12}$. When $x = -6$, [1] becomes $\frac{3}{-6}=\frac{-6}{12}$.

If you don't like the shortcut, you can rewrite [1] as

$$\frac{36}{12x}=\frac{x^2}{12x}$$

and conclude that $36 = x^2$.

We started with [1], which is a fractional equation, and converted it to [2], which is a quadratic equation — one we already know how to solve.

Example 9

For what value of x does

$$\frac{x+3}{x+4}=\frac{6}{5} \qquad [3]$$

Equation [3] is true only when $x = -9$.
  If you cross multiply, you get

$$5(x + 3) = 6(x + 4) \qquad [4]$$

Though [3] is a fractional equation, [4] is a linear equation. You may use the distributive property to rewrite it as

$$5x + 15 = 6x + 24 \qquad [5]$$

As shown in the margin, you can solve this linear equation easily to get the correct answer.

  Again, the problems may become more complicated, but the principle stays the same.

CHECK

$$\frac{-9+3}{-9+4}=\frac{-6}{-5}=\frac{6}{5}$$

$$\begin{aligned}5x + 15 &= 6x + 24\\ -5x &\quad -5x\\ 15 &= x + 24\\ -24 &\quad -24\\ -9 &= x\end{aligned}$$

Example 10

For what value(s) of x does

$$\frac{x^2+3}{x-4}=28 \qquad [6]$$

There are two values: x may be 5 or 23.

CHECK
With $x = 5$,

$$\frac{x^2+3}{x-4}=\frac{25+3}{5-4}=28$$

With $x = 23$,

$$\frac{x^2+3}{x-4}=\frac{529+3}{23-4}=\frac{532}{19}=28$$

To help working with fractions, rewrite 28 as $\frac{28}{1}$. In this way, [6] becomes

$$\frac{x^2 + 3}{x - 4} = \frac{28}{1}$$

If you cross multiply, you get

$$1(x^2 + 3) = 28(x - 4)$$
$$x^2 + 3 = 28x - 112$$
$$x^2 - 28x + 115 = 0 \qquad [7]$$

Equation [7] may then be written in factored form as

$$(x - 5)(x - 23) = 0 \qquad [8]$$

From [8] you see that $x - 5 = 0$ or $x - 23 = 0$. That is, $x = 5$ or $x = 23$.

> The key point is the fact that a fractional equation [6] has been replaced by a more familiar quadratic equation [7].

Sometimes we have to apply two or more principles to solve the given equation. In such cases, there may be more than one proper system to use.

Example 11
For what value of x does

$$\frac{x}{3} + \frac{x}{5} = 40 \qquad [9]$$

> CHECK
> $$\frac{75}{3} + \frac{75}{5} = 25 + 15 = 40$$

Equation [9] is satisfied only when $x = 75$.
Let's solve this problem in two different ways.

METHOD 1
A common denominator for the left side of [9] is 15. That is,

$$\frac{x}{3} + \frac{x}{5} = \frac{5x}{15} + \frac{3x}{15}$$
$$= \frac{5x + 3x}{15} = \frac{8x}{15} \qquad [10]$$

If you use [10], [9] becomes

$$\frac{8x}{15} = 40 \qquad \text{or} \qquad \frac{8x}{15} = \frac{40}{1} \qquad [11]$$

You can now use cross multiplication in [11] to get

$$\qquad 1(8x) = 15(40)$$
$$\text{or} \qquad 8x = 600 \qquad [12]$$

Then you divide both sides of [12] by 8 to get $x = 75$.

> We may be better off to leave it as $8x = 15(40)$. Then, when we divide both sides by 8, we get
> $$x = \frac{15(40)}{8}$$
> If we divide 8 into 40, we can reduce the equation to $x = 15(5)$, or 75. This point isn't that important, but it does show us that if we want to, we can avoid multiplying large numbers.

METHOD 2
If you look at [9], you see that all the denominators are divisors of 15. So you can multiply both sides of [9] by 15. That is,

$$15\left(\frac{x}{3} + \frac{x}{5}\right) = 15(40) \qquad [13]$$

With the distributive property, [13] becomes

$$15\left(\frac{x}{3}\right) + 15\left(\frac{x}{5}\right) = 15(40)$$
$$5x + 3x = 15(40)$$
$$8x = 15(40) \qquad \text{or} \qquad x = 75 \qquad [14]$$

> Remember that $15\left(\frac{x}{3}\right)$ is the same as $\frac{15}{1}\left(\frac{x}{3}\right) = \frac{15x}{3} = 5x$.

Section 15.1 Fractions Revisited 401

Equations [12] and [14] are equivalent. The difference is that Method 2 helps you to avoid extensive use of fractions.

In the exercise set, there is ample drill on using the rules for common fractions in solving problems involving algebraic fractions. Once you have acquired the necessary skills, the rest of the lesson will show you various situations in which these skills are useful.

PRACTICE DRILL

Solve each of the following fractional equations.

1. $\frac{8}{x} = 4$ 2. $\frac{1}{x} = 4$ 3. $\frac{3}{x} = \frac{4}{5}$ 4. $\frac{x+2}{x} = \frac{2}{3}$ 5. $\frac{x}{x+1} = 7$

Write as a single term.

6. $\frac{4}{x} + \frac{5}{x}$ 7. $\frac{x}{4} + \frac{x}{5}$ 8. $\frac{x}{4} + \frac{5}{x}$ 9. $\frac{2}{x+1} + \frac{3}{x+2}$

Answers
1. $x = 2$ 2. $x = \frac{1}{4}$
3. $x = \frac{15}{4} = 3\frac{3}{4}$ 4. $x = -6$
5. $x = -\frac{7}{6}$ 6. $\frac{9}{x}$
7. $\frac{9x}{20}$ 8. $\frac{x^2 + 20}{4x}$
9. $\frac{5x+7}{(x+1)(x+2)} = \frac{5x+7}{x^2+3x+2}$

CHECK THE MAIN IDEAS

1. Because $x + 1$ appears in the denominator, you call $\frac{x}{x+1}$ a _____ expression.

2. Similarly, you call $\frac{x}{x+1} = 7$ a fractional _____.

3. By multiplying both sides of a fractional equation by a common denominator, you can change it into a polynomial equation. By multiplying sides of $\frac{x}{x+1} = 7$ by $x + 1$, you get the linear equation $x =$ _____.

4. The solution of this equation is given by $x =$ _____. One point of this section is to show that you can use algebra to reduce new equations (such as fractional) to equations you already know how to solve (such as linear).

5. In general, you deal with fractional expressions just as you would with ordinary fractions. You must be careful to pick values of the variable that prevent the _____ of any fraction from being zero.

6. For example, since $x^2 - 4x + 3 = 0$ when either $x = 1$ or $x = 3$, you do not let x have the values of either _____ in the expression $\frac{3x}{x^2 - 4x + 3}$.

7. Other than that, you deal with fractional expressions using the same rules that you use for numerical fractions. For instance, the sum of $\frac{2}{x+1}$ and $\frac{5}{x+1} = \frac{7}{(\underline{})}$. That is, you keep the common denominator and add the two numerators.

8. If you want to add $\frac{2}{x+1}$ and $\frac{3}{x+2}$, you use _____ as a common denominator.

9. You rewrite $\frac{2}{x+1}$ as _____ and $\frac{3}{x+2}$ as $\frac{3(x+1)}{(x+1)(x+2)}$.

10. You then add the fractions to obtain $\frac{(\underline{})}{(x+1)(x+2)}$. That is, just as with numerical fractions, you find a common denominator, add numerators, and keep the common denominator.

Answers
1. fractional 2. equation
3. $7(x+1)$, or $7x + 7$ 4. $-\frac{7}{6}$
5. denominator 6. 1 or 3
7. $x + 1$ 8. $(x+1)(x+2)$
9. $\frac{2(x+2)}{(x+1)(x+2)} = \frac{2x+4}{(x+1)(x+2)}$
10. $2(x+2) + 3(x+1)$, or $5x + 7$

Section 15.2
Ratio and Proportion

OBJECTIVE
To apply the ideas of fractional equations to ratio and proportion.

In the previous section, we studied fractional equations. In this section, we shall see some simple situations in which we may want to use fractional equations.

Example 1

Two people are partners. Their agreement is that for every dollar of profit the junior partner receives, the senior partner receives two dollars. If the profit for a year is $30,000, how much does each partner get?

The junior partner gets $10,000, and the senior partner gets $20,000.
If you want to use algebra, you let j be the amount the junior partner gets. Then $2j$ is the amount the senior partner gets.
The equation becomes

$$j + 2j = 30,000$$
$$3j = 30,000 \qquad [1]$$

[1] is a linear equation whose solution is $j = 10,000$.

There is no need to make Example 1 more difficult, but we should observe that there are at least two other ways of solving the problem.
One way is by ordinary arithmetic. If one partner gets $1 to each $2 the other gets, then each $3 is split so that one partner gets $1 and the other, $2. That is, one partner gets 1 of each 3 dollars, and the other gets 2 of each 3 dollars. In common fractions, one partner gets one third of the profit, and the other gets two thirds.
The other way is to use algebra to apply the one-to-two ratio to the entire profit. We may still let j denote the junior partner's share. But now we let $30,000 - j$ denote the senior partner's share. The two shares must still be in the ratio of 1 to 2.
We use fraction notation to write this as

$$\frac{j}{30,000 - j} = \frac{1}{2} \qquad [2]$$

As a check notice that $\frac{2}{3}$ of 30,000 is 20,000 while $\frac{1}{3}$ of 30,000 is 10,000. This agrees with the answer we got by using algebra.

Notice that there is no one right way to set up a solution.

Equation [2] is read as, "j is to $30,000 - j$ as 1 is to 2."

Equation [2] is a *fractional equation*.
According to what we learned in the previous section, we can solve [2] by cross multiplying to obtain the linear equation

$$2j = 1(30,000 - j)$$

which we can solve to verify that $j = 10,000$.
For Example 1, it may seem more logical to use [1] than [2]. But there are some cases where the reverse is true. For example, the ratio may not be quite so simple.

No matter how we do the problem, we must get the same answer if we haven't made any mistakes.

Example 2

In time, the two partners from Example 1 decide to change their contract. They decide that the junior partner deserves a larger share of the profits. They agree to share the profits so that the senior partner gets $3 for each $2 the junior partner gets. If the profit is still $30,000, how much does the junior partner get?

This time the junior partner gets $12,000.
The ratio is now 2 to 3. You can let j be the amount the junior partner gets. Then $30,000 - j$ is the amount the senior partner gets. The difference is that now

the ratio must be 2 to 3 rather than 1 to 2. Therefore, the equation becomes

$$\frac{j}{30{,}000 - j} = \frac{2}{3} \qquad [3]$$

Equation [3] is a fractional equation. You can solve it by cross multiplying to obtain

or
$$\begin{aligned}
3j &= 2(30{,}000 - j) \\
3j &= 60{,}000 - 2j \\
+2j & \quad\quad\quad +2j \\
\hline
5j &= 60{,}000
\end{aligned} \qquad [4]$$

Don't lose sight of the main objective. [3] is a fractional equation derived from a real situation and [4] is the linear equation into which [3] has been converted.

From [4] you see that $j = 12{,}000$.

Of course, there are other ways to solve Example 2. We can still use arithmetic. Since one partner gets $2 to each $3 the other gets, then out of each $5 (3 + 2), one partner gets 3 and the other, 2. This means that one partner gets $\frac{2}{5}$ of the profit, and the other gets $\frac{3}{5}$. One-fifth of 30,000 is 6,000; so $\frac{2}{5}$ of $30,000 is $12,000 and $\frac{3}{5}$ of $30,000 is $18,000.

It may be easier for many of us to do this problem by arithmetic than by algebra. We're trying to show how fractional equations may occur, and in either method we have to know how to work with fractions.

NOTE

When we say that the ratio is 2 to 3, we mean that one partner gets $\frac{2}{5}$ and the other, $\frac{3}{5}$. In this case, when we write 2 to 3 as $\frac{2}{3}$ (as we did in [3]), the fraction is the ratio of one partner's share to the other's. It is *not* that one partner gets $\frac{2}{3}$ of the whole amount.

If one partner got two thirds of the whole amount, the ratio would be 2 to 1. That is, out of each $3, one partner gets $2, and the other gets $1, which is what happened in Example 1.

Example 3

The recipe for an extremely dry martini calls for seven parts of gin to one part of vermouth. How many ounces of each must be used to make 200 ounces of martini?

You need 25 ounces of vermouth and 175 ounces of gin.

If you let v stand for the amount of vermouth, then $200 - v$ is the amount of gin. Since the ratio of vermouth to gin is 1 to 7, the equation is

$$\frac{v}{200 - v} = \frac{1}{7} \qquad [5]$$

You solve the fractional equation [5] by cross multiplying to obtain the linear equation

$$\begin{aligned}
7v &= 1(200 - v) \qquad [6] \\
7v &= 200 - v \\
+v & \quad\quad +v \\
\hline
8v &= 200 \\
v &= 25
\end{aligned}$$

We could let $7v$ denote the amount of gin and obtain the same result from the equation

$$v + 7v = 200 \qquad [7]$$

If we want to use arithmetic, the ratio of 1 to 7 means that since $1 + 7 = 8$, 1 of every 8 parts is vermouth and 7 of every 8 parts is gin. Hence, $\frac{1}{8}$ of the 200 ounces (or 25 ounces) is vermouth, and the rest is gin.

Example 4

A different martini calls for 2 parts of vermouth to 3 parts of gin. How many ounces of vermouth do you need in order to make 200 ounces of martini?

Now you need 80 ounces of vermouth.
Since the ratio is 2 to 3, the algebraic equation [5] is modified to become

$$\frac{v}{200-v} = \frac{2}{3} \qquad [8]$$

If you cross multiply, the fractional equation [8] becomes

$$3v = 2(200 - v) \qquad [9]$$

and the solution follows.

$$3v = 400 - 2v$$
$$+ 2v \qquad + 2v$$
$$5v = 400$$

So $v = 80$.

If we want to express the gin in terms of v, we have to use a relationship like $\frac{v}{g} = \frac{2}{3}$, or $2g = 3v$, or $g = \frac{3}{2}v$. We could then write

$$v + \frac{3}{2}v = 200 \qquad \text{or} \qquad \frac{5}{2}v = 200$$

So any method requires the ability to handle fractions.

The solutions of Examples 3 and 4 give us a hint that there is a connection between ratios and linear relationships.

For example, when we say that the ratio of gin to vermouth is 7 to 1, we can write this as

$$\frac{g}{v} = \frac{7}{1} \qquad \text{or} \qquad g = 7v$$

When we say that the ratio of gin to vermouth is 3 to 2, we can write this as

$$\frac{g}{v} = \frac{3}{2} \qquad \text{or} \qquad g = \frac{3}{2}v$$

These relationships are special forms of linear relationships: there is no constant added to the right side. We give such a relationship a special name.

DEFINITION
The variables x and y are said to be **directly proportional** if there exists a non-zero constant m such that $y = mx$.

Direct proportions exist almost everywhere.

Arithmetic still works very well. The ratio of 2 to 3 tells us that the vermouth is $\frac{2}{5}$ of the mixture and gin is $\frac{3}{5}$ of the mixture.

REVIEW
The relationship between x and y is said to be linear if there exist constants m and b such that $y = mx + b$.

That is, we have the form $y = mx$. Geometrically, linear relationships graph as straight lines. If the form is $y = mx$, the line passes through the origin.

NOTE
$y = mx$ is equivalent to $\frac{y}{x} = m$. This is another way of defining a direct proportion. That is, two variables are in direct proportion if their quotient is constant.

Example 5

If 30 gallons of oil cost $17.10, what is the cost of 60 gallons of oil, assuming that the price per gallon stays the same?

60 gallons cost $34.20.
There are several ways of solving this problem. The "tried-and-true" arithmetic way is to find the cost per gallon and then to multiply by the number of gallons. If you divide $17.10 by 30, you get $0.57, or 57¢ per gallon. If you multiply this by 60, you get the correct answer.
But there is a much simpler way. Since 60 gallons is twice 30 gallons, the cost of 60 gallons is twice the cost of 30 gallons. Twice $17.10 is $34.20.

The assumption that the price per gallon stays the same is crucial. Sometimes you get a better price by buying in larger amounts. When this happens the cost is no longer said to be directly proportional to the amount you buy.

This method is summarized by

$$\text{cost} = \frac{\text{cost}}{\text{gal}} \times \text{gal}$$

Section 15.2 Ratio and Proportion

The idea behind Example 5 is the connection between direct proportion and ratios. Since the oil costs 57¢ per gallon, we see that the cost in dollars (C) of g gallons of oil is given by

$$C = 0.57g \qquad [10]$$

If we divide both sides of [10] by g, we get $\frac{C}{g} = 0.57$. This tells us that there is a fixed ratio between the cost and the number of gallons. The cost is to the number of gallons as 57 is to 100.

Example 6

A car traveling at a constant speed goes 200 miles in 6 hours. How far will it go in 9 hours?

In 9 hours it goes 300 miles.

If you let x denote how far it goes in 9 hours, you may use the ratio

$$\frac{x}{200} = \frac{9}{6} \qquad [11]$$

This fractional equation becomes

$$6x = 9(200) = 1{,}800 \quad \text{or} \quad x = 300$$

In less elaborate language, 9 hours is 50% more than 6 hours, so the distance the car travels is 50% more than 200 miles.

If you want to use arithmetic, you would divide 200 by 6 and find that the speed of the car is $33\frac{1}{3}$ (or $\frac{100}{3}$) miles per hour. The direct proportion is

$$d = \frac{100}{3}t$$

where d is the distance in miles and t is the time in hours.

> There is no one best way to set up a proportion. You may find it easier to write
>
> $$\frac{200 \text{ miles}}{6 \text{ hours}} = \frac{x \text{ miles}}{9 \text{ hours}}$$
>
> The answer will be the same.

Our only goal is to see how fractional equations may rise, so additional examples are left for the exercise set.

PRACTICE DRILL

1. A recipe requires that you use twice as much vinegar as oil.
   (a) How much vinegar should you use if you use 6 ounces of oil?
   (b) How much oil should you use if you use 6 ounces of vinegar?
   (c) If you use 60 ounces of the mixture, how much should be oil?

2. A recipe requires that you use two parts of oil to three parts of vinegar.
   (a) How much vinegar should you use if you use 6 ounces of oil?
   (b) How much oil should you use if you use 6 ounces of vinegar?
   (c) If you use 60 ounces of the mixture, how much should be oil?

3. Without finding the price per pound, find the price of 150 pounds of meat if 80 pounds cost $128.

4. If meat costs $1.60 per pound, write a linear relationship that expresses the cost of meat in dollars (C) for p pounds.

5. Use the formula in Problem 4 to find the cost of 150 pounds of meat.

Answers
1. (a) 12 ounces (b) 3 ounces (c) 20 ounces   2. (a) 9 ounces (b) 4 ounces (c) 24 ounces
3. $240   4. $C = 1.60p$
5. $240

CHECK THE MAIN IDEAS

1. The variables x and y are directly proportional if there exists a constant k such that $y =$ _____.
2. In a direct proportion, you can find the value of k by knowing one value of x and the corresponding value of y. Suppose you know that
$$y = kx \qquad [1]$$
and that $x = 2$ and $y = 8$. You replace x by 2 and y by 8 in [1] to find that $8 =$ _____, so that $k = 4$.
3. You replace k by 4 in [1] to obtain
$$y = \text{_____} \qquad [2]$$
4. From [2] you can find the value of y for any given value of x. If $x = 16$, you replace x by 16 in [2] to find that $y =$ _____.
5. In a direct proportion, you never have to find the value of the constant _____ if you don't want to.
6. Because k is a constant and $y = kx$, then for every pair of x and y values, you have that $\dfrac{y}{x} =$ _____.
7. Since k is a constant, each pair of x and y values has the same ratio. That is, $\dfrac{y}{x}$ is constant. Therefore, if you know that when $x = 2$, $y = 8$, and you want to find the value of y when $x = 16$, you write the ratio
$$\frac{y}{16} = \text{_____} \qquad [3]$$
8. Equation [3] is a _____ equation.
9. One way to solve [3] is to cross _____.
10. This tells you that $y =$ _____, which agrees with the answer to Problem 4. While this introduces a form of fractional equation, it also shows you how direct proportion is related to linear relationships.

Answers
1. kx
2. $k2$, or $2k$
3. $4x$
4. $4(16)$, or 64
5. k
6. k
7. $\dfrac{8}{2}$, or 4
8. fractional
9. multiply
10. 64

Section 15.3
Inverse Proportions and Other Variations

Sometimes when one quantity gets bigger, another quantity gets smaller. This happens quite obviously when we increase the denominator of a common fraction.

In other words, the more pieces we divide something into, the smaller each piece is.

OBJECTIVE
To see how algebraic fractions are used when we deal with inverse proportions.

> **DEFINITION**
> The variables x and y are said to be **inversely proportional** if there exists a constant k such that
> $$y = \frac{k}{x} \qquad [1]$$

In a direct proportion, $y = kx$. That is, in the direct proportion, the constant is multiplied by x. In the inverse proportion, the constant is *divided* by x.

We often write [1] in the form $xy = k$. Thus, Example 1 implies that $xy = 6$.

Section 15.3   Inverse Proportions and Other Variations

Example 1

In Formula [1], find the value of k if $y = 2$ when $x = 3$.

$k = 6$.

You replace x and y by their given values and solve the resulting equation. If you replace x by 3 and y by 2, equation [1] becomes

$$2 = \frac{k}{3} \quad [2]$$

Multiply both sides of [2] by 3 to get that $k = 6$.
  Once you know that $k = 6$, you can rewrite formula [1] as

$$y = \frac{6}{x} \quad [3]$$

Example 2

You are given the formula

$$y = \frac{k}{x} \quad [1]$$

and you know that when $x = 3$, $y = 2$. What is the value of y when $x = 6$?

When $x = 6$, $y = 1$.
  There are a few ways to do this problem. One way is to realize that this is a special case of the result in Example 1. In Example 1, you found out that under the given conditions, k must equal 6. That is shown in [3]. Knowing that $y = \frac{6}{x}$, you can replace x by 6 to find out that $y = \frac{6}{6}$, or $y = 1$.

Let's find the inverse proportion in this example. When x is 3, y is 2, but when we increase x to 6, y decreases to 1. In other words, as x increases in size, y decreases in size.

NOTE

We are usually given Example 2 and solve Example 1 as a first step. That is, we need one x value and a corresponding y value to determine k. Once we know the value of k, we can find x for any given value of y or y for any given value of x.

k is usually called the constant of proportionality.

  Let's try some more examples.

Example 3

Variables x and y are related by the formula

$$y = \frac{k}{x} \quad [1]$$

When $x = 5$, $y = 6$. What is the value of y when $x = 3$?

When $x = 3$, $y = 10$.
  You know that when $x = 5$, $y = 6$, so you begin by replacing y by 6 and x by 5 in formula [1]. You get

$$6 = \frac{k}{5} \quad [4]$$

Then you multiply both sides of [4] by 5 to see that $30 = k$.
  Once you know that k is 30, you replace k by 30 in [1] to obtain

$$y = \frac{30}{x} \quad [5]$$

The fact that k is a constant means that if k is 30 when $x = 5$ and $y = 6$, it's always 30 in the problem.

Then you replace x by 3 in [5] to obtain

$$y = \frac{30}{3} \quad \text{or} \quad y = 10 \qquad [6]$$

[6] Formula [1] is the same in both examples. The difference in the examples is that the given information determines different values for k. But once we find the value for k, it remains the same throughout that problem.

If we multiply both sides of Formula [1] by x, we get

$$xy = k \qquad [7]$$

Remember that k is a constant. Hence [7] tells us that in this problem the value of x times y is always the same. Equation [5] tells us that in Example 3, xy must always equal 30.

This checks with the fact that when $x = 5$, $y = 6$ ($5 \times 6 = 30$), and that when $x = 3$, $y = 10$ ($3 \times 10 = 30$).

Example 4

Using the information in Example 3, find the value of y when $x = 15$.

When $x = 15$, $y = 2$.
You know from Example 3 that

$$y = \frac{30}{x} \qquad [5]$$

All you have to do is replace x by 15 in [5] and solve for y. That is,

$$y = \frac{30}{15} = 2 \qquad [8]$$

Another method is to use the fact that in this problem x and y are related by the equation

$$xy = 30 \qquad [9]$$

So if $x = 15$, then $y = 2$.

[9] Equation [9] is just a rewriting of [5].

It doesn't make much difference whether we use [5] or [9]: they are different ways of saying the same thing. But if we invert the situation by giving a value for y and asking for the corresponding value of x, see what happens.

Example 5

Using the information in Examples 3 and 4, find the value of x when $y = 7$.

When $y = 7$, $x = \frac{30}{7}$, or $4\frac{2}{7}$.
If you use [9], you replace y by 7 to obtain

$$7x = 30 \quad \text{or} \quad x = \frac{30}{7}$$

If you replace y by 7 in [5], you get the fractional equation $7 = \frac{30}{x}$, which you convert to $7x = 30$ by cross multiplication.

Both direct proportions and inverse proportions lead us to fractional equations, but the two types of proportion are quite different.

We should also realize that sometimes we use more complicated proportions. For example, suppose we want to paint a square region. The amount of paint we use is proportional to the area of the square. If P denotes the amount of paint and A the area of the square, we have

$$P = kA \qquad [10]$$

If s denotes the length of the side of the square, $A = s^2$. If we replace A by s^2 in [10], we get

$$P = ks^2 \qquad [11]$$

Formula [11] says that P is proportional to the *square* of s.

Example 6

Given the formula

$$P = ks^2 \qquad [11]$$

find the value of k if $P = 18$ when $s = 3$.

The value of k is 2.

All you do is replace P by 18 and s by 3 in [11]. This gives you

$$18 = k(3)^2 = 9k \qquad [12]$$

Now you divide both sides of [12] by 9 to find that $k = 2$.

REMARK
If we use graphs, the difference is quite vivid. For example, $y = 6x$ is the straight line that passes through $(0,0)$ and $(1,6)$, as shown in the figure.

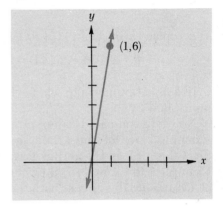

But look at $y = \dfrac{6}{x}$. We get the chart

x	$y \left(= \dfrac{6}{x} \right)$
1	6
2	3
3	2
4	$\dfrac{6}{4} = \dfrac{3}{2}$
6	1
12	$\dfrac{6}{12} = \dfrac{1}{2}$

These points are shown in the next figure. We see that the curve they lie on is more complicated than a line.

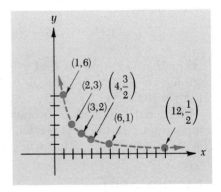

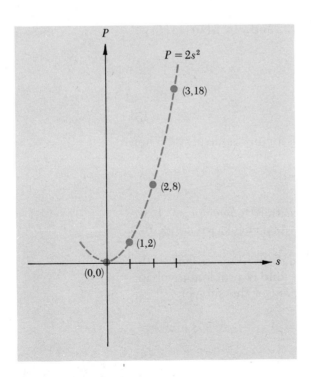

NOTE
[11] indicates a quadratic relationship between s and P. Therefore, the graph of the relationship will be a parabola.

s	s^2	$p\ (= 2s^2)$
0	0	0
1	1	2
2	4	8
3	9	18

Example 7

Using the information in Example 6, find the value of P when $s = 12$.

When $s = 12$, $P = 288$.
  You already know from Example 6 that $k = 2$. So the formula is
$$P = 2s^2 \qquad [13]$$
You replace s by 12 in [13] to get
$$P = 2(12)^2 = 2(144) = 288$$

  In a direct proportion, when we multiplied one variable by 4 the other was also multiplied by 4.
  Now let's compare Example 7 with Example 6. In Example 6, we let $s = 3$. In Example 7, we let s be 4 times as large: that is, we let $s = 12$. But in Example 6, the value of P is 18, and in Example 7, the value of P is 288. In going from Example 6 to Example 7, the value of P was multiplied by 16, not 4. The square of 4 is 16, and this checks with the fact that in [11] P is proportional to the *square* of s.

Example 8

Suppose that x and y are related by
$$y = kx^3 \qquad [14]$$
and that when $x = 2$, $y = 24$. What is the value of y when $x = 4$?

When $x = 4$, $y = 192$.
  First you find the value of k. To do this, replace x by 2 and y by 24 in [14] to get
$$24 = k(2^3) = 8k$$
This tells you that $k = 3$.
  Then you replace k by 3 in [14] to get
$$y = 3x^3 \qquad [15]$$
  Now you can use equation [15] to find the value of y for any value of x. When $x = 4$, you have
$$y = 3(4^3) = 3(64) = 192$$

  Equation [15] tells us that y is proportional to the third power of x. If we double x, y changes by a factor of 8 (that is, $2^3 = 8$). As a check, we notice that $24 \times 8 = 192$.
  Additional examples of proportions, both direct and inverse, are left to the exercise set. In the next section, we'll study another kind of problem involving fractional equations. That will complete our treatment of the subject.

PRACTICE DRILL

Find the value of k in each of the following.
1. $y = \dfrac{k}{x}$ and $x = 4$ when $y = 7$.
2. $y = \dfrac{k}{x}$ and $x = 7$ when $y = 4$
3. $y = kx^2$ and $y = 75$ when $k = 5$
4. Suppose $y = \dfrac{k}{x}$ and $x = 7$ when $y = 8$. What is the value of y when $x = 28$?

Answers
1. 28   2. 28   3. 3
4. 2

Section 15.4 "Sharing the Work" Problems

CHECK THE MAIN IDEAS

1. It is true that x and y are inversely proportional if there exists a constant k such that $y = $ _____.
2. To find the value of k, you need to know only one value of x and the corresponding value of y. Suppose you are given that

$$y = \frac{k}{x} \quad [1]$$

   and that $x = 7$ when $y = 8$. To find the value of k in [1], you replace x by 7 and y by 8 to get:

$$8 = \underline{} \quad [2]$$

3. From [2] you may cross multiply to find that $k = $ _____.
4. Once you know that $k = 56$, you may rewrite [2] as

$$\underline{} = \frac{56}{x} \quad [3]$$

5. From [3] you can now find the value of y for any given value of x. If x is 28, you replace x by 28 in [3] to find that $y = $ _____.
6. By cross multiplying in [3], you see that x and y are always related by the relationship $xy = $ _____.
7. The main point of this lesson is to study [3] because it is a form of a _____ equation. This is true because x appears in the denominator.
8. There are many other variations in which x and y are neither directly nor inversely proportional. For example, you may have

$$y = kx^3 \quad [4]$$

   In this case, as x increases, so does y, but the increase is not _____.
9. Using [4], suppose $y = 48$ when $x = 2$. From this information, you find that $48 = $ _____, or that $k = 6$.
10. In this case, [4] can be rewritten as

$$y = 6x^3 \quad [5]$$

   If you let $x = 4$ in [5], $y = 6(4)^3$ or _____. Comparing Problems 9 and 10, you see that though x only doubled, y increased by a factor of 8. That is, as x increased, so did y, but the increase was not linear.

Answers
1. $\frac{k}{x}$ 2. $\frac{k}{7}$ 3. 56
4. y 5. $\frac{56}{28} = 2$ 6. 56
7. fractional   8. linear
9. $k(2^3) = 8k$ 10. $6(64)$, or 384

Section 15.4
"Sharing the Work" Problems

OBJECTIVE
To learn how to develop and to solve more complex fractional equations.

Sometimes problems come up that cause us to make silly mistakes, but the mistakes don't seem silly at first. In fact, often they don't seem like mistakes until after we've had time to think about them.

Let's examine a mistake that many students make in a certain type of problem. The problem reads like this: "A large pool is being drained by two pipes. One of the pipes can drain the pool by itself in 6 hours. The other pipe can drain the pool by itself in 12 hours. How long does it take the two pipes, working together, to drain the pool?"

A typical impulse is to take the average of the two times. The average of 6 and 12 is 9, so the typical answer is 9 hours. Can you see why that answer is silly? Even if you don't know the right answer, isn't it logical to assume that if one of the pipes working by itself can empty the pool in 6 hours, the two pipes together do the job in less than 6 hours?

This mistake teaches two valuable lessons. One is that just because we know a formula (in this case, finding an average), the formula is not necessarily the right one for that problem. The second lesson is that we must think about the logic involved and to make an estimate. For example, without any other knowledge, we should know that the answer *has to be* less than 6 hours.

This problem is significant because the correct way to solve it makes use of fractional equations in a somewhat different way from the problems in the previous sections. Let's look at this problem piece by piece.

1. The pipe that empties the pool in six hours empties one-sixth of the pool in one hour.
2. The other pipe empties the pool in 12 hours, so it empties one-twelfth of the pool in one hour.
3. Therefore, in one hour the two pipes empty a total of $\frac{1}{6} + \frac{1}{12}$ of the pool.
4. But $\frac{1}{6} + \frac{1}{12} = \frac{2}{12} + \frac{1}{12} = \frac{3}{12} = \frac{1}{4}$.
5. So working together, the two pipes empty $\frac{1}{4}$ of the pool in one hour.
6. Therefore, working together, the two pipes will take 4 hours to empty the pool.

The following example gives us some practice with this technique.

In this entire discussion, we assume that the pipes work at a steady pace and that they don't interfere with one another.

These assumptions aren't so obvious as they may seem. For example, the pipes may be set up to empty the pool rapidly at the start and to slow down as the pool becomes less full. Or it may be that one pipe has an effect on the other (like two boys painting a fence but with only one brush).

In the solution of any real-life problem, we always make certain assumptions. Our answers, therefore, depend on the assumptions we make.

Example 1

One pipe can drain a pool in 6 hours. Another pipe can drain the same pool in 3 hours. How long does it take for both pipes working together to drain the pool?

It takes them 2 hours if they work together.

One pipe drains $\frac{1}{6}$ of the pool in an hour. The other drains $\frac{1}{3}$ in an hour. So together, in 1 hour they drain $\frac{1}{6} + \frac{1}{3}$ or $\frac{1}{2}$ of the pool. The whole pool is drained in two hours.

Sometimes the answer is not a whole number, but there's no reason why it should be.

Before you start this problem, do you realize that the correct answer must be less than 3 hours?

Example 2

One pipe drains a pool in 6 hours. Another pipe drains the same pool in 9 hours. How long does it take both pipes, working together, to drain the pool?

It takes them 3 hours and 36 minutes $\left(3\frac{3}{5} \text{ hours}\right)$.

The method remains the same. In one hour, one pipe empties $\frac{1}{6}$ of the pool and the other empties $\frac{1}{9}$ of the pool. So together, in one hour the pipes empty $\frac{1}{6} + \frac{1}{9}$ or $\frac{5}{18}$ of the pool. Therefore, it will take $\frac{18}{5}$ hours to empty the whole pool. This is $3\frac{3}{5}$ hours; and since one-fifth of an hour is 12 minutes, the total time can be expressed as 3 hours and 36 minutes.

Without least common denominators, we'd get

$$\frac{1}{6} + \frac{1}{9} = \frac{9+6}{54} = \frac{15}{54}$$

This has the same value as $\frac{5}{18}$.

Section 15.4 "Sharing the Work" Problems

NOTE

It should be clear that this type of problem involves using fractions. If you didn't know how to get $\frac{18}{5}$ from the $\frac{5}{18}$, use trial and error. For example, if $\frac{5}{18}$ is emptied in one hour, then twice that amount, $\frac{10}{18}$, is emptied in 2 hours, $\frac{15}{18}$ is emptied in 3 hours, and $\frac{20}{18}$ is emptied in 4 hours. But $\frac{20}{18}$ is more than the whole amount. So you know that it takes between 3 and 4 hours to complete the job.

Example 3

One pipe empties a pool in $3\frac{1}{2}$ hours, and the other pipe empties the pool in $5\frac{1}{3}$ hours. How long does it take them to empty the pool if they work together?

It takes them $\frac{112}{53}$ hours, or approximately 2 hours and 7 minutes.

Start by using common fractions rather than mixed numbers. The first pipe takes $\frac{7}{2}$ hours, so in one hour it empties $\frac{2}{7}$ of the pool. The second pipe empties the pool in $\frac{16}{3}$ hours, so in one hour it empties $\frac{3}{16}$ of the pool.

Working together, in one hour the two pipes empty $\frac{2}{7} + \frac{3}{16}$, or $\frac{2(16) + 7(3)}{7(16)}$, or $\frac{53}{112}$ of the pool. Therefore, the whole pool is emptied in $\frac{112}{53}$ hours.

So far in this section, we have not used any algebra. Our aim has been to feel comfortable with this type of example. But there is a formula that we can use. If f stands for the time it takes the first pipe to empty the pool, and s stands for the time it takes the second pipe to empty the pool, then the total time t it takes for both pipes working together to empty the pool is given by

$$\frac{1}{f} + \frac{1}{s} = \frac{1}{t} \qquad [1]$$

By its very form, equation [1] is a fractional equation. Depending on how the question is worded, Equation [1] can be very helpful.

Example 4

One pipe takes twice as long as another to empty a pool. Working together, the two pipes empty the pool in 6 hours. How long would it take each pipe separately to empty the pool?

One pipe would do it in 9 hours, and the other in 18 hours.

Let f stand for the time it takes the faster pipe. Then $2f$ is the time it takes the other pipe. Since the pool is emptied in 6 hours, $t = 6$. Using equation [1], you have

$$\frac{1}{f} + \frac{1}{2f} = \frac{1}{6} \qquad [2]$$

To see that the answer is the reciprocal of $\frac{5}{18}$, it may be helpful to write

$$\frac{\frac{5}{18} \text{ pool}}{1 \text{ hour}} = \frac{1 \text{ pool}}{x \text{ hours}}$$

$$\frac{5}{18} = \frac{1}{x}$$

If numbers like $3\frac{1}{2}$ and $5\frac{1}{3}$ bother you, make approximations. For example, pretend the times are 3 hours and 5 hours. This answer would be less than the true time. Then use 4 hours and 6 hours, and this answer would be more than the true time. But getting the exact answer isn't too hard if you understand Example 3.

In Example 1, f is 6 and s is 3. So equation [1] becomes

$$\frac{1}{6} + \frac{1}{3} = \frac{1}{t}$$

or

$$\frac{1}{2} = \frac{1}{t}$$

Then by cross multiplying or by any other method, we find that $t = 2$.

In the first three examples, however, it seemed easier to use arithmetic rather than to depend on equation [1].

Our whole discussion centers around [2]. In the last section, we learned how to solve an equation like [2]. In this section, we've tried to show where an equation like [2] may come from.

Since f, $2f$, and 6 are all divisors of $6f$, you can multiply both sides of [2] by $6f$ to obtain

$$6f\left(\frac{1}{f}+\frac{1}{2f}\right) = 6f\left(\frac{1}{6}\right)$$
$$6f\left(\frac{1}{f}\right) + 6f\left(\frac{1}{2f}\right) = 6f\left(\frac{1}{6}\right)$$
$$6 + 3 = f$$

If $f = 9$, then $2f = 18$.

As a check, note that the faster pipe empties $\frac{1}{9}$ of the pool in one hour. The slower one empties $\frac{1}{18}$ of the pool in one hour. Together they empty $\frac{1}{9}+\frac{1}{18}=\frac{1}{6}$ of the pool in one hour. That is, it takes them 6 hours to empty the pool.

We don't use Equation [1] just for draining-the-pool problems. There are many situations in which two or more people or objects share a work load. By chance, Example 1 was stated in terms of two pipes draining a pool. But it could have been two painters painting a house, or two children whitewashing a fence. The situations solved by equation [1] are indeed numerous.

Example 5

An experienced typist types at three times the speed as a beginning typist. Working together, they complete a typing assignment in 6 hours. How long would it take the typists to do the job if each works alone?

It would take the experienced typist 8 hours, and the beginner 24 hours.
 Let f stand for the time it takes the faster typist. Then the time for the slower typist is $3f$. Since the whole job takes 6 hours, t is 6. Equation [1] becomes

$$\frac{1}{f}+\frac{1}{3f}=\frac{1}{6} \qquad [3]$$

Equation [3] is a fractional equation that can be made into a linear equation by multiplying both sides by $6f$. You get

$$6f\left(\frac{1}{f}+\frac{1}{3f}\right) = 6f\left(\frac{1}{6}\right)$$
$$6 + 2 = f$$

The variations of [1] are almost endless. Sometimes [1] leads us to quadratic equations, depending on how the question is worded.

Example 6

It takes one typist 8 hours longer to do a typing project than it takes another typist. Working together, the two typists complete the project in 3 hours. How long would it take the faster typist to finish the project working alone?

It would take the faster typist 4 hours to do the job.
 Let f denote the time it takes the faster typist. Then $f + 8$ is the time it takes the slower typist. They complete the job in 3 hours, so $t = 3$. Hence equation [1] becomes

$$\frac{1}{f}+\frac{1}{f+8}=\frac{1}{3} \qquad [4]$$

To clear [4] of fractions, multiply both sides of the equation by $3f(f+8)$. That is,

$$3f(f+8)\left(\frac{1}{f}+\frac{1}{f+8}\right) = 3f(f+8)\left(\frac{1}{3}\right)$$
$$\frac{3f(f+8)}{f}+\frac{3f(f+8)}{(f+8)} = \frac{3f(f+8)}{3}$$
$$3(f+8)+3f = f(f+8)$$
$$3f+24+3f = f^2+8f$$
$$0 = f^2+2f-24$$
$$0 = (f+6)(f-4)$$

This tells you that either $f=4$ or $f=-6$ (which is impossible, because the time must be nonnegative).

Every course has to end somewhere, and our course ends here. The material in this lesson summarizes what we've learned in the course and how the material is used in more advanced courses.

The particular topics we've studied are not that important. The important point is that we've learned a structured way to tackle new problems in terms of old ones. We believe that this type of reasoning can be extended to any decision-making process we may need.

> We could also work on the left side of the equation to get
>
> $$\frac{1(f+8)+f(1)}{f(f+8)}$$
>
> or
>
> $$\frac{2f+8}{f(f+8)}$$
>
> Equation [4] becomes
>
> $$\frac{2f+8}{f(f+8)} = \frac{1}{3}$$
>
> By cross multiplying, we get
>
> $$3(2f+8) = f(f+8)$$
> $$6f+24 = f^2+8f$$
> $$0 = f^2+2f-24$$
>
> The main point is not how we solve [4] but that we realize it can be rewritten as a quadratic equation.

PRACTICE DRILL

1. If one pipe can empty a pool in two hours and a second pipe can empty the same pool in three hours, then how long will it take for both pipes, working together, to empty the pool?
2. If one typist can do a project in 3 hours and a second typist can do the same project in 12 hours, then, working together, how long will it take them to do the project?
3. Suppose one typist types twice as fast as another, and that working together they can complete a project in three hours. How long would it take each of them, working alone, to do the project?

> **Answers**
> 1. $\frac{6}{5}$ hours, or 1 hour and 12 minutes   2. $\frac{12}{5}$ hours, or 2 hours and 24 minutes   3. $4\frac{1}{2}$ hours and 9 hours

CHECK THE MAIN IDEAS

1. In this section, you see another place in which _____ equations occur.
2. The theory is that if it takes n hours to do a job, then you do $\frac{1}{n}$ of the job in _____ hour.
3. This assumes that the work is done at the _____ rate of speed throughout.
4. Under this assumption, if a pipe can drain a pool in 2 hours, then in one hour it drains _____ of the pool.
5. If a second pipe can drain the same pool in _____ hours, then in one hour it drains one-third of the pool.
6. If the two pipes work together, in one hour they drain a total of $\frac{1}{2}+\frac{1}{3}$ or _____ of the pool.
7. It will take _____ hours to drain the whole pool. You find this answer by taking the reciprocal of the answer to the previous question.

> **Answers**
> 1. fractional   2. one
> 3. same   4. one half $\left(\frac{1}{2}\right)$
> 5. three (3)   6. $\frac{5}{6}$
> 7. $\frac{6}{5}$ or $1\frac{1}{5}$ 8. one third $\left(\frac{1}{3}\right)$
> 9. $2x$ 10. $6x$

8. Algebra enters the picture when you know how long a joint project takes, and you want to find out the individual rates. For example, if two typists finish a job in 3 hours, then in one hour they do _____ of the job.

9. Suppose that one typist works at twice the speed of the other. If you assume that the faster typist takes x hours, then the slower typist takes _____ hours.

10. The equation would be $\frac{1}{x} + \frac{1}{2x} = \frac{1}{3}$. To clear this equation of fractions, you multiply both sides by _____ to get $3 + 6 = 2x$, from which it follows that $x = 4\frac{1}{2}$ and $2x = 9$. Thus you have solved a fractional equation to determine that the faster typist requires $4\frac{1}{2}$ hours to do the job working alone, and the slower typist requires 9 hours for the same job.

EXERCISE SET 15 (Form A)

Section 15.1

1. Give the values (if any) of x for which each of the following is undefined, and then write each expression in lowest terms.

   (a) $\dfrac{36x}{45x}$ (b) $\dfrac{12(x-2)}{18(x-2)}$ (c) $\dfrac{12(x-1)}{18(x-2)}$

2. Reduce to lowest terms.

   (a) $\dfrac{2(x+5)(x+3)}{4(x+3)(x+7)}$ (b) $\dfrac{2(x^2+8x+15)}{4(x^2+10x+21)}$

3. Evaluate $\dfrac{2(x^2+8x+15)}{4(x^2+10x+21)}$ when $x = -6$.

Give the values of x (if any) for which each of the following expressions is undefined, and then write each expression as a quotient of polynomials in lowest terms.

4. $\dfrac{2}{x^2+1} + \dfrac{x^2-1}{x^2+1}$

5. $\dfrac{2}{x^2-1} + \dfrac{x-1}{x+1}$

6. $\dfrac{2}{x+1} - \dfrac{x+5}{x^2+4x+3}$

7. $\left(\dfrac{x+3}{x-1}\right)\left(\dfrac{x+1}{x+3}\right)$

8. $\left(\dfrac{x^2-9}{x^2+5x+6}\right)\left(\dfrac{x+2}{x-3}\right)$

9. $\dfrac{4x}{x-4} \div \dfrac{8x^2}{x^2-6x+8}$

10. Find the value of each of the following expressions when $x = 1.2$.

    (a) $\dfrac{2}{x^2+1} + \dfrac{x^2-1}{x^2+1}$ (b) $\left(\dfrac{x+3}{x-1}\right)\left(\dfrac{x+1}{x+3}\right)$ (c) $\dfrac{4x}{x-4} \div \dfrac{8x^2}{x^2-6x+8}$

Solve for x in each of the following.

11. $\dfrac{x+1}{3} - \dfrac{x}{4} = \dfrac{1}{2}$

12. $\dfrac{2x}{x-1} - 3 = \dfrac{7-3x}{x-1}$

13. $\dfrac{x+1}{x+2} + \dfrac{1}{x+3} = \dfrac{19}{20}$

14. $\dfrac{8}{x} = \dfrac{x}{18}$

15. $\dfrac{2x}{x-1} - 3 = \dfrac{2}{x-1}$

Answers: Exercise Set 15
(Form A)

1. (a) $\dfrac{4}{5}$, $x \neq 0$ (b) $\dfrac{2}{3}$, $x \neq 2$

   (c) $\dfrac{2x-2}{3x-6}$, $x \neq 2$

2. (a) $\dfrac{(x+5)}{2(x+7)}$, $x \neq -3$ or -7

   (b) $\dfrac{(x+5)}{2(x+7)}$ or $\dfrac{x+5}{2x+14}$

3. $-\dfrac{1}{2}$

4. 1

5. $\dfrac{x^2-2x+3}{x^2-1}$, $x \neq -1$ or 1

6. $\dfrac{1}{x+3}$, $x \neq -1$ or -3

7. $\dfrac{x+1}{x-1}$, $x \neq 1$ or -3

8. 1, $x \neq -3, -2,$ or 3

9. $\dfrac{x-2}{2x}$, $x \neq 2, 0,$ or 4

10. (a) 1 (b) 11 (c) $-\dfrac{1}{3}$

11. $x = 2$ 12. $x = 2$ $(x \neq 1)$

13. $x = -7$ or $x = 2$

14. $x = -12$ or 12

15. no solution 16. $7\dfrac{1}{2}$ or 7.5

17. $26\dfrac{2}{3}$ 18. 870

19. 120 20. 2,000

21. 10 22. (a) $u = k\sqrt{v}$

(b) 9 23. (a) $y = \dfrac{kuv}{w^2}$

(b) 62.5 24. (a) 22

(b) 22 hours 25. (a) $x = 3$ or $x = -2$ (b) 6 hours

Exercise Set 15 (Form B)

Section 15.2

16. Suppose that $y = kx$ and that $x = 6$ when $y = 8$. For what value of x will $y = 10$?
17. If $\frac{x}{y} = \frac{2}{3}$, for what value of x will $y = 40$?
18. In a certain school, the ratio of boys to girs is 3:5 (3 to 5). How many boys are enrolled in the school if there are 1,450 girls enrolled?
19. For what value of x will $\frac{200-x}{x} = \frac{2}{3}$?

Section 15.3

20. In the formula $y = \frac{k}{x^3}$, $y = 2$ when $x = 5$. For what value of y will $x = \frac{1}{2}$?
21. There is a formula in chemistry that states $P = k\frac{T}{V}$. Suppose that $P = 20$ when $T = 300$ and $V = 600$. Find the value of P when $T = 200$ and $V = 800$.
22. Suppose that u varies directly as the square root of v.
    (a) Express u in terms of v, letting k denote the constant of variation.
    (b) If $u = 8$ when $v = 4$, for what value of v will $u = 12$?
23. (a) Translate into an equation, using k as the constant of variation: y varies directly as the product of u and v, and inversely as the square of w.
    (b) Suppose that $y = 60$ when $u = 6$, $v = 5$, and $w = 4$. For what value of v will $y = 100$, $w = 10$, and $u = 5$?

Section 15.4

24. (a) For what value of x does $\frac{1}{x} + \frac{1}{2x} + \frac{1}{3x} = \frac{1}{12}$?
    (b) It takes pipe B twice as long to empty a pool as it takes pipe A. It takes pipe C three times as long to empty a pool as it takes pipe A. Working together, the three pipes empty the pool in 12 hours. How long will it take pipe A to empty the pool by itself?
25. (a) For what values of x does $\frac{1}{x} + \frac{1}{x+3} = \frac{1}{2}$?
    (b) It takes Tom 3 hours longer to paint a fence than it takes Betty to paint the same fence. Working together, they paint the fence in two hours. How long would it take Tom to paint the fence if he painted it by himself?

EXERCISE SET 15 (Form B)

Section 15.1

1. Give the values of x (if any) for which each of the following is undefined, and then write each expression in lowest terms.
    (a) $\frac{81x}{90x}$ (b) $\frac{8(x-3)}{4(x-3)}$ (c) $\frac{8(x-2)}{4(x-3)}$

2. Reduce to lowest terms.
    (a) $\frac{8(x+3)(x+4)}{2(x+4)(x+1)}$ (b) $\frac{8(x^2+7x+12)}{2(x^2+5x+4)}$

3. Evaluate $\frac{8(x^2+7x+12)}{2(x^2+5x+4)}$ when $x = -2$.

Give the values of x (if any) for which each of the following is undefined, and then rewrite each expression as a quotient of polynomials in lowest terms.

4. $\dfrac{3}{x+1} + \dfrac{x-2}{x+1}$

5. $\dfrac{3}{x^2-1} + \dfrac{x+1}{x-1}$

6. $\dfrac{2}{x+2} - \dfrac{x+4}{x^2+5x+6}$

7. $\left(\dfrac{x+5}{x+1}\right)\left(\dfrac{x+1}{x-5}\right)$

8. $\left(\dfrac{x^2-16}{x^2+6x+8}\right)\left(\dfrac{x+2}{x-4}\right)$

9. $\dfrac{5x}{x-4} \div \dfrac{10}{x^2-9x+20}$

10. Find the value of each of the following when $x = 5.1$.

   (a) $\dfrac{3}{x+1} + \dfrac{x-2}{x+1}$
   (b) $\left(\dfrac{x+5}{x+1}\right)\left(\dfrac{x+1}{x-5}\right)$
   (c) $\dfrac{5x}{x-4} \div \dfrac{10}{x^2-9x+20}$

For what values of x is each of the following true?

11. $\dfrac{x+3}{5} - \dfrac{x}{6} = \dfrac{2}{3}$

12. $\dfrac{2x}{x-3} + 5 = \dfrac{6x-5}{x-3}$

13. $\dfrac{x+1}{x+3} + \dfrac{1}{x-2} = \dfrac{13}{12}$

14. $\dfrac{12}{x} = \dfrac{x}{48}$

15. $\dfrac{2x}{x-3} + 5 = \dfrac{6(x-2)}{x-3}$

Section 15.2

16. Suppose that $y = kx$ and that $x = 7$ when $y = 9$. For what value of x will $y = 14$?

17. If $\dfrac{x}{y} = \dfrac{4}{9}$, for what value of x will $y = 14$?

18. In a certain school, the ratio of boys to girls is 4:7. How many boys are in the school if there are 2,240 girls in the school?

19. For what value of x will $\dfrac{400-x}{x} = \dfrac{9}{11}$?

Section 15.3

20. In the formula $y = \dfrac{k}{x^3}$, y is 5 when x is 2. For what value of y will $x = \dfrac{1}{2}$?

21. A chemistry formula states that $P = k\dfrac{T}{V}$. Suppose that when $T = 500$ and $V = 400$, $P = 30$. What is the value of P when $V = 300$ and $T = 400$?

22. Suppose that u varies directly as the square root of v.

   (a) Express u in terms of v, letting k denote the constant of proportion.
   (b) If $u = 6$ when $v = 9$, for what value of v will $u = 12$?

23. (a) Translate into a formula, using k as the constant of variation: y varies directly with the square of w and inversely with the product of u and v.
   (b) Suppose that $y = 6$ when $u = 8$, $v = 12$, and $w = 4$. What is the value of v if $y = 288$, when $w = 2$, and $u = 0.1$?

Exercise Set 15 (Form B)

Section 15.4

24. (a) For what value of x does $\dfrac{1}{x} + \dfrac{1}{3x} + \dfrac{1}{4x} = \dfrac{1}{12}$?

   (b) It takes pipe B three times as long to empty a pool as it takes pipe A. It takes pipe C four times as long to empty a pool as it takes pipe A. Working together, the three pipes empty the pool in 12 hours. How long would it take pipe A to empty the pool by itself?

25. (a) For what values of x does $\dfrac{1}{x} + \dfrac{1}{x+8} = \dfrac{1}{3}$?

   (b) It takes Mary 8 hours longer to set up the lab than it takes Helen. Working together, they set up the lab in 3 hours. How long would it take Helen to set up the lab by herself?

Appendix

Arithmetic Review

The Arithmetic of Common Fractions

(1) To add two common fractions that have a common denominator, we add the numerators and keep the common denominator. For example,

$$\frac{3}{8} + \frac{2}{8} = \frac{3+2}{8} = \frac{5}{8}$$

That is, 2 *eighths* + 3 *eighths* is 5 *eighths*.

(2) If we multiply both numerator and denominator by the same nonzero number, we get an equal fraction. For example, $\frac{2}{5} = \frac{6}{15}$ since we get from $\frac{2}{5}$ to $\frac{6}{15}$ by multiplying both numerator and denominator by 3. Likewise, we can say that we get from $\frac{6}{15}$ to $\frac{2}{5}$ by dividing both numerator and denominator by 3. It's like saying that at 2 for 5¢ the cost per item is the same as buying 6 for 15¢. (*Note:* We can't just do the same thing to numerator and denominator. For example, if we add 6 to both the numerator and the denominator of $\frac{2}{5}$, we get $\frac{2+6}{5+6}$, or $\frac{8}{11}$. But $\frac{2}{5}$ and $\frac{8}{11}$ are not equal. For example, 2 for 5¢ is not the same price as 8 for 11¢.)

(3) To add two common fractions that have different denominators, we replace each fraction by equal ones that have the same denominator and then proceed as in (1). For example, to add $\frac{2}{5}$ and $\frac{3}{7}$, we see that 35 — that is, 5 × 7 — is a common denominator. So we multiply numerator and denominator of $\frac{2}{5}$ by 7 and numerator and denominator of $\frac{3}{7}$ by 5. This gives us: $\frac{2}{5} + \frac{3}{7} = \frac{14}{35} + \frac{15}{35}$, or $\frac{29}{35}$. (*Note:* It is incorrect to add numerators and denominators. For example, $\frac{1}{2} + \frac{1}{3}$ is not the same as $\frac{2}{5}$. If you work for a half hour (30 minutes) and then for another third of an hour (20 minutes), you've worked 50 minutes or $\frac{5}{6}$ of an hour. But $\frac{2}{5}$ of an hour is only 24 minutes.)

(4) To subtract two common fractions, we rewrite them so that they have common denominators. Then we subtract numerators and keep the common denominator. For example, $\frac{1}{2} - \frac{1}{3} = \frac{3}{6} - \frac{2}{6} = \frac{1}{6}$. That is, if we take 2 sixths from 3 sixths, we have 1 sixth left.

(5) To multiply common fractions, we multiply numerators and we multiply denominators. For example, $\frac{2}{5} \times \frac{3}{7} = \frac{6}{35}$. Notice that we do not need common denominators. What we are doing is taking $\frac{2}{5}$ *of* $\frac{3}{7}$. To get a number divisible by 5, we replace $\frac{3}{7}$ by $\frac{3 \times 5}{7 \times 5}$,

or $\frac{15}{35}$. Now 1 fifth of 15 thirty-fifths is 3 thirty-fifths, so 2 fifths of 15 thirty-fifths is twice 3 thirty-fifths or 6 thirty-fifths, which we write as $\frac{6}{35}$. The shortcut would be to multiply the numerators to get 6 and to multiply the denominators to get 35. Notice that the shortcut is based on our knowing the right answer. It is not a logical device that should be obvious to you.

(6) To divide common fractions, remember that division is the inverse of multiplication. For example, $\frac{2}{5} \div \frac{3}{7}$ means the number we must multiply by $\frac{3}{7}$ to get $\frac{2}{5}$. If we were to multiply $\frac{3}{7}$ by $\frac{7}{3}$, we would get 1; and if we then multiplied 1 by $\frac{2}{5}$, we would get $\frac{2}{5}$. That is, $\left(\frac{2}{5} \times \underbrace{\frac{7}{3}\right) \times \frac{3}{7}}_{1} = \frac{2}{5}$.

A quick way to get from $\frac{2}{5} \div \frac{3}{7}$ to the correct answer $\frac{2}{5} \times \frac{7}{3}$ (or $\frac{14}{15}$) is by the "invert and multiply" rule. Leave the first number alone, change division to multiplication, and invert the second number (that is, take its reciprocal). This is a shortcut for getting the correct answer. The point is that $\frac{14}{15} \times \frac{3}{7} = \frac{42}{105}$, or $\frac{2}{5}$. (*Note:* Don't invert both or don't invert the first instead of the second. If you do, the number you get is not the number that when multiplied by $\frac{3}{7}$ equals $\frac{2}{5}$.)

The Arithmetic of Decimal Fractions

(1) When we use decimal fractions, we can multiply by 10 simply by moving the decimal point one place to the right. For example, $3.24 \times 10 = 32.4$. If there is no decimal point, we may assume it's after the digit furthest to the right. That is, 34. and 34 name the same number. In this case $34 \times 10 = 34. \times 10 = 340.$, where the 0 tells us that the decimal point has been moved one place to the right. So, for example, since $1{,}000 = 10 \times 10 \times 10$, we multiply by 1,000 by moving the decimal point one place to the right three times — that is, 3 places to the right.

(2) To divide by 10 using decimal fractions, we move the decimal point one place to the left. For example, $342 \div 10 = 34.2$.

(3) To multiply two decimal fractions, first multiply as if the decimal points were not there. Then count the total number of digits to the right of the decimal point in each factor and move the decimal point that number of places to the left. For example, to multiply 3.14 by 2.7, first multiply 314 by 27 to get 8,478. Next count the total number of digits in both factors to the right of the decimal points (that is, 3.14 and 2.7). Then move the decimal point in 8,478. three places to the left to get 8.478. (*Note:* The reason this works is that 3.14×2.7 is the same as $\frac{314}{100} \times \frac{27}{10}$, or $\frac{314 \times 27}{1{,}000}$. This means that we first multiply 314×27 and then divide by 1,000 — which means that we move the decimal point three places to the left. But you don't have to memorize this rule. Use the fact that 3.14 is between 3 and 4, while 2.7 is between 2 and 3. Therefore, 3.14×2.7 is between 2×3 and 3×4. But if the product is between 6 and 12, the decimal point in 8478 must be between the 8 and the 4. For example, 84.78 is greater than 12, while 0.8478 is less than 6.

(4) To divide two decimal fractions, look at both the dividend and the divisor and see how many places the decimal point must be moved in each to make both numbers whole numbers. Move the decimal point in *both* the same amount as the number with the most places to the right of the decimal point. Since we have moved the decimal point the same

amount, we have multiplied the dividend and the divisor by the same number and haven't changed the ratio. We then divide the resulting whole numbers to obtain the quotient. For example, to compute $0.86 \div 0.2$, we see that if we move the decimal point two places to the right, 0.86 becomes the whole number 86. If we move the decimal point one place to the right, 0.2 becomes 2; and if we move it two places to the right, it becomes 20. To keep the ratio the same, we move the decimal point two places to the right in both to get $86 \div 20$ or $43 \div 10$ or 4.3 (*Note:* It sometimes comes as a surprise that when we divide two small numbers like 0.86 and 0.2, we get a quotient as great as 4.3, but the reason for this is that the ratio is a relative size. That is, when we divide 0.86 by 0.2, we want the number we must multiply by 0.2 to get 0.86 and this number is 4.3.)

(5) To add decimal fractions, line them up in a column so the decimal points are matched up. (We do this to make sure we're adding like denominations.) Then place the decimal point in the sum directly below where it appears in the addends, and add as if the decimal points were not there. For example to find the sum of 2.38 and 4.19;

$$\text{write} \quad \begin{array}{r} 2.38 \\ +4.19 \\ \hline \end{array} \quad \text{and then add as with whole numbers} \quad \begin{array}{r} 2.38 \\ +4.19 \\ \hline 6.57 \end{array}$$

That is, 238 hundredths + 419 hundredths = 657 hundredths or 6.57.

(6) To subtract two decimal fractions, we proceed as we did to add them, but this time we subtract the bottom number from the top. For example, $4.19 - 2.38$ means 419 hundredths − 238 hundredths = 181 hundredths or 1.81.
That is,

$$\begin{array}{r} 4.19 \\ -2.38 \\ \hline 1.81 \end{array}$$

(7) We do not change the value of a decimal fraction by putting 0's *after* the digit furthest to the *right* of the decimal point. For example, 8.3 and 8.3000 name the same number — both indicate 8 ones and 3 tenths. But we change the value if we put the 0 between two digits. For example, in 8.3 the 3 means 3 tenths, but in 8.03 the 3 means 3 hundredths. If we think in terms of money, $8.30 and $8.03 are not equal amounts. So if we wanted to subtract 8.49 from 13.1, instead of writing

$$\begin{array}{r} 13.1 \\ -8.49 \\ \hline \end{array}$$

we could rewrite 13.1 as 13.10 to get

$$\begin{array}{r} 13.10 \\ -8.49 \\ \hline 4.61 \end{array}$$

But it would be wrong to replace 13.1 by 13.01 since these two decimal fractions do not name the same number.

The Arithmetic of Percents

(1) Just think of "percent" (written as the symbol, %) as "divided by a hundred."

(2) To change a percent to a common fraction, we divide by one hundred (shown by writing the number over 100). For example, 27% of 1,800 would mean $\frac{27}{100} \times 1{,}800$ or 27×18 or 486. (It helps to remember that in the present context, "of" means "times".)

(3) To change a percent to a decimal fraction, we move the decimal point two places to the left. So if we use decimal fractions, 27% of 1,800 becomes $0.27 \times 1{,}800$, or 486. (*Note:* We

write 0.27 rather than .27 simply to highlight the presence of the decimal point.)

(4) Whether we use decimal fractions or common fractions, 27% of 1,800 is 486. The key point is to replace "percent" by "divided by a hundred" and "of" by "times".

(5) Another way of thinking of percent is "per hundred." In this sense, 27% means that we take 27 per 100. If we were to use money, this would mean 27 cents of each dollar. Then, since 1,800 is a hundred 18 times, we take 27 eighteen times or 27×18, which is still 486.

(6) An entire amount is called 100%. So if we take one-third of an amount, we're taking one-third of 100%. That is, $\frac{1}{3}$ means $\left(\frac{1}{3} \times 100\right)\%$, which is $33\frac{1}{3}\%$. So when we convert from a fraction to a percent, we *multiply* by a hundred.

(7) Since percent means to divide by a hundred, we multiply by a hundred to change a fraction back to a percent. This is easy to remember if we recall that multiplication and division are inverses.

The Arithmetic of Mixed Numbers

(1) A mixed number such as $6\frac{2}{3}$ is read as 6 *and* $\frac{2}{3}$, which means $6 + \frac{2}{3}$. To rewrite the sum using common denominators, we replace 6 by $\frac{18}{3}$ ($\frac{18}{3}$ means 18 divided by 3, which is 6) and get $6 + \frac{2}{3} = \frac{18}{3} + \frac{2}{3}$ or $\frac{20}{3}$. This rule for converting from a mixed number to a common fraction can be expressed by saying that we multiply the whole number (in this case, 6) by the denominator (in this case, 3) to get 18; then we add the numerator (in this case, 2) to get 20; and then we write the result over the denominator to get $\frac{20}{3}$. As another example, to write $45\frac{3}{4}$ as a common fraction, we multiply 45 by 4 to get 180, add 3 to get 183, and then put 183 over 4 to get $\frac{183}{4}$. This says that $45\frac{3}{4} = 45 + \frac{3}{4} = \frac{45}{1} + \frac{3}{4} = \frac{180}{4} + \frac{3}{4} = \frac{183}{4}$.

(2) If we start with a common fraction whose numerator is greater than the denominator, we divide the numerator by the denominator to get the mixed number. For example, suppose we want to write $\frac{655}{7}$ as a mixed number. We have

$$\frac{655}{7} = \begin{array}{r} 93 \text{ R } 4 \\ 7\overline{)655} \\ -63 \\ \hline 25 \\ -21 \\ \hline 4 \end{array} \quad \text{or} \quad 93\frac{4}{7}$$

That is, if we divide 655 by 7, we get 93 and have 4 of the 7 needed to get another "bundle" of 7. That is, we have 93 and $\frac{4}{7}$.

(3) To convert a decimal fraction to a mixed number, read the decimal point as "and." For example, 34.59 means 34 and 59 hundredths or $34 + \frac{59}{100}$, or $34\frac{59}{100}$.

(4) To do arithmetic using mixed numbers, we can always rewrite the mixed numbers as common fractions and then use the arithmetic of common fractions. For example, to add

The Arithmetic of Mixed Numbers

$3\frac{1}{4}$ and $2\frac{1}{3}$, we can write $3\frac{1}{4}$ as $\frac{13}{4}$ and $2\frac{1}{3}$ as $\frac{7}{3}$. So the sum is $\frac{13}{4} + \frac{7}{3}$. If we now use 12 as a common denominator, we get

$$\frac{13 \times 3}{4 \times 3} + \frac{7 \times 4}{3 \times 4} = \frac{39}{12} + \frac{28}{12} = \frac{67}{12} = 5\frac{7}{12}$$

(*Note:* Since $3\frac{1}{4}$ means $3 + \frac{1}{4}$ and $2\frac{1}{3}$ means $2 + \frac{1}{3}$, we can find the sum by first adding 2 and 3 to get 5 and then adding $\frac{1}{4}$ and $\frac{1}{3}$ to get $\frac{7}{12}$, which tells us that the sum is, $5 + \frac{7}{12}$. Either method is acceptable.)

CAUTION

Don't think that we always combine the whole numbers and the fractions separately. While this works for addition, it doesn't work for multiplication. For example, suppose we want the value of $3\frac{1}{2} \times 4\frac{1}{3}$. Certainly $3 \times 4 = 12$ and $\frac{1}{2} \times \frac{1}{3} = \frac{1}{6}$ but $3\frac{1}{2} \times 4\frac{1}{3}$ is not equal to $12\frac{1}{6}$! In fact, since $4\frac{1}{3}$ is more than 4, $3\frac{1}{2} \times 4\frac{1}{3}$ is more than $3\frac{1}{2} \times 4$; but $3\frac{1}{2} \times 4 = \frac{7}{2} \times 4 = 14$ — and this is already more than $12\frac{1}{6}$. To get the correct answer, you may write $3\frac{1}{2}$ as $\frac{7}{2}$ and $4\frac{1}{3}$ as $\frac{13}{3}$. Then compute $\frac{7}{2} \times \frac{13}{3}$ to get $\frac{91}{6}$, or $15\frac{1}{6}$.

(5) If you want to multiply two mixed numbers without converting them to fractions (and this might well be the case with a mixed number like $234\frac{879}{922}$), the rule is quite complicated. Let's use $3\frac{1}{2} \times 4\frac{1}{3}$ as the example.

(a) Multiply the two whole numbers ($3 \times 4 = 12$).

(b) Multiply the two fractions $\left(\frac{1}{2} \times \frac{1}{3} = \frac{1}{6}\right)$.

(c) Multiply the first whole number by the second fraction $\left(3 \times \frac{1}{3} = 1\right)$.

(d) Multiply the second whole number by the first fraction $\left(4 \times \frac{1}{2} = 2\right)$.

(e) Add the results in (a) through (d) $\left(12 + \frac{1}{6} + 1 + 2, \text{ or } 15\frac{1}{6}\right)$.

(f) (e) is the product $\left(3\frac{1}{2} \times 2\frac{1}{3} = 15\frac{1}{6}\right)$.

(6) We could avoid the problem of using this complicated rule for multiplication if we use common fractions. That is, since $3\frac{1}{2}$ is $\frac{7}{2}$ and $4\frac{1}{3}$ is $\frac{13}{3}$, $3\frac{1}{2} \times 4\frac{1}{3} = \frac{7}{2} \times \frac{13}{3} = \frac{91}{6} = 15\frac{1}{6}$. The same applies to division. If we want to compute $3\frac{1}{2} \div 4\frac{1}{3}$, we write $\frac{7}{2} \div \frac{13}{3} = \frac{7}{2} \times \frac{3}{13} = \frac{7}{13}$.

(7) We can also convert mixed numbers to common fractions for subtraction. But if we keep the mixed number form, we must be careful if the subtraction involves borrowing. For example, suppose we want to subtract $2\frac{3}{4}$ from $4\frac{1}{6}$. We can't take $\frac{3}{4}$ from $\frac{1}{6}$. So we borrow one from the four in $4\frac{1}{6}$. But remember that in terms of sixths, 1 is 6 sixths. That is,

we write $4\frac{1}{6}$ as $3 + 1 + \frac{1}{6} = 3 + \frac{6}{6} + \frac{1}{6} = 3 + \frac{7}{6}$ $\left(\text{or } 3\frac{7}{6}\right)$. This gives us $3\frac{7}{6} - 2\frac{3}{4}$. Using common denominators, we get $3\frac{28}{24} - 2\frac{18}{24} = 1\frac{10}{24} = 1\frac{5}{12}$. If n is the denominator of the fractional part of the mixed number, we must remember that if we borrow 1, we write it as $\frac{n}{n}$.

(*Note:* The shortcut for borrowing 1 in an expression like $7\frac{3}{5}$ is to replace the whole number 7 by 6 and the numerator of the fraction by the *sum* of the numerator and denominator. That is, $7\frac{3}{5} = 6\frac{8}{5}$. The reason in this case is that 1 is $\frac{5}{5}$ and $\frac{5}{5} + \frac{3}{5} = \frac{8}{5}$.)

Arithmetic Glossary

Addends. The numbers that are added. For example, in the expression $3 + 2 + 6$, the addends are 3, 2, and 6. Addends are also called *terms*.

Borrowing. The inverse (or, opposite) of "carrying." It is used when we subtract and indicates that we have exchanged one of a place (denomination) for ten of the next lower place. For example, when we subtract 47 from 63, we can't take 7 ones from 3 ones. So we "borrow" one of the 6 tens in 63 and exchange it for 10 ones. That is, we rewrite 63 as $50 + 10 + 3$ or $50 + 13$.

We write it as $\overset{5}{\cancel{6}}\,^1 3$. In other words,

$$\begin{array}{r} 63 \\ -47 \\ \hline \end{array} \quad \text{means} \quad \begin{array}{r} \overset{5}{\cancel{6}}\,^1 3 \\ 4\ 7 \\ \hline 1\ 6 \end{array}$$

Carrying. A term used when we add. It means that we exchange ten of a place for one of the next greater place. For example, when we add 67 and 85 we write

$$\begin{array}{r} \overset{1}{}67 \\ +\ 85 \\ \hline 152 \end{array}$$

We have added 5 ones and 7 ones to get 12 ones, which we write as a ten and 2 ones. We indicate the ten by "carrying" 1 to the tens place.

Change-making. A form of subtraction in which we use addition to subtract. This is usually how we make change. If a person owes you $2.39 and pays you with a $5 bill, you often make change by adding onto $2.39 the amount necessary to make $5. For example:

$$\begin{array}{rr} \$2.39 & \\ .01 & .01 \\ \hline \$2.40 & \\ .10 & .10 \\ \hline \$2.50 & \\ .50 & .50 \\ \hline \$3.00 & \\ 2.00 & \$2.00 \\ \hline \$5.00 & \$2.61 \text{ change} \end{array}$$

Common Denominator. Refers to two or more common fractions having the same (that is, a common) denominator. To add common fractions, we must use a common denominator. Just as 2 nickels and 3 nickels are 5 nickels or 2 dimes and 3 dimes are 5 dimes, 2 of any denomination and 3 of that same (common) denomination are 5 of the common denomination. But while 2 nickels and 3 dimes are 5 coins, they are neither 5 nickels nor 5 dimes.

Common Fraction. A special numeral used to denote a rational number. For example, a common fraction that names the rational number $3 \div 5$ is $\frac{3}{5}$. It is read as "three over five" or "three-fifths". 3 is called the *numerator* and 5 is called the *denominator*. The denominator tells us the size (denomination) and the numerator *enumerates* (that is, counts) how many of that size.

Common Multiple (of two or more numbers). A common multiple of two or more whole numbers is any whole number that is divisible by each of the numbers. For example, 54 is a common multiple of 9 and 6 because 54 is divisible by both 9 and 6. One way to get a common multiple (as we did here) is to multiply each of the given numbers.

Decimal Fraction. A place-value numeral used to name a rational number. For example, since $\frac{1}{2}$ and $\frac{5}{10}$ name the quotient $1 \div 2$, the decimal fraction that names this quotient is 0.5 (which we read as "point five" or "five-tenths").

Difference. The correct answer when we subtract two numbers. For example, when we write $5 - 3 = 2$, 2 is called the difference of 5 and 3. Some people call 5 the minuend and 3 the subtrahend. That is,

$$\text{minuend} - \text{subtrahend} = \text{difference}$$

Digits. In place value, the name given to the ten numerals: 0, 1, 2, 3, 4, 5, 6, 7, 8, and 9. Any place-value numeral is a combination of these digits. For example, we refer to both 862 and

253 as three-digit numerals. We are allowed to use only one digit per place in place value.

Dividend. In an expression like $f \div s$, the dividend is the name given to f. For example, when we say "6 divided by 3," 6 is called the dividend and 3 is called the divisor. When the quotient is written as a common fraction, the dividend is the numerator and the divisor is the denominator. If we use the form

$$3 \overline{)6}^{\,2}$$

the divisor, 3, is the "outside" number and the dividend is the "inside" number.

Division. The inverse of multiplication. For example, $6 \div 3$ means the number we must multiply by 3 to get 6, and since this number is 2, we write $6 \div 3 = 2$. When we read this as "six divided by three is two," we lose the connection between multiplication and division. But, notice that the usual check for division is to multiply the quotient by the divisor to see if we get the dividend. That is,

dividend ÷ divisor = quotient   means that
quotient × divisor = dividend

Divisor. See *Dividend*.

Equal. The term used to indicate that two numerals name the same number. For example, when we write $5 = 3 + 2$, we do not mean that the numerals 5 and $3 + 2$ look alike. What we do mean is that 5 and $3 + 2$ name the same number (five). In a similar way, if we write $\frac{4}{2} = \frac{6}{3}$, we mean that these two different common fractions name the same rational number. That is, if $4 is equally divided between two people, each would get the same amount as three people dividing $6 would get.

Factor. The name given to each of the numbers that are multiplied. For example, in the expression 2×3, 2 and 3 are called factors. (Some people prefer to call the first factor the multiplier and the second factor the multiplicand, but since 2×3 is equal to 3×2, the terms are, for all practical purposes, interchangeable.)

Fraction. A numeral used to name a rational number. To be more exact, see either *Decimal Fraction* or *Common Fraction*.

Irrational Number. Any number that isn't rational (that is, *not* the quotient of two whole numbers). If we represent numbers with decimals, rational numbers are those decimals that either terminate (come to an end), such as 0.5 or 0.34, or else repeat the same cycle of digits endlessly beyond a certain place, such as $0.3\overline{3}\ldots$ or $0.13\overline{53}\ldots$, where the "bar" tells us the repeating cycle. On the other hand, an irrational number is represented by a non-ending decimal that never repeats the *exact* cycle endlessly, such as $0.28288288828888\ldots$, where each time we add one more 8 to the cycle (so that the same cycle never repeats, since each time the new cycle has one more 8 than the previous cycle).

Least Common Multiple (of two or more numbers). The smallest *natural* number that is divisible by each of the given numbers. For example, the least common multiple of 9 and 6 is 18. Any other common multiple of 9 and 6 must be a multiple of 18; that is, 36, 54, 72, 90, and so forth. We exclude 0 from being the least common multiple because then 0 would always be the least common multiple. That is, when 0 is divided by 9, 6, or any other non-zero number, the quotient is 0.

Minus. The symbol used to indicate subtraction. We read $5 - 3$ as "five minus three" or "five take away three" or "subtract three from five."

Mixed Number. A combination of a whole number and a common fraction. Any common fraction in which the numerator is greater than the denominator can be made into a mixed number. For example, $215 \div 7$ is 30 with a remainder of 5. This is the same as $30 + \frac{5}{7}$, and when we write this as $30\frac{5}{7}$, the result is called a mixed number—mixed because it's part whole number and part fraction.

Natural Numbers. The numbers such as one, two, three, and so on that are used in counting. They are also called *counting numbers*. The term "natural" is used to indicate that the most natural use of numbers was for counting.

Numerals. Symbols that name or represent numbers. For example, X is a Roman numeral used to stand for the number ten. Other more modern numerals for naming ten are, for example, 10, 5×2, $6 + 4$, $12 - 2$, and $30 \div 3$. (*Note:* 5 and 2 are two digits but 5×2 is *a* numeral that names the number ten.)

Percent. Literally, "per hundred" or "divided by 100." So if we say 24% of 400, we mean 24 hundredths of 400. In decimal form this is 0.24×400 and in common fraction form it's $\frac{24}{100} \times 400$. In either case, 24% of 400 is 96.

Place Value. A numeral system in which the denomination (place) of a digit depends on the placement of that digit. For example, in 372 the digit 3 means 3 *hundred*, but in 834 it means 3 *tens*. The "face value" of 3 is always three; but the "place value" tells us three of what denomination.

Plus. The name of the symbol used to denote addition. That is, $3 + 5$ is read as "three plus five". It is also called the sum of 3 and 5.

Product. The name given to the correct answer when we multiply two or more numbers. For example, when we write $2 \times 3 = 6$, we say that the product of 2 and 3 is 6.

Quotient. The name given to the correct answer when we divide one number by another. For example, we read $6 \div 3 = 2$ as "six divided by three is two"; and we say that when 6 is divided by 3, the quotient is 2. (*Note:* When we divide, the order makes a difference. For example, when we divide 3 by 6, the quotient is not 2 but $\frac{3}{6}$ or $\frac{1}{2}$.)

Rational Number. The quotient of two whole numbers, provided only that the divisor is not zero (see definition of *Division*). For example, $6 \div 2$ is a rational number because it is the quotient of two whole numbers. In this case, the quotient is also a whole number. In fact, if n denotes any whole number, n is also a rational number — because it equals $n \div 1$, which is the quotient of two whole numbers, n and 1. But the quotient of two whole numbers doesn't have to be a whole number to be a rational number. It is also the quotient of two whole numbers. For example, if we divide $5 between 2 people, each gets $2.50, which is a "real" amount of money but not a whole number of dollars.

Reciprocal (of a number). The number we multiply a given number by to get 1. In terms of common fractions, the reciprocal is obtained by interchanging numerator and denominator. For example, the reciprocal of $\frac{2}{3}$ is $\frac{3}{2}$, since $\frac{2}{3} \times \frac{3}{2} = \frac{6}{6}$, or 1. (*Note:* 0 has no reciprocal because any number times 0 is still 0.)

Remainder. When we divide the dividend by the divisor, the quotient may not be a whole number. For example, when we divide 17 by 3 (think of dividing 17 objects into groups of 3, for example) we get 5 with 2 *remaining*. That is, $17 = (3 \times 5) + 2$. The number 2 is called the remainder. The remainder can be as small as 0 (which is the case when the quotient is a whole number — for example, when we divide 15 by 3, we get 5 with no remainder) to as much as one less than the divisor (for example, when we divide 17 by 3, we get 5 "bundles" of 3 with 2 remaining).

Rounding off. A form of estimating numbers that is very useful in checking arithmetic. For example, 3.82 is between 3 and 4 but closer to 4. If we replace 3.82 by 4, we say that we've rounded off 3.82 to the nearest whole number. Similarly, since 3.82 is between 3.8 and 3.9 but closer to 3.8, if we replace 3.82 by 3.8, we say that we've rounded off 3.82 to the nearest tenth. By rounding off to the nearest whole number, we could replace, for example, 3.82×7.98 by 4×8, which is 32. This indicates that the product of 3.82 and 7.98 is "reasonably close to" 32 and helps us guard against serious errors. (*Note:* The exact value of 3.82×7.98 is 30.4836. The fact that we know the answer is near 32 warns us not to misplace the decimal point and write 304.836 since 304.836 hardly is "reasonably close to" 32.)

Subtraction. Often called the inverse of addition. That is, by $5 - 3$ we mean the number we must *add* to 3 to get 5. Since this number is 2, we see that $5 - 3 = 2$. Because we often use the wording "five take away three is two," we may not realize that subtraction is a form of addition. But "minuend − subtrahend = difference" means the same thing as "difference + subtrahend = minuend."

Sum. The correct answer to a problem involving the addition of two or more numbers. For example, because $3 + 2 + 6$ is 11, we say that the sum of 3, 2, and 6 is 11.

Times. The name given to the multiplication symbol. That is, 3×5 is read as "three times five". In dealing with rational numbers, the word "times" is often replaced by "of". For example, we read $\frac{1}{2} \times \frac{1}{3}$ as "one-half *of* one-third." (*Note:* Since one-half of any non-zero amount is less than the original amount, $\frac{1}{2} \times \frac{1}{3}$ is less than $\frac{1}{3}$ even though we are multiplying.)

Whole Numbers. The name given to the collection consisting of the natural numbers together with 0. Note that 0 is the only whole number that is not a natural number. One reason 0 is not included among the natural numbers is that it's not "natural" to begin counting until there's at least one to count. But place value makes the digit 0 necessary to hold the "place" of a missing denomination. For example, in the numeral 203 the 0 tells us that we have no 10's, so that the 2 means 2 *hundreds*.

SQUARES AND SQUARE ROOTS

No.	Sq.	Sq. Root	No.	Sq.	Sq. Root
0.1	0.01	0.3162	5.1	26.01	2.2583
0.2	0.04	0.4472	5.2	27.04	2.2804
0.3	0.09	0.5477	5.3	28.09	2.3022
0.4	0.16	0.6325	5.4	29.16	2.3238
0.5	0.25	0.7071	5.5	30.25	2.3452
0.6	0.36	0.7746	5.6	31.36	2.3664
0.7	0.49	0.8367	5.7	32.49	2.3875
0.8	0.64	0.8944	5.8	33.64	2.4083
0.9	0.81	0.9487	5.9	34.81	2.4290
1.0	1.00	1.0000	6.0	36.00	2.4495
1.1	1.21	1.0488	6.1	37.21	2.4698
1.2	1.44	1.0954	6.2	38.44	2.4900
1.3	1.69	1.1401	6.3	39.69	2.5100
1.4	1.96	1.1832	6.4	40.96	2.5298
1.5	2.25	1.2247	6.5	42.25	2.5495
1.6	2.56	1.2649	6.6	43.56	2.5690
1.7	2.89	1.3038	6.7	44.89	2.5884
1.8	3.24	1.3416	6.8	46.24	2.6077
1.9	3.61	1.3784	6.9	47.61	2.6268
2.0	4.00	1.4142	7.0	49.00	2.6458
2.1	4.41	1.4491	7.1	50.41	2.6646
2.2	4.84	1.4832	7.2	51.84	2.6833
2.3	5.29	1.5166	7.3	53.29	2.7019
2.4	5.76	1.5492	7.4	54.76	2.7203
2.5	6.25	1.5811	7.5	56.25	2.7386
2.6	6.76	1.6125	7.6	57.76	2.7568
2.7	7.29	1.6432	7.7	59.29	2.7749
2.8	7.84	1.6733	7.8	60.84	2.7928
2.9	8.41	1.7029	7.9	62.41	2.8107
3.0	9.00	1.7321	8.0	64.00	2.8284
3.1	9.61	1.7607	8.1	65.61	2.8460
3.2	10.24	1.7889	8.2	67.24	2.8636
3.3	10.89	1.8166	8.3	68.89	2.8810
3.4	11.56	1.8439	8.4	70.56	2.8983
3.5	12.25	1.8708	8.5	72.25	2.9155
3.6	12.96	1.8974	8.6	73.96	2.9326
3.7	13.69	1.9235	8.7	75.69	2.9496
3.8	14.44	1.9494	8.8	77.44	2.9665
3.9	15.21	1.9748	8.9	79.21	2.9833
4.0	16.00	2.0000	9.0	81.00	3.0000
4.1	16.81	2.0248	9.1	82.81	3.0166
4.2	17.64	2.0494	9.2	84.64	3.0332
4.3	18.49	2.0736	9.3	86.49	3.0496
4.4	19.36	2.0976	9.4	88.36	3.0659
4.5	20.25	2.1213	9.5	90.25	3.0822
4.6	21.16	2.1448	9.6	92.16	3.0984
4.7	22.09	2.1679	9.7	94.09	3.1145
4.8	23.04	2.1909	9.8	96.04	3.1305
4.9	24.01	2.2136	9.9	98.01	3.1464
5.0	25.00	2.2361	10.0	100.00	3.1623

METRIC CONVERSIONS

		English to Metric		Metric to English
LENGTH	1 inch	= 2.54 cm (centimeters)		
		= 25.4 mm (millimeters)	1 mm (millimeter)	= 0.039 inches
	1 foot	= 30.48 cm (centimeters)	1 cm (centimeter)	= 0.033 feet
	1 yard	= 0.914 m (meters)	1 m (meter)	= 1.094 yards
	1 mile	= 1.609 km (kilometers)	1 km (kilometer)	= 0.621 miles
WEIGHT (MASS)	1 ounce	= 28.35 (grams)	1 g (gram)	= 0.035 ounces
	1 pound	= 0.454 (kilograms)	1 kg (kilogram)	= 2.205 pounds
	1 short ton (2,000 pounds)	= 0.908 metric tons	1 t (metric ton) (1000 kg)	= 1.102 short tons
VOLUME	1 teaspoon	= 5 ml (milliliters)	1 ml (milliliter)	= 0.2 teaspoons
	1 tablespoon	= 15 ml (milliliters)		= 0.067 tablespoons
	1 fluid ounce	= 30 ml (milliliters)		= 0.034 ounces
	1 cup	= 237 ml (milliliters)		= 0.004 cups
	1 pint	= 437 ml (milliliters)	1 l (liter)	= 4.226 cups
	1 quart	= 0.946 l (liters)		= 2.113 pints
	1 gallon	= 3.785 l (liters)		= 1.057 quarts
				= 0.264 gallons
AREA	1 square inch	= 6.452 cm² (square centimeters)	1 cm² (square centimeter)	= 0.155 square inches
	1 square foot	= 929.030 cm² (square centimeters)	1 m² (square meter)	= 10.764 square feet
	1 square yard	= 0.836 m² (square meters)		= 1.196 square yards
	1 square mile	= 2.590 km² (square kilometers)	1 km² (square kilometer)	= 0.405 square miles
	1 acre	= 0.405 ha (hectares)	1 ha (hectare)	= 2.471 acres

TEMPERATURE To convert from degrees Fahrenheit (°F) to degrees Celsius (°C):

$$(°F - 32) \times \frac{5}{9} = °C$$

To convert from degrees Celsius (°C) to degrees Fahrenheit (°F):

$$\left(°C \times \frac{9}{5}\right) + 32 = °F$$

INDEX

Absolute value, 46
Addition
　associative property of, 68–71, 110, 286
　commutative property of, 71–73, 117, 282, 286, 366
　of equations, 5, 30
　of fractions, 397–98
　and grouping symbols, 68–71, 73, 78, 88–89
　of polynomials, 285–88
　of signed numbers, 47–51
Additive identity, 79–80, 83, 110
Additive inverse, 79–83, 84–85, 199
"Add the opposite" rule, 53–55, 289–91. See also Signed numbers; Subtraction
Algebraic expressions, 67, 71, 396–401
Algebraic fractions, 395–410
Algorithm, 316–21
Areas
　of rectangles, 9–10, 76, 96, 125, 350
　of squares, 11, 14
Arithmetic, 3–7, 275, 375
　vs. algebra relationship, 25, 30, 37–38, 333–36, 403–404
　glossary, 427–29
　and quadratic equations, 328–36
　review, 421–26
　and solving equations, 34, 37, 38
　and square roots, 364–68
Associative property
　for addition, 68–71, 110, 286
　vs. commutative property, 78
　definition of, 68
　for multiplication, 74–75, 109
Average, 6
Axis, coordinate, 157–61, 165, 173–77, 192

Balance, 5
Base, 248, 255, 260–61
Brace, 196, 233
Brackets, 50–51, 58, 93–94

Cancellation property for fractions, 17, 199
Cartesian coordinate system, 156–61, 194
Change-making method, 288–90, 427
Coefficient(s)
　definition of, 76–78
　negative, in equalities, 224–25
　numerical, 279–80
　of simultaneous equations, 199–204
　of square roots, 365
Common denominator(s), 305, 365, 396–400, 427
Commutative property
　for addition, 71–73, 117, 282, 286, 366
　for multiplication, 76–78, 292, 294
Completing the square method, 369–72, 380, 383, 386–88
　for solving quadratic equations, 373–79
Complex fractions, 411–15
Complex number(s), 161, 362
Consecutive integers, 153, 350–51
Constant(s), 31–33, 280
　in direct proportions, 404
　and inequalities, 222–23
　in inverse proportions, 406–10
　in linear relationships, 103–107, 108–12, 404
　of proportionality, 407
Constant speed problems. See Motion problems
Continuity, 332
Conversion, 17–20
　table, 430
Coordinates
　Cartesian, 156–61
　and inequalities, 227–30
　of linear relationships, 163–67
　of quadratic relationships, 381–84
Cross multiplication, 398–402, 408, 413
Curved graph, 165, 381–85

Decimal(s)
　fractions, 5, 9, 128, 253, 422–23
　place, 262–65, 266–69
　system, 17–18
Decreasing order, 282–83, 295, 306, 316. See also Order; Usual order
Degree
　of monomial, 279–80, 344
　of polynomial, 282, 341–45
Denominator, 140, 303–304, 396–401
　common, 305, 365, 396–98, 400, 427
Dependent equations, 207
Dependent variables, 32
Deposit, 5
Descartes, René, 157
Difference, 44–45
Directed distance, 154–55
Direct proportion, 404–406, 408
Distributive property, 92–97, 111–12, 117, 167, 197, 223, 364–65, 376, 399
　of polynomials, 283, 289, 292–93, 295–96, 308–309
Dividend, 303, 316–21
Division, 29, 34, 75, 78, 398
　with exponents, 251–53, 257–58
　of fractions, 398
　and grouping symbols, 74, 89–90
　of monomials, 302–304
　of polynomials, 302–306, 312–21
　of signed numbers, 58–60
　by zero, 302–303, 337–39
Divisor, 303, 316–21
　of zero, 337–39

Elimination method, 201, 203
English system, 17–20
　conversion table, 430
Equality, 27, 29–30, 36, 218
　vs. inequalities, 235
Equation(s)
　algebraic, 25–40, 127–43, 396–401
　constant terms in, 31–33
　definition of, 29–30
　dependent, 207
　equivalent, 115, 141, 200, 206
　as false statements, 117
　fractional, 395, 398–405, 409, 411–15
　graphs of, 163–71, 180–83, 225–30, 235–36, 380–85, 404, 409
　identity, 76, 117–18
　inconsistent, 116–18, 206–207
　linear, 102, 108–12, 119–21, 125–27, 163–71, 180–83, 345, 359, 369, 399–401
　linear in two variables, 191–96
　linear in three variables, 208
　as open sentences, 26–30
　polynomial, 338–39, 345
　quadratic, 327–32, 340–53, 359–63, 369–72, 373–79, 399–401, 409
　simultaneous, 191–96, 197–201, 202–208
　solving, 28–29, 34–37, 38–40, 60–62, 95–96, 113–16, 118, 211, 225–30, 235–36, 344–53, 373–79, 396
　straight-line, 173–78
　as true statements, 76, 116–18, 206
　variable terms in, 31–33
Equivalent equations, 115, 141, 200
Evaluating expressions, 36, 38–39, 280, 284
　with signed numbers, 60–61, 69–70
Exponential notation, 246–49, 367
Exponents, integral
　decimal fraction, 253
　definition of, 248, 250, 255, 259
　division of, 251–53, 257–58
　multiplication of, 254–57, 260–64, 268–69
　negative, 250–53, 258–59
　order of, 248, 282–83, 295, 306, 316
　positive, 246–49
　raising to a power, 259–61
　rules of, 255, 258
　zero, 248, 250, 255, 259
Expression
　definition of, 36
　linear, 103–107
　simplified, 69–71, 75

Factor(s), 35, 246, 254, 428
　product of, 76–77, 338–39
Factoring, 301, 307–12, 340–44
　definition of, 307
　polynomials, 312–15

433

quadratic equations, 344-49, 350-53, 373-74, 380
    solving equations by, 344-53, 373-74, 380
Factorization, prime, 367
Formula(s), algebraic, 12-13, 25, 62
    constant terms in, 31-33
    for coordinates of a point, 163
    definition of, 8-10
    for linear equations, 125-27
    for motion problems, 136-38, 213
    for percent problems, 140-41
    for polynomial expressions, 281-82
    for solving equations, 28-30, 36-37
Fractions
    addition of, 397-98
    algebraic, 395-410
    cancellation property for, 17, 199
    common, 29, 303-305, 366, 396-401, 421-22, 427
    completing the square, 372
    complex, 411-15
    decimal, 5, 9, 253, 427
    division of, 398
    least common denominator in, 412
    multiplication of, 398
    in percent problems, 140-41
    reducing, 307, 311
    square roots, 365, 375-76
    subtraction of, 398
    from word problems, 128, 140-41
Fractional equations, 395, 398-401, 408
    as quadratic equations, 415
    and ratio and proportion, 402-410
    solution of, 95-96, 411-15
Fractional notation, 402

Galileo, Galilei, 12
Geometry, 151, 152, 155, 193-94
Graphs
    definition of, 165
    and inequalities in two variables, 225-30
    of linear equations, 163-71, 175, 177, 178, 180-83, 404
    of quadratic equations, 380-85, 409
    and solution of simultaneous equations, 235-36
Greater than, 218-21
Gross pay, 140
Grouping symbols, 93, 129, 134, 282
    in addition, 68-71, 73, 78, 88-89
    braces, 196, 233
    brackets, 50-51, 58
    in division, 74, 89-90
    in multiplication, 74-75, 77-78, 89-90, 280
    parenthesis, 4, 7, 13, 33, 51, 68, 88-89, 93-94, 111, 337
    removing, 70, 88-91, 94-96, 311
    in subtraction, 69-71, 74

Half-line, 236

Half-plane, 229, 231-34
Hypotenuse, 351-52

Identity, 116-18
    additive, 79
    definition of, 76
    multiplicative, 84-85
Imaginary number(s), 344-49
Inconsistent equations, 116-18, 206-207
Independent variables, 32
Inequalities
    linear in one variable, 221-25
    linear in two variables, 225-30, 233
    and measurement, 266-69
    signed numbers, 220
    simultaneous, 231-36
    symbols, 219-21, 222, 302
Integer(s), 45-46, 360-61
    consecutive, 153, 350-51
    definition of, 45, 152-53
    even, 351
    as exponents, 246-53, 257-59
    negative, 250-53, 258
    nonnegative, 277-78, 303-304
    positive, 246-49, 253, 255, 258
    powers of, 248, 253, 259-61
    product of, 55
    quotient of, 302, 396
Intercept(s), 165-70, 177, 192-93, 198
Intersection of points, 194-96, 235
Inverse, 75, 78, 424
    additive, 79-83
    process of, 119-21
    proportion, 406-10
    multiplicative, 84-87
"Invert and multiply" rule, 75, 78, 119-21, 422, 424
Irrational number(s), 428

Least common denominator, 412
Less than, 44-46, 218-21
Like terms, 283, 293
Linear equation(s), 102, 108-12
    definition of, 108, 345
    equivalent, 115
    from formulas, 125-27
    from fractional equations, 395, 399-404, 414
    graphs of, 163-71, 175, 177-78, 404
    identity, 117-18
    inconsistent, 116-18
    inverting, 119-21
    from mixture problems, 132-35
    from motion problems, 136-39
    from percent problems, 140-43
    solving, 113-16, 118
    solving with graphs, 180-83
Linear equations in two variables, 191-96
    solving, 197-207
    solving word problems with, 210-13
Linear equations in three variables, 208
Linear expressions, 103-107

Linear inequalities, 221-25
    solving, 231-36
Linear inequalities in two variables, 225-30
Linear polynomials, 340-45
    square of, 369-72
Linear relationships, 163-71, 276, 404
Line segment, 236. See also Straight line

Magnitude, of signed numbers, 46, 48-50, 55-58, 84, 152-53
Mathematical relationships, 7-10
Measurement, 266-69
Metric system, 17-20
    conversion table, 430
Minus sign(s), 89, 91
Mixed number(s), 9, 424-26
Mixture problems, 132-35
Monomials, 276-80
    definition of, 278, 345
    degree of, 279-80
    division of, 302-304
    product of, 294
Motion problems, 136-39, 212-13
More than, 44-46, 218-21
Multiplication, 5
    associative property of, 74-75
    commutative property of, 76-78
    with exponents, 254-57, 260-64, 268-69
    of fractions, 398
    and generalized distributive property, 96-97
    and grouping symbols, 74-75, 77-78, 88-90, 280
    of linear inequalities, 222-24, 229
    of monomials, 294
    of polynomials, 292-97
    of signed numbers, 55-58, 84
Multiplicative identity, 84-86, 109
Multiplicative inverse(s), 84-87, 109. See also Reciprocal

Natural number(s), 428
Negative intergral exponents, 250-53, 258-59
Negative numbers, 44, 232
    in graphs, 167
    in inequalities, 220-21, 224
    multiplication of, 335
    on a number line, 154-55
    products of, 55-58, 342-43
    of signed numbers, 48-50, 62
    square of, 363
    square root of, 362
Negative slope, 170
Net pay, 140
Nonlinear equations, 182-83
Nonnegative integers, 277-78, 303-304
Nonnegative number(s), 58, 153, 335, 338, 350, 366
Notation
    decimal, 261-65

exponential, 246-49, 367
scientific, 261-65, 269, 277
Number line, 152-55, 157
Number(s). *See also* Negative numbers; Positive numbers; Signed numbers
  irrational, 428
  mixed, 9, 424-26, 428
  natural, 428
  vs. points, 153-54, 160-61
  rational, 45, 429
  real, 58, 153, 335, 338
  whole, 143-46, 429
Numerals, 266-69, 428
Numerator, 303-305, 396-401
Numerical
  coefficient, 279-80
  fractions, 362
  method, 396

Open sentence, 26-30, 34, 223
Opposites, 52-53, 81-82, 84
Order
  in addition, 71-74, 78
  of exponents, 248, 282-83, 295, 306, 316
  vs. grouping, 72
  in multiplication, 76, 78
  of polynomials, 282-83, 285, 289, 295, 306
Ordered pairs, 156-61

Parabola, 383
Parenthesis, 4, 7, 13, 33, 51, 68, 88-89, 93-94, 110, 337
Percent
  definition of, 141, 423-24
  problems, 140-43
Perfect square, 346, 361, 371
Plus signs, 88-89, 91
Points, 192-96
  coordinates of, 156-61, 163-67, 170-71
  for inequalities, 227-30, 235-36
  on a number line, 153-54
  set of, 165
  on a straight line, 170, 175, 176
Polynomials, 275, 281-85, 338-49
  addition of, 285-88
  definition of, 284, 345
  degree of, 282, 341-45
  division of, 302-306, 312-21
  factoring, 311-15, 344-49
  multiplication of, 292-97
  subtraction of, 288-91
Positive integral exponents, 246-49, 252, 255, 258
Positive numbers
  in inequalities, 220-21
  product of, 55-58
  of signed numbers, 47-50, 54
  square roots of, 360-61, 363, 368, 370

Power(s), 248, 253, 259-65
Prime factorization, 367
Problem solving, 124, 127-35, 136-39, 140-43, 210-13, 231. *See also* Word problems
Product(s), 74, 76, 307
  of integers, 245-46
  of monomials, 294
  of polynomials, 292-97, 340-44
  of signed numbers, 55-59
  as zero, 337-39
Proportion
  direct, 404-405
  inverse, 406-10
Pythagorean Theorem, 351-52, 362

Quadratic equations, 327-32
  definition of, 336, 344
  from fractional equations, 395, 399-401, 414-15
  graphs of, 409
  and Pythagorean Theorem, 352
  solutions by arithmetic, 328-39, 333-36
  solutions by completing the square, 373-79, 380, 386-88
  solutions by factoring, 344-53, 380
  solutions by graphs, 380-85
  standard form, 347
  word problems, 350-53, 386-88
Quotient, 44-45, 302-304, 404
  of polynomials, 312-21, 396
  of signed numbers, 58-60

Radical, 366
Radicand, 366
Ratio, 402-405
Rational expression, 396
Rational number, 45, 429
Real number(s), 58, 153, 335, 338
Reciprocal, 85, 413. *See also* Multiplicative inverse
Rectangles, area of, 9-10, 76, 96, 125, 350
Remainder, 319, 321
Right triangle, 351-52
Roots, 346, 378, 383

Scientific notation, 261-65, 269, 277
Set(s) of points, 165, 227-30, 235-36
"Sharing the work" problems, 411-15
Signed numbers, 43-64, 152-53
  addition of, 47-51
  additive inverse of, 81, 84
  definition of, 46
  division of, 58-60
  inequalities, 220
  multiplication of, 55-58, 84
  solving equations with, 60-62
  square of, 58
  subtraction of, 52-55
  and zero, 44-46, 152, 161, 220, 232
Significant figures, 266-69

Simplifying, 69, 75, 94, 112
  and additive inverse, 82, 85
  complex fractions, 411-15
  square roots, 366-67
Simultaneous inequalities, 231-36
Simultaneous linear equations, 191-96, 199
  solving, 191-208
  solving word problems with, 210-13
Slope, 168-70, 198, 226
Solutions
  of equations, 34-40, 337-39, 395-401
  of linear equations, 113-21, 180-83
  of linear inequalities, 221-25, 225-30
  of open sentence(s), 26-30, 34
  of polynomials, 338-39
  of quadratic equations, 327-36, 344-53, 359-79
    by completing the square, 373-80, 386-88
    by factoring, 344-53, 380
    by graphs, 380-85
  of simultaneous inequalities, 231-36
  of simultaneous linear equations, 191-208
  of word problems, 127-43, 210-13
Solving equations
  by addition, 5, 30
  by completing the square, 373-80, 386-88
  by factoring, 344-53, 380
  by guessing, 38-40
  by multiplication, 29, 34, 37
  with signed numbers, 60-62
  by substitution method, 211
  and "undoing" method, 30, 34-37, 61, 79, 86
Square
  definition of, 11
  of negative numbers, 363
  perfect, 346, 361, 371
  of real numbers, 58, 153, 335, 338
Square roots, 14-16, 360-68, 369-79, 387-89
  simplifying, 366-67
  table of, 431
Squaring a number, 11-13, 15, 369-70
  definition of, 11, 15, 361
  a signed number, 58
Standard form
  of linear equations, 111-12, 175, 177
  of linear expressions, 105
  for quadratic equations, 347
Straight lines, 163, 165, 173-78, 180, 381, 404
Substitution method, 211
Subtraction, 5
  "add the opposite" rule, 53-55, 289-91
  and associative property, 73, 75
  and equations, 30, 34
  of fractions, 398
  of polynomials, 288-91

435

of signed numbers, 52–55, 69, 155, 258, 289
Superscript, 246
Symbols, 31–33. *See also* Terms
  equal, 219
  grouping, 93, 129, 134, 280, 282
  inequality, 219–22
  superscript, 246
  unequal, 302
Symmetry, 366

Terms, 31–33, 88–90, 279, 282–85
  like, 283
Triangles, right, 351–52
Trinomial, 371

"Undoing" method, 29–30, 34–37, 61, 79, 86, 109–11, 328–29
Usual order, 282–83, 285, 289, 295, 306

Variable(s), 31–33, 103–107, 108–16, 119–21, 125–27
  dependent, 32
  independent, 32
Vertex, 383

Whole number(s), 5, 9, 43–46, 164, 312, 341, 360, 365
  coordinates, 164
  exponents, 259
Withdrawal, 5
Word problems, 124, 127–43, 210–13, 231, 411–15

and quadratic equations, 350–53, 386–88

x-axis, 157–61, 165, 174–76
x-intercepts, 165

y-axis, 157–61, 165, 173–75
y-intercepts, 165

Zero
  addition by, 79–81
  division by, 302–303, 337–39
  exponents, 250, 280
  multiplication by, 84–85, 87
  products, 222, 279–80, 308, 337
  and signed numbers, 44–46, 152, 161, 220, 222
  square root of, 363